Introduction to
Plant Pathology

Introduction to Plant Pathology

Richard N. Strange
University College London

WILEY

Copyright © 2003 by John Wiley & Sons Ltd,
The Atrium, Southern Gate, Chichester,
West Sussex PO19 8SQ, England

National   101243 779777
International   (+44) 1243 779777

e-mail (for orders and customer service enquiries):
cs-books@wiley.co.uk

Visit our Home Page on http://www.wiley.co.uk
or http://www.wiley.com

Other Wiley Editorial Offices

John Wiley & Sons, Inc., 605 Third Avenue,
New York, NY 10158-0012, USA

Wiley-VCH Verlag GmbH, Pappelallee 3,
D-69469 Weinheim, Germany

John Wiley & Sons (Australia) Ltd, 33 Park Road, Milton,
Queensland 4064, Australia

John Wiley & Sons (Asia) Pte Ltd, 2 Clementi Loop #02-01,
Jin Xing Distripark, Singapore 0512

John Wiley & Sons (Canada) Ltd, 22 Worcester Road,
Rexdale, Ontario M9W 1L1, Canada

Wiley also publishes its books in a variety of electronic formats. Some content that appears in
print may not be available in electronic books.

Library of Congress Cataloging-in-Publication Data
Strange, Richard N.
  Introduction to plant pathology / Richard N. Strange.
     p. cm.
Includes bibliographical references and index.
  ISBN 0-470-84972-X (Cloth : alk. paper) – ISBN 0-470-84973-8 (Paper : alk. paper)
  1. Plant diseases. I. Title.
  SB731-S868 2003
  632′-3–dc22   I0034994 6x   2003016087

British Library Cataloguing in Publication Data

A catalogue record for this book is available from the British Library

ISBN 0 470 84972 X (Hardback)
      0 470 84973 8 (Paperback)

Typeset in 10.5/12.5 pt Times by Kolam Information Services Pvt. Ltd, Pondicherry, India
Printed and bound in Great Britain by TJ International Ltd., Padstow, Cornwall
This book is printed on acid-free paper responsibly manufactured from sustainable forestry,
in which at least two trees are planted for each one used for paper production.

To my dear wife, Lilian,
with love and appreciation for her patience
and support over the last several months.

Contents

Introduction to Plant Pathology by Richard Strange
© 2003 John Wiley & Sons, Ltd   ISBN 0 470 84972 X (cased)   ISBN 0 470 84973 8 (pbk)

Preface

Pathogenic agents which infect plants are many, both in kind and number, and they are in direct competition with us for our food and our cash crops. They are therefore a cause of malnutrition, starvation and poverty. They also deplete our food reserves by spoiling stored produce and ruin our environment by destroying vegetation.

'Know your enemy' is a good adage in most walks of life and is particularly apposite to that of the plant pathologist. The range of organisms which cause plant disease and the techniques for their identification are therefore introduced in the first two chapters of this book together with accounts of some of the worst epidemics they have caused. Those with an historical inclination may know that a disease of potato was the cause of a great famine in the 1840s in which Ireland lost over a quarter of its population through starvation and emigration. Similarly, death by starvation of an estimated two million people living in Bengal (now the Province of West Bengal in India and Bangladesh) is thought to have been caused by a disease of rice in the 1940s. As an example of the devastation of a cash crop, the growing of coffee in Sri Lanka became uneconomic in the last part of the 19th century owing to coffee rust; in consequence, the farmers turned to growing tea – and the British, obligingly, to drinking it! With regard to the environment, parents and grandparents of today's students may remember that elm trees were a distinctive and attractive feature of parts of the USA and the southern part of England but that in the 1970s they rapidly died out owing to 'Dutch Elm Disease'. Finally, of the many organisms that cause spoilage of produce in storage, fungi of the *Aspergillus flavus* group are perhaps the most notorious; not only are the infected or infested plant products such as maize and peanuts toxic, they are also highly carcinogenic.

Once the identity of a pathogen is known and there are adequate techniques for its rapid identification (Chapters 1 and 2), the study of its dissemination and the losses it causes become practicable (Chapter 3 and 4, respectively). Chapter 5 addresses the control of inoculum of the pathogen since prevention is not only better than cure but is also, apart from growing resistant plants, virtually the only feasible option for poorer parts of the world. Richer countries may be able to afford pesticides but concerns about their adverse effects on human health and the environment have led to several of them being phased out. Emphasis is

Introduction to Plant Pathology by Richard Strange
© 2003 John Wiley & Sons, Ltd   ISBN 0 470 84972 X (cased)   ISBN 0 470 84973 8 (pbk)

therefore placed on biological control as a potentially less hazardous and more environmentally friendly way of reducing inoculum.

Plant pathogens have evolved many ingenious ways of locating and effecting entrance into their hosts, often involving the recognition of host topography or chemical signals. In some instances, entrance is achieved through natural openings and wounds, but in others direct penetration of the cell wall is achieved by the secretion of enzymes or the development of high turgor pressure in special organs (appressoria). Enzymes are also required by many pathogens for the colonization of the host and the release of metabolizable products and their action may give rise to a considerable amount of cell death (necrosis). Location, penetration and colonization are the topics discussed in Chapter 6.

Some pathogens subvert the metabolism of their hosts by altering hormone levels, the ultimate level in sophistication in this regard being species of *Agrobacterium*. These organisms are genetic engineers since they insert genes into plants that encode enzymes for the synthesis of two important plant hormones, as well as other genes that harness the metabolism of infected plants to produce compounds which can only be used by the pathogen (Chapter 7). Some pathogens influence the metabolism of their hosts and may kill them by the production of phytotoxic compounds (Chapter 8).

Faced with this onslaught from pathogens, it is not surprising that plants have developed mechanisms for fighting back. Some of these are constitutive (Chapter 9) while others are induced by the challenging organism (Chapter 11). The induced responses are often triggered by recognition of the invader, the genetic elements responsible on the part of the plant and the pathogen being termed resistance and avirulence genes, respectively. This gene-for-gene relationship is discussed in Chapter 10 together with structures of the genes involved and the means by which their products may be brought together and interact.

Finally, in Chapter 12, the material of the six previous chapters is viewed from the perspective of developing control measures that are environmentally friendly.

A number of people deserve my thanks for help in producing this book, those who sent illustrations and who are acknowledged in the legends, Andy Slade and his team at John Wiley and Sons Ltd., and especially John Mansfield who kindly read an earlier version of the manuscript. Any mistakes that remain are mine.

Richard Strange
January, 2003

1 The Causal Agents of Plant Disease: Identity and Impact

Summary

Eleven groups of organisms cause plant disease: parasitic angiosperms, fungi, nematodes, algae, Oomycetes, Plasmodiophoromycetes, trypanosomatids, bacteria, phytoplasmas, viruses and viroids. Pathogenicity is normally established by the application of Koch's postulates, modified in the cases where culture of the causal organism is not feasible. Individual members of all 11 groups cause severe diseases of important crops with consequent significant economic and social impact. Their ability to reproduce prolifically and survive transportation and long periods of inactivity as dormant propagules has ensured that they continue to present a formidable threat to the health of our crops and therefore to the world's food security.

1.1   Introduction

Plants, through their ability to fix carbon dioxide by photosynthesis, are the primary producers of the food that feeds the world's human population as well as the many animals and other organisms that are heterotrophic for carbon compounds. It is not surprising, therefore, that among the latter there is a considerable number which, in order to have first call on these rich pickings, have adopted the parasitic mode of life. They range from higher plants themselves, the parasitic angiosperms, to viroids, naked fragments of nucleic acid, in some instances less than 300 nucleotides in length. Between these extremes of size, there are plant pathogenic organisms among the fungi, nematodes, algae, Oomycetes, Plasmodiophoromycetes, trypanosomatids, bacteria, phytoplasmas and viruses. In almost all of these categories there are organisms that cause catastrophic plant diseases, affecting the lives of millions of people by competing for the plant products on which they depend for food, fibre, fuel and cash. In this chapter all 11 classes of plant pathogenic agent will be introduced and those that are particularly destructive will be highlighted together with the impact that they have had on

Introduction to Plant Pathology by Richard Strange
© 2003 John Wiley & Sons, Ltd   ISBN 0 470 84972 X (cased)   ISBN 0 470 84973 8 (pbk)

the people who have been most seriously affected. However, the first imperative of a plant pathologist is to establish unequivocally the cause of disease.

1.2 Establishing the Cause of Disease

The correct diagnosis of a plant disease and its cause is not always an easy task. In the first instance symptoms may be ill defined which make their association with any organism problematic (Derrick and Timmer, 2000) and, secondly, plants grow in environments which are notably non-sterile. In particular, besides supporting a microflora on their aerial parts, the phylloplane, they are rooted in soil which may contain in excess of 1 million organisms per gram. The plant pathologist is therefore faced with trying to determine which, if any, of the organisms associated with the diseased plant is responsible for the symptoms. This is normally achieved by the application of the postulates of Robert Koch, a German bacteriologist of the 19th century, which for plant pathogens may be stated as follows:

(1) The suspected causal organism must be constantly associated with symptoms of the disease.

(2) The suspected causal organism must be isolated and grown in pure culture.

(3) When healthy test plants are inoculated with pure cultures of the suspected causal organism they must reproduce at least some of the symptoms of the disease.

(4) The suspected causal organism must be reisolated from the plant and shown to be identical with the organism originally isolated.

Clearly, these criteria can only be met with organisms that can be cultured, ruling out all obligate pathogens which include a number of important fungi, many phytoplasmas and all viruses and viroids. Establishing these organisms as causal agents of disease usually involves purification of the suspected agent rather than culture and the demonstration that these purified preparations reproduce at least some of the disease symptoms.

1.3 The Range of Organisms that Cause Plant Disease

1.3.1 Parasitic angiosperms

The parasitic angiosperms are higher plants which are parasitic on other higher plants. They number more than 3000 species and are found in nine families (Stewart and Press, 1990). *Striga* and *Orobanche* are two prominent genera which

Figure 1.1 (a) *Striga hermonthica* flowering on sorghum (reproduced courtesy of Dr Chris Parker and CAB International); (b) *Orobanche foetida* flowering on *Vicia faba* (reproduced courtesy of Dr Diego Rubiales, University of Cordoba). A colour reproduction of this figure can be seen in the colour section

are said to be parasitic weeds. On germination of their seeds, they attach themselves to the roots of their host plants by a structure termed a haustorium (Section 6.6) and abstract nutrients, causing losses of yield which may be total (Figure 1.1(a)) (see plates between pages 32–33 and 240–241). The use of the term weed is perhaps unfortunate since, until relatively recently, it seems to have diverted the attention of plant pathologists from this important group of plant pathogens! Species of *Striga* are particularly notorious as they infect more than two-thirds of the 73 million hectares of cereals and legumes grown on the African continent, affecting the lives of over 100 million people living in 25 countries. Losses may be total and infestation of some areas may be so great that continued crop production becomes impossible (Estabrook and Yoder, 1998). It seems likely that these diseases will continue to exact a high toll of crops in Africa since the pathogens responsible seed prolifically and the seed remains viable in the soil for at least 10 years. However, with regard to *S. gesnerioides*, which infects cowpea, two advances were made in 1996: the discovery of races of the pathogen and the clustering of the races in discrete geographical areas in West Africa (Lane *et al.*, 1996). These exciting results raise the possibility of breeding resistant cultivars and distributing them to areas where their resistance will be effective.

Species of *Orobanche* cause serious losses of plants belonging to the families Solanaceae, Fabaceae, Apiaceae and Asteraceae. They produce large numbers of

Figure 1.2 (a) *Viscum album* subsp. *album* on *Robinia pseudoacacia* (courtesy of Dr Doris Zuber, Swiss Federal Institute of Technology, Zurich); (b) *Cuscuta reflexa* on an unidentified shrub (reproduced courtesy of Dr Chris Parker and CAB International). A colour reproduction of this figure can be seen in the colour section

seed but relatively small numbers are required to harm the crop. For example, fewer than 1000 and 250 seeds/kg of soil reduced yields of faba beans and peas, respectively, to zero (Linke, Sauerborn and Saxena, 1991; Bernhard Lensen and Andreasen, 1998). As a consequence, in some areas such as North Africa, farmers have been forced to abandon growing these crops (Figure 1.1(b)).

Other genera of parasitic angiosperms such as mistletoe (*Viscum* spp.), dwarf mistletoe (*Arceuthobium* spp.) and dodder (*Cuscuta* spp. and *Cassythia* spp.) also cause diseases which have profound economic and social consequences (Figure 1.2). In the Intermountain Region of the USA, *Arceuthobium americanum* was estimated to cause losses of ca. 500 000 m^3 of lodgepole pine per annum, equivalent to more than 80 per cent of the annual harvest (Hoffman and Hobbs, 1985) and a more recent estimate puts the annual loss at more than 3.2 billion 'board feet' of lumber in the USA as a whole (Parker and Riches, 1993).

1.3.2 Fungi

There are at least five reasons why fungi may cause catastrophic plant disease.

(1) They sporulate prolifically, the spores providing copious inoculum which may infect further plants.

(2) Their latent period, i.e. the time between infection and the production of further infectious propagules, usually spores, may be only a few days.

(3) The spores, if they are wettable, may be spread as high-density inoculum in surface water or in droplets by rain-splash. Alternatively, non-wettable spores may be carried long distances by the wind.

(4) They may produce compounds that are phytotoxic and/or a battery of enzymes that destroy the plant's structure.

(5) Biotrophic pathogens, such as the rusts and mildews, draw nutrients away from the economically valuable part of the plant by the production or induction of growth regulators such as cytokinins and consequently depress yields.

These points will be reviewed in later chapters.

Fungal taxonomy and identification was traditionally based on morphology but, since many fungi are poorly endowed with distinctive morphological features, this approach has led to error and confusion. For example, Marasas and co-workers (1985) reported that about 50 per cent of a collection of 200 *Fusarium* strains reported to be toxigenic in the literature were incorrectly named on the basis of the taxonomic concepts current at the time. Fortunately, DNA sequence data are becoming the norm in fungal classification (see Section 2.5.5) and although these often uphold traditional classifications for higher taxa they frequently do not do so at the species level. All data support the view that fungi may be separated into the four classes originally defined on the basis of the morphology of their sexual structures and the sporulating organs formed after sexual reproduction. These are the Chytridiomycetes, Zygomycetes, Ascomycetes and Basidiomycetes. A further class, the Deuteromycetes, alternatively known as the Fungi Imperfecti, is reserved for those fungi for which no sexual phase is known. Organisms that are highly destructive to plant life are found in all five classes.

Oomycetes, which contain such important plant-pathogenic genera as *Phytophthora, Peronospora* and *Pythium*, have, until recently, been classified as fungi but are more properly placed well away from the fungi in the kingdom, Chromista (Cavalier-Smith, 1998; Baldauf *et al.*, 2000) and are considered separately (Section 1.3.5). Similarly, the Plasmodiophoromycetes, which are sometimes referred to as slime moulds, are not fungi and are therefore also given a section of their own Section 1.3.6). In the following paragraphs we shall only be concerned with plant pathogens which are true fungi.

Chytridiomycetes

Synchytrium endobioticum is the cause of potato wart disease (Figure 1.3). It is an obligate pathogen consisting only of two types of sporangia, winter and summer. Winter sporangia germinate in the spring and release zoospores, which are propelled by a single flagellum, enabling them to move in soil water and reach the living host. Penetrated host cells enlarge and the fungus forms within them the evanescent summer sporangium. Numerous zoospores are rapidly discharged from the summer sporangium and reinfect surrounding cells. Under favourable conditions further summer sporangia are produced, inducing swelling of penetrated cells and those surrounding them and giving rise to the cauliflower-like appearance which is characteristic of the disease (Figure 1.3). Under conditions of stress, zoospores fuse in pairs and form a zygote from which the thick-walled

Figure 1.3   Cauliflower-like symptoms of potato wart caused by the Chytridiomycete fungus, *Synchytrium endobioticum* (reproduced courtesy of Dr Hans Stachewicz and CAB International). A colour reproduction of this figure can be seen in the colour section

winter sporangium develops. These are released from rotting warts and may remain viable for at least 30 years, providing a long-term source of inoculum for succeeding crops.

Other members of this class of fungi, while causing little damage themselves, may serve as vectors for viruses (Section 3.3.2). For example, *Olpidium brassicae* may transmit lettuce big vein virus (LBVV) and tobacco necrosis virus (TNV, Figure 1.4).

Zygomycetes

Rhizopus is one example of a genus in the Zygomycetes which causes significant losses of many disparate crop plants such as cassava, groundnuts, sorghum and cucurbits. *Rhizopus stolonifer* is an important cause of postharvest disease in soft fruit (Michailides and Spotts, 1990) and sweet potato (Holmes and Stange, 2002) and *R. oryzae* causes damping off of cotton (Howell, 2002).

Ascomycetes

Many plant pathogenic fungi are Ascomycetes and several fungi that were originally classified as Deuteromycetes are now placed in this class owing to the discovery that they form asci (singular ascus). These are the characteristic sack-like structures of the group which are the result of sexual reproduction, each ascus usually containing eight ascospores, the products of meiosis and one round of mitosis. Normally the Deuteromycete name is retained but the Ascomycete name is also used, particularly when sexual reproduction, leading to the produc-

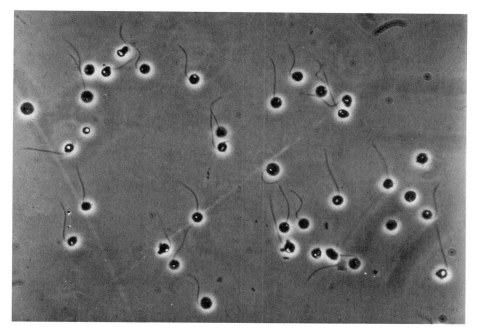

Figure 1.4 Zoospores of *Olpidium brassicae*; these may act as vectors for virus diseases such as lettuce big vein virus and tobacco necrosis virus (courtesy of Ian Macfarlane and Rothamsted Research, Harpenden, Herts., UK)

tion of asci, is common. The Deuteromycete form is known as the anamorph and the Ascomycete form as the teleomorph. For example, the greatest recorded epidemic of a plant disease was that caused by a fungus originally referred to as *Helminthosporium maydis*, the anamorph name. The teleomorph of the same organism is *Cochliobolus heterostrophus* (Figure 1.5).

The fungus was first found on ears and stalks of samples from a seed field in Iowa, USA, in October 1969 where it caused a greyish-black rot but it was not until the next year that the epidemic struck. Then it caused a severe epidemic, particularly in the Midwest and South of the USA, with some areas reporting losses of 50 to 100 per cent. In monetary terms at today's (2003) prices this represents about $ 4 \times 10^9$.

The reason for the devastation lay in the widespread use of a gene in the crop for breeding purposes that had the unfortunate pleiotropic effect of conferring acute susceptibility to a variant of the fungus which, up to that time, had caused little damage. The gene concerned was cytoplasmically inherited and conferred male sterility. As it originally came from Texas it was known as the *T*exas cytoplasmic *m*ale *s*terile gene or the *Tcms* gene for short. Plant breeders used the gene in developing high-yielding hybrids as it prevented self-pollination and obviated the need to remove the pollen-bearing structures, the tassels, from the plant destined to be the female parent, a tedious and costly process. A field

Figure 1.5 Symptoms of Southern Corn Leaf Blight (SCLB) caused by *Cochliobolus hetero-strophus* (reproduced courtesy of Dr M. Listman, CIMMYT). A colour reproduction of this figure can be seen in the colour section

planted exclusively with *Tcms* maize sets no seed as there is no pollen capable of fertilizing the ovules. To overcome this problem, the farmer either buys a blend, made of *Tcms* seed and seed with normal cytoplasm, or seed which is a hybrid between a *Tcms* parent and a pollen parent containing a restorer gene. Such a gene, as its name implies, restores the fertility of the pollen to the hybrid. The advantages of using *Tcms* in hybrid seed production ensured that the gene became widely distributed and, by 1970, it was present in about 85 per cent of the American maize crop. Ironically, therefore, a technique used to broaden the genetic base of the crop by the production of hybrids had resulted in the uniformity of one part of the cytoplasmic genome – that containing a gene conferring male sterility. Remarkably, this gene had the pleiotropic effect of conferring acute susceptibility to a disease which had previously been of minor importance. The virulent pathogen, which was the cause of the disease, was a variant of *C. heterostrophus* and was specifically aggressive for maize with *Tcms*. It was therefore designated race T to distinguish it from the less virulent race O. Subsequent work showed that race T produced a family of compounds which are specifically toxic to maize bearing the *Tcms* gene. These toxins and the reasons for their specific toxicity to *Tcms* maize are discussed in Chapter 8 (Section 8.4.2).

Claviceps is one of the most notorious plant pathogenic genera in the Asco-mycetes. The fungus parasitizes the developing cereal grain and supplants it by a sclerotium, called an ergot, consisting of a hard mat of fungal hyphae, usually somewhat larger than the grain itself. The sclerotia are rich in a number of alkaloids which cause the contraction of smooth muscle. This property has been exploited in childbirth in which the mother is often given an injection of one of the compounds, ergometrine, to aid parturition and to promote the return of the uterus to its normal size after birth. However, in uncontrolled doses, as may occur when grain contaminated with ergot is consumed, horrific symptoms ensue. In the Middle Ages these were regarded as divine punishment for misdeeds and were known as 'St Anthony's Fire'. They have been graphically described by Large (1940) as follows: 'Ergot, through its constrictive action on blood vessels, not only caused abortion in women, it cut off the blood supply to the extremities of the body; hands and feet became devoid of sensation and then rotted most horribly away. In the progress of the ergot gangrene whole limbs fell off at the joints, before the shapeless trunk was released from its torments.'

Sorghum, *Sorghum bicolor*, is the world's fifth most important cereal crop. Since the 1960s the development and use of F_1 hybrid seed has boosted yields to between 3 and 5 tonnes/hectare under high-input farming conditions but much lower yields, averaging less than 1 tonne/hectare, are obtained when farming inputs are low. Unfortunately, production of F_1 hybrid seed is jeopardized by ergot caused by the fungi, *Claviceps sorghi* and *C. africana*. They enter the host principally through the pistils of unfertilized gynoecia. When pistils are fertilized the fungus grows slowly or fails to grow altogether. The production of hybrids from male sterile lines is therefore particularly hazardous since pollen from the restorer line (which restores fertility to the progeny) may not synchronize with flowering of the male sterile line, thus leaving the pistil unfertilized and susceptible to infection. Production of copious amounts of honeydew oozing from florets is an early symptom of infection (Figure 1.6). Later in the season sclerotia are produced which contain alkaloids but their presence and toxicity has not been as well-established as those of *C. purpurea* (Bandyopadhyay *et al.*, 1998).

A third Ascomycete, *Cochliobolus miyabeanus* (teleomorph name; anamorph = *Helminthosporium oryzae*), is a destructive disease of rice which, under favourable conditions, causes severe losses (Figure 1.7). In 1942, an epidemic occurred in Bengal (now the Province of West Bengal in India and Bangladesh), an area that was normally deficient in both wheat and rice. In 1942, the winter rice crop, reaped in November and December, was exceptionally poor with late cultivars being particularly badly affected. Yield reductions recorded at research stations varied from nearly 40 per cent to over 90 per cent (Padmanabhan, 1973). As a consequence of the shortages, prices escalated early in 1943, putting them beyond the means of ordinary people. Many, who lived in rural areas, left their villages and travelled to the larger cities in search of work and rice. There, 'finding neither, they slowly died of starvation' (Padmanabhan, 1973). They numbered two million. However, some scientists remain unconvinced that *C. miyabeanus* was the cause of the disease in West Bengal and suggest that high levels of iron or high aluminium may have been responsible while others suspect that the problem was actually a brown planthopper outbreak (R. S. Zeigler, personal communication).

Figure 1.6 Honeydew on sorghum caused by infection with the fungus *Claviceps africana* (courtesy of Dr F Workneh, Texas Agricultural Experiment Station). A colour reproduction of this figure can be seen in the colour section

Basidiomycetes

Some members of this class of fungi such as the rusts and smuts are highly destructive plant pathogens. Rusts are obligate pathogens and therefore cannot normally be grown in pure culture as demanded by Koch's second postulate. However, since they often produce massive numbers of spores, usually of the rust colour that gives these fungi their name and which, at least under the stereoscan electron microscope, are distinctive, their causal relation to disease is relatively easy to prove – the spores may simply be collected and used as inoculum on test plants. Production of the same type of spores on these plants is clear evidence that the fungus is the cause of the disease. A more stringent test would be to adopt one or more of the serological or nucleic acid procedures detailed in Chapter 2 (Sections 2.5.4 and 2.5.5).

The rusts and smuts are highly specialized with regard to their host ranges but their prolific production of aerially borne spores ensures their continuation as threats to our crops (Figure 1.8). Although the rusts produce small, although abundantly sporulating lesions on their hosts, other members of the Basidiomycetes produce their prolific numbers of spores from large structures known as sporopohores. These are often tree pathogens such as *Ganoderma* and, because of

Figure 1.7 Lesions on rice leaves caused by *Cochliobolus miyabeanus*, often cited as the cause of rice crop failures in 1943 that led to the great Bengal famine in which an estimated 2 million people died (courtesy of Dr. R. S. Zeigler). A colour reproduction of this figure can be seen in the colour section

the shape of their sporophores, they are commonly known as bracket fungi (Figure 1.9).

Deuteromycetes

The Deuteromycetes contain many important pathogens such as species of *Alternaria, Fusarium* and *Helminthosporium*, some of which produce powerful toxins which kill plant cells (Chapter 8) and mycotoxins, compounds of fungal origin that are toxic to people or domestic animals. For example, *Alternaria alternata* f. sp *lycopersici* causes stem canker of tomato on susceptible lines of the plant and produces a host selective toxin (AAL toxin) which is essential for pathogenicity. Remarkably, a number of *Fusarium* species synthesize compounds known as the fumonisins which are structurally related to AAL toxin and are mycotoxins (Section 8.4.1). Other Deutermoycetes produce mycotoxins, the most notorious being those of the *Aspergillus flavus* group which synthesize aflatoxins. Aflatoxin B_1 has high mammalian toxicity and is also extremely carcinogenic,

Figure 1.8 Stripe rust of wheat caused by *Puccinia striiformis*; the orange colour of the lesions on the leaves is given by the copious numbers of spores which they contain. A colour reproduction of this figure can be seen in the colour section

causing liver tumours (Figure 1.10). *A. flavus* and its toxins are frequently found in maize and peanuts (groundnuts).

1.3.3 Nematodes

Nematodes are an important group of plant parasitic organisms, causing crop losses directly by their parasitic activities on the plants they infect and also indirectly by acting as vectors for plant viruses. There are 17 orders of nematodes but only two contain plant pathogens, the Tylenchida and the Dorylaimida with virus vectors found only in the latter (Wyss, 1982). As an example, *Ditylenchus dipsaci* attacks over 450 different plant species, including many weeds (Goodey, *Franklin and Hooper*, 1965) and is one of the most devastating nematode species

Figure 1.9 Sporophore of *Ganoderma boninense* at the base of an oil palm (courtesy of Paul Bridge). A colour reproduction of this figure can be seen in the colour section

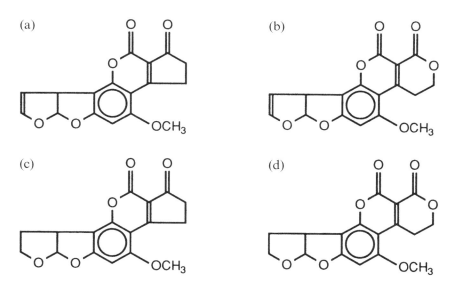

Figure 1.10 The chemical structures of the four major aflatoxins, potent mycotoxins from fungi of the *Aspergillus flavus* group which contaminate many crops, including ground nut and maize: (a) aflatoxin B_1; (b) aflatoxin G_1; (c) aflatoxin B_2; (d) aflatoxin G_2

Figure 1.11 Dwarfing symptoms on oats (left) compared with control (right) caused by the nematode *Ditylenchus dipsaci* (courtesy of Dr S Taylor, South Australian Research and Development Institute). A colour reproduction of this figure can be seen in the colour section

(Figure 1.11). Important economic hosts include onion, leek, pea, oats and maize as well as ornamentals such as *Narcissus* and tulip. When soil is heavily infested losses may be total. The nematode is also associated with other pathogens and has been claimed to transmit the bacterium *Corynebacterium insidiosum* (= *Clavibacter michiganensis* subsp. *insidiosus*) to lucerne plants (Hawn, 1963).

1.3.4 Algae

Cephaleuros virescens has been associated with disease symptoms in over 280 species and cultivars of higher plants and in one study up to 98 per cent of the leaves of Tahiti lime, *Citrus latifolia*, were found to be infected by an alga of this genus (Holcomb, 1986; Marlatt, Pohronezny and McSorley, 1983). More recently, Holcomb, Vann and Buckley (1998) reported the isolation of the alga from blackberry canes with symptoms consisting of stem cracking, tissue discoloration beneath the bark and the presence of an orange, velvet-like growth which was identified as the alga. However, data for losses do not seem to be available. Other species of algae, which have been implicated in plant disease, belong to the genera *Chlorochytrium, Rhodochytrium* and *Phyllosiphon.*

1.3.5 Oomycetes

The Oomycetes were long regarded as fungi but a number of their features such as the absence of chitin from their cell walls, their predominantly diploid karyotype and their biflagellate zoospores, do not accord with true fungi. Owing to nucleic acid and protein sequence data (Baldauf *et al.* 2000) we now know that these organisms are more closely related to the golden-brown algae. Although they share the first four characteristics of fungi that make them such dangerous pathogens of plants (Section 1.3.2) another important point about them is that they have evolved a plant parasitic life style independently of true fungi. Therefore some plant defences such as the production of saponins and chitinases which specifically target the fungal attributes, membrane sterols and chitin, respectively, are ineffective (see Chapter 9, Section 9.3.1 and 9.3.2 and Chapter 11, Section 11.3.6).

The Oomycetes contain many destructive plant pathogens, among them some species of the genera *Phytophthora, Pythium* and *Peronospora* stand out; in fact, the name *Phytophthora*, means 'plant destroyer'. In the 19th century the organism now known as *Phytophthora infestans* was the cause of abject human suffering in Ireland owing to its aggressive parasitism of potato (Large, 1940; Strange 1993; Figure 1.12). The potato was introduced into Europe in about 1570 but had remained free of the pathogen until 1845. Since potato was the

Figure 1.12   Stem lesions and wilted leaves on potato infected with *Phytophthora infestans* (reproduced courtesy of William E. Fry and CAB International). A colour reproduction of this figure can be seen in the colour section

predominant source of nourishment of a large section of the Irish population the arrival of the blight organism spelt famine. About a million people died of starvation and a further 1.5 million emigrated, many to North America, although about a quarter of these did not survive the voyage owing to their poor health.

The catastrophe of the Irish potato famine initiated the scientific investigation of plant disease and may be thought of as the key event that led to the establishment of Plant Pathology as a scientific discipline. Pride of place as the originator must be given to the Rev. Miles J. Berkeley who contended, against considerable opposition, that the fungal-like organism associated with the disease was its cause, not its consequence.

Other species of the genus *Phytophthora* cause diseases of many economically important plant species such as cocoa, cantaloupe melons and pigeonpea. In Australia, the widespread destruction of the jarrah forests and the susceptibility of about half the flora there to *Phytophthora cinnamomi* suggest that this pathogen was a recent introduction to the continent (Figure 1.13).

1.3.6   Plasmodiophoromoycetes

These are a group of soil protozoa known as the Myxomycetes or slime moulds, whose vegetative body is a plasmodium rather than a mycelium. The Plasmodiophoraceae are found here and several members are important plant pathogens. For example, the causal agent of the very damaging club root of brassicas, *Plasmodiophora brassicae*, is a member of this family (Figure 1.14). Spores of this organism may remain viable in the soil for many years and the characteristic swelling of roots is attributed to the higher concentrations of cytokinins in the infected tissue (Section 7.3.1). *Spongospora subterranea* f. sp. *subterranea* is the causal agent of powdery scab of potato (Figure 1.15) but species of another genus of the family, *Polymyxa*, cause more damage to plants indirectly than directly since they serve as vectors for viral diseases. These include wheat mosaic virus and barley yellow mosaic virus, transmitted by *P. graminis* and beet necrotic yellow vein virus, the cause of rhizomania disease, transmitted by *P. betae* (Figure 1.16).

1.3.7   Trypanosomatids

This is another group of plant pathogenic protozoa. Members of the group are common in the latex, phloem, fruit sap, seed albumin and nectar of many plant families (Camargo, 1999). Although they were originally described in the early years of the last century it was only in 1976 that they were recognized as the causal agents of serious plant disease when they were connected with two important disorders in palms – coconut hartrot and palm marchitez (Figure 1.17). McCoy and Martinez-Lopez (1982), working in South America, found *Phytomonas staheli* in dwarf coconut palms with lethal wilt and in mature African oil palms affected with sudden wilt disease (marchitez sopresiva) as well as in immature African oil palms with case nine (caso nueve) syndrome. A wilt disease of coffee is now ascribed to *Phytomonas leptovasorum* which may kill trees within 2 months of the appearance of the first symptoms. The disease is characterized by necrosis of the phloem and this symptom has given the disease its common name – phloem necrosis.

Several species of *Phytomonas* have been differentiated on the basis of rDNA sequences (Dollet, Sturm and Campbell, 2001; see Section 2.5.5). Two main groups were separated by their non-transcribed spacer sequences, the second of which was further subdivided into subgroups designated IIa and IIb. The genus could also be separated by differences in the spliced leader RNA gene (Dollet, Sturm and Campbell, 2001).

1.3.8 Bacteria

Bacteria causing plant disease were originally classified in five genera – the Gram-positive *Corynebacterium* and the Gram-negative *Agrobacterium, Erwinia, Pseudomonas* and *Xanthomonas* – but to these must be added the Actinomycetes. In the last two decades the classification has begun to be extensively revised. For example, the plant pathogenic coryneform bacteria are generally classified in the genera *Curtobacterium, Arthrobacter, Rhodococcus* and *Clavibacter* although some authors still retain the old nomenclature (Davis, 1986).

Coryneform bacteria

Species of *Corynebacterium, Curtobacterium, Arthrobacter, Rhodococcus* and *Clavibacter* cause diseases on a number of crop plants. For example, potatoes are infected by *Clavibacter michiganensis* subsp. *sepedonicus*. Symptoms consist of wilting of leaves which become pale green to yellow and develop necrotic areas but these are easily confused with other wilts and foliage diseases, and natural senescence. In tubers vascular tissue becomes discoloured and corky-brown tissue sometimes surrounds hollows that develop in the vascular ring (Figure 1.18). Where infection levels are high, losses may be total but certification of seed potato with a zero tolerance level (see Section 5.6.4) has meant that losses are usually only sporadic. Another subspecies of *Clavibacter michiganensis, C. michiganensis* subsp. *michiganensis* was found in tomato fields in Cyprus and losses were reported as heavy (Ioannou, Psallidas and Glynos, 2000). Some species of *Clavibacter* are nutritionally fastidious and limited to the xylem of their hosts but can, nevertheless, cause serious diseases in a number of crop plants (Raju and Wells, 1986). For example, ratoon stunt of sugar cane caused by *Clavibacter* (=*Leifsonia*) *xyli* subsp. *xyli* caused cane losses of 14 per cent in the first year of cultivation but this increased to 27 per cent in the third year (Grisham, 1991).

(continues)

Figure 1.14 Club root of cabbage caused by *Plasmodiophora brassicae* (courtesy of Richard Cooper, University of Bath). A colour reproduction of this figure can be seen in the colour section

Rhodococcus fascians is the cause of malformations in a wide variety of host plants, the most severe being a leafy gall composed of multiple shoots and shoot primordia (Vereecke *et al.*, 2000; Figure 1.19). Pathogenicity is determined by a linear plasmid which carries several virulence loci since strains that are cured of the plasmid are non-pathogenic. Outbreaks of the disease on ornamental plants may cause serious financial losses.

Agrobacterium

Agrobacterium tumefaciens is the causal agent of crown gall in well over 200 dicotyledonous species and a related bacterium *A. rhizogenes* causes 'hairy root'

Figure 1.13 (a) Impact of *Phytophthora cinnamomi* on vegetation in an infested part of the jarrah (*Eucalyptus marginata*) forest of Western Australia; (b) a lesion caused by *P. cinnamomi* on a young *E. marginata* plant – the outer bark has been cut away to show the blackened diseased tissue; (c) *Xanthorrhoea preissii* killed by *P. cinnamomi* in the jarrah forest of Western Australia; (d) purple lesion of *P. cinnamomi* on the stem of a *Banksia grandis* plant – members of this genus are highly susceptible to the pathogen. Figures kindly supplied by Keith McDougall, NSW National Parks and Wildlife Service, PO Box 2115, Queanbeyan, NSW, Australia, 2620. A colour reproduction of this figure can be seen in the colour section

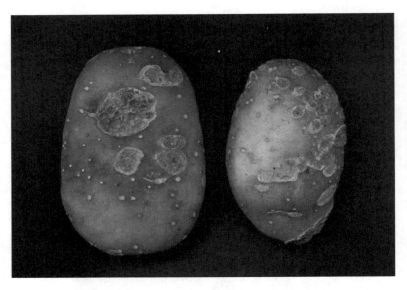

Figure 1.15 Symptoms of powdery scab of potato caused by *Spongospora subterranea* f. sp. *subterranea* – note the individual and coalesced raised lesions (courtesy of Herbert Torres). A colour reproduction of this figure can be seen in the colour section

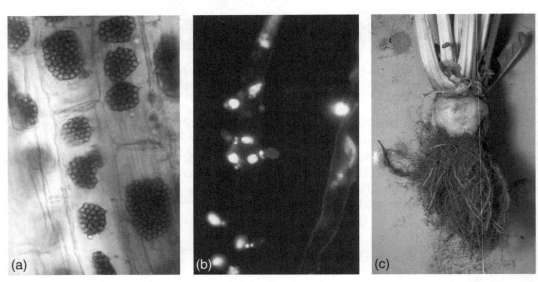

Figure 1.16 (a) Resting spores of *Polymyxa betae* in epidermal cells of sugarbeet rootlets; (b) zoospores of *Polymyxa betae* attaching and infecting sugarbeet root hairs; (c) Sugarbeet infected with beet necrotic yellow vein virus via its vector *Polymyxa betae*, showing abnormal root proliferation (rhizomania). (Kindly provided by Mike Asher of Broom's Barn Research Station, Higham, Bury St Edmunds, Suffolk 1P28 6NPR). A colour reproduction of this figure can be seen in the colour section

Figure 1.17 An 18-year old oil palm infected with *Phytomonas sp.*, demonstrating early symptoms of Marchitez disease consisting of browning and drying of the leaves (courtesy of Michel Dollet, CIRAD, Montpellier, France). A colour reproduction of this figure can be seen in the colour section

disease (Figure 1.20). Three other species have also been distinguished, *A. rubi, A. vitis* and *Agrobacter radiobacter* but the last of these is synonymous with *Agrobacterium tumefaciens*. However, based on comparative 16S rDNA analyses, these species together with *Allorhizobium undicola* form a monophyletic group with all *Rhizobium* species. A recent proposal is therefore to keep *Agrobacterium* as an artificial genus comprising plant-pathogenic species but to form the new combinations *Rhizobium radiobacter, Rhizobium rhizogenes, Rhizobium rubi, Rhizobium undicola* and *Rhizobium vitis* (Young *et al.*, 2001).

Galls or, as they are often called, tumours caused by *A. tumefaciens* are frequently free of the bacterium. They can be grown in tissue culture and do not require the plant growth substances, auxin and cytokinin, which are normally essential for the *in vitro* growth of plant tissue. The tumorous growth was therefore viewed as a plant cancer and a search was made for the tumour-inducing principle, TIP. This was finally identified as a plasmid (Kerr, 1969; Chilton *et al.*, 1977). The mystery as to how the tumorous state was initiated and maintained was solved when it was discovered that a portion of the plasmid, containing the genes for auxin and cytokinin synthesis, is transferred to the host

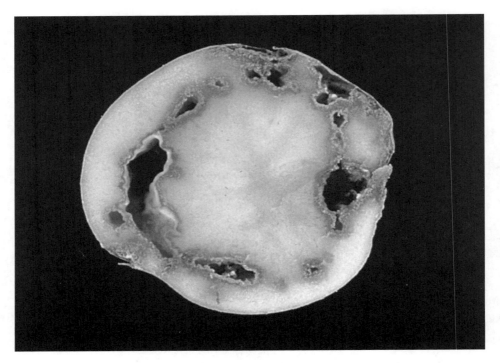

Figure 1.18 Internal symptoms of *Clavibacter michiganensis* subsp. *sepedonicus* on potato. Note the destruction of the vascular tissue and the surrounding creamy to brown coloured lesions (reproduced courtesy of Dr S. H. DeBoer and CAB International). A colour reproduction of this figure can be seen in the colour section

Figure 1.19 *Rhodococcus fascians* causing malformations on tobacco (courtesy of Carmem-Lara de O. Manes). A colour reproduction of this figure can be seen in the colour section

Figure 1.20 Crown gall on *Euonymus* (reproduced courtesy of Jennifer Smith, University of Missouri-Columbia). A colour reproduction of this figure can be seen in the colour section

plant and stably incorporated into its genome (Section 7.4). *Agrobacterium*, with the genes for auxin and cytokinin synthesis deleted and replaced with genes of interest, is now routinely used to transform plants (Section 12.6).

Erwinia

Species of *Erwinia* are responsible for blights, wilts and soft rots of a large number of economically important plants. *Erwinia amylovora* is the cause of fireblight of apples and pears (Figure 1.21). It is sporadic, destructive outbreaks occurring when climatic conditions are favourable to the disease and where susceptible genotypes are grown. Examples include infections of cider apples in the UK in 1980 and 1982 (Gwynne, 1984), in pears and apples in France in 1978 and 1984 respectively (Lecomte and Paulin, 1989) and in apple and pear flowers in Switzerland in 1995 (Mani, Hasler and Charriere, 1996). Pear varieties grown in the Crimea are susceptible and losses of 60 to 90 per cent of flowers and buds have been recorded in some years with yields reduced by a factor of 8–10 (Kalinichenko and Kalinichenko, 1983). A particularly severe epidemic occurred in south-west Michigan apple orchards in 2000 following unusually warm, humid, wet weather in May. Estimates of the number of trees that will die lie between 350 000 and 450 000 and the total economic loss in the region will be about $42 million (Longstroth, 2000).

   E. chrysanthemi pv. *zeae* causes a highly destructive disease of maize in tropical and subtropical countries, particularly under conditions of high temperature and humidity (Reifschneider and Lopes, 1982; Saxena and Lal, 1984; Sah,

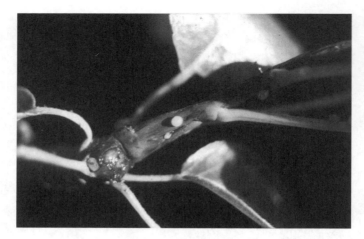

Figure 1.21   Fireblight of pears caused by *Erwinia amylovora* – note exudate and extending lesions (reproduced courtesy of Dr J. P. Paulin and CAB International). A colour reproduction of this figure can be seen in the colour section

1991). Plants are most susceptible when they are 40–60 days old and symptoms consist of withered leaves and brown, soft and water-soaked stems. Infected plants usually emit an unpleasant odour and, when the disease is advanced, collapse.

E. carotovora subsp. *carotovora* causes soft rot of potatoes and a wide range of fruits and vegetables. The bacterium may limit potato storage, particularly in warm environments and may cause total loss of seed and ware potatoes during shipment. Part of the difficulty in controlling the post-harvest disease syndrome is that infected potatoes may appear healthy, the infection being present but latent.

Pseudomonas, Burkholderia and Ralstonia

Many serious plant diseases are caused by species and pathovars of *Pseudomonas* but some of these bacteria are now classified as members of the genera *Burkholderia* and *Ralstonia*. For example, the preferred name of bacterial wilt of potato and tomato is *Ralstonia solanacearum* rather than *Pseudomonas solanacearum* and a disease of rice is now known as *Burkholderia glumae* rather than *Pseudomonas glumae*. The many pathovars of *Pseudomonas syringae* cause a variety of symptoms on a wide range of hosts. Symptoms include discoloured or necrotic spots on various organs of many plants, dieback of shoots and cankers on twigs and branches. Water soaking is often an initial symptom of a susceptible reaction but when resistance genes are present in the host, which are matched by complementary genes for avirulence in the pathogen, a hypersensitive response occurs that is associated with resistance (see Chapter 10 and Figure 10.2).

Xanthomonads

Some species and pathovars of *Xanthomonas* are devastating. For example, black rot caused by *X. campestris* pv. *campestris*, is considered to be the most important disease of crucifers (the cabbage family) world-wide, severe epidemics developing from low levels of seed infestation (Schaad, Sitterly and Humaydan, 1980; Figure 1.22). Also citrus canker, caused by *X. campestris* pv. *citri*, is regarded as a sufficiently serious disease for 20 million trees to have been destroyed in the citrus groves of Florida in an effort to eliminate it (Figure 1.23; Schoulties *et al.*, 1987).

Another Xanthomonad, *Xylella fastidiosa*, is a fastidious xylem-limited bacterium that causes a range of economically important diseases, such as Pierce's disease of grape (Figure 1.24), mulberry leaf scorch, plum leaf scald, phoney peach, coffee leaf scorch and citrus variegated chlorosis. The complete genome of a strain that causes the last of these has been sequenced (Simpson *et al.*, 2000). It consists of a 52.7 per cent GC-rich 2 679 305-base-pair (bp) circular chromosome and two plasmids of 51 158 bp and 1285 bp. There are 2904 predicted coding regions and putative functions have been assigned to almost half of these. Those concerned with pathogenicity and virulence involve toxins, antibiotics and ion sequestration systems. Orthologues of genes encoding some of these proteins

Figure 1.22   Black rot of cabbage caused by *Xanthomonas campestris* pv. *campestris* (courtesy of Said Massomo). A colour reproduction of this figure can be seen in the colour section

Figure 1.23 Citrus canker, caused by *Xanthomonas campestris* pv. *citri* (courtesy of Gerry Saddler, Scottish Agricultural Science Agency). A colour reproduction of this figure can be seen in the colour section

have been identified in animal and human pathogens, suggesting that they are constituents of pathogenicity islands (Hacker and Kaper, 2000, and see Section 10.5.2). These regions of the genome of pathogens are conserved and independent of the host and therefore imply horizontal gene transfer. This is corroborated by the finding of at least 83 genes in the genome of *X. fastidiosa* that are bacteriophage-derived and which include virulence-associated genes from other bacteria.

Rickettsias

Davis and co-workers (1998) have recorded a rickettsia or closely related organism as a possible cause of bunchy top of papaya (Figure 1.25). The bacterium was found in lactifers and although it could not be cultivated, its phylogeny was established by isolating and sequencing portions of genes corresponding to those for 16S rRNA, the flavoprotein subunit of succinate dehydrogenase (SdhA), citrate synthase (GltA), and the 17-kDa rickettsial common antigen.

Figure 1.24 Symptoms of Pierce's disease of grape-vine caused by *Xylella fastidiosa* (reproduced courtesy of Dr Alexander H. Purcell and CAB International). A colour reproduction of this figure can be seen in the colour section

The bacterium was detected by polymerase chain reaction in diseased, but not healthy, papaya tissues and in the leafhopper vector, *Empoasca papayae*.

Streptomycetes

All phytopathogenic actinomycetes, with the exception of *Nocardia vaccinii*, which causes galls and bud proliferation in blueberry, belong to the genus *Streptomyces* (Locci, 1994). *Streptomyces* species are gram positive, filamentous prokaryotes which are dispersed by spores and cause diseases of underground parts of a range of plant species (Loria *et al.*, 1997). The most widespread and economically important of these is Common Scab of potato caused by *S. scabies*. Although the loss of potatoes in terms of weight is small, the loss to the grower in financial terms is considerable since badly scabbed potatoes are not accepted by consumers. A similar disease of carrot has also been attributed to a streptomycete but the causal agent differs from that of Common Scab of potato by possession of echinulate rather than smooth spores (Janse, 1988; Hanson, 1990). *S. ipomoeae* causes a fibrous root rot of sweet potato (*Ipomoea batatas*) resulting in the reduction of both yield and quality owing to distortions and necrotic regions on the storage roots.

Figure 1.25 Bunchy top of papaya caused by a rickettsia-like organism (reproduced courtesy of Dr Michael J. Davis, University of Florida). A colour reproduction of this figure can be seen in the colour section

1.3.9 Phytoplasmas

Yellows diseases of plants have been recognized since the early 1900s and were originally assumed to be caused by viruses, although viruses could neither be consistently isolated from diseased plants nor visualized in them (Lee, Davis and Gundersen-Rindal, 2000). In 1967, Doi and co-workers (Doi *et al.*, 1967) showed that ultra-thin sections of the phloem of plants affected by such diseases contained particles that resembled animal and human mycoplasmas and are therefore actually bacteria. Since then phytoplasmas have been associated with several hundred diseases of plants including many that are important agriculturally. For example, coconut lethal yellowing disease has virtually destroyed the coconut industry in Ghana.

Although phytoplasmas are often rounded with average diameters ranging from 200 to 800 μm, others, such as *Spiroplasma citri*, the causal agent of 'stubborn' disease of citrus, are helical. They lack rigid cell walls and are surrounded by a single unit membrane.

Disease symptoms are consistent with the disturbance of plant hormone balance. For example, infected plants may be sterile or stunted or develop witches' brooms owing to the release of axillary buds from apical dominance. Internally, infected plants often have extensive phloem necrosis.

Phytoplasmas require vectors for transmission and are normally spread by sapsucking insects belonging to the families Cicadellidea (leafhoppers) and Fulgoridea (planthoppers).

1.3.10 Viruses

There are over 700 known plant viruses, many of which cause catastrophic diseases and have wide host ranges. They have been classified into three families and 32 groups (Martelli, 1992). These are based on morphology, the type of nucleic acid they contain (RNA or DNA), whether the nucleic acid is single- or double-stranded, whether it exists as a single unit or is divided, and the means of transmission. For example, furoviruses are fungal-transmitted, rod-shaped, single-stranded RNA viruses with divided, typically bipartite genomes (Rush and Heidel, 1995). Increasingly, serological and nucleic acid techniques are being used to establish the identity or relatedness of plant viruses (Sections 2.5.4 and 2.5.5).

Barley yellow dwarf viruses (BYDV) are luteoviruses which are world-wide in distribution and infect over 150 species of the Poaceae (grasses), including all the major cereals – wheat, barley, oats, rye, rice and maize. They are transmitted by at least 25 species of aphid in a persistent, circulative but non-propagative manner and there is a high degree of specificity between virus and vector (see Chapter 3, Section 3.3.2). Despite the ubiquity of these viruses and the importance of their hosts there have been surprisingly few studies of the losses inflicted, although these are deemed to be serious. Wangai (1990) reported losses of 47 per cent in wheat experimentally inoculated with the PAV strain of the virus and El Yamani and Hill (1990) gave figures of around 11–12 per cent for natural infections in Morocco. In oats the PAV strain was reported to cause an approximately 4.5 per cent loss for each 10 per cent increase in virus incidence (Bauske, Bissoriette and Hewings, 1997).

Cassava is the most important locally-produced food in a third of the world's low-income, food-deficit countries and the fourth most important source of carbohydrates for human consumption in the tropics, after rice, sugar, and maize. World production from 1994 to 1996 averaged 166 million tons/year and was grown on 16.6 million hectares, just over half of it on the African continent (Bellotti, Smith and Lapointe, 1999). Unfortunately, infection by a virus disease causing mosaic symptoms is extremely high on the African continent, ranging from 80–100 per cent and losses have been estimated at about 50 per cent (Figure 1.26). It is now clear that this disease is usually caused by one or other of two whitefly-transmitted viruses, African Cassava Mosaic Virus (ACMV) or East African Cassava Mosaic Virus (EACMV). However, in the late 1980s an unusually severe form of the disease was reported from Uganda.

Figure 1.26 Symptoms of African Cassava Mosaic Virus–note the yellow mosaic symptoms and puckering of the leaves. A colour reproduction of this figure can be seen in the colour section

Zhou and co-workers (1997) have shown that this resulted from double infection by a recombinant virus derived from ACMV and EACMV termed (UgV) and one of the parental strains, ACMV. The area affected has now expanded to cover virtually all of Uganda and large parts of Kenya, Tanzania, Sudan and the Democratic Republic of Congo. Local cultivars are generally so sensitive and the losses so great that many farmers have abandoned cassava cultivation, consequently destabilizing food security in East Africa (Legg, 1999).

   Cassava mosaic viruses are not known in South America, the centre of origin of the plant, suggesting that, once cassava was imported into Africa during the 16th century, the viruses were acquired from indigenous plants via their whitefly vector.

1.3.11 Viroids

Viroids are small (246–375 nucleotides), single-stranded, covalently closed circular, unencapsidated RNAs and are characterized by a highly base-paired, rod-like secondary structure (Diener, 2001). A viroid etiology is established for at least a dozen plant diseases, including such economically important disease agents as potato spindle viroid, citrus exocortis viroid and the viroid that causes cadang-cadang (meaning dying-dying) disease of coconut palms. The last, although consisting of less than 300 nucleotides, has destroyed over 30 million palm trees in the Philippines (Hanold and Randles, 1991). Infected palms pass through three well-defined disease stages over a period of 7–15 years and these are associated

Figure 1.27   Tinangaja, a viroid disease of coconut palms – note dying fronds and deformed coconuts (courtesy of George Wall). A colour reproduction of this figure can be seen in the colour section

with distinct changes in the molecular structure of the viroid (Haseloff, Mohamed and Symons, 1982). A similar disease in Guam is caused by the closely related coconut tinangaja viroid (CtiVd; Figure 1.27). These two viroids are the only ones known to affect monocotyledonous plants and to kill their hosts.

Koch's second postulate, that the causal organism be grown in pure culture, clearly cannot be applied to viruses or viroids since they cannot replicate in the absence of their hosts. Establishment of these particles as agents of disease therefore relies upon their constant association with disease symptoms and the use of purified virus or viroid as inoculum. Reproduction of at least some of the symptoms of the disease in healthy hosts by inoculation with highly purified preparations leaves little doubt that the virus or viroid is the cause of the disease.

Figure 1.1 (a) *Striga hermonthica* flowering on sorghum (repro-
duced courtesy of Dr Chris Parker and CAB International); (b)
Orobanche foetida flowering on *Vicia faba* (reproduced courtesy of
Dr Diego Rubiales, University of Cordoba)

Figure 1.2 (a) *Viscum album* subsp. *album* on *Robinia pseudoacacia* (courtesy of Dr Doris
Zuber, Swiss Federal Institute of Technology, Zurich); (b) *Cuscuta reflexa* on an unidentified
shrub (reproduced courtesy of Dr Chris Parker and CAB International)

Figure 1.3 Cauliflower-like
symptoms of potato wart
caused by the Chytridiomycete fungus, *Synchytrium
endobioticum* (reproduced
courtesy of Dr Hans Stachewicz and CAB International)

Figure 1.7 Lesions on rice leaves caused by *Cochliobolus miyabeanus*, often cited as the cause of rice crop failures in 1943 that led to the great Bengal famine in which an estimated 2 million people died (courtesy of Dr. R. S. Zeigler)

Figure 1.5 Symptoms of Southern Corn Leaf Blight (SCLB) caused by *Cochliobolus heterostrophus* (reproduced courtesy of Dr M. Listman, CIMMYT)

Figure 1.6 Honeydew on sorghum caused by infection with the fungus *Claviceps africana* (courtesy of Dr F Workneh, Texas Agricultural Experiment Station)

Figure 1.8 Stripe rust of wheat caused by *Puccinia striiformis*; the orange colour of the lesions on the leaves is given by the copious numbers of spores which they contain

Figure 1.9 Sporophore of *Ganoderma boninense* at the base of an oil palm (courtesy of Paul Bridge)

Figure 1.12 Stem lesions and wilted leaves on potato infected with *Phytophthora infestans* (reproduced courtesy of William E. Fry and CAB International)

Figure 1.11 Dwarfing symptoms on oats (left) compared with control (right) caused by the nematode *Ditylenchus dipsaci* (courtesy of Dr S Taylor, South Australian Research and Development Institute)

Figure 1.14 Club root of cabbage caused by *Plasmodiophora brassicae* (courtesy of Richard Cooper, University of Bath)

Figure 1.13 (a) Impact of *Phytophthora cinnamomi* on vegetation in an infested part of the jarrah (*Eucalyptus marginata*) forest of Western Australia; (b) a lesion caused by *P. cinnamomi* on a young *E. marginata* plant – the outer bark has been cut away to show the blackened diseased tissue; (c) *Xanthorrhoea preissii* killed by *P. cinnamomi* in the jarrah forest of Western Australia; (d) purple lesion of *P. cinnamomi* on the stem of a *Banksia grandis* plant – members of this genus are highly susceptible to the pathogen. Figures kindly supplied by Keith McDougall, NSW National Parks and Wildlife Service, PO Box 2115, Queanbeyan, NSW, Australia, 2620

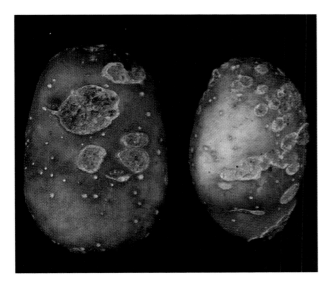

Figure 1.15 Symptoms of powdery scab of potato caused by *Spongospora subterranea* f. sp. *subterranea* – note the individual and coalesced raised lesions (courtesy of Herbert Torres)

Figure 1.17 An 18-year old oil palm infected with *Phytomonas sp.*, demonstrating early symptoms of Marchitez disease consisting of browning and drying of the leaves (courtesy of Michel Dollet, CIRAD, Montpellier, France)

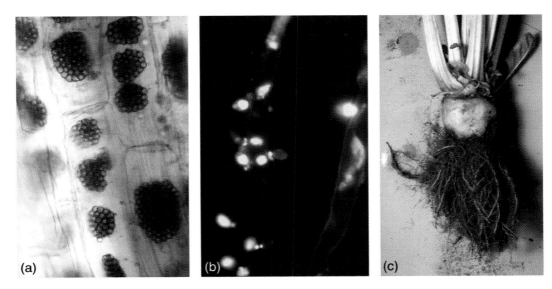

Figure 1.16 (a) Resting spores of *Polymyxa betae* in epidermal cells of sugarbeet rootlets; (b) zoospores of *Polymyxa betae* attaching and infecting sugarbeet root hairs; (c) Sugarbeet infected with beet necrotic yellow vein virus via its vector *Polymyxa betae*, showing abnormal root proliferation (rhizomania). (Kindly provided by Mike Asher of Broom's Barn Research Station, Higham, Bury St Edmunds, Suffolk 1P28 6NPR)

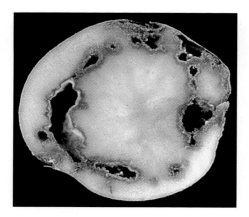

Figure 1.18 Internal symptoms of *Clavibacter michiganensis* subsp. *sepedonicus* on potato. Note the destruction of the vascular tissue and the surrounding creamy to brown coloured lesions (reproduced courtesy of Dr S. H. DeBoer and CAB International)

Figure 1.20 Crown gall on *Euonymus* (reproduced courtesy of Jennifer Smith, University of Missouri-Columbia)

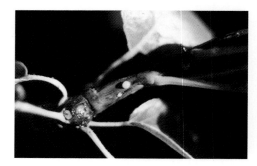

Figure 1.21 Fireblight of pears caused by *Erwinia amylovora* – note exudate and extending lesions (reproduced courtesy of Dr J. P. Paulin and CAB International)

Figure 1.19 *Rhodococcus fascians* causing malformations on tobacco (courtesy of Carmem-Lara de O. Manes)

Figure 1.22 Black rot of cabbage caused by *Xanthomonas campestris* pv. *campestris* (courtesy of Said Massomo)

Figure 1.25 Bunchy top of papaya caused by a rickettsia-like organism (reproduced courtesy of Dr Michael J. Davis, University of Florida)

Figure 1.23 Citrus canker, caused by *Xanthomonas campestris* pv. *citri* (courtesy of Gerry Saddler, Scottish Agricultural Science Agency)

Figure 1.24 Symptoms of Pierce's disease of grape-vine caused by *Xylella fastidiosa* (reproduced courtesy of Dr Alexander H. Purcell and CAB International)

Figure 1.26 Symptoms of African Cassava Mosaic Virus–note the yellow mosaic symptoms and puckering of the leaves

Figure 1.27 Tinangaja, a viroid disease of coconut palms – note dying fronds and deformed coconuts (courtesy of George Wall)

Figure 2.2 *Rafflesia tuan-mudaes* flowering (courtesy of Sandra Patiño)

Figure 3.3 A field of chickpea (*Cicer arietinum*) showing foci of infection of chickpea by *Ascochyta rabiei* (courtesy Walter Kaiser); the foci are likely to have arisen from the planting of infected seed

2 The Detection and Diagnosis of Plant Pathogens and the Diseases They Cause

Summary

Attention is usually first attracted to a plant disease by symptoms visible to the naked eye. Since they are rarely sufficiently specific for a confident diagnosis to be made, other techniques are required. These often include the culture of the causal organism, which may require selective media, and in some instances, the establishment of its host range, as well as the identification of the pathogen. Many techniques are available for pathogen identification. They include substrate metabolism, fatty acid profiles (FAME analysis), protein analysis and serological techniques, usually as one of the formats of the enzyme linked immunoabsorbent assays (ELISA). Increasingly, techniques that involve nucleic acid hybridization such as molecular beacons and amplification by the polymerase chain reaction (PCR) are being exploited to detect nucleic acid sequences of pathogens at levels as low as the femtogram range (10^{-15}g). The convenience, objectivity, speed, accuracy and sensitivity of serological and nucleic acid techniques will ensure that their use becomes ever more widespread. A diagram of the techniques currently available for disease and pathogen identification is presented in Figure 2.1.

2.1   Introduction

Inferior genetic material and unfavourable environmental conditions such as drought stress and mineral deficiency may cause poor plant performance, but these are not considered in this book, except insofar as they interact with the diseased condition. Our subject is the biotic causes of plant disease and in this chapter we shall be concerned with their detection and identification and the diagnosis of the diseases for which they are responsible. Detection is fundamental to quarantine procedures (Waterworth, 1993; Section 5.6.4) and identification

Introduction to Plant Pathology by Richard Strange
© 2003 John Wiley & Sons, Ltd   ISBN 0 470 84972 X (cased)   ISBN 0 470 84973 8 (pbk)

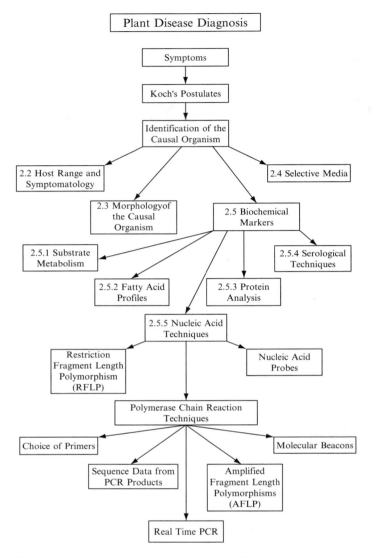

Figure 2.1 Diagram to show the range of techniques available for the detection and diagnosis of plant diseases and plant pathogens (the numbers denote sections of the chapter in which the techniques are discussed)

allows the experimenter to draw on the published experiences of other workers with the pathogen or its close relatives. Although a single technique may be sufficient for a confident identification, often two or more are used. These may vary from determination of host range to sequence analysis of the nucleotides from part of the pathogen's genome. Diagnosis on the basis of symptoms alone is hazardous unless these are specific to the disease. For example, many plant pathogens cause chlorosis, necrosis and wilts so these symptoms, although they may be dramatic, are generally of little diagnostic use. In contrast, the occurrence of unorganized tumorous tissue at the crown of a plant strongly suggests that *Agrobacterium* may be the cause.

Comparatively recently the CD-ROM has become a most welcome aid to the plant pathologist called upon to identify an outbreak of disease. For example, the American Phytopathological Society has a collection of CD-ROMs with digital images of subjects such as diseases of vegetables and small fruits. More ambitiously, CAB International (http://pest.cabweb.org) produced its third edition of the Crop Protection Compendium Global Module in 2002. Not only does this monumental work provide ready access to pictures of disease symptoms and advice on diagnosis but also provides a wealth of information on many other topics such as host range of the pathogen, geographic distribution, biology and ecology, economic impact, phytosanitary significance, morphology and control. Thus a plant pathologist equipped with a portable personal computer loaded with this CD-ROM (and a sufficiently charged battery!) will be in a position to offer valuable advice to farmers in the most remote areas of the world.

2.2 Host range and symptomatology

Plant pathologists are seldom required to diagnose a disease on an unknown plant. The host plant is therefore an immediate given and a useful starting point. For example, the crop may be one for which a compendium of diseases is published by the American Phytopathological Society. Alternatively, reference may be made to the CAB International Crop Protection Compendium mentioned in the previous section. For example, under barley in this CD-ROM there is a long list of pests and diseases, including nearly 50 pathogens for which barley is an important host. The search may be narrowed down by entering the symptoms option and then looking up the possible candidates in the pathogen files.

Although the primary host of the pathogen in question may be known, its host range may not, despite the fact that such information may be vital for epidemiological studies (see Chapter 3). Host range may also be a valuable aid to identification. In particular, some fungal and bacterial pathogens are given trinomials, the third name designating the host. For example, isolates of *Fusarium oxysporum* are classified into many formae speciales according to the hosts that they infect; *F. oxysporum* f. sp. *lycopersici* infects tomatoes and *F. oxysporum* f. sp. *cubense* infects bananas, causing Panama wilt. Similarly, isolates of *Xanthomonas campestris* are classified into many pathovars: *X. campestris* pv. *citri* causes citrus canker (see Figure 1.23) whereas *X. campestris* pv. *malvacearum* infects cotton. However, it is in the determination of physiological races (often referred to simply as races) of an organism that host-range studies are of particular importance and were, until recently, the only way in which a race could be identified. Physiological races, although morphologically indistinguishable, differ in their virulence towards a range of genotypes of the host plant, causing severe symptoms on some but not others. For example, some isolates of *Xanthomonas campestris* pv. *vesicatoria* cause bacterial spot of tomato and others cause a similar disease in pepper; these have been designated the tomato and pepper races, or the A and B groups, respectively. More recent work has classified the tomato race or A group as *Xanthomonas axonopodis* pv. *vesicatoria* and the pepper race or B group as *X. campestris* pv. *vesicatoria*. However, the pathogens may be differentiated still further by their virulence or avirulence for different genotypes of these hosts. Originally, two tomato races were recognized, one of which, T1, induced the hypersensitive response (a resistant reaction; see Section 11.3.1) on the tomato genotype Hawaii 7998 while the other, T2, was compatible with this genotype (i.e. the plant was susceptible) and did not induce the hypersensitive response. Jones and co-workers (1995) studied a number of strains from tomato fields in Florida which were classed as T2 but found that, unlike authentic T2 strains, these induced the hypersensitive response on three further genotypes of tomato. These strains were therefore designated tomato race 3. This example illustrates the importance of the choice of hosts in such tests, often referred to as differential cultivars, since the omission of one or more may lead to a failure to differentiate races in samples of pathogen isolates.

When the pathogen is international in distribution, such as stem rust of wheat, caused by *Puccinia graminis* f. sp. *tritici*, an agreed set of cultivars is usually used, but these should be supplemented with local ones. A further refinement, well-illustrated by *P. graminis* f. sp. *tritici*, is the classification of symptoms according to severity (Table 2.1.). With the increased knowledge of the complementary genetic systems of such obligate pathogens and their hosts (Section 10.3) it is now

Table 2.1   Infection types of *Puccinia graminis* f. sp. *tritici*

Infection type[*]	Designation	Symptoms
0	Low	No uredia or other sign of infection
;	Low	No uredia, but hypersensitive necrotic or chlorotic flecks of varying size present
1	Low	Small uredia often surrounded by necrosis
2	Low	Small to medium uredia often surrounded by chlorosis or necrosis
X	Low	Random distribution of variable-sized uredia on a single leaf with a pure culture
Y	Low	Ordered distribution of variable-sized uredia with larger uredia at leaf tip
Z	Low	Ordered distribution of variable-sized uredia with larger uredia at leaf base
3	Low	Medium-sized uredia that may be associated with chlorosis or rarely necrosis
4	High	Large uredia without chlorosis or necrosis

[*]These may be modified with various suffixes such as: = lower limit of size, − small, + large, and ++ extra large. N denotes more necrosis than usual. Infection types 2, X, Y, Z and 3 are considered to have inadequate levels of resistance for commercial use and 4 is extremely susceptible (Roelfs and Martens, 1988).

possible to test isolates of the pathogen against cultivars with known resistance genes. The results from such tests quickly show if the resistance gene is effective or not and if this information is obtained early enough it may influence the choice of cultivars planted subsequently (Section 5.6.1).

In Koch's first postulate (Section 1.2) the aim is to show a consistent association of symptoms of a disease with a suspected causal organism. Usually diseases are first noticed only when symptoms have become severe since earlier and more subtle effects, which may be discovered later, are liable to be missed initially. Disease symptoms in plants are diverse and range from biochemical perturbations of a few cells to death of the whole plant. In the case of parasitic angiosperms, the pathogen itself is often one of the most distinctive signs of infection. *Rafflesia*, in fact, has the largest flowers in the entire plant kingdom (Figure 2.2) and the flowering shoots of *Striga* and *Orobanche* are prominent features of heavily parasitized crops (Figures 1.1(a) and (b)). Mistletoes, themselves, are also frequently conspicuous features of infected trees (Figure 1.2(a)).

Symptoms caused by other groups of plant pathogens are usually less easy to attribute to specific organisms since, although the symptoms caused may be obvious, the pathogen itself is likely to be microscopic or sub-microscopic in size and may be completely embedded in the tissues of the host. In particular, diseases caused by viruses are especially prone to erroneous diagnosis when this is made entirely on the basis of symptoms. As Bock (1982) pointed out, errors may result from a failure to distinguish between a 'new disease' and the virus causing it which may or may not be 'new' as well as naming viruses as 'new' on insufficient evidence.

Plant virologists have made considerable use of indicator plants which often react to mechanical inoculation with distinctive local lesions. Sap from plants

Figure 2.2 *Rafflesia tuan-mudaes* flowering (courtesy of Sandra Patiño). A colour reproduction of this figure can be seen in the colour section

showing symptoms is rubbed onto the leaf of the indicator plant, often with an abrasive, and any symptoms that develop observed. Choice of the indicator plant is important as, for example, the work of Van Dijk, Vandermer and Piron (1987) has shown. They examined accessions of Australian *Nicotiana* species and found that susceptibility and sensitivity was more common in the sections *Acuminatae, Bigelovianae* and *Suaveolentes* than in other sections of the genus. In particular, they recommended *N. benthamiana*-9, *N. miersii*-33 and *N. occidentialis*-37B for routine inoculation tests.

More specifically, Damsteegt and co-workers (1997) have reported that *Prunus tomentosa* is a suitable host for the diagnosis of plum pox virus (PPV) and three other *Prunus* viruses, prunus necrotic ringspot virus (PNRSV), prune dwarf virus (PDV) and sour cherry green ring mottle virus (GRMV). Moreover, they found that representatives of two serogroups of PPV, Marcus and Dideron, could be distinguished; the former produced strong chlorotic, vein-associated patterns quickly followed by necrotic flecking and vein-associated necrotic patterns while the latter produced only vein-associated necrotic patterns.

Indicator plants have also been used to detect the presence of viruses in insect vectors. Allen and Matteoni (1991) tested a range of plants, including three species of tobacco, as monitors for the detection of tomato spotted wilt virus (TSWV) in western flower thrips (*Franklinella occidentalis*). Petunia was found to sustain the greatest number of feeding wounds, the highest percentage of infected plants and the greatest number of viral lesions.

Once an informed guess as to the nature of the causal agent of a disease has been made from the symptoms caused, confirmation should be sought by application of one or more of the techniques described below.

2.3 Morphology of the causal organism

Parasitic angiosperms and some fungal pathogens of plants are sufficiently large and distinct for diagnosis to be made on the basis of morphology seen with the naked eye. As mentioned above (Section 2.2), the flowers of *Rafflesia* are hard to miss since they are the largest in the plant kingdom and the appalling smell of decaying flesh that it exudes is a further prompt (Figure 2.2)! The sporophores of *Ganoderma boninense*, a fungal pathogen of African oil palm (*Elaies guineensis*), coconut (*Cocos nucifera*) and betelnut palm (*Areca catechu*) are also distinctive (Figure 1.9). Other pathogens will probably require the aid of the light microscope or the higher magnifications obtainable with the scanning or transmission electron microscope.

Several reference books for the identification of plant pathogens with keys are available. For parasitic angiosperms, the monograph of Kuijt (1969) still seems to be the standard text. Fungi often prove difficult to identify on the basis of morphology as seen under the light microscope since, when cultured on agar media, they frequently refuse to yield anything more interesting than wefts of mycelium. Various culture treatments can sometimes induce these species to sporulate but even then a skilled mycologist is often required to make a definite identification. Moreover, their morphology may be influenced by the medium. For example, Adaskaveg and Hartin (1997) found that the morphology of conidia of isolates of *Colletotrichum acutatum* varied according to whether they were grown on potato dextrose agar, pea straw agar or almond fruit. For a guide to the literature on the identification of plant pathogenic fungi the reader is referred to Rossman, Palm and Spieman (1987). Nematodes are classified on the basis of their morphology as seen under the light microscope (Siddiqi, 1986) but for the other groups of plant pathogens, light microscopy is seldom sufficient. Even if additional morphological information from transmission and stereoscan electron microscopy is available, a positive identification usually requires data from the supplementary techniques which are described in the remainder of this chapter.

2.4 Selective media

Some plant pathogenic organisms may be cultured *in vitro*. For these, placing infected material on a suitable medium and inspecting the resulting colonies of the pathogen may be sufficient for a positive identification. However, the choice of medium is important. All media are selective and considerable effort has been made for some pathogens or groups of pathogens to adapt substrates so that only the organisms of interest will grow. In early work, Kado and Heskett (1970) published details of five media which were selective for the five genera of plant pathogenic bacteria recognized at the time: *Agrobacterium*, *Corynebacterium*, *Erwinia*, *Pseudomonas* and *Xanthomonas*. Both their selectivity and plating efficiency have been improved by further work. For example, Fatmi and Schaad (1988) published a semi-selective medium for the isolation of *Clavibacter* (formerly *Corynebacterium*) *michiganense* subsp. *michiganense* from tomato seed which, besides nutrients, contained nalidixic acid, potassium tellurite and cycloheximide. The medium gave recovery rates of 85–132 per cent when compared with nutrient-broth yeast-extract medium and allowed the detection of a single contaminated seed containing 50 colony-forming units in samples of 10 000 seeds. Similarly, Gitaitis and co-workers (1997) have published a diagnostic medium which was effective for the semi-selective isolation and enumeration of *Pseudomonas viridiflava*, the causal agent of bacterial streak and rot of onion. The medium contained tartrate as a carbon source and the antibiotics, bacitracin, vancomycin, cycloheximide, novobiocin and penicillin G. Selectivity was enhanced by incubation at 5°C rather than higher temperatures.

Many media selective for particular fungi have been published. Here the problem is often to suppress fast-growing, saprophytic fungi and bacteria associated with diseased plant material, which would otherwise swamp the slower-growing pathogens, while allowing these to grow out of the infected plant material. For example, Manandhar, Hartman and Wang (1995) found that a semi-selective medium consisting of one-quarter strength potato dextrose agar and seven antibiotics, fenarimol, vinclozolin, choramphenicol, erythromycin, iprodione, neomycin and tetracycline was significantly less inhibitory to *Colletotrichum capsicum* and *C. gloeosporioides* than to *Alternaria* spp. and *Fusarium* spp. As a result *C. gloeosporioides* was detected more frequently in pepper seeds than when they were placed on moist filter paper, the normal means of detection. Similarly, Duffy and Weller (1994) supplemented diluted potato dextrose agar with rifampicin and tolclofosmethyl for isolation of *Gaeumannomyces graminis* var. *tritici*, the causal agent of take-all, an important disease of wheat. Isolation of this organism is difficult since roots of the plant often harbour several other pathogenic fungi. Moreover, lesions caused by *G. graminis* var. *tritici* are rapidly invaded by many opportunistic soil fungi, which displace the pathogen. Another difficulty with the identification of *G. graminis* var. *tritici* is that it does not sporulate in culture, necessitating pathogenicity and molecular tests (Sections 2.2, 2.5.4 and 2.5.5). However, the medium of Duffy and Weller (1994) was also of value in identification since the fungus altered its orange colour, imparted by rifampicin, to purple.

Often the most selective medium of all is the host plant itself and, in some instances, this has been exploited in a technique known as baiting. For example, Jeffers and Aldwinckle (1987) used apple cotyledons to detect *Phytophthora cactorum* in naturally infested soil. The soil to be tested was air-dried and remoistened before flooding several days later with water and adding the apple cotyledons. In soils that were infested with the Oomycete, the colonized cotyledons turned brown and abundant sporangia, characteristic of the fungus, could be seen with the aid of a microscope.

Baiting may also be used to detect viruses in soil samples. Fillhart, Bachand and Castello (1998) used *Chenopodium amaranticolor* to detect tobamoviruses in forest soils by growing the plant in the soils and assaying their roots for the viruses using DAS-ELISA (see Section 2.5.4).

2.5 Biochemical markers

All organisms have distinctive biochemical features and these can be used for diagnosis. Some characteristics are shared by large groups while, at the opposite end of the scale, others are unique to individual populations. The choice of character is therefore paramount in determining the taxonomic level to which an organism is defined.

Bacteria have long been identified by their metabolic functions, such as their ability to metabolize certain substrates and, more recently, by analysis of their fatty acid profiles. Additionally, soluble protein analysis by gel electrophoresis has been adopted both for bacteria and fungi. All three methods rely upon gene expression and, since this may be regulated by environmental factors, care has to be taken to standardize these.

Serological techniques were initially used for virus identification but are now applied to several of the other groups of plant pathogenic organisms (Section 2.5.4). Finally, some of the methods currently used for analysing nucleic acids provide exquisite sensitivity in the detection and quantitative analysis of plant pathogens independent of environmental conditions (Section 2.5.5). These molecular technologies have been reviewed by Martin, James and Levesque (2000). More specifically, Louws, Rademaker and de Bruijn (1999) have reviewed the application of the polymerase chain reaction to the analysis of the diversity of bacterial plant pathogens and their detection as well as the diagnosis of the plant diseases they cause.

2.5.1 Substrate metabolism

Goor and co-workers (1984) were among the first to investigate the applicability of the 'galleries' of biochemical tests contained in the API (Appareils et Procédés d'Identification) systems to the identification of phytopathogenic strains of *Erwinia* and *Pseudomonas*. Those useful for distinguishing strains of *Erwinia* were API 20E, API 50CHE and the oxidase, zym and aminopeptidase systems. *Pseudomonas* strains were differentiated using the API auxanographic systems 50CH, 50AO and 50AA. The authors concluded that, provided an adequate reference system was available, the tests provided useful alternatives to conventional procedures

Biolog is an alternative to the API system and gave better differentiation of 204 bacterial pathogens associated with a sheath rot complex and grain discoloration of rice in the Philippines (Cottyn *et al.*, 1996). Using this system, Cottyn and coworkers (1996) found that all the reported strains of *Pseudomonas fuscovaginae* were positive for the production of 2-ketogluconate but strains of *Acidovorax avenae* and *Burkholderia glumae* were negative. In contrast, *B. glumae* was positive for the production of acid from inositol but negative for the production of 2-ketogluconate and *A. avenae* was negative for both these reactions.

de Laat, Verhoeven and Janse (1994) identified the causal agent of bacterial leaf rot of a species of aloe (*Aloe vera*) as *Erwinia chrysanthemi* biovar 3 on the basis of its ability or failure to metabolize a number of substrates as well as its agglutination by an antiserum prepared against a defined strain of the organism. Similarly, Pernezny and co-workers (1995) were able to define the bacterial species causing a severe outbreak of bacterial spot in lettuce fields in Florida as *Xanthomonas campestris* on the basis of substrate utilization, the pathovar being defined as *vitians* by its fatty-acid profile (see Section 2.5.2).

In some instances organisms may be identified by their production of unusual metabolites. For example, strains of *Aspergillus flavus* that were aflatoxigenic (Section 1.3.2) were recognized by their production of volatile $C_{15}H_{24}$ compounds such as alpha-gurjunene, *trans*-caryophyllene and cadinene. These compounds were not produced by non-toxigenic strains (Zeringue, Bhatnagar and Cleveland, 1993).

2.5.2   Fatty acid profiles (FAME analysis)

Identification of bacterial pathogens of plants by fatty acid methyl ester analysis is usually performed on pure cultures of the organism. About 40 mg of wet cells are saponified and methylated. The fatty acid methyl esters (FAME) are extracted in an ether–hexane mixture and analysed by gas chromatography. Areas of the resulting peaks on the chromatograms are calculated and compared with profiles of known reference strains by computer programs (Roy, 1988). For example, the organism responsible for an outbreak of bacterial spot of lettuce was defined as *Xanthomonas campestris* pv. *vitians* as the fatty acid profiles of the strains collected from the field matched this pathovar most closely (Pernezny *et al.*, 1995). In a more extensive study, Wells, Vanderzwet and Hale (1994a) and Wells, Vanderzwet and Butterfield (1994b) were able to differentiate the five species of *Erwinia* of the 'amylovora' group as well as the four species of the 'herbicola' group.

Norman and co-workers (1997) differentiated three species of *Xanthomonas, X. albilineans, X. fragariae* and some pathovars of *X. campestris* as well as *Stenotrophomonas maltophilia* using fatty acid analyses with 100 per cent accuracy. Other pathovars of *X. campestris* were more diverse and therefore FAME analysis was not suitable for them.

2.5.3   Protein analysis

Electrophoresis of soluble proteins from plant pathogens often gives rise to complex patterns and these can be used for identification purposes. Instead of using a general protein stain, such as Coomassie Blue, particular proteins, which for example might have enzymic activity, may be revealed by appropriate staining methods. MacNish and co-workers (1994), by staining for pectic enzymes, were able to place 4250 Australian isolates of *Rhizoctonia solani* in 10 groups which they termed zymograms. However, a major disadvantage of the technique is that, compared with immunological or PCR methods (see Sections 2.5.4 and 2.5.5), relatively large quantities – 50–100 mg – of the pathogen must be grown in order to obtain the enzymes either from culture filtrates or from extracts of the organism itself (Bonde, Micales and Peterson, 1993).

In contrast, peptide profiling may be performed on comparatively small amounts of pathogen and this technique has been used to define virus strains. Kittipakorn and co-workers (1993) investigated virus isolates from groundnut which caused a wide variety of symptoms varying from mild mottle to systemic necrosis and stunting. Virion proteins (150–190 µg) were isolated, digested with trypsin and separated by high-performance liquid chromatography (HPLC). Despite the disparate symptoms all were found to have identical HPLC profiles and were identified as strains of peanut stripe potyvirus.

2.5.4 Serological techniques

Serological techniques rely on the specificity of antigen–antibody binding and a sensitive method for detection of the resulting complex. They have been used for the detection of plant viruses for many years but more recently they have been applied to the diagnosis of fungi, Oomycetes, phytoplasmas and bacteria as well. In order to obtain reliable results, the antibody preparation used, whether polyclonal or monoclonal, must have the required specificity and high affinity for an appropriate epitope of the pathogen.

Polyclonal antibody preparations are obtained by injecting an animal with antigen and, as their name implies, contain a population of different antibodies. As pointed out previously (Strange, 1993) these are likely to be heterogeneous with regard to the antigen with which they bind, the affinity of the binding and their suitability for the attachment of reagents used to detect them. Purification of the antigen to a high level will eliminate some of this variation. In contrast monoclonal antibodies, although more difficult to obtain, may be screened for specificity to a single epitope and will provide a stable source of antibody since the hybridoma may be kept in perpetuity by culturing and storage in the frozen state (Ball *et al.*, 1990). In some instances the extreme specificity of a monoclonal antibody can be a disadvantage since it may not detect all members of a taxonomic group. Here a mixture of monoclonal antibodies may be used to overcome this problem.

Advances in molecular immunology have allowed the development of specific recombinant monoclonal antibodies. One such reagent has been used to detect *Ralstonia solanacearum* (formerly *Pseudomonas solanacearum*) race 3, a pathogen of potato, tomato and eggplant for which many countries have quarantine regulations (Griep *et al.*, 1998; Section 5.6.4). The antibody is specific to the lipopolysaccharide of the bacterium and holds promise for being sufficiently reliable to obviate the need for time-consuming bioassays, which are normally required to provide confirmation of positives detected by polyclonal antibodies.

Antibody binding may be recorded and quantified by a variety of techniques but enzyme-linked immunosorbent assay (ELISA) in one of its variants is the one that is used predominantly. The most common of these are double-antibody sandwich ELISA (DAS-ELISA) and indirect ELISA (I-ELISA; Figure 2.3).

In DAS-ELISA, antibody is bound to a solid support, such as the wells of a microtitre test plate and the test sample, containing the antigen of interest, added. This is 'sandwiched' by adding another antibody which has been conjugated with an enzyme such as alkaline phosphatase. When substrate for the enzyme is added such as *p*-nitrophenyl phosphate a coloured reaction product is obtained. Normally the intensity of the colour is determined spectrophotometrically to give a measure of the amount of antigen, using a plate reader.

In I-ELISA, antibody is bound to a solid support, as in DAS-ELISA, and this binds the antigen. However, the bound antigen is recognized by an antibody raised in a second animal and this second antibody is recognized by a further antibody, which is conjugated to an appropriate enzyme. The advantage of

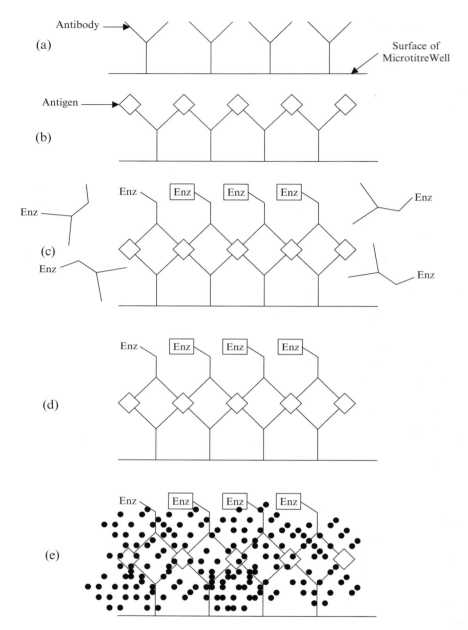

Figure 2.3 Double antibody sandwich-enzyme linked immunosorbent assay (DAS-ELISA): (a) antibody bound to the wells of a microtitre plate; (b) sample containing antigen ◇ of interest added and wells washed well with buffer; (c) antibody conjugated with an enzyme added; (d) wells washed with buffer; (e) substrate for enzyme added and coloured product ● measured – the intensity of the colour developed is proportional to the amount of antigen

I-ELISA is that the antibody used to recognize the antigen of interest is not constrained by conjugation with an enzyme and is likely to bind more completely. For this reason it may be less strain specific than DAS-ELISA. Another variant of the technique involves trapping the antigen to the $F(ab')_2$ fragments of specific IgG on a solid support and detecting it by intact specific immunoglobulin (IgG) to which enzyme has been conjugated in the Fc region (Miller and Martin, 1988).

Dot ELISA and dipsticks are two further developments of the ELISA technique. In both, antibodies are bound to a membrane and the membrane is incubated with a preparation suspected to contain the antigen. After washing, the membranes are treated with enzyme-conjugated antibody and incubated with the enzyme substrate. The membrane is then dried and the intensity of colour measured, for example, with a hand-held reflectometer.

Cahill and Hardham (1994a) describe a development of the dipstick technique which they combined with baiting to detect *Phytophthora cinnamomi* from soil (for examples of the damage done by this pathogen see Figure 1.13). Soil samples (20g) were flooded with distilled water and baited with cotyledons from *Eucalyptus sieberi*. After incubation for 3 days the samples were cold shocked at 4°C for 20–30 min to induce the release of zoospores and three dipsticks were floated, membrane side downwards, on the water for 1.5 h. Some of the released zoospores encysted on the dipsticks and were detected immunologically as follows. The dipsticks were treated with a murine monoclonal antibody specific for an antigen of *P. cinnamomi* located on the cyst periphery; the antibody was recognized by an anti-mouse antibody conjugated to alkaline phosphatase and the conjugate was visualized by treatment of the dipsticks with a mixture of 4-chloro-2-methylbenzene-diazonium salt and naphthol AS-MX phosphate to give a red insoluble dye, which was easily visible under a hand-held lens or dissecting microscope.

Other techniques for detecting bound antigens include gold-conjugated antibody, antibody conjugated with fluorescent dyes and, for biotinylated antibodies, avidin conjugates, since the glycoprotein avidin binds strongly to biotin. Milne and co-workers (1995) used immunogold labelling successfully to detect three unrelated phytoplasmas in plant tissue both before and after embedding and Jones, D.A.C. and co-workers (1994) used fluorescence labelling to detect *Erwinia carotovora* subsp. *atroseptica* in seed potato. This organism is responsible for blackleg in cool temperate regions and the incidence of the disease is directly related to the incidence of contamination of the seed potato, the threshold for disease development being about 10^3 cells per tuber. Juice was extracted from the peel of potatoes and placed in the wells of 24-well tissue culture plates together with an agarized medium containing polygalacturonic acid. After incubation for 48 h the agar in the wells was dried to a thin film in a hot-air oven at 50°C and stained with fluorescein isothiocyanate conjugated with antiserum specific to the bacteria. The films were washed before photographing under a fluorescence microscope and colonies of the bacterium were enumerated from the negatives using a commercial imaging system.

Immunofluorescence was also used by Chittaranjan and de Boer (1997) to detect *Xanthomonas campestris* pv. *pelargonii* in greenhouse nutrient solution. A murine monoclonal antibody, 2H5, was used and binding was detected by an antimouse antibody conjugated with indocarbocyanine.

Blotting is a very convenient technique and may be adapted for field applications. For example, Lin, Hsu and Hsu (1990) were able to detect several viruses and phytoplasmas in plant material by blotting freshly cut surfaces of the hosts onto nitrocellulose membranes and treating the membranes with biotinylated antibodies specific to the pathogens. Antibody binding was visualized in these instances by treatment of the membranes with avidin–enzyme conjugates and the appropriate enzyme substrates.

2.5.5 Nucleic acid techniques

The genome of every organism consists of a unique sequence of nucleotides but the degree of conservation of different domains is variable. Highly-conserved regions are found among many organisms which are distantly related whereas some regions such as satellite DNA sequences, which are repetitive and usually non-transcribed, are very variable. In consequence, a considerable degree of discrimination is possible. For example, Piotte and co-workers (1994, 1995) found two highly reiterated but different satellite DNA sequences with monomeric units of 295 bp and 169 bp in the nematodes, *Meloidogyne incognita* and *M. hapla*, respectively. The monomers were cloned and used as probes. That from *M. incognita* was non-specific since it hybridized with all populations belonging to *M. incognita, M. arenaria* and *M. javanica*. In contrast, the monomer from *M. hapla* was specific and, because of its variability, could be used to differentiate three populations of this species as well as to distinguish it from sympatric populations of M. *chitwoodi* and *M. incognita*. Moreover, because of the high reiteration – about 15 000 monomers per haploid genome – identification of the organism was possible by direct hybridization with as little material as one squashed female nematode, obviating the need to extract DNA. For non-reiterative DNA, the selection of the sequence to be analysed is critical and, in practice, this means the choice of appropriate restriction enzymes or nucleic acid sequences for the probes or primers in the techniques described below.

A further advantage of nucleic acid analyses is that, unlike the techniques described so far, they are not dependent on gene expression and therefore are independent of environmental factors. In the past few years a large number of techniques has been published, many of which are extremely ingenious and which are revolutionizing the detection and assay of plant pathogens. In order to give some coherence to this rapidly developing field the techniques are described roughly in order of specificity.

Gel fractionation

Nucleic acids fragments are separated according to size by gel electrophoresis. Agarose gels are normally used for separating fragments which are 300 nucleotides in length or more and polyacrylamide gels for shorter lengths. The resulting bands are visualized by staining with ethidium bromide, which is often included in the gel, and viewing under ultraviolet light. Usually the technique is used in conjunction with other procedures such as cutting extracted DNA with restriction enzymes or amplifying specific domains by the polymerase chain reaction. However, Hodgson, Wall and Randles (1998) were able to obtain preliminary identification of a viroid associated with the lethal Tinangaja disease of coconuts present on the island of Guam (see Figure 1.27 for symptoms of this disease) by two-dimensional polyacrylamide gel electrophoresis of RNA extracted from palms showing symptoms.

Restriction fragment length polymorphism (RFLP)

RFLP analysis was one of the first techniques to be exploited in determining subtle differences in the DNA from two or more organisms. DNA preparations from the organisms of interest are digested with restriction enzymes and the resulting fragments run on a gel. Different organisms will give fragments of different size according to the distribution of their restriction sites. In some cases the restriction enzymes used may delineate repetitive DNA, in which case banding patterns, visualized by staining with ethidium bromide and viewing under ultraviolet light, may be clear without any further treatment. More often, however, the differences are obliterated by a smear of thousands of differently sized DNA fragments. In these instances it is necessary to use a probe to pick out the fragments of interest.

Nucleic acid probes

Probes are selected on the basis of their complementarity to unique nucleotide sequences of the organism to be detected and will therefore bind wherever these sequences occur. Recognition that binding has occurred is achieved by labelling the probe. Frequently the label is ^{32}P or ^{33}P which enables binding to be recognized by autoradiography but non-radioactive probes are being developed which may have greater applicability in areas where elaborate laboratory facilities are not available (Miller and Martin, 1988). For example, biotin-labelled adenine may be introduced into the probe either by nick translation or chemically with photobiotin. The probe may then be recognized by avidin conjugated to an enzyme in an ELISA assay since avidin binds tightly to biotin (see Section 2.5.4; Bertaccini et al., 1990).

Various methods for the preparation of probes have been used. For plant viruses, labelled complementary DNA (cDNA) can be prepared directly from the virus RNA using the enzyme reverse transcriptase. Alternatively, screening of a cDNA library of the organism of interest may allow the identification of a number of sequences which vary in their degree of conservation. The more highly conserved sequences will be common to related groups while strain variation may be detected among less conserved sequences. For example, Waterhouse, Gerlach and Miller (1986) found both general and luteovirus serotype specific probes in a cDNA library of barley yellow dwarf virus.

Three other examples will further illustrate the importance of choosing the appropriate fragment of DNA as a probe and the different techniques that have been used to produce them. Coronatine is a toxin produced by a number of pathovars of *Pseudomonas syringae* and is an important virulence factor (Section 8.5.1). It would therefore be useful to be able to determine specifically the prevalence of toxigenic isolates. Cuppels and Elmhirst (1999), using the probe TPRI, derived from the *Pseudomonas syringae* pv. *tomato* gene cluster controlling production of the toxin, in conjunction with a semi-selective medium, were able to trace natural populations of the pathogen on tomato plants from just before planting to harvest.

Magnaporthe grisea, the cause of rice blast disease, was thought to be genetically unstable and this was deemed to be an explanation of the apparent instability of pathotype. Levy and co-workers (1991) were able to disprove this view by using the highly repetitious sequence (termed MGR for *Magnaporthe grisea* repeat), discovered by Hamer and co-workers (1989) as a probe. When the MGR probe was used with *Eco*R1 digests of strains of the fungus, which had been separated by agarose gel electrophoresis, the resulting MGR-DNA fingerprints distinguished and accurately identified the major pathotypes collected in the USA over 30 years. The fingerprints also defined the clonal lineages within and among pathotype groups.

Bacterial spot of tomatoes and peppers is caused by *Xanthomonas axonopodis* pv. *vesicatoria* and *Xanthomonas vesicatoria*, corresponding to the groups A and B of the former taxon, *X. campestris* pv. *vesicatoria* (Section 2.2). Diagnosis of the diseases and identification of the pathogens is complicated by the occurrence of epiphytic xanthomonads which are non-pathogenic. Kuflu and Cuppels (1997) have developed a probe for the pathogens by genomic subtraction. DNA from a pathogenic strain was labelled with digoxigenin and denatured. The labelled DNA was hybridized with denatured unlabelled DNA from three non-pathogenic strains. After one round of subtraction the preparation was found to hybridize only with Southern blots of restriction enzyme-digested DNA from pathogenic but not non-pathogenic strains, allowing the identification of a fragment which preferentially hybridized to DNA from only pathogenic strains.

Probes that are either radio, chemically or antigenically labelled may be used in a method which is analogous to dot ELISA (Section 2.5.4). The nucleic acid to be tested is spotted and baked onto a solid matrix such as a nitrocellulose membrane and the remaining sites on the membrane are blocked with heterologous DNA, such as that from calf thymus, and protein (e.g. bovine serum albumin). The probe is then hybridized with the test nucleic acid preparation under carefully controlled conditions. Variation of these conditions alters the stringency of the test (i.e. the amount of mismatching of bases) and hence the degree of binding of the probe. For example, Herdina-Roget (2000) has published an assay using a radioactive probe that detects DNA levels of *Gaeumannomyces graminis* var. *tritici* at less than 30 pg, 30–50 pg and more than 50 pg in 0.1 g soil organic matter corresponding to low, moderate and high levels of the disease, respectively. The assay allows farmers to have soil samples assessed before sowing and can be used to predict the potential yield loss.

Probes may also be used to recognize pathogens in tissue blots. For example, Duran-Vila, Romero-Durban and Hernandez (1996) pressed freshly cut stems or leaves of chrysanthemum on polyvinylidene membranes and were able to detect the presence of chrysanthemum stunt viroid with a digoxigenin labelled probe.

Polymerase chain reaction techniques

The polymerase chain reaction (PCR) has facilitated advances in many areas of biology and the diagnosis of plant disease is no exception. Essentially the technique provides a method for amplifying *in vitro* specific nucleotide sequences from the organisms of interest, usually corresponding to 0.5–2.5 kDa, although sequences in excess of 10 kDa have been obtained. The basic procedure is now so well known that it hardly requires any introduction. Specificity of the region to be amplified is determined by the primers used to delineate the two ends of the sequence and the amplification itself is carried out in a thermocycler. Normally target DNA for amplification is given a preliminary heating to ca. 94°C in order to separate the strands. It is then mixed with the other constituents of the reaction – the primers, dNTPS (the deoxynucleotide triphosphates ATP, TTP, GTP and CTP), buffer, $MgCl_2$ and a thermostable DNA polymerase. Usually this last is *Taq* polymerase from *Thermus aquaticus* but other heat stable polymerases are now available such as *Vent* which has the advantages of greater fidelity and products that are blunt ended (*Taq* DNA polymerase has a tendency to add an extra A residue to the 3'-end of each strand). The reaction mixture is then cycled around 30 times through a temperature régime of 94°C, 45–55°C and 72°C. At 94°C the DNA strands separate and at 45–55°C primer annealing to the separated strands occurs, the precise temperature depending on primer composition and the degree of stringency required by the experimenter. DNA complementary to the separated strands is synthesized at 72°C. The time at each temperature varies according to the experimental material but 30 s at the lowest and highest temperatures and 1 min at the intermediate temperature are commonly used. Allowing for ramping time among the three temperatures, 30 cycles of such a régime can be achieved in less than 3 h. After the reaction, the PCR products may be visualized by running them on a conventional agarose gel and staining with ethidium bromide.

Choice of primers   The detection of variation within pathogen populations and the detection of specific pathogens are usually the applications of PCR of most interest to plant pathologists. For the former, Random Amplified Polymorphic DNA (RAPD) is often suitable since it does not presume any knowledge of DNA sequences of the organism. In this technique, rather than selecting primers of known complementarity to the pathogen, arbitrary sequences of about 10 nucleotides are used. Separation of the products by agarose gel electrophoresis often reveals polymorphisms among isolates. For example, Serrano, Camargo and Teixeira (1999) were able to separate 48 isolates of *Phytomonas* species, of which 31 were obtained from plants and 17 from insects into five main clusters and show that they were grouped according to geographic origin.

RAPDS are not so suitable for the diagnosis of pathogens for at least two reasons. First, the pattern of bands on the gels is contingent upon very precise conditions of amplification and it is difficult to reproduce these from one thermocycler to another. Even in the same instrument, the heating block may not be uniform. Second, in complex substrates such as plants or soil, DNA from these sources may interfere. In these instances, primers that are known to recognize only the organisms of interest are required.

One method for choosing primers is to design them on the basis of the sequence of specific amplicons obtained by RAPD analysis. For example, Parry and Nicholson (1996) found two RAPD amplicons that were common to all isolates of *Fusarium poae* tested. These were cloned and used to probe Southern blots of DNA from a range of isolates pathogenic to seeds and stem bases of wheat. One fragment was partially sequenced and the sequence data used to design primers. When tested, these were found to amplify DNA from all isolates of *F. poae* but not from a range of other fungal species associated with diseases of cereal ears or seed. Similarly, Schilling, Moller and Geiger (1996) identified sequences which enabled them to design primers which gave rise to amplicons in a PCR reaction that allowed differentiation of *Fusarium culmorum* and *Fusarium graminearum*.

In fungi, sequences from ribosomal genes are particularly favourable targets for detection and may also provide useful data concerning the phylogenetic relations of isolates. This is because ribosomal genes are present in high copy number (around 60–200 per haploid genome) and there is a wide disparity in the degree of conservation of the genes and their spacer regions. Thus, although the rDNA genes themselves are conserved, there is considerable heterogeneity in the internal transcribed spacer (ITS) regions between the genes, more in the transcribed external spacer (ETS) regions and still more in the non-coding or intergenic spacer (IGS) units within rDNA repeat units (Figure 2.4). As a result, it is possible to use universal primers, based on sequences of the conserved genes, which give rise to amplicons that span regions of heterogeneity. One method of identifying differences among amplicons is to digest them with restriction enzymes and run gels to reveal differences in banding patterns. For example, Balesdent and co-workers (1998) investigated the *Leptosphaeria maculans* complex which is the cause of blackleg disease of oilseed rape. They were able to propose a scheme for the large-scale identification of species of the complex

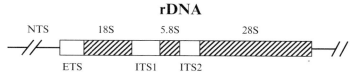

NTS = Non-Transcribed Spacer (InterGenic Spacer)
ETS = External Transcribed Spacer
18S, 5.8S and 28S = Ribosomal Genes
ITS1 and ITS2 = Internal Transcribed Spacers

Figure 2.4 Diagram of eukaryotic ribosomal DNA: a single repeat unit is shown which may exist in copy numbers as high as 200 making conserved regions easy targets for Polymerase Chain Reaction (PCR) primers; when these span less conserved regions such as the Internal Transcribed Spacers, ITS1 and ITS2, these will also be amplified, allowing their sequences to be used for diagnosis either by restriction enzyme digestion and gel electrophoresis of the product or sequencing

based on amplification of the ITS1–5.8S–ITS2 region of the fungi involved and digestion of the amplicons with four restriction enzymes.

More specifically, sequence data from the ITS1–5.8S–ITS2 region may be used to design primers for specific amplification of sequences from the organism of interest. For example, Schilling, Moller and Geiger (1996) found that there was sufficient sequence variation, especially in ITS2, to construct primers for the specific amplification of *F. avenaceum* but that polymorphism between the ITS sequences of *F. culmorum* and *F. graminearum* was insufficient to design primers specific for these species.

In bacteria, three unrelated families of repetitive DNA sequences, known as BOX, ERIC (enterobacterial repetitive intergenic consensus) and REP (repetitive extragenic palindromic) are distributed throughout the genome. Thus, outwardly facing primers complementary to these sequences allows the amplification of the intervening regions giving rise to many amplicons of different sizes which may be separated on agarose gels. Using this technique, Louws and co-workers (1998) were able to separate *Clavibacter michiganensis* into five subspecies which corresponded to their host ranges.

Other workers have designed primers which recognize sequences required for virulence of pathogens. For example, bacteria such as *Pseudomonas syringae* pv. *savastanoi*, *Agrobacterium tumefaciens* and *Erwinia herbicola* form galls on their hosts as a result of the expression of genes which code for the plant-growth regulating compounds indole acetic acid (IAA) and cytokinins (see Chapter 7). Manulis and co-workers (1998) designed primers which recognized non-conserved sequences of the IAA and cytokinin biosynthesis genes that enabled them to detect specifically gall forming as opposed to saprophytic strains of *Erwinia herbicola* pv. *gysophilae*. Moreover, sensitivity of detection was boosted 100 fold by nested PCR in which, after one round of PCR with primers which recognized cytokinin biosynthesis genes and gave an amplicon of 607 bp, a second round was performed with primers which were complementary to sequences internal to this amplicon and gave a product of 522 bp.

Toxins are important pathogenicity or virulence factors (see Chapter 8). For example, *Pseudomonas syringae* pv. *syringae* produces small cyclic lipodepsipeptides which are phytotoxic and antifungal (Section 8.5.1). Production of these compounds is dependent on the expression of the gene *SyrB*. Use of primers specific to sequences of the *SyrB* gene allowed the polymerization of a 752 bp fragment only from toxigenic isolates of the bacterium (Sorensen, Kim and Takemoto, 1998)

For RNA viruses the RNA is reverse-transcribed into cDNA before amplification (RT-PCR). For example, Haber and co-workers (1995) developed a technique for detecting double-stranded RNA associated with flame chlorosis of cereals in small amounts of tissue and candidate fungal vectors. Total RNA was extracted from milligram quantities of test tissue, reverse transcribed, and amplified by PCR. Two primer pairs were used which were predicted to yield 358- and 347-bp DNA fragments from a consensus 821-bp sequence covering an open reading frame.

Sequence data from PCR products   *Colletotrichum* is a notoriously 'difficult' fungal genus with taxa that have been recorded on a wide range of plant species (Sutton, 1992). In particular, *C. gloeosporioides* has been described as a 'group species' and the opinion expressed that 'no progress in the systematics and identification of isolates belonging to this complex is likely to be made based on morphology alone' (Sutton, 1992). Accordingly, Sherriff and co-workers (1994 and 1995) used rDNA sequence analysis to distinguish members of this genus. Their work revealed new species groupings and confirmed the distinction between *C. graminicola*, a pathogen of maize and *C. sublinoleum*, a pathogen of sorghum and *Rottboellia*. Also, Moses and co-workers (1996), using the same technique, were able to confirm that *C. gloeosporioides* was the cause of a dieback disease of cassava that is prevalent in West Africa.

Quantitative PCR   PCR may also be used to quantitate target DNA sequences. One way of doing this is to include in the reaction mixture as an internal standard a known amount of a sequence that is heterologous to the target and differs in size while retaining the sequences that are recognized by the primers. On the assumption that both the standard and target sequences are amplified equally, the amount of the target sequence can be calculated from the amount of initial standard sequence and the ratio between the amounts of amplified products of both target and standard sequences. For example, Mahuku, Goodwin and Hall (1995) were able to quantify *Leptosphaeria maculans* during the development of blackleg symptoms in oilseed rape using a template derived from *Leptosphaeria korrae* as the internal standard. Alternatively, quantitative results may be obtained by the separate amplification of members of a dilution series. Amplification of target DNA will take place at the higher but not the lower concentrations and the amount present in the initial sample may be calculated by the

Poisson distribution (Sykes *et al.*, 1992). However, the results should still be treated with caution since inhibitors of the PCR reaction may be present and interfere.

As an alternative, PCR may be carried out on a dilution series of the template DNA in the presence of digoxigenin labelled UTP, allowing its incorporation into the PCR product. The PCR product is denatured and hybridized to a biotin labelled capture probe which has been designed to anneal to an internal sequence of the PCR product. The hybridization products are immobilized on streptavidin coated microtitre plates and peroxidase-conjugated anti-digoxigenin antibodies are used to detect the hybrids. The technique is 10–100 fold more sensitive than fluorescent staining of PCR products on agarose gels and is capable of discriminating point mutations since the stringency of the binding of the capture probe can be increased by increasing the temperature of hybridization (Bonants *et al.*, 1997; Figure 2.5).

Real time PCR   In the last few years, a sophisticated piece of equipment has appeared on the market which gives PCR results in real time and obviates the need for running gels in order to visualize amplicons. The PCR reaction is performed in a 96-well microtitre plate with the addition of a probe, which recognizes the amplicon and which is labelled with a fluorochrome and a quencher. When polymerization occurs, the 5′ exonuclease activity of *Taq* polymerase dissociates the fluorochrome and its quencher and the resulting fluorescence is recorded, the excitation and emission light being transmitted to and from the wells by fibre optics (Figure 2.6). Mumford and co-workers (2000) have used this technique for the simultaneous detection in a multiplex assay of potato mop

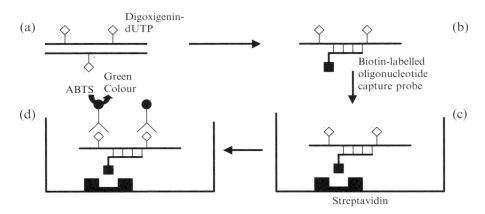

Figure 2.5   A sensitive PCR-ELISA system for detecting plant pathogens: (a) PCR is performed with digoxigenin-dUTP which is incorporated into the PCR product; (b) a sample of the PCR product is denatured and hybridized with a biotin-labelled oligonucleotide capture probe; (c) the hybridization products are immobilized on a streptavidin coated microtitre plate; (d) peroxidase-conjugated anti-digoxigenin antibodies and ABTS substrate are used to detect the hybrids (redrawn from Boehringer Mannheim advertising material)

Figure 2.6 An ABI 7700 quantitative PCR machine (photograph courtesy of David Bacon, Imperial College, London)

top virus (PMTV) and tobacco rattle virus (TRV), the causal agents of a syndrome known as spraing or corky ringspot of potatoes. The symptoms of the disease are brown necrotic arcs or lines in the flesh of the tuber and, although not causing severe reductions in yield, can lead to rejection of crops. For the assay, primers were used that recognized sequences within different conserved domains of the two viruses and specific probes were designed to recognize these domains. In order to improve specificity, the probes had melting temperatures 10°C higher than the primers. The quencher, tetra-methylcarboxyrhodamine (TAMRA), was the same for both probes but the fluorochrome for TRV was VIC and for PMTV, FAM (6-carboxyfluorescein). The assay allowed the experimenters to replace the two separate tests, reverse transcription PCR for TRV and ELISA for PMTV, with a single tube format; moreover the new assay was 100 and 10 000 times more sensitive, respectively, than the assays it replaced.

Cullen and co-workers (2002) also used quantitative real-time PCR to detect *Colletotrichum coccodes*, the cause of black dot disease in potato tubers and soil samples. They showed a high correlation ($r = 0.978$) between the threshold cycle, defined as the cycle number at which a statistically significant increase in reporter fluorescence occurs, and the logarithm of the starting quantity of DNA down to 1 femtogram, the lowest level tested (Figure 2.7). This enabled the experimenters to detect as few as three spores of the fungus/g of soil, equivalent to 0.06 microsclerotia.

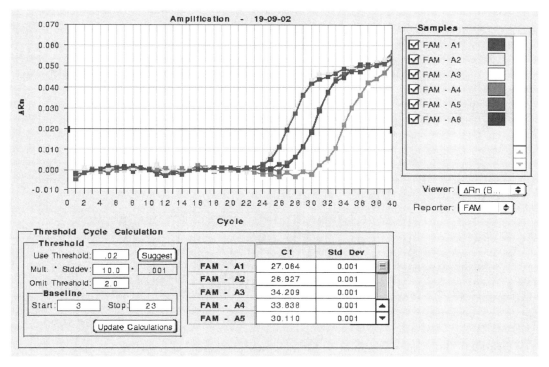

Figure 2.7 Print-out of screen from a computer controlling a real time PCR machine, showing samples reaching the set threshold of 0.02 at different cycles according to the amount of template DNA

Amplified Fragment Length Polymorphisms (AFLP)

This DNA fingerprinting technique which, like RAPD, does not require sequence knowledge, was introduced by Vos and co-workers (1995). The DNA is first digested with restriction enzymes and then ligated to double-stranded adapters. These consist of a core sequence and an enzyme-specific sequence and serve as primer sites for amplification of the restriction products. The number of fragments detected can be tuned by addition of one or more nucleotides to the 3' end of the primers. For example, Bragard and co-workers (1997) used the restriction enzymes *Apa*I and *Taq*I and the corresponding primers, each with an additional 3'-terminal guanine in their analysis of the diversity of *Xanthomonas translucens* from small grains.

Scoring the bands can be a difficulty with the AFLP technique. One way around this is to use GeneScan. Here one of the primers is labelled with a fluorochrome and a fluorescent ladder is included in the samples when they are separated by gel electrophoresis. The gel is read by a laser and the data computerized to give electropherograms and precise read outs of the size of the fragments (Figure 2.8).

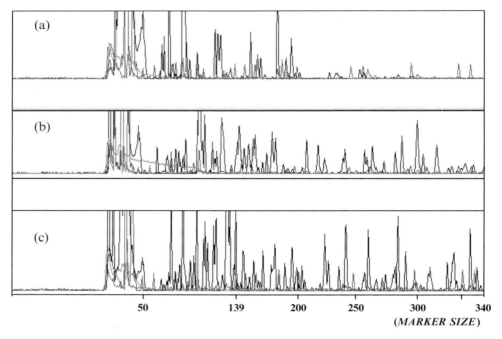

Figure 2.8 Electropherograms of AFLP fragments, ranging in size from 0 to 350 nucleotides, obtained from three isolates of *Colletotrichum gloeosporioides* from different geographic locations (a) South Pacific (b) Hainan Island, China and (C) Ghana.

Molecular beacons

Molecular beacons are another technique that employs fluorochromes and quenchers which was originally developed by Tyagi and Kramer (1996). The beacon consists of single-stranded DNA with a sequence complementary to that of the target sequence and a stem loop formed by the annealing of the 5′ and 3′ arm sequences. A fluorochrome is attached to the 5′ end and a quencher to the 3′ end of the beacon. The beacon is added to a solution of target DNA, heated to 80°C and allowed to cool. When the beacon anneals with the complementary DNA the fluorochrome and its quencher are moved apart resulting in fluorescence (Figure 2.9). Since different fluorochromes may be used in the same reaction mixture it is possible to detect multiple target sequences simultaneously. For example, using this technique, Eun and Wong (2000) were able to detect both Cymbidium mosaic virus and Odontoglossum ringspot virus in orchids. Since these are RNA viruses it was necessary to perform RT-PCR on specific sequences of the two viruses first and use the molecular beacons to detect these cDNA molecules.

(a)

(b)

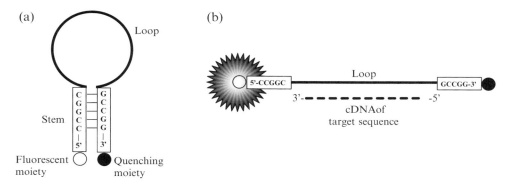

Loop

Stem

Fluorescent moiety Quenching moiety

cDNA of target sequence

Figure 2.9 Diagram of a molecular beacon: (a) the beacon before annealing to the complementary DNA showing the fluorescent and quenching moieties held in close proximity by the complementary bases of the stem loop; (b) the beacon annealed to a complementary target DNA sequence showing the displacement of the fluorescent and quenching moieties (image kindly provided by Dr Wong and reproduced with permission of the American Phytopathology Society)

2.5.6 Choice of diagnostic techniques

With such a large number of diagnostic techniques available, the experimenter may find difficulty in deciding which one is most suitable for a particular application. Usually the requirements fall into one or more of the following categories.

(1) identification of an unknown pathogen;

(2) detection of a known pathogen;

(3) quantitative assessment of infection.

For identification purposes, sequence data of an appropriate part of the pathogen's genome is unequivocal. Often the sequences of the ITS regions of rDNA are sufficient but in some instances they have not separated isolates that differ significantly in other respects. For example, three species of *Seiridium, S. cardinale, S. cupressi* and *S. unicorne*, associated with cypress canker were not separated by their ITS sequences. Barnes and co-workers (2001) therefore analysed the histone and partial β-tubulin sequences of 14 isolates of the genus from cypress trees. The sequence data separated the isolates into two clades, one of which contained the less virulent *S. unicorne* and the other the more virulent *S. cardinale*.

Detection may be required for quarantine purposes and in this instance the most sensitive technique available would probably be the one to choose. A combination of techniques may be better still. For example, Schaad and co-workers (1995) used a combined biological and enzymatic amplification technique to detect *Pseudomonas syringae* pv. *phaseolicola*. Bean seeds were soaked in water overnight containing a low concentration of detergent and aliquots plated on media. Colonies were washed from the plates and used directly for nested PCR with primers designed to amplify a domain of the *tox* region of the organism's DNA responsible for the synthesis of phaseolotoxin (Section 8.5.1). The detection threshold was 1–2 cells.

Another way to boost the sensitivity of a test is to use immunocapture before PCR. For example, Moury and co-workers (2000) found that they could increase the sensitivity of detection of prunus necrotic ringspot virus (PNRSV) in rose tissue about 100 times over that of DAS-ELISA. In brief, microtubes were coated with anti-PNRSV antigen and incubated with plant extract. After washing and removing moisture, tubes were preheated at 65°C for 15 min before adding the RT-PCR mix containing primers complementary to the coat protein gene of the virus. Amplicons were detected by agarose gel electrophoresis.

Quantitative assays may be required to indicate the level of resistance of a plant to a pathogen. Here the convenience of the many kits available on the market for ELISA may influence the researcher. Quantitative PCR with an internal standard may give satisfactory results but for those fortunate enough to be able to afford the equipment for real time PCR, this is clearly the technique of choice.

2.5.7   Ralstonia solanacearum – a case study

R. solanacearum was originally named *Pseudomonas solanacearum*. It is a pathogen with a wide host range and may be devastating on potato. Owing to its importance many of the techniques discussed above – symptoms, morphology, selective media and biochemical markers – have been used in its diagnosis, making it an ideal subject for a case study. The following descriptions refer mainly to bacterial wilt of potato.

Symptoms

In bacterial wilt of potato, caused by *R. solanacearum*, wilting of leaves at the ends of branches during the day with recovery at night is the first visible symptom. Epinasty of petioles may also occur and leaves become bronzed. As the disease progresses, plants fail to recover at night and die. Stems may develop brown streaks 2.5 cm or more above the soil line and, when cut, a white, slimy mass of bacteria exudes from the vascular bundles. If such stems are placed in water, the exudate forms threads characteristic of the bacterium. These are not formed by other bacterial pathogens of potato, making the test useful for field diagnosis.

Morphology of the causal organism

R. solanacearum is a Gram-negative rod, 0.5–1.5 μm in length, with a single polar flagellum. The bacterium may be distinguished from *Erwinia* species by a positive staining reaction for poly-β-hydroxybutyrate granules with Sudan Black B or Nile Blue and stains heavily at the poles with carbol fuchsin. Colonies on agar are initially smooth, shining and opalescent, but become brown with age (Crop Protection Compendium, 2001).

Selective media

Several semi-selective media have been published such as that of Elphinstone and co-workers (1996) which was used by Pradhanang, Elphinstone and Fox (2000a, b) to detect the bacterium in soil. These media are often used as a means of enrichment prior to more specific techniques employing the polymerase chain reaction.

Biochemical markers

Caruso and co-workers (2002) were able to detect the bacterium routinely in symptomless potato tubers by enrichment involving shaking samples in Wilbrink broth and double-antibody sandwich indirect enzyme sorbent assay using the specific monoclonal antibody 8B-IVIA. A variety of techniques was used by Timms-Wilson, Bryant and Bailey (2001) to characterize isolates of the bacterium from different geographic areas. These included the phenotypic characters sodium dodecyl sulphate polyacrylamide gel electrophoresis (SDS-PAGE) of proteins, fatty acid methyl ester (FAME) analysis, growth profiles and exopolysaccharide production. Genotypes were analysed using restriction fragment length polymorphism (RFLP) of 16S rRNA, amplified ribosomal DNA restriction analysis and sequence analysis of 16S–23S rRNA ITS and flanking regions.

Weller and co-workers (2000) have developed a real time PCR assay using two different fluorogenic probes, one which detected all biovars of *R. solanacearum* and the other which was specific for biovar 2A. The assay detected 100 cells or fewer in pure culture but was less sensitive for potato extracts.

3 Epidemiology

Summary

Three types of pathogen behaviour have been recognized in the development of plant-disease epidemics: monocyclic, in which there is only one pathogen generation each cropping season; polycyclic, when there is more than one generation each cropping season; and polyetic in which the pathogen is monocyclic but the incidence of diseased plants increases over several seasons. Epidemics may originate from discrete areas, often as a result of planting infected seed. Such areas are termed foci and give distinct patches of disease as opposed to the more general infection which may originate from a spore shower. Fitting disease progress curves to epidemiological data has emphasized the importance of the capacity of the pathogen to multiply and spread to other plants.

Inoculum may arise from many different sources, the most important generally being host material, either living or dead, e.g. crop debris. However, inoculum may long survive the host from which it originated and may persist in the soil at levels that preclude obtaining an economic return from subsequent crops for many years. Viruses and viroids require vectors for dissemination and these are most frequently insects, but some fungi and nematodes may also fulfil this role. Severe epidemics usually occur only when an aggressive pathogen with a high reproductive rate and a large population of susceptible hosts are present. In addition, environmental conditions which favour the pathogen, such as high moisture and appropriate temperature, must occur over sufficient time for penetration and colonization of the plant to take place as well as reproduction and dissemination of the pathogen.

3.1 Introduction

Epidemiology is the study of the spread of disease over time and space. For an epidemic to occur there must be:

(1) a source of inoculum,

(2) availability of susceptible hosts,

(3) appropriate environmental conditions.

Introduction to Plant Pathology by Richard Strange
© 2003 John Wiley & Sons, Ltd   ISBN 0 470 84972 X (cased)   ISBN 0 470 84973 8 (pbk)

Inoculum of a pathogen may be endemic but fail to bring about disease on a serious scale owing to its avirulence for the crops under cultivation. Mutation to virulence may alter the situation dramatically, the virulent variant soon establishing itself on the susceptible crop as the dominant constituent of the pathogen population. As a result, a damaging epidemic may ensue, especially if the crop is genetically uniform and grown over a wide area (see for example Southern Corn Leaf Blight; Section 1.3.2).

Alternatively, serious epidemics may occur as a result of importing inoculum. Among Oomycetes, examples include *Phytophthora infestans*, the cause of potato blight and *P. cinnamomi* (Section 1.3.5 and Figures 1.12 and 1.13). The latter has not only caused dieback of the jarrah (*Eucalyptus* species) forests in Australia but is also virulent for about half the native flora of that continent. As mentioned previously (Section 1.3.5), the lack of resistance in such a wide range of plants strongly suggests that the pathogen is a recent introduction. *Cryphonectria parasitica*, the fungus that causes chestnut blight is also thought to have been imported into the USA where it destroyed essentially all the chestnut trees.

Serious epidemics may arise when plants are introduced to new areas where they succumb to indigenous pathogens. Candidates for this explanation include African Cassava Mosaic Virus, East African Cassava Mosaic Virus (Section 1.3.10) and the virus causing swollen shoot of cocoa. Both cassava and cocoa were imported to Africa from South America where they are free of infection by these agents.

Appropriate environmental conditions are the third factor that is necessary for development of an epidemic. These must occur over a sufficient period of time to allow infection, development of the pathogen within the host, reproduction and dissemination.

Fry (1982) has summarized the three factors, host, pathogen and environment which need to operate over a period of time in an equation as follows:

$$D_t = \Sigma_{i=0}^t f(p_i,\ h_i,\ e_i) \tag{3.1}$$

where D_t is a measure of disease at time t. p_i, h_i and e_i are all the pathogen, host and environmental factors, respectively, that contribute to an increase in disease. f is a factor that relates the interaction of p, h and e over the period $i = 0$ to t to the amount of disease at time t.

Theories of epidemic development will first be described in this chapter followed by a consideration of the roles of pathogen, host and environment. How these are affected by control measures will be reserved for Chapters 5 and 12.

3.2   Theories of epidemic development

Originally, three types of pathogen behaviour were recognized: monocyclic, polycyclic and polyetic. As we shall see, these terms are primarily descriptive of the development of disease in time. More recently, increased attention has been paid to the development of disease in space with the result that the aggregate rather than the uniform nature of disease has become better appreciated (Madden and Hughes, 1995; Ristaino and Gumpertz, 2000). This 'patchiness' is strongly influenced by the distribution of susceptible plants, the distribution of inoculum and many environmental factors, such as the direction of the wind, the occurrence of free-standing water and a host of others.

The space–time properties of epidemiology have led to numerous attempts to model disease with the aims of obtaining a greater understanding of disease progress and the critical factors that control it. As these models increase in accuracy they will become of increasing utility in the prediction of disease and the evaluation of control measures.

3.2.1   Development of disease in time

Monocyclic pathogens

Pathogens that are restricted to one generation per cropping season are said to be monocyclic. Such pathogens may have a life cycle that occupies about one season and have no repeating reproductive stages of shorter duration. Many smut fungi fall into this category, rusts which do not have a repeating stage and some nematodes. Alternatively, there may only be a limited time frame in which the plant is susceptible. For example, pearl millet is only susceptible to infection by *Claviceps fusiformis* for a few days since pollination causes withering of the stigmas through which invasion normally occurs and prevents the progress of the fungal hyphae (Thakur and Williams, 1980). Similarly, headblight of wheat caused by *Fusarium graminearum* is primarily a monocyclic disease since anthesis is the most susceptible period (Fernando *et al.*, 1997: Strange and Smith, 1971).

The behaviour of monocyclic pathogens is described by the following equation:

$$x_t = x_0 rt \tag{3.2}$$

where x_t is the proportion of the crop which is diseased at time t, x_0 is the initial amount of disease or inoculum, r is a measure of the rate of disease increase and t is the time over which the host and pathogen interact. It follows that:

$$\frac{dx}{dt} = x_0 r \tag{3.3}$$

i.e. the increase in disease over time is a function of initial inoculum (x_0) and its rate of increase (r). As an epidemic develops, the probability that inoculum will be deposited on plant material that is already infected increases. Such inoculum is unlikely to cause an increase in disease. Therefore a value of 1 is ascribed to the uninfected crop and x, where x is any value from 0 to 1, to the proportion already diseased. The equation may then be written as follows:

$$\frac{dx}{dt} = x_0 r (1 - x) \tag{3.4}$$

This may be rearranged:

$$\frac{dx}{(1 - x)} = x_0 r dt \tag{3.5}$$

And solved to give:

$$x = 1 - e_0^{-xrt} + x_0 e^{-x_0 rt} \tag{3.6}$$

where x_0 is the amount of disease at $t = 0$.

Polycyclic pathogens

Many pathogens produce new batches of propagules several times during the growing season of a crop plant, the time between infection and reproduction of some pathogens being only a few days, e.g. 5.5 days for blight of chickpea caused by *Ascochyta rabiei* (Trapero-Casas and Kaiser, 1992) and 8–10 days for *A. fabae* which infects faba bean (Wallen and Galway, 1977). Such pathogens, under appropriate conditions, may cause explosive epidemics. For example, blight of chickpea caused by *A. rabiei* develops rapidly under cool and damp conditions and may completely destroy a crop. Bacteria have even shorter generation times and bacterial diseases of plants can consequently develop quickly. *Pseudomonas tabaci*, for example, spread through tobacco fields in Virginia so rapidly that the disease it causes was given the name wildfire! Nematodes may also be polycyclic but their generation times are seldom fewer than 20 days. For example, the generation time of *Paratrichodorus minor*, a pathogen of more than 100 hosts, is 30 days.

The inoculum of polycyclic pathogens, unlike that of monocyclic pathogens, increases during the season. An equation that takes this into account is as follows:

$$\frac{dx}{dt} = x r (1 - x) \tag{3.7}$$

where, as before, x is the amount of disease on a scale of 0–1, r is the exponential rate of disease increase and t is the time under consideration during which host and pathogen have interacted.

Since the proportion of diseased plants early in an epidemic caused by a polycyclic pathogen is usually very small it can be ignored and the expression may then be simplified as follows:

$$\frac{dx}{dt} = xr \tag{3.8}$$

This equation may be rearranged:

$$\frac{dx}{x} = rdt \tag{3.9}$$

which on integration becomes:

$$\ln x = rt + k \tag{3.10}$$

Taking antilogarithms, the amount of disease x at time t is:

$$x = x_0 e^{rt} \tag{3.11}$$

where, as before, x_0 is the amount of disease at $t = 0$, e is the base of the natural logarithm and r and t are as described for Equation (3.7). However, just as in monocyclic diseases, when the proportion of diseased material becomes significant the rate of disease progress declines. An equation that takes this into account can be obtained by solving Equation (3.7) to give:

$$x = \frac{x_0 e^{rt}}{1 - x_0 + x_0 e^{rt}} \tag{3.12}$$

Lesion expansion is another component that should be taken into account in deriving models for epidemics of plant disease (Berger, Bergamin and Amonim, 1997). As these authors point out, disease intensifies as a consequence of lesion expansion under conditions in which fresh infections cannot take place. Moreover, in contrast to new infections in which there is a period of time before the pathogen becomes reproductive, lesion expansion in many diseases immediately makes a greater area available for the production of propagules of the pathogen.

Polyetic pathogens

In some parts of the world there is a definite break between cropping seasons and this may be sufficient to prevent carry over of inoculum into the following season. For example, Kocks and co-workers (1998) found that if cabbage plants were chopped and the resulting plant debris rotovated into the soil there was little carry over of *Xanthomonas campestris* pv. *campestris* into the next season. Alternatively, if the inoculum is hardy, as in the case of *Ceratocystis* species causing Dutch elm disease, carry over can be significant and may show multiplication from season to season. Such pathogens are said to be polyetic.

3.2.2 Development of disease in space

Spatial analysis of epidemics may give valuable information about the mechanisms of dispersal of pathogen propagules and the physical and biological factors that are critical for their dissemination (Ristaino and Gumpertz, 2000). For a plant disease to spread, infective propagules must be released, transported and deposited on new susceptible plants, which they successfully infect. All of these processes are probabilistic and, given the existence of susceptible plants, are contingent on environmental conditions and numbers of infective propagules. Where these are unfavourable to the pathogen it may die out before many plants are infected (Shaw, 1994). Alternatively, under favourable conditions an asymptote is reached when an infected individual fails to infect any more healthy plants (Shaw, 1994). In agricultural terms, this normally constitutes a level of infection of the crop that is unacceptably high.

Transport by vectors as well as release and deposition will be discussed later (Sections 3.3.2 and 3.5.2, respectively). Transport by water splash and wind turbulence will be considered here since they have been described mathematically.

Splash dispersal

Splash dispersal by rain or irrigation water probably occurs for most pathogens that produce aerial propagules and is more important, relative to dispersal by air currents, for pathogens with large propagules or those that are aggregated in mucilage (Mundt, 1995). Dispersal is influenced by many factors such as drop size and velocity, duration of rainfall, physical characteristics of the propagule, propagule substrate and crop canopy but generally occurs over distances of less than 1 m (Mundt, 1995). In consequence, disease gradients are steep and can often be expressed accurately by an exponential model:

$$y = ae^{-bx} \tag{3.13}$$

where y is the number of propagules deposited or the number of infections initiated per unit area at x units of distance from the source of inoculum, a is the number of propagules or infections per unit area at the source, e is the base of natural logarithms and b is a parameter that describes the steepness of the gradient (Kiyosawa and Shiyomi, 1972).

Genetic diversity of pathogens that are spread by splash dispersal is a consequence of the steepness of the disease gradient. For example, Morjane and co-workers (1994) found 12 haplotypes of *Ascochyta rabiei*, a fungus that reproduces by pycnidiospores during the growing season of chickpea, in a single field of chickpea in the Béja region of Tunisia. This and further data obtained by genetic finger printing of other pathogens strongly suggest that many plant pathogen populations are not clones but metapopulations, i.e. a collection of evanescent individual populations of varying sizes which frequently come into existence and as frequently vanish (Shaw, 1994).

Dispersal by turbulent air currents

Many plant pathogens are dispersed by turbulent air currents and these, too, form disease gradients which, in contrast to those of splash dispersed pathogens, are better described by an inverse power function:

$$y = ax^{-b} \qquad\qquad (3.14)$$

where y, x and b are as in Equation (3.13) and a is the number of propagules or infections per unit area at one unit distance from the source.

The importance of this model is that it predicts that a significant proportion of the inoculum will be deposited far from the source, a prediction that is borne out by many observations, e.g. the rapid spread of new virulence types of the cereal pathogen, *Blumeria graminis* in Europe over long distances.

Focus expansion

If a pathogen arrives at a crop as a spore shower, then its distribution is likely to be fairly even. In contrast, if inoculum were to arrive in the seed of the crop as a few infected individuals these could provide local sources of inoculum. Epidemics arising from these local infected points are said to be focal.

Zadoks and Van den Bosch (1994) have reviewed the expansion of foci of infection, making a telling analogy with the outer ripple or wave caused by throwing a stone into a still pond. After the initial build-up of the focus, the velocity of the wave becomes almost constant and the numbers of plants in its path rise exponentially. As the wave passes over the crop, infected plants pass through a latent period during which they are not infectious, an infectious period, termed the 'time kernel', when further propagules are produced, and a second non-infectious period when plants have either been killed or no longer support reproduction of the pathogen (Figure 3.1). Numbers of lesions generated by a single mother lesion fall off exponentially with distance and this relationship was called 'contact distribution' (Figure 3.2).

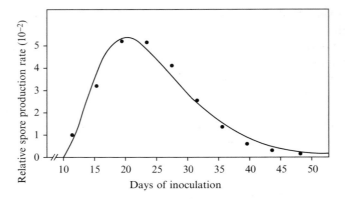

Figure 3.1 Graph showing the 'time kernel' during which active spore production of *Puccinia striiformis* occurs on primary leaves of winter wheat in relation to date of inoculation (from van den Bosch, 1988c)

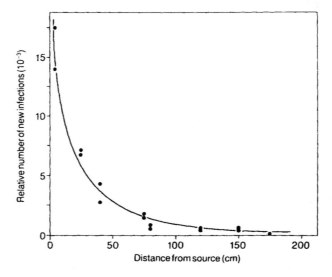

Figure 3.2 The spatial distribution ('contact distribution') of daughter lesions of *Puccinia striiformis* on winter wheat around a single mother lesion; each point represents the average of four compass directions (from van den Bosch, 1988c)

The numbers of propagules produced by a lesion that successfully infect further plants was termed 'gross reproduction' and, when the first few generations of pathogen reproduction do not overlap, this may be estimated by dividing the number of plants infected in a given generation by that of the preceding one. Van den Bosch and co-workers (1988c) used these three concepts of the time kernel, contact distribution and gross reproduction to construct models of focus expansion by stripe rust (*Puccinia striiformis*) of wheat and mildew (*Peronospora farinosa*) of spinach which fitted the experimental data well.

The theory of focus expansion was further elaborated by Zawolek (1993) in computer simulations which included wind and combined short- and long-distance dispersal of spores. Stochastity (randomness) was introduced by randomization of the numbers of lesions resulting from long-distance spore dispersal. The model simulated the real-life phenomena of a non-circular central focus owing to wind and a ragged front and daughter foci owing to long-distance spore dispersal.

Van den Bosch and co-workers (1990) have also considered the effect of mixing resistant and susceptible plants on focus expansion and, from theory, demonstrated that when the mix was homogeneous focus expansion was proportional to the square root of the area occupied by the susceptible plants. However, when actual data for bean rust, caused by *Uromyces appendiculatus*, was analysed, the rate of focal expansion was found to be linearly related to the logarithm of the fraction of susceptible plants in the mixture rather than the square root (Assefa *et al.*, 1995).

In a more recent paper van den Bosch, Metz and Zadoks (1999) presented a model of pandemics of focal plant diseases. Pandemics are diseases that spread over large areas such as a continent. For example, blue mould of tobacco, caused by *Peronospora tabacina* was recorded in Australia in the 19th century and was reported in the USA in 1921 where a pandemic occurred between the years 1931 and 1950. The speed at which the disease spread there was about 50 km/year. In contrast, the disease, which was reported in England in 1959, spread over mainland Europe and Mediterranean countries at a rate of 130 km/year. The model included variables, functions and parameters such as numbers of foci initiated per unit of time, the probability that a spore leaving the canopy is redeposited in the field of origin and that it initiates a new focus, age of focus and pandemic dispersal distribution. The authors suggest that the model may be used to compare pandemics at various places in the world with different cropping or climatic conditions.

3.2.3 Fitting disease progress curves to epidemiological data

Van der Plank (1963), a pioneer of plant disease epidemiology, used a monetary metaphor to explain the importance of the capacity to multiply and spread to other plants when he stated that, 'A high rate of interest is more important than money in the bank.' In other words, if r in the equations given earlier in the chapter is high then the pathogen is capable of producing explosive epidemics. Delay in the onset of the epidemic and the carrying capacity of the crop (i.e. the maximum disease level obtainable) are two other important factors to be considered. All three factors have been considered by Gilligan (1990) both from a theoretical and practical point of view. The theoretical models confirmed the importance of r (β in Gilligan's notation) and confirmed that, until the carrying capacity of the crop was reached, less disease resulted when the onset of an epidemic was delayed. When this upper maximum was set at a low level it is obvious that there is less disease but this has important implications for disease control (Chapter 12).

Often plant diseases do not fit neatly into the descriptions given above of monocyclic, polycyclic and polyetic and there may be overlap. For example, Rekah, Shtienberg and Katan (2001) found that crown and root rot of tomato caused by *Fusarium oxysporum* f. sp. *radicis-lycopersici* consisted of overlapping monocyclic and polycyclic phases. The monocyclic phase resulted from over-summering inoculum and was the source of primary infection whereas the polycyclic phase resulted from root-to-root contact and was the source of the secondary infection. Also as pointed out previously (Section 3.2.1), whether a pathogen is considered to be monocyclic or polyetic depends on the time period over which the disease is considered.

3.3   The role of the pathogen

For disease to occur the pathogen must be present with sufficient 'inoculum potential', defined by Garrett (1956) as 'the energy available for infection of a host at the surface of the infection court'. In epidemics of plant disease, the origin of such inoculum, its multiplication and its dissemination are crucial factors. Inoculum usually originates and multiplies on infected plant material but relies on other agents such as the soil, the atmosphere or vectors for transmission. Once a plant has been infected, the extent to which it provides inoculum with sufficient potential to infect other host plants is crucial, particularly in the cases of polycyclic pathogens. With these, a short generation time and massive multiplication in susceptible hosts can rapidly give rise to devastating epidemics (Section 3.2).

3.3.1   Sources of inoculum

Infected seed

In some diseases, infected seed is an important source of inoculum, particularly where the pathogen is polycyclic and has efficient means of transmission. For example, an outbreak of *Ascochyta* blight of chickpea at many foci in Idaho State in the USA was attributed to the planting of infected seed (Derie *et al.*, 1985). Since the latent period of this fungus is 5.5 days (Trapero-Casas and Kaiser, 1992) and spore production is abundant, such foci can quickly lead to a build-up of the disease (Figure 3.3). Bacterial diseases, such as black rot of crucifers caused by *Xanthomonas campestris* pv. *campestris* may also have short latent periods and rapidly multiply, giving rise to serious epidemics (Schaad, Sitterly and Humay-dan, 1980).

Figure 3.3 A field of chickpea (*Cicer arietinum*) showing foci of infection of chickpea by *Ascochyta rabiei* (courtesy Walter Kaiser); the foci are likely to have arisen from the planting of infected seed. A colour reproduction of this figure can be seen in the colour section

About 18 per cent of plant viruses are seed-transmitted and this figure is expected to rise to one third as the result of further research (Johansen, Edwards and Hampton, 1994; Stace-Smith and Hamilton, 1988). Transmission rates for cucumber mosaic virus (CMV) ranged up to 8.8 per cent in clover and in lupin they were 24.5 per cent when plants were inoculated at the mid-vegetative stage, about 90 days post-emergence, later inoculations at the beginning of flowering giving lower rates of transmission (Jones and McKirdy, 1990; Geering and Randles, 1994).

Epidemics may occur as a result of planting infected seed even if the incidence of infection is low. For example, Schaad, Setlerly and Humaydan (1980) showed that serious epidemics of black rot of crucifers caused by *Xanthomonas campestris* pv. *campestris* could arise from as few as three infected seed in 10 000 owing to its short latent period and rapid multiplication already noted. Such observations have led to interest in determining the thresholds of infection below which an epidemic is unlikely to result. These have only been set for a few diseases and are contingent not only on the pathogen and its means of transmission but also environmental conditions (Mcgee, 1995). For example, Shah, Bergstrom and Ueng (1995) found that in a season that was mildly conducive to blotch of wheat caused by *Septoria* (= *Stagnospora*) *nodorum*, epidemics resulted from 1 per cent seed infection and, in a conducive season, 0.5 per cent infection, both the lowest levels tested (Figure 3.4). Evidence that the infected sown seed was indeed responsible for the epidemics, rather than inoculum from another

Figure 3.4 Symptoms of *Stagnospora nodorum* (= *Septoria nodorum*) on wheat leaves and ears (copyright Peter Scott). A colour reproduction of this figure can be seen in the colour section

source, was obtained by DNA fingerprinting. Fingerprints for the pathogen from both sown seed and the harvested crop were identical.

Jones (2000) has pointed out that despite advances in testing for seed-borne viruses (see Chapter 2) there are very few virus diseases for which threshold values have been set. In order to determine these it is necessary to perform field experiments in which infected seed is sown and the consequences are followed in terms of virus spread, yield losses and infection of newly produced seed. Jones (2000) suggests that such experiments should be continued over several years and at diverse sites in order to determine variation in the risk of economic loss. He also describes how such field experiments and surveys were used to define threshold values for CMV in lupin (*Lupinus angustifolius*) subterranean clover (*Trifolium subterraneum*) as well as alfalfa mosaic virus (AlMV) in annual burr medic (*Medicago polymorpha*). Despite the difficulties in obtaining the necessary data, thresholds for some seed-borne viruses have been set. For example, the tolerance of seed infected with lettuce mosaic virus (LMV) is 0 in 2000 in the Netherlands but 0 in 30 000 in California. There are two reasons for this difference in tolerance: in the Netherlands the disease cycle is broken by the planting of other crops whereas in California lettuce is grown under continuous cultivation; secondly, the cooler climate of the Netherlands supports a lower population of the aphid vector of the virus than in California (Kuan, 1988).

Infected pollen

Mink (1993) has reviewed the role of pollen in the transmission of viruses and quoted 19, the figure of Manandhar and Gill (1984), as the number of viruses transmitted by this means. Most are transmitted vertically through seed but Mink (1993) listed nine that were transmitted horizontally from plant to plant. Referring to his own work (Mink, 1982) in which the earlier work of the 1960s and 1970s was re-examined a new hypothesis was presented, i.e. that spread of ilaviruses from plant to plant involves both thrips and pollen (Mink, 1992). Pollen as a source of inoculum is particularly serious where perennial species are concerned since the virus may be transmitted not only vertically to the seed of the next generation but also horizontally. For example, Sour Cherry Yellows can be spread to uninfected trees by this route (Gilmer, 1965). Bulger, Stacesmith and Martin (1990) have confirmed earlier work by Murant, Chambers and Jones (1974) which showed that Raspberry Bushy Dwarf Virus (RBDV) was pollen transmitted in red raspberry, *Rubus idaeus*. Additionally, they suggested that bumblebees and honeybees could play important roles as vectors since they were very common in raspberry fields, carried large quantities of pollen and damaged flower parts while foraging. Blueberry Leaf Mottle Virus is another viral agent that is disseminated by foraging honeybees carrying infected pollen (Childress and Ramsdell, 1987).

Mink (1993) has also reviewed the transmission of viroids through pollen. Six were found to be transmitted through seed and in three cases the viroids were found to invade mother plants, suggesting the possibility of horizontal spread although this has not been demonstrated under field conditions.

Infected vegetative material

A number of important crop plants are multiplied by vegetative propagation and are subject to progressive infection, particularly by viruses. 'Seed' potato may contain several viruses such as Potato Virus X (PVX) and Potato Virus Y (PVY) and their titres can build up over successive generations to give severe yield reductions. Plants and bulbs of garlic (*Allium sativum*) may also harbour a complex of viruses which may depress yields by as much as 50 per cent (Verbeek, Vandijk and Vanwell, 1995).

Viruses may be spread by grafting and this phenomenon led to the early definition of plant viruses as infectious agents which could be transmitted by grafting but were not retained by bacteriological filters (Matthews, 1981).

Reservoir plants

Although healthy material may be planted, in some instances it rapidly becomes infected, although there are no similar species in the vicinity which show signs of infection. For example, the source of fireblight epidemics caused by *Erwinia amylovora* in the pear orchards of Kent, UK, following stormy weather, was traced to infected hawthorn trees which had been planted as windbreaks around the orchards (Glasscock, 1971). In contrast, de Mackiewicz and co-workers (1998) did not find any evidence that herbaceous weeds served as reservoirs of *Erwinia tracheiphila*, the causal agent of cucurbit wilt. Although they were able to detect the bacterium by ELISA in six species of weed plants 3 weeks after artificial inoculation, they were not able to isolate it and concluded that the immunoassays detected dead cells.

A number of viruses have wide host ranges and the source of crop infections can sometimes be traced to weed plants infected with the same virus. For example, Franz, Makkout and Vetten (1997) found 29 plant species as new hosts of faba bean necrotic yellows virus (FBNYV) in Egypt and Syria, bringing the total number of hosts of the virus to 58. They also showed that from September to August, when no major crops susceptible to the virus are grown, that the virus may survive in perennial weeds such as *Melilotus officinalis, Onobryches* spp. and some members of the Amaranthaceae.

Movement of host plants to new areas may result in their succumbing to viruses present in indigenous hosts as discussed in Chapter 1 (Section 1.3.10) for infection of cassava with mosaic viruses. Cocoa (*Theobroma cacao*) is another example of a plant which succumbed to a devastating virus disease, swollen shoot, when transported to the African continent from South America (Thresh and Owusu, 1986).

Infected volunteer plants

Self-sown (volunteer) plants of a number of crops often occur as a result of the shedding of seed at or before harvest. In wheat, for example, up to 3 per cent of the grain may be lost in this way (Schuler, Rodakowski and Kucera 1977). Such plants may provide a 'green bridge' of susceptible material that can be important in maintaining and increasing the inoculum of a pathogen. In an experiment designed to gain evidence for this view, Brassett and Gilligan (1990) found that the incidence and severity of take-all disease of wheat was increased if sowings in soil infested by the fungus were preceded by a preliminary sowing of the plant. In consequence they suggested that the 'green bridge' provided by volunteer seedlings could promote the development of severe autumn epidemics. The 'green bridge' phenomenon has also been invoked as a reason for epidemics of rice tungro bacilliform virus (RTBV) in southern India. Here changes in cropping patterns and irrigation practices have provided a summer 'green bridge' for the virus and its leafhopper vector (Nagarajan, 1993).

Eversmeyer and Kramer (2000), reviewing the epidemiology of wheat leaf and stem rust in the USA, pointed out that overwintering of these fungi in volunteer or newly-sown plants affords the pathogens the earliest opportunity to begin their polycyclic development resulting in the maximum levels of infection.

Infected crop debris

Debris, the remains of a crop left on the soil surface after harvest, may be an important source of inoculum for succeeding crops if it is infected with a pathogen. For example, *Alternaria brassicae* and *Alternaria brassicicola* produced viable spores on oilseed rape leaves left outside on the soil for 6 weeks in winter and 8 weeks in summer. On cabbage, sporulation continued for even longer periods, 8 weeks in winter and 12 weeks in summer but in both cases these periods corresponded with the time that the leaves remained intact. Stems, which decomposed more slowly, supported spore production for 23 weeks (Humpherson-Jones, 1989). Such sources of inoculum could be of considerable significance in the survival of the pathogens and their transmission to the newly-planted crop.

Ascochyta rabiei, the cause of blight of chickpea (see this section above and Figure 3.3), also survives in crop debris. Navas-Cortés, Trapero-Casas and Jimenez-Diaz (1995) found that the fungus remained viable for 2 years (the length of time of the experiment) on the surface of soil but survival was severely curtailed by burial. The fungus also reproduced sexually on debris but mature asci were only produced at 10°C under the controlled conditions of the experiment.

Burial of crop debris was also found to destroy inoculum of common blight of bean caused by *Xanthomonas campestris* pv. *campestris*. Although contaminated seed is the primary source of inoculum for this pathogen (see above), the disease was still found on occasions in fields planted with certified disease-free seed, suggesting that inoculum could have originated on crop debris. Gilbertson, Rand and Hagedorn (1990) found that the organism was able to survive on bean debris in no-tillage plots from October to May in two successive years but not beyond February in either year if the fields were disked or ploughed.

3.3.2 Vectors

The spread of plant disease from one host plant to another requires that infectious propagules should be produced in the parent plant and should be transported to another susceptible plant where it is able to establish a parasitic relationship. Many fungal pathogens, which produce a profusion of spores on the surface of their hosts and have an array of sophisticated mechanisms for breaching plant cell walls and entering their new hosts, need nothing more than air currents for transport. In contrast, phytoplasmas, viruses and viroids need the help of a third party to exit the parent plant, to be transported to the new plant and to effect entry. In many instances the third party is an insect vector but in some cases fungi or nematodes fulfil these roles.

Virus transmission by insects

Aphids are the most successful group of insects to exploit plants as a source of food and are also the most important group of virus vectors (Matthews, 1981). Other plant-feeding insects include leafhoppers and planthoppers (Auchenorrhynca), mealybugs (Coccoidea and Pseudococcoidea), whiteflies (Aleyrodidae), bugs (Miridae and Piesmatidae), thrips (Thysanpotera) and mites (Arachnida).

The relations of plant viruses with their insect vectors have been divided into four classes:

(1)   non-persistently transmitted, stylet-borne viruses,

(2)   semi-persistently transmitted, foregut-borne viruses,

(3)   persistently transmitted, circulative viruses,

(4)   persistently transmitted, propagative viruses (Nault, 1997).

Non-persistently transmitted, stylet-borne viruses.   Most of the 250 viruses that are known to be transmitted by aphids are non-persistent. After the acquisition of virus particles the aphid is immediately infective, i.e. there is no latent period, but infectivity falls off rapidly and does not survive a moult. These viruses may also be transmitted easily by mechanical inoculation, which is usually achieved by rubbing a leaf with abrasive and inoculum.

There is a range of specificity in the relationships of non-persistent viruses and their vectors. *Myzus persicae* transmits at least 70 viruses whereas other species are known to transmit only one (Figure 3.5). Such specificity is highlighted in transmission studies of plants infected by more than one virus. For example, both cauliflower mosaic virus (CaMV) and turnip mosaic virus (TuMV) are transmitted by *M. persicae* from doubly-infected crucifers but *M. ornatus*, feeding on the same plants, only transmits cauliflower mosaic virus. In a study of the acquisition and transmission of CaMV by six aphid species, Markham and co-workers (1987) found that two clones of *Aphis fabae* were unable to transmit CaMV but the other species, i.e. *Acyrosiphon pisum, Brevicoryne brassicae, Megoura viciae, M. persicae* and *Rhopalosiphum padi*, were able to do so. Furthermore, some strains of virus which are normally stylet-borne are not transmitted by this means. These examples of specificity lead to the conclusion that the geometry of virus receptors on the aphid stylets is precise and that this precision must be mirrored by a complementary precision of the virus itself. A further complication with caulimoviruses, of which CaMV is one, is that not only are they transmitted non-persistently from stylets during probes of epidermal cells of host but also semi-persistently from the foregut when phloem feeding (Nault, 1997).

Some non-persistent viruses require helper components for transmission. These are encoded by the virus but are not part of the virion (Pirone and Blanc, 1996). They act by forming a bridge between the virus and the vector, allowing retention of the virus in the vector's alimentary tract whence it

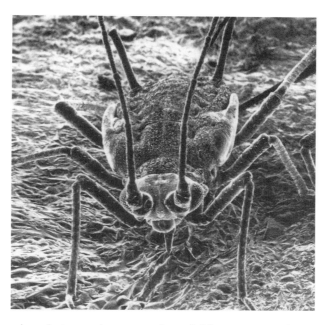

Figure 3.5 Scanning electron microscope view of *Myzus persicae* (courtesy Rothamsted Research, Harpenden, Herts, UK)

can be egested to initiate infection. Helper component proteins may be distinct for different viruses. For example, Thornbury and Pirone (1983) raised antisera to the helper components of two potyviruses, tobacco vein mottling virus (TVMV) and potato virus Y (PVY). Treatment with the homologous antiserum drastically reduced transmission but this was not so if heterologous sera were used. A study of the acquisition of CaMV has shown that there are two viral encoded proteins P2 and P3 that are necessary for aphid transmission (Drucker *et al.*, 2002). P2 binds to the aphid stylets and P3 forms the bridge between P2 and virions. The authors showed by immunogold labelling that P2:P3 and P3:virion complexes were predominantly spatially separated in the infected plant and suggest that the complete P2:P3:virion complex, necessary for transmission, is formed in the aphid.

Semi-persistently transmitted, foregut-borne viruses. Apart from caulimo-viruses which, as already mentioned, are transmitted bimodally, semi-persistent viruses are not acquired by insects during brief probes. Rather, their acquisition time reflects the time it takes for the aphids to reach the phloem (Nault, 1997). Electron microscope studies have shown in a number of cases that virus or virus-like particles in the vector are found attached to the linings of the cibarium and the pharynx.

Persistently transmitted circulative viruses. Viruses transmitted by vectors such as aphids or leafhoppers, which remain viruliferous after a moult when stylets, skin, foregut and hindgut are shed, are said to be persistent. They often undergo a latent period immediately after acquisition but may be transmitted weeks or months later. All gemini viruses are thought to be circulative with latent periods in leafhopper vectors ranging from 4 to 19 h (Nault and Ammar, 1989). Examples include beet curly top virus (BCTV), maize streak virus (MSV) and tomato yellow leaf curl virus (TYLCV). This last is a complex of geminiviruses which infects tomatoes world-wide and is transmitted by *Bemisia tabaci*, not only from plant to plant but also from insect to insect by sexual activity (Ghanim and Czosnek, 2000)!

Persistently transmitted, propagative viruses. There is now substantial evidence from various types of experiment that several viruses which cause plant disease multiply in their insect vectors. The first successful experiments were done by Black and Brakke (1952). They used wound-tumour virus (WTV) and its vector, *Agallia constrictor*. The leafhoppers were maintained on alfalfa, which is immune to WTV, and injected with extracts from viruliferous hoppers. After the seventh passage through the leafhoppers, the dilution factor was estimated at 10^{18}, much beyond the dilution endpoint of the original inoculum. Replication of the virus must therefore have occurred in the vector. Other techniques have also been used to establish virus replication in their vectors. These include transovarial passage in which succeeding generations were shown to be viruliferous, increase of virus antigen in the vector, enumeration of virus particle counts in the electron microscope and replication of the virus in vector-cell monolayers. In this last procedure, viral inoculum is incubated with cultured cells of the vector and virus-infected cells revealed by binding of a fluorochrome-tagged antibody to the virus.

Vector-plant relations

In addition to vector–virus interactions, successful transmission depends on the distribution of virus in the plant and the vector–plant interaction. Virus distribution and concentration may vary markedly from tissue to tissue and viruses may be differentially available in different cell types thus affecting the efficiency of transmission.

Aphids themselves are rather specific as to the type of plants that they feed on. For example, potato plants infected by potato leaf roll virus (PLRV) exude volatile compounds which attract alatae of *Myzus persicae*, the principal vector of PLRV, causing preferential colonization of infected plants (Eigenbrode *et al.*, 2002). In some instances the identity of the compounds involved has been established. The mustard oil, sinigrin, found in brassicas, for instance, is a feeding stimulant for *Brevicoryne brassicae* (Nault and Styer, 1972) while, conversely, *Solanum berthaultii* wards off aphids by release of the alarm pheromone E-β-farnesene (Gibson and Pickett, 1983). One very practical example of the importance of this phenomenon is Bud Necrosis Virus (BNV) of groundnut which is transmitted by the thrip vector, *Frankliniella schultzei*. The cultivar Robut 33-1 is susceptible to the virus but repels the vector. Consequently, field infection levels with the virus are low (Amin, 1985).

Virus transmission by zoosporic vectors

Two species of Chytridiomycetes, *Olpidium brassicae* and *O. bornovanus*, and three species of Plasmodiophoromycetes, *Polymyxa graminis, P. betae* and *Spongospora subterranea*, are vectors of plant viruses (Campbell, 1996).

The *Olpidium* species are soil-inhabiting obligate pathogens that seem to do relatively little harm to their hosts in which they form resting spores but they transmit several viruses that cause important diseases. These include cucumber necrosis virus (CNV), melon necrotic spot virus (MNSV) and tobacco necrosis virus (TNV). On disintegration of the host, zoospores are released that may infect other plants (Figure 1.4). Several experiments have strongly implicated the capsid protein as the site which binds the virus. For example Kakani, Sgro and Rochon (2001) have studied the transmission of CNV by zoospores of *O. bornovanus*. They found that for transmission to be successful, virus particles must attach to the surface of zoospores before they encyst on host roots. When a population of CNV was screened for mutants, which were deficient in transmission, six were found. All of the mutants bound less efficiently to zoospores than the wild type and all the mutations involved the capsid protein.

Polymyxa graminis vectors several important cereal viruses. These include soilborne wheat mosaic virus (SWMV), rice stripe necrosis virus (RSNV), barley yellow mosaic virus (BYMV) and wheat spindle streak mosaic virus (WSSMV). In contrast, *P. betae* is reported as the vector of only two viruses, beet soil-borne virus (BSBV) and beet necrotic yellow vein virus (BNYVV), the complex of BNYVV and vector giving rise to a severe disease of sugarbeet known as rhizomania (Figure 1.16(c)).

Two subspecies of *Spongospora subterranea* vector viruses. Potato mop-top virus (PMTV) is transmitted by *S. subterranea* ssp. *subterranea* and two watercress diseases, watercress yellow spot and watercress chlorotic leaf spot agent are vectored by *S. subterranea* ssp. *nasturtii*.

Transmission by nematodes

Hewitt, Raski and Goheen (1958) were the first to show that nematodes vectored plant viruses. Indications that they might be fulfilling this role include:

(1)   correlation of patches of diseased plants with the distribution of the nematode (as long as virus symptoms can be distinguished from those caused by the nematode);

(2)   healthy plants grown in the same container as diseased become infected only
       when the appropriate nematode is added;

(3)   addition of nematodes from infected plants to pots in which in which
       indicator plants have been grown-development of symptoms in the indicator
       plants suggests that the nematodes are acting as vectors;

(4)   treatment of the soil with nematocides which do not affect the virus directly-
       failure of the virus to spread strongly suggests that nematodes serve as
       vectors for the virus;

(5)   observation of virus particles associated with the nematode by electron
       microscopy.

Vector specificity is determined by the degree of adsorption of virus particles to
the cuticular lining of the buccal capsule or the oesophagus. For example,
Scottish and English forms of raspberry ringspot virus are serologically related
but they are vectored by different species of *Longidorus* (Harrison, 1977). Ploeg,
Zoon and Maas (1996) investigated the transmission of five strains of tobacco
rattle virus (TRV) or pea early-browning virus (PEBV) tobraviruses by *Paratrichodorus teres*. Virus-free nematodes were allowed to feed on roots of infected
Petunia hybrida and *Nicotiana tabacum* cv. 'White Burley' plants and were
subsequently allowed to feed on virus-free plants. Two strains of TRV differing
in serotype were efficiently transmitted but two other strains of TRV and one
strain of PEBV with serotypes similar to the transmitted ones were not transmitted. Visser and Bol (1999) have further investigated the transmission of TRV but
using *Paratrichodorus pachydermus* rather than *P. teres* as vector. They found
that a 40 kDa protein coded by RNA 2 of the virus was required for transmission
and that a 32.8 kDa protein may be involved with transmission by other species.

3.4   The role of the host

It is axiomatic that the availability of susceptible plants is of critical importance
in the development of epidemics of plant disease. It is also necessary that they
should support the multiplication of the pathogen.

3.4.1   Host-plant distribution

Providing environmental conditions are favourable and the pathogen is present,
growing uniformly susceptible plants at high density over a wide area provides an
ideal situation for the development of plant-disease epidemics. Often, perceived
commercial pressures have ensured that such uniformity is the norm. Lower plant

density and dilution of the crop with resistant plants both reduce disease. Growing plants at reduced density is not usually an option in modern agriculture but dilution of the susceptible crop with resistant plants is gaining acceptance. This can take the form of intercropping, much practised in Developing Countries, or planting multilines, crop plants which have similar genetic backgrounds but differing in their resistance to the prevailing pathogens. Their adoption has sometimes given dramatic control and increased yields and is treated further in Chapter 5 (see Section 5.6.1). Another advantage of plant diversity is that it promotes acquired resistance (Sections 11.3.7 and 12.4.3). Here, although inoculum may be produced on susceptible members of the intercropped species or multiline, much of it may alight on resistant members. Resistance is often expressed as a hypersensitive response (HR) in which a few cells die (see Chapter 11, Section 11.3.1) but the plant as a whole survives. Not only is the pathogen contained but resistance throughout the plant is induced that is non-specific. Thus, a pathogen, which is normally virulent for the plant, is likely to be resisted.

Intercropping, however, may not always be a solution since some pathogens such as *Colletotrichum gloeosporioides* have wide host ranges. For example, this fungus was isolated from both coffee and cassava (which had been grown as a shade crop for the coffee!) growing on Hainan Island, China. Identity of the pathogen in both crops was established by DNA sequencing and AFLP (Section 2.5.5; Aiyere and Strange, unpublished data). It therefore seems entirely probable that intercropping exacerbated the disease situation in this instance, both plants providing inoculum not only for themselves but each other.

3.4.2 The effect of host resistance on inoculum multiplication

Although the initial inoculum level is important in establishing an epidemic in both monocyclic and polycyclic diseases, it is the rate of inoculum production in polycyclic diseases that is important. Where this is high and the disease cycle short (i.e. the time between infection and production of new propagules) explosive epidemics may occur (Van der Plank, 1963; Gilligan, 1990). Low levels of inoculum and delay in its production are components of plant resistance and are considered further in Chapters 9–11. Suffice it to reiterate here the exponential nature of many plant diseases (Section 3.2.1) and therefore the importance of resistance.

3.5 The role of the environment

Plants are normally rooted in soil but maintain light-capturing organs and other structures in the atmosphere. Both environments are subject to variation which may profoundly affect the initiation and development of disease.

3.5.1 The soil

Many important classes of plant pathogens such as the parasitic angiosperms, *Striga* and *Orobanche*, nematodes and fungi are soil-borne and their resting structures, i.e. seeds, cysts and spores, respectively, may survive in the soil for many years. In addition, fungi and nematodes may also act as vectors of plant viruses (Section 3.3.2). These make the economic cultivation of important crops difficult or impossible in some circumstances and place a high premium on plant resistance or soil treatments that reduce pathogen populations.

Many factors affect disease development in infested soils. These include moisture content, temperature and pH. In some instances their overriding influence is on the pathogen while in others it is on the host. A further important factor is the presence of other members of the microbial flora; there is now much interest in this area since some of the species are antagonistic to plant pathogens and efforts are being made to exploit them in biocontrol (see Section 5.5).

Soil moisture, aeration and temperature

Soil moisture and aeration are interactive in that high moisture levels generally mean low aeration and vice versa. Moreover, temperature is likely to influence soil moisture levels since at high temperatures soils are likely to dry out more quickly than at low temperatures. All three factors may profoundly affect plant and pathogen and hence alter the inception and progression of disease.

Few plant pathogens are able to infect their hosts when the soil is dry and some pathogens are killed by prolonged periods of drought. However, dry conditions are usually unfavourable to the host as well and may predispose it to infection. For example, drought stress has long been known to enhance the invasion of groundnuts by *Aspergillus flavus*, possibly by depressing their ability to synthesize phytoalexins (Wotton and Strange, 1985, 1987 and see Section 11.3.1). In contrast, high moisture exacerbated the development of root rot of citrus by *Phytophthora citrophthora* (Feld, Menge and Stolzy, 1990). One reason for this was that the high soil moisture promoted the production and release of zoospores of the pathogen but another was that it also reduced the host's ability to produce new roots. This inability meant that the whole root system was liable to become rotted and the tree to die, whereas at lower moisture levels, new roots were formed which compensated for those affected by the pathogen.

Café and Duniway (1995) and Café, Duniway and Davis (1995) have investigated the effect of irrigation schedules on the incidence of disease of pepper (*Capsicum annuum*) and squash (*Cucurbita pepo* var. *melopepo*) caused by *Phytophthora capsici* (Figure 3.6) In one year, the yield of plots of pepper infested with the fungus and irrigated every 21 days did not differ from that of control plots without the pathogen. In contrast, yields from plots irrigated at 7- and 14-day intervals were only 45 and 83 per cent of the controls. However, when the

Figure 3.6 Pepper showing roots infected by *Phytophthora capsici* (courtesy of Dr Jean Beagle Ristaino, Department of Plant Pathology, North Carolina State University). A colour reproduction of this figure can be seen in the colour section

experiment was repeated the following year, disease levels were higher and no effective control was obtained by increasing the intervals between irrigation. The discrepancy between the years was most likely explained by differences in pre-incoulation soil temperatures. In the first year the temperature increased slowly and was continuously in the optimum range for peppers while in the second year it oscillated widely and lay outside the optimum for the plant for long periods. Results with squash were consistent in both years and showed that less frequent irrigation controlled the disease.

Since drought stress and elevated temperatures often go hand in hand it is usually difficult to separate their effects. In peanuts, for example, the canopy recedes in dry conditions allowing more direct solar radiation to reach the soil. This in turn causes a rise in the temperature of the geocarposphere. In environ-mentally controlled plots, Cole and co-workers (1985) were able to separate these effects on the invasion of peanuts by *Aspergillus flavus* and the accumulation of aflatoxin. When the mean geocarposphere temperature was 24.6°C aflatoxin concentrations were low in undamaged groundnut kernels from drought-stressed plants but reached values as high as 3100 parts per billion in kernels from plants in which the mean geocarposphere temperature was 29.6°C. The percentage infection of plants grown at the elevated temperature was also greater, although here it would have been useful if the authors had measured the amount of fungus as a parameter of susceptibility.

pH

Blaker and MacDonald (1983) found that pH values less than 4.0 reduced sporangium formation, zoospore release and motility of zoospores of *Phy-tophthora cinnamomi*. In contrast, raising the pH of soil by liming has long been used as a control measure for clubroot of brassicas caused by *Plasmodio-phora brassicae* (Section 5.2.5).

Antagonists

Oyarzun, Gerlagh and Zadoks (1998) measured the effect of a number of abiotic and biotic factors on 'soil receptivity', a term which they defined as the capacity of the soil to allow soil-borne pathogens to produce disease. They used three pathogens of pea as their models: *Thielaviopsis basicola, Fusarium solani* and *Aphanomyces euteiches*. Although several relationships were shown including one with pH, experiments with soil sterilization demonstrated that soil biota were the main factor responsible for the inhibition of the pathogens investigated. Certainly there are many reports of organisms in the soil that are antagonistic to plant pathogens. For example, over 100 species of fungi trap and prey on nematodes (Jatala, 1986) and many fungi are hyperparasites of other fungi (Adams, 1990). Among these are species of *Trichoderma* that secrete lytic enzymes which are active against fungal cell walls (Sivan and Chet, 1989), *Talaromyces flavus* which is able to attack the sclerotia of *Sclerotinia sclerotiorum* and *Verticillium dahliae* (McLaren *et al.*, 1989) and *Sporidesmium sclerotiorum* which is able to attack the sclerotia of five important plant pathogens (Whipps, 2001; Section 5.5.3).

Some soils suppress soil-borne plant pathogens, one of the earliest leads to the discovery of this phenomenon being take-all decline. Here the growth of wheat continuously in successive years first led to an increase in the disease caused by the fungus, *Gaeumannomyces graminis* var. *tritici* and then to its decline. Further experimentation showed that the population of *Pseudomonas* spp. which produced the antibiotic diacetylphloroglucinol increased to at least 2×10^5 per g root, levels that were double that required to suppress the disease (Raaijmakers and Weller, 1998). Other ways in which soil microorganisms may suppress plant pathogens are competition and parasitism, subjects to which we shall return in Chapter 5 (Section 5.4.2).

3.5.2   The atmosphere

It is not surprising that aerially borne pathogens are strongly influenced by weather conditions. In the first instance, their propagules must become detached from the host plant. Although many fungal pathogens have mechanisms for releasing their spores into the air beyond the still boundary layer surrounding the plant, in some instances rain splash is critical. The flight of the spore is entirely dependent on prevailing wind currents or, in the case of wettable spores, the flight path of water droplets emanating from rain drops splashing onto sporulating lesions. During the flight the spore may encounter adverse radiation in the form of ultraviolet light which may prevent it being infectious on alighting on a susceptible plant. On arrival at a suitable host the spore must have the appropriate conditions over a sufficient time period for it to germinate and effect entry. These usually mean high relative humidity and temperatures appropriate to the particular pathogen. Rain, wind, temperature and humidity are the substances of weather forecasts and these have long been used to warn of the potential for outbreaks of plant disease (Section 5.4.1).

Generation of inoculum

Vloutoglou and co-workers (1995) found that the greatest number of conidia of *Alternaria linicola* in the air above linseed crops occurred between 12:00 h and 13:00 h, when the relative humidity was lowest and, on a seasonal basis, the period of most prolific conidial production was from July to early September. At this period not only were weather conditions favourable for sporulation but there was an increase in the incidence of the disease in the senescing crop.

In contrast, many fungi require high humidity or even free water for spore release. For example, Dennis and co-workers (1992) carried out an extensive study of spore release from the basidiome of *Oncobasidium theobromae*. They found that spore release was highly correlated with the number of consecutive hours of basidiome wetness with a threshold of 5 h and a maximum of 12 h. Spore release was also highly correlated with the number of rainy days but low numbers were also released in response to dew. Studies on basidiomes in the laboratory showed that spore release required alternating periods of light and dark but that release only occurred during the dark period.

Dissemination of inoculum

There have been numerous studies on the effects of the atmosphere on the dissemination of propagules of plant pathogens. Air movements are clearly important not only for dislodging spores of some fungi from the parent plant but also for transporting them. Aerial parts of plants maintain a relatively static layer of air adjacent to them, referred to as the boundary layer. In order to escape this, pathogens have had to develop a number of mechanisms. For example, the aecidiospores of *Puccinia graminis* f. sp. *tritici* round up suddenly, generating sufficient momentum for them to escape (see Figure 5.2 for a diagram of the life cycle of this pathogen). In some species of *Peronospora*, a rapid decline in the humidity of the atmosphere is associated with spore discharge. Here it is thought that as leaves of the host plant dry they develop a charge and this repels the similarly charged spores. Other pathogen propagules rely on removal from the host plant environs by wind gusts and rain splash (Aylor, 1990).

Most investigations have shown that the majority of spores fall within a short distance of the parent plant, plots of the logarithm of spore numbers deposited versus the logarithm of the distance often approximating to a straight line (Figure 3.7 and cf. the 'contact distribution' of van den Bosch *et al.*, 1988, Figure 3.2 and Section 3.2.2). However, some escape and may travel intercontinentally before being brought down by air currents or washed out by rain (Nagarajan and Singh, 1990). For example, *Hemileia vastatrix* (the cause of coffee rust) is thought to

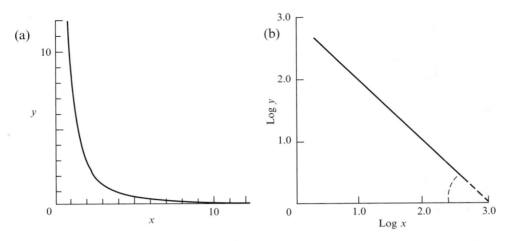

Figure 3.7   Typical disease gradients from a point source: (a) numbers of infections (*y*) plotted against distance (*x*) on linear scales; (b) plotted on double logarithmic scales (after Gregory, 1973)

have spread across the Atlantic ocean from Angola on the west coast of Africa to Brazil by means of wind transport (Bowden, Gregory and Johnson, 1971).

Rain splash is an important dispersal mechanism for many Oomycetes, fungi and bacteria. Sporangia of *Phytophthora cactorum*, causal agent of leather rot of strawberries, are dispersed by rain splash (Reynolds *et al.*, 1988). Strawberry plants are usually mulched but the type of mulch has marked effects on the incidence of disease. Madden and Ellis (1990) showed that fruit from plants mulched with plastic had the highest disease rating (ca. 82 per cent), presumably owing to the splashing that occurred when droplets hit this material, whereas those with straw mulch had the least (ca. 15 per cent).

As previously mentioned, outbreaks of fireblight caused by *Erwinia amylovora*, is spread from neighbouring hawthorn trees during stormy weather (Section 3.3.1 and Figure 1.21). More recently, McManus and Jones (1994) showed that the bacterium becomes aerosol-borne during rain and, when these conditions are accompanied by wind, the organism is readily disseminated within an apple orchard from infected budwood.

The role of atmospheric conditions in penetration

The critical roles of duration of appropriate temperatures and moisture levels for successful penetration become obvious when plotted as a three-dimensional graph or response surface. For example, Shaw, Adaskaveg and Ogawa (1990) developed such a graph for predicting the number of lesions caused by *Wilsonomyces carpophilus*, the causal agent of shot-hole of almond. From this it is apparent that the highest lesion numbers developed when temperatures were 15°C or above and the wet period lasted more than 24 h (Figure 3.8).

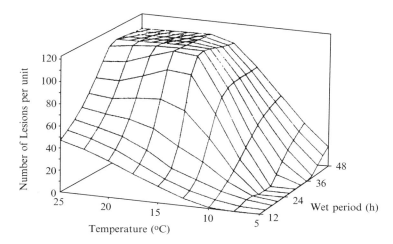

Figure 3.8 A response surface predicting the numbers of shot-hole lesions caused by *Wilsonomyces carpophilous* on leaves of almond under varying temperatures and duration of wetness (courtesy of D.A.Shaw, University of California, San Diego and the American Society of Phytopathologists)

Fungal pathogens are profoundly affected by moisture since their spores normally require high relative humidity or free water for germination and it is usually imperative that the germ tube is not subject to desiccation before it has penetrated the haven of the host. For example, conidia of *Monilina fructigena* only germinated at 97 per cent humidity or more but germination was greatest at 100 per cent relative humidity or in free water (Xu, Guerin and Robinson, 2001). It was also rapid, more than 70 per cent of viable conidia producing germ tubes within 2 h, at 20 and 25°C.

Uredospores of rusts are examples of fungal spores that require free water for germination as they contain inhibitors which must be leached from the spore before germination can occur. This does not happen in the mother lesion as the spores are hydrophobic and, when present *en masse* in a pustule, are not easily wetted. Once disseminated, however, they may become associated with dewdrops that are of sufficient size to leach the inhibitor. Germination then occurs rapidly, usually in less than 90 min for uredospores of *Puccinia graminis* f. sp. *tritici*. Presumably this behaviour is an adaptation to the comparatively short period between dew deposition and the drier conditions that ensue once the sun has come up as these would be expected to arrest germ tube growth and penetration of stomata (see Section 6.6.4). The inhibitors of the uredospores of both

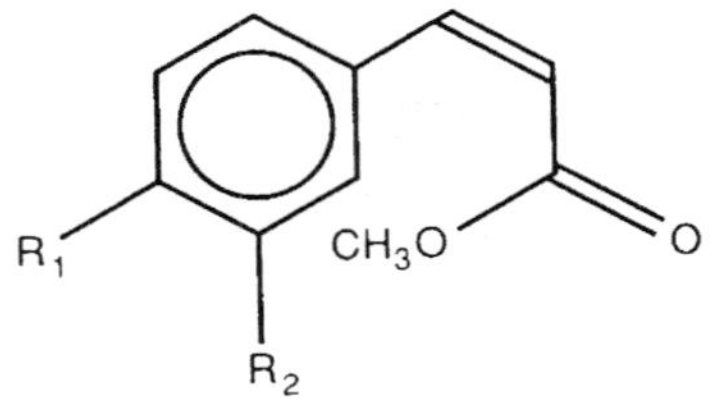

Figure 3.9 Endogenous germination inhibitors of rust uredopores: R_1 = OH, R_2 = CH_3O methyl ferulate (methyl 4-hydroxy-3-methoxy cinnamic acid), the germination inhibitor of *Puccinia graminis* f. sp. *tritici*. R_1 = R_2 = CH_3O (methyl 3,4-dimethoxy cinnamic acid) the germination inhibitor of *Uromyces phaseoli*

P. graminis f. sp. *tritici* and *Uromyces phaseoli* have been identified as the *cis* methyl esters of cinnamic acid derivatives (Allen, 1971; Figure 3.9). They are thought to act by inhibiting the dissolution of the plug occupying the germ-tube pore but once this has been digested the addition of inhibitor is ineffective in preventing elongation of the germ tube.

4 The Measurement of Inoculum and Disease Severity and their Effects on Crop Yields

Summary

Yield is affected by inoculum levels of pathogens and the severity of the diseases they cause. The measurement of all three factors is therefore necessary in order to set priorities for control, predict yield losses and evaluate control measures. Disease pressure can be measured by comparing the performance of plants grown in plots under standard conditions with those in which pathogens have been excluded by either physical or chemical means. Pathogen populations in the soil may be measured by a variety of techniques depending on the organism of interest and those in the air by suction traps. The extent of fungal colonization of plants may be measured by chitin or ergosterol analysis. Fungi, as well as other plant pathogens *in planta*, may also be measured by ELISA or nucleic acid probes. Disease severity is often measured visually with the help of keys although image analysis has been successfully used in a few instances. A number of formulae that relate disease severity to yield have been derived empirically. These use disease severity data recorded at a single time or several times during the growth of the plant. The area under the disease progress curve (AUDPC) is generally regarded as the most successful predictor of yield. Recent developments have included attempts to couple disease measurement with plant growth in order to predict yields more accurately and economic models in which disease severity is related to the return a grower can expect from a crop.

Introduction to Plant Pathology by Richard Strange
© 2003 John Wiley & Sons, Ltd   ISBN 0 470 84972 X (cased)   ISBN 0 470 84973 8 (pbk)

4.1   Introduction

Plant disease may be measured not only in epidemiological terms as described in the last chapter but also in terms of its effects on both individual plants and communities of plants, especially when they are grown as crops. Measurement of the severity of plant disease and its effects on crop yield, quality and value are crucial both for the establishment of priorities for control and for the evaluation of any control measures that may be instituted. In particular, plant breeders are often faced with the problem of screening thousands of lines in the field in order to select the most resistant. Frequently, in these circumstances, plants are scored on a scale such as 1–5 or 1–9. Although there is generally little disagreement among scorers at either end of the scale (a plant cannot be more sick than dead!) the central part of the scale, where often the critical decision as to whether to discard the plant or not in a breeding programme is taken, is subject to wide variation. Notably this occurs when the scale is purely descriptive and the scorer has no visual prompt. Reliable disease severity data are also required when the need arises to relate such information to failure of the crop to reach attainable yields (Gaunt, 1995). Ideally, one would like to be able predict the yield and quality of a crop in the absence of disease and in its presence at several levels of severity. In addition, one would like to predict the probability of the disease progressing through these levels of severity during the growing season.

Yield and quality affect value. An insufficient harvest causes price increases while a glut depresses prices. Blemished and unsightly produce can often be sold at only a small fraction of the price of an apparently healthy product or may be unsaleable. The Great Bengal Famine was an outstanding example of price rises being driven out of reach of the poorer members of the community, owing to low yields caused by the attack of *Helminthosporium oryzae* (= *Cochliobolus miyabeanus*: Chapter 1, Section 1.3.2 and see Figure 1.7). Nevertheless, the farmer who was able to salvage part of his crop benefited from the high price he was able to charge for it. Yield and quality are therefore causally linked to the profitability of the crop and not only affect the farmer's pocket but also decisions as to whether it would be economic to include disease-control measures as part of the process of crop husbandry. A fuller discussion of the economic and sociological effect of plant disease is given in the review by Zadoks (1985). In this chapter we shall be concerned primarily with the more restricted subject of the relationships among the presence of inoculum, disease severity and the extent to which these govern failure to reach attainable yield.

Determination of the effect of a pathogen on yield is fraught with difficulties. One method is to exclude it from test plots and compare the yield from these with that obtained from plots affected by the disease, an approach that may be feasible for seed and soilborne pathogens but not for airborne ones unless they are grown in a greenhouse that is proof against propagules of the pathogen, e.g. spore proof. Generally, where the pathogen is endemic control measures have to be applied in order to obtain a disease-free crop but the difficulty here is in finding a treatment that has no effect other than eliminating the target organism – an impossible mission. For example, a fungicide may affect the ability of the plant to form mycorrhizal associations or may itself affect the plant directly and adversely, a phenomenon known as iatrogenic disease.

An alternative approach is to measure inoculum or disease severity as accurately as possible and attempt to relate this to yield. Inoculum density and disease severity will normally be measured before harvest and may be used to predict yields. Although, in some instances, quite accurate forecasts of yield have been made on the basis of measurements made at one point in time, accuracy is generally increased when multiple measurements are made over a period.

The initiation and development of disease are influenced by many factors. These include the incidence and reproductive capacity of the pathogen, the distribution and degree of susceptibility of the plants in the crop, the nature of the soil, its flora and fauna, and the prevailing weather conditions. Variation in any one of these can profoundly affect the course of an epidemic and the damage it causes.

4.2 Parameters of disease and their measurement

The presence of the pathogen in the soil or aerial environment is the first indication that a crop may suffer disease. Therefore the methods available for estimating plant pathogens in these two environments will be reviewed in this section. Additionally, techniques for measuring pathogens in plants will be discussed as such measurements provide one of the more reliable estimates of the virulence of the invader and the susceptibility of the host.

4.2.1 Measurement of disease pressure in the soil

In order to determine the disease pressure in the soil for a given crop it is necessary to remove the influence of potential pathogens in test plots and compare them with controls. One way of doing this is by treating plots to kill pathogens and comparing the yields of plants in these plots with those of comparable untreated plots. A problem with this approach is that, as mentioned above (Section 4.1), no treatment is exclusively selective for plant pathogens. Therefore such experiments are confounded by other changes brought about by the treatment such as the killing of other members of the soil microflora and changes in the physical and chemical properties of the soil. Nevertheless, such experiments can be informative. For example, Rovira (1990) reported that fumigation of soil before planting wheat increased yields by between 70 per cent and 300 per cent. The pathogens responsible for causing decreases in yields were identified as *Gaeumannomyces graminis* var. *tritici* (take-all), *Rhizoctonia solani* (bare patch, purple patch), *Heterodera avenae* (cereal-cyst eelworm), *Pythium* spp. (root rot), *Bipolaris sorokiniana* (common root rot) and *Fusarium grami-nearum* group 1 (crown rot).

An alternative approach to assessing the risk of soil-borne disease is to grow plants of the crop in test samples of the soil and determine the disease severity or yield. As Oyarzun (1993) points out, disease severity results from:

(1)   density of the pathogens,

(2)   virulence of the pathogens,

(3)   physical and chemical characteristics of the soil,

(4)   competitiveness of the pathogens in relation to the other soil microflora,

(5)   environmental conditions,

(6)   susceptibility of the test plant.

The first two factors make up the 'inoculum potential' of the pathogen, which was defined by Garrett (1956) as 'the energy available for infection of a host at the surface of the infection court'. This is influenced by the biotic and abiotic factors of the soil to give the 'inoculum potential of the soil (IPS)'. Environmental conditions and the susceptibility of the plant determine the actual amount of disease generated by a given IPS. An important property of such assays is that they do not rely on the prior identification of the causal agent of disease.

Oyarzun (1993) applied these concepts to the formulation of a method for predicting the severity of root rot of pea. In the autumn or early spring, preceding the planting of the pea crop, an area of 1 ha from each prospective field was selected for sampling on the basis of homogeneity and at a distance of at least 10 m from the border. Fifty samples were extracted with an auger, mixed and sieved, before distributing to four large pots (2.6l). Seeds were planted in the pots and the moisture level brought to field capacity. After the plants had been allowed to grow to the green-bud stage (13th leaf) under standardized conditions, they were re-moved, the roots carefully washed free of soil and scored on a 0–5 scale: 0 = healthy and 5 = roots 100 per cent rotten. From the results a root disease index (RDI) was calculated for each plant in which disease of the epicotyl and root was weighted more heavily (0.35) than that of the cotyledons and xylem (0.20 and 0.10 respectively). From these results the RDI per sample, which is a measure of the IPS of the soil, was calculated as an average of all the plants in the sample.

4.2.2   Measurement of the pathogen populations in the soil

Representatives of many of the classes of organisms discussed in Chapter 1 are important soil-borne pathogens. These include *Striga* and *Orobanche* among the parasitic angiosperms, fungi, nematodes and bacteria. In addition, viruses may be vectored by nematodes and fungi in the soil.

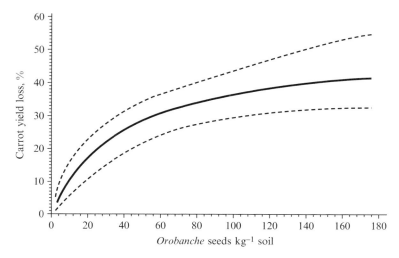

Figure 4.1 Estimated yield loss of carrot in per cent (solid curve) and 95 per cent bootstrap prediction intervals (dashed curves) in relation to *Orobanche* sp. populations in the soil (after Bernhard, Jensen and Andreasen, 1998)

Once the causal agent of a disease has been identified more specific techniques may be adopted for its detection and, importantly, since disease severity is normally related to incidence of pathogen propagules, enumeration. For example, Bernhard, Jensen and Andreasen (1998) were able to predict the losses in carrots to *Orobanche crenata* and *O. aegyptica* by counting the numbers of seeds of each pathogen in soil samples since the relation of yield loss to seed numbers fitted a rectangular hyperbolic model (Figure 4.1). They also found that there was a linear relation between the logarithm of seed numbers and the number of *Orobanche* plants that developed.

A flotation method to separate the microsclerotia of *Verticillium dahliae* from soil samples using caesium chloride or sucrose solutions has been used (Termorshuizen *et al.*, 1998). In this pathosystem, like that of *Orobanche* and carrots, a linear relationship was found between the $\log_{10}$ -transformed numbers of microsclerotia g^{-1} of soil and the $\log_{10}$ – transformed numbers of colonies of the pathogen growing from roots on ethanol agar medium (Nagtzaam and Bollen, 1994). However, an inter-laboratory study has shown just how difficult it can be to quantify microsclerotia of this pathogen. In a blind trial of 14 soil samples by 13 research groups from seven countries there was a 118-fold difference between groups with the highest and lowest mean estimates (Termorshuizen *et al.*, 1998)!

Recently, monoclonal antibody techniques have been investigated as a means of quantifying soil-borne pathogens but as Dewey and co-workers (1997) conclude many problems remain to be solved with regard to the extraction of antigens from the soil before reliable estimates of biomass of specific fungal pathogens can be obtained. For example, Otten, Gilligan and Thornton (1997) showed that clay soil reduced the sensitivity of a monoclonal antibody-based ELISA assay for *Rhizoctonia solani* by more than 1000-fold.

Fluorescent *in situ* hybridization (FISH) is a relatively recent technique that has been used to identify and quantify bacteria and fungi. A complementary probe to DNA or RNA sequences of the organism of interest is made and labelled directly or indirectly with a fluorochrome. rRNA is often the target for probes as ribosomes are usually numerous and contain both conserved and variable regions, allowing the specificity of the probe to be tailored from the universal to the subspecies level (Li, Spear and Andrews, 1997).

Because of the difficulties of direct enumeration of soil-borne fungal and bacterial pathogens an amplification procedure is often included. Traditionally, this has involved culturing the pathogen itself, but more recently it has become feasible to amplify DNA sequences of the organism of interest by the polymerase chain reaction (PCR). Quantitative PCR procedures (see Section 2.5.5) allow the determination of the original numbers of the organisms of interest.

The dilution plate method is one way of visualizing the number of propagules of a pathogen. A suspension of the soil is made in sterile distilled water and serially diluted. Samples of the original suspension and the dilutions (usually 0.1 ml) are spread on Petri dishes of a medium which is selective for the pathogen of interest and incubated. Results are expressed as colony-forming units (cfu)/g of soil. For example, Conn and co-workers (1998) developed a medium for the selective isolation of *Streptomyces* spp. from soil. An additional advantage of the medium was that it allowed the colonies to express a distinctive smooth and non-sporulating phenotype which aided the identification of pathogenic strains. Moreover, colonies with this phenotype also produced thaxtomin, a toxin produced by all pathogenic strains (see Chapter 8, Section 8.5.1).

The Baermann-funnel technique is usually used to enumerate nematodes (Figure 4.2). Soil samples are contained in a cloth bag or put on a gauze sieve and placed in the funnel which is closed off with a piece of tubing and a clip. The funnel is filled with enough water to reach the sample and the apparatus left overnight or longer. After this time, the clip is opened and about 5 ml of the water, which contains most of the nematodes, is allowed to flow out of the bottom of the funnel. Nematodes in the sample may then be examined in a counting dish. The method may also be adapted for counting the numbers of nematodes in infected plants. Usually the plant tissue is finely divided or homogenized before being enveloped in muslin or paper and placed on top or in the funnel as described for soil samples.

Soil-borne pathogens are seldom distributed homogeneously, either in space or time, and therefore present a sampling problem. Where a crop is already in the field it is important to sample populations both within the rhizosphere and outside it during the growth cycle. Taylor A.L. (1971) recommended that samples should be taken in a systematic manner over the whole area of interest and should be repeated for at least 3 years. The size and number of these is critical and, for example, in estimating nematode numbers the total weight of soil from 1 ha should not be less than 10–20 kg. More recent work with specific nematode

Figure 4.2 Baermann funnels – a simple apparatus for enumerating nematodes; on the left with a cloth bag containing the sample and on the right with a gauze sieve (courtesy of C. C. Doncaster, Rothamsted Research)

species suggests that the size of samples may be substantially reduced. For example, Turner (1993) has reviewed experiments made by the Department of Agriculture for Northern Ireland in order to optimize the detection of potato cyst-nematodes. Factors evaluated were the efficiency and ease of use of augurs for sampling, numbers of samples and patterns of sampling. It was concluded that a composite sample, consisting of 30 to 36 probes taken from the perimeter of a field was satisfactory for detection of potato cyst-nematodes for seed certification purposes (see Section 5.6.4).

4.2.3 Measurement of pathogen populations in the air

Many fungal pathogens produce copious numbers of spores which may be disseminated by air currents or rain splash. Also many vectors of mycoplasmas, viruses and viroids are air-borne.

Spore traps may be used to sample air-borne pathogens (Figure 4.3). The main features of the one illustrated are a pump, a vane, a small orifice and a rotating

Figure 4.3   The Burkard spore trap used for trapping spores of plant pathogenic fungi in the air (courtesy of G. Wili of the Burkard Manufacturing Co. Ltd)

drum covered with transparent tape and an adhesive. The vane maintains the orientation of the apparatus so that the orifice, through which air is drawn by the pump, is always into the wind. Any solids that are present, such as fungal spores, are impacted onto the drum. Modern spore traps, such as the one illustrated, allow the air spora to be sampled over a period of 1 week before the tape and adhesive with the impacted spores requires replacing. The tape itself may then be examined. Here the principal difficulty of the technique becomes apparent – the identification of any organism present to an appropriate level. Simple microscopy may be inadequate and supplementary procedures may be necessary such as stereoscan electron microscopy and fluorescent antibody technique.

Suction traps may also be used to trap insect vectors of viruses and their efficiency may be increased by the use of pheromones where these are known (Taylor, L.R., 1971).

4.2.4   Measurement of pathogens in the plant

The extent of colonization of a plant by a pathogen is a useful measure of disease and several techniques are available which rely on estimating the amounts of pathogen components that are not represented in the plant. For example, chitin and ergosterol are prominent constituents of the cell walls and membranes,

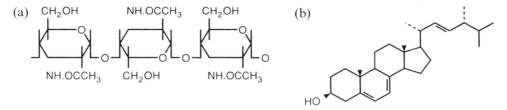

Figure 4.4 The chemical structures of (a) chitin and (b) ergosterol, constituents of the cell walls and membranes, respectively, of many fungi

respectively, of fungi but not plants (Figure 4.4). Chitin may be measured by a colorimetric technique involving deamination and depolymerization of the compound in concentrated alkali followed by treatment with nitrous acid to give an aldehyde. This reacts with methyl benzothiazolone hydrazone in the presence of ferric ions to give a blue formazan dye, which may be determined spectrophotometrically. The amount of chitin in a given fungus may be determined from standard curves of chitin and the assay performed on the fungus grown *in vitro*. However, when using samples of diseased material, care must be taken to allow for inflated readings resulting from interfering compounds containing hexosamine residues that may be present in the uninfected plant (Ride and Drysdale, 1972). Also, chitin is refractory and persists after the death of the fungus and therefore does not give an estimate of living hyphal mass.

In contrast to chitin, ergosterol is unstable and is therefore potentially a measure of viable hyphal mass. This supposition has been confirmed for ectomycorrhizas where the two assays have been used in combination to estimate both total and living fungal biomass (Ekblad, Wallander and Nasholm, 1998). Alcoholic potassium hydroxide is the most efficient solvent for the extraction of ergosterol. The compound is subsequently partitioned into pentane and after evaporation of the solvent dissolved in methanol before injection into an HPLC. Effluent from the column is monitored at 282 nm and the ergosterol content estimated by comparison of peak areas with standards (Padgett and Posey, 1993). Gretenkort and Ingram (1993) have also published an ergosterol method but in this case the compound was quantified by a thin-layer chromatography (TLC) method. Spots on TLC plates corresponding to ergosterol were eluted and the absorbance of eluates, which were dissolved in ethanol, read in a spectrophotometer at 278 nm.

Although the speed and ease of use of ergosterol assays have much to commend them, Bermingham, Maltby and Cooke (1995) have questioned their use since they found that in only three out of nine species of aquatic hyphomycetes was there a significant regression between ergosterol content and biomass. In contrast, Newton (1989) found that sterol analysis as a means of quantifying mildew infections of barley compared favourably with colony numbers and area of sporulation as assessed with an image analyser (Section 4.3). Newton (1990) also characterized the sterol as (3β)-ergosta-5,24(28)dienol and showed that it could be used to detect partial resistance to mildew. Although Gunnarsson and co-workers (1996) found that the ergosterol content of *Bipoaris sorokiniana* varied considerable according to age and nutrient conditions in culture it was, nevertheless, superior to visual ratings of disease in terms of sensitivity and objectivity.

Neither the chitin nor the ergosterol technique is specific and so results from them may be confounded if the load of epiphytic fungi on the plant is high. One way of countering this problem is to use serological techniques such as ELISA (Section 2.5.4). For example, Harrison and co-workers (1990) were able to estimate the amounts of mycelium of *Phytophthora infestans* in potato leaf tissue using either plate-trapped antigen or the $F(ab')_2$ antibody-fragment technique. In a different approach, Oliver and co-workers (1993) have described the use of transformants of fungal pathogens constitutively expressing β-glucuronidase. The enzyme is stable and may be quantified by a fluorometric assay using the substrate 4-methylumbelliferyl-β-D-glucuronide or detected histochemically in infected plants using 5-bromo-4-chloro-3-indolyl-β-D-glucuronide. Although these are formidable advantages which may find wide application, for example in the rapid screening of germ plasm, it is important to establish that the virulence of the transformants has not been compromised and that the reporter gene is expressed throughout the phase of the life cycle of the pathogen of interest.

As discussed in Chapter 2 (Section 2.5.5), various protocols and instrumentation have recently become available for quantifying target RNA or DNA sequences by PCR and the reader is referred to these.

Bacteria are most simply enumerated by homogenizing the infected plant and plating the homogenate and its dilutions on a suitable selective medium (see Section 2.4). ELISA or nucleic acid probes (Sections 2.5.4 and 2.5.5) are usually used to quantify viruses and these techniques may be used for all classes of plant pathogen if appropriate antibodies or probes are available.

In the last few years a number of companies have developed kits for the detection and quantitation of plant pathogens, some of which may be used in the field and give results in minutes. Recent experience by the author in the use of some of these kits in West Africa showed that they were robust and, in the case of fungal and bacterial pathogens, circumvented extended procedures of cultivation and the risk of misidentification.

4.3   Measurement of symptoms

Infection of plants by pathogens gives rise to a variety of symptoms which also vary according to the severity of the attack. As Gaunt (1995) points out, the measurement of these faces the plant pathologist with a dilemma. Should the disease be measured by methods that allow comparison of epidemics such as the number and size of sporulating lesions or by methods that are related to host productivity and yield such as remaining green leaf area? Also some pathogens cause systemic symptoms such as stunting and wilting which are more difficult to quantify. Most methods rely upon direct observation by trained observers. *Puccinia graminis* f. sp. *tritici* causes well-defined lesions on aerial parts of wheat (Chapter 2, Table 2.1) although there is a continuum of lesion sizes which may, on occasions, cause difficulty in assigning symptoms to a given infection type. In contrast, *Ascochyta rabiei* causes a variety of symptoms on chickpea. These include areas of chlorosis and necrosis on all aerial parts of the plant which may contain concentrically arranged pycnidia. Stem and petiole breakage are also prominent features of severe attacks. Several scales have been proposed and one of these, a 1–9 scale, is given in Table 4.1. Unfortunately, there is wide variation among different workers in the scores that they allot, particularly in the middle of the scale. This part of the scale is crucial since breeders are inclined to discard material that scores a 6 but keep material that scores a 5. When 11 delegates at a coordination meeting of an international project on Ascochyta blight of chickpea were asked to score 66 photographs of diseased plants standard deviations more than 1 (highest 1.75) were recorded for 32 of the plants. For a further discussion of the problems associated with visual scoring, the reader is referred to the review by Nilsson (1995).

One solution is to use actual measurements. For example Riahi and co-workers (1990) proposed the linear infection index for Ascochyta blight of chickpea which is calculated as follows:

$$\frac{NL \times ALL}{SL} \times 100 \tag{4.1}$$

where NL = number of lesions, ALL = average lesion length and SL = stem length. This method gives an objective estimate of disease severity but it is laborious and not suited to the screening of the large numbers of genotypes often encountered in breeding programmes. Instead reliance is usually placed on visual assessment by trained observers.

Table 4.1   A scale for scoring infection of chickpea by *Ascochyta rabiei* (Reddy, Singh and Nene, 1981)

Infection type	Symptoms
1	No lesions visible
2	Lesions on some plants, usually not visible
3	Few, scattered lesions, usually seen after careful examination
4	Lesions and defoliation on some plants, not damaging
5	Lesions common and easily observed on all plants, but defoliation and/or damage not great
6	Lesions and defoliation common, some plants killed
7	Lesions very common and damaging, 25% of plants killed
8	All plants with extensive lesions causing defoliation and drying of branches, 50% of plants killed
9	Lesions extensive on all plants, defoliation and drying of branches, more than 75% of plants killed

The main problem with visual assessment, as described above, is that it is inaccurate. Part of the reason for this is that the eye obeys the Weber–Feckner law which states that visual discrimination is a function of the logarithm of the intensity of the stimulus. The eye can therefore only assess accurately very low or very high levels of disease (Horsfall and Barratt, 1945). To overcome this problem there has been interest in determining the relationship between disease incidence and disease severity (Seem, 1984). Incidence may be defined as the proportion of diseased entities within a sampling unit. Such entities may be the proportion of plants in a given area, diseased leaves on a plant or diseased fruit on a tree. Severity is the amount of disease affecting the diseased entities. For example, Edwards and co-workers (1999) investigated peppermint rust, a disease that when not combated with fungicide causes the plant to die out within four seasons. They found that the relationship between disease incidence and severity could be described by the following equation:

$$Y = 0.26 \times 10^{0.02X} \tag{4.2}$$

where Y = disease severity (%) and X = disease incidence (%).

One way in which operator error may be reduced is to use hand-held pictorial keys which are available for many of the more important diseases (James, 1971). Falloon and co-workers (1995) have published disease severity keys for powdery mildew (*Erysiphe pisi*) and downy mildew (*Peronospora viciae*) of pea and powdery scab (*Spongospora subterranea*) of potato. Here the keys were constructed from hand drawings of pea leaflets or potato tubers with colonies or lesions drawn to resemble typical infections of the respective diseases. Image analysis was used to confirm the surface area of lesions and the illustrations were scanned into a computer for manipulation into illustrated keys for publication (Figure 4.5). Such keys, if referred to at the time of making disease assessments, can improve precision considerably.

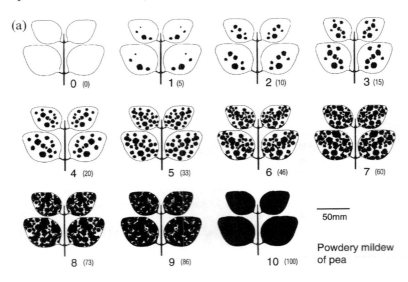

(a)

0 (0)   1 (5)   2 (10)   3 (15)

4 (20)   5 (33)   6 (46)   7 (60)

8 (73)   9 (86)   10 (100)

50mm

Powdery mildew of pea

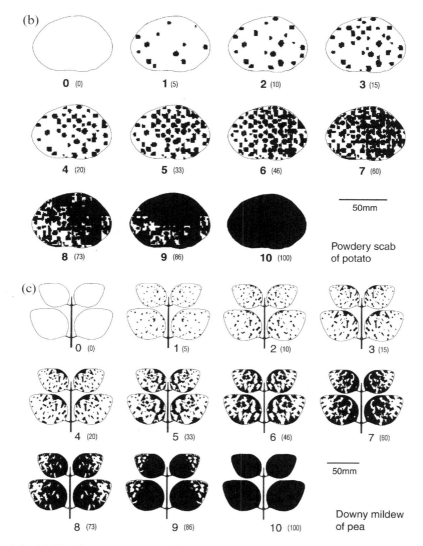

Figure 4.5 (a) Key for measurement of *Peronospora* on pea; (b) key for scoring powdery scab of potato, and (c) key for scoring downy mildew of pea (Falloon, R.E. *et al.* (1995): Disease severity keys for powdery and downy mildews of pea, and powdery scab of potato. *New Zealand Journal of Crop and Horticultural Science* **23**: 31–37; reprinted by permission of RSNZ Publishing, courtesy of R.E. Falloon and the New Zealand Journal of Crop Protection)

On a larger scale, plant stress, including disease, may be detected by cameras carried by low-flying aircraft or radio-piloted model planes. Such data can be valuable in determining factors which affect the amount of disease in a crop. In pioneer work, Brenchley (1966, 1968) was able to identify the early stages of potato blight and monitor its spread during the growing season. Areas of infection appeared as dark patches on the photographs that fanned out asymmetrically from the primary foci in a manner typical of wind-borne diseases. These data allowed Brenchley (1968) to trace the source of infection to waste potato dumps and old clamps.

More recently, Kobayashi and co-workers (2001) determined the spectral regions most sensitive to panicle blast infection of rice caused by *Magnaporthe grisea* and found that reflectance increased at the dough stage, with the greatest sensitivity lying in the visible region around 485 and 675 nm. The authors concluded that air-borne multispectral scanners may be effective in detecting the occurrence of panicle blast at this stage using band combinations of 530 to 570 nm and 650 to 700 nm.

The use of the radiometers on satellites such as the American Landsat series and the European SPOT (Système Probataire pour l'Observation de la Terre) is a logical extension to those located closer to the Earth and they have provided a wealth of data including information on plant stresses, diseases and pests (Nilsson, 1995). However, to be really useful to plant pathologists, particularly epidemiologists, further refinement of spectral and geometric resolution will have to be achieved. There is also a problem of timing of satellite orbits. For example, satellites pass over Sweden early in the morning when on some days there is dew on the vegetation and on others it has dried, profoundly affecting spectral reflectance. There is the further factor of weather, cloud cover often obstructing measurements in many parts of the world (Nilsson, 1995).

4.4 Measurement of yield and quality

Yields of plant products are generally non-controversial and are usually recorded in terms of weight or number. Cereal yields, for example, are usually measured in terms of 1000-grain weight, spikelets per tiller, numbers of tillers and kg or metric tonne per hectare. For example, Adhikari and co-workers (1999) measured losses of rice caused by *Xanthomonas oryzae* pv. *oryzae* as reductions in the number of tillers, grains per panicle and 1000-grain weight.

Quality is highly prized and so, unfortunately, is uniformity. As a result, produce which has a few superficial lesions sells for very much less than produce that is apparently completely sound; also supermarkets are keen to show off piles of uniform fruit and vegetables that seem to be without blemish. To a plant pathologist, the genetic uniformity and probable use of biocides that these mountains of perfectly identical and apparently perfect produce imply give pause for thought.

4.5 Establishing the relation between disease and yield

So far in this Chapter we have been more concerned with the measurement of the pathogen than the symptoms it causes. However, it is the disease symptoms that cause the failure of plants to attain their yield potential. They do this by reducing the stand of the crop or its photosynthetic rate, accelerating senescence, stealing light, appropriating assimilates, consuming tissues or reducing turgor. All of these factors would be expected to affect both radiation interception and its efficient use (Gaunt, 1995) and they will receive further treatment in Chapters 7, 8 and 11. In this section we shall be concerned with relating the measurements available, whether pathogen or symptoms or a combination of both to yield.

Various models have been proposed that relate some estimate of disease such as incidence of pathogen or severity of disease to failure of a crop to reach its attainable yield. These models are usually differentiated on the basis of how many estimates are made and how they are used.

The critical point model is the simplest as there is only one estimate of disease. Although this may give an accurate estimate of yield in some circumstances, it is of limited usefulness when environmental conditions vary unpredictably after the estimate has been made as these are likely to alter the progress of the disease, exacerbating it when they favour the pathogen and moderating it when they favour the plant. However, one particularly valuable application of the critical point model is when it is used to define a damage threshold rather than disease severity at a particular growth stage of the crop. Damage threshold may be defined as the lowest pathogen population density or injury level for which at least some damage is projected and may provide the warning needed to take remedial action. For a full definition of thresholds and other terms relevant to the concepts of yield and crop loss the reader is referred to the paper by Nutter, Teng and Royer (1993).

A solution to the problem of varying conditions affecting the development of a disease is to make several estimates over time and/or space. The data may then be treated by multiple regression analysis. A variation of this model is to integrate disease severity over time to give the area under the disease progress curve (AUDPC) and relate this to yield.

A further factor that should be taken into account is the growth of the plant since disease affects plant growth and plant growth affects disease. Finally, and of critical importance to the farmer, there is the problem of relating any model to the price that the crop will fetch.

4.5.1 Critical-point models

Critical-point models attempt to relate yield to disease assessment at a specific time. Usually the model is linear and takes the form:

$$L = cX \qquad (4.3)$$

where L = loss, c = a coefficient which varies according to the crop and disease under discussion and X = the proportion of disease at a particular growth stage of the crop. It is unusual for the regression line to go through the origin and so a constant must be added or subtracted accordingly.

Diseases caused by soil-borne organisms are particularly favourable for application of critical-point models as they can be applied to the presence of inoculum in the soil rather than diseased plants. For example, Brown and Sykes (1983) found that there were direct relationships between the numbers of eggs of the potato-cyst nematodes (*Globodera pallida* and *G. rostochiensis*) in the soil and potato yield. Yield losses were 6.25 t/ha for each 20 eggs/g soil in the 0–40 eggs/g range and 1.67 t/ha/20 eggs/g for populations in the 40–160 eggs/g range. Determination of the population size of soil-borne pathogen propagules can therefore be useful where it correlates with disease as areas of high disease pressure can be avoided or planted with cultivars that are resistant.

As described earlier (Section 4.2.2), Bernhard, Jensen and Andreasen (1998) found that the number of *Orobanche* plants in carrot fields was proportional to the number of seeds extracted from the soil by an elutriation process. However, yield losses caused by *O. crenata* and *O. aegyptica* stabilized at about 50 per cent for moderate infestations of about 200 seed/kg soil whereas infestation at this level caused virtually complete loss of peas. Unfortunately, the soil sampling and seed extraction methods used to count seeds of the pathogens are laborious, precluding their use by farmers.

Other workers have related yield to severity of symptoms at a specific growth stage of the crop. For example, Roberts and co-workers (1995) found that severity of bacterial blight of peas caused by *Pseudomonas syringae* pv. *pisi* at growth stage 208 of the plant was the best predictor of yield. When this was combined with terms for environmental effects and bird damage a disease severity of 1, equivalent to 5 per cent leaf area affected, predicted a loss of 0.98 t/ha.

As described in the previous Chapter, some diseases are polyetic, building up over successive seasons in a predictable manner. For example, Bélair and Boivin (1988) found that there was a significant logarithmic relation between galls on carrot caused by *Meloidogyne hapla*, measured as a gall index, one year and that of the succeeding year. Losses were described by the equation:

$$L = 3.895 + 21.7078X \qquad (4.4)$$

where L = percentage loss and X = a gall index describing the number of galls on carrot roots caused by the nematode. Thus, when the gall index was high, planting carrots the following year could be avoided or the field treated in some way to reduce the nematode population.

4.5.2 Multiple-point models and area under the disease progress curve

Intuitively one would expect yield predictions to be more accurate when disease estimates are taken several times during the growing season rather than at a single time (James and Teng, 1979). For example James and co-workers (1972)

estimated losses caused by potato blight at nine weekly intervals according to the
following equation:

$$L = b_1 X_1 + b_2 X_2 + b_3 X_3 + \ldots \tag{4.5}$$

where L = percentage loss in yield, b_1, b_2 etc. = coefficients for each week relating
yield loss to disease severity and X_1, X_2 etc. = weekly estimates of disease
severity. In nine out of 10 cases the difference between the estimated loss and
the actual loss was less than 5 per cent. The choice of coefficient is important as
yield is affected differently at different growth stages (see Section 4.5.3).

Determination of the area under the disease progress curve (AUDPC) is a
natural progression from multiple point models. Again, the time selected is of
importance and should reflect the period that is critical in determining yield.
Shaner and Finney (1977) defined AUDPC as follows

$$\text{AUDPC} = (\sum_{i=1}^{n}(X_{i+1} + X_i)/2)(t_{i+1} - t_i) \tag{4.6}$$

where X_i is a measure of disease severity at the ith observation, t is a measure of
time (usually in days) and n = the total number of observations.

As mentioned in Section 4.3, the parameter chosen to measure disease is
critical since failure to attain full yield may be more closely related to impaired
plant functioning rather than measures of disease severity (Rouse, 1988; Gaunt
1995). For example, there was a high correlation between the integral of necrosis
over time and loss of kernel weight in infections of wheat by *Septoria tritici*
(Forrer and Zadoks, 1983). Reciprocally, yield may be related to the amount of
tissue remaining healthy as in mildewed barley (Oerke and Schönbeck, 1990).
Here, yield was more closely related to photosynthetically active leaf material
($r = +0.96$ and $+0.93$ for the cultivars Catinka and Tapir, respectively) than
disease intensity, measured as percentage area covered by the fungus ($r = -0.32$
and -0.85, respectively). Similarly, Bergamin and co-workers (1997) found that
yields of *Phaseolus vulgaris* infected with *Phaeoisariopsis griseola*, causing angular leaf spot, were generally not related to disease severity or AUDPC but to the
duration of healthy leaf area and healthy leaf area absorption.

4.5.3 Coupling disease progress with plant growth

Unless the plant is killed swiftly by the pathogen, it will continue to grow as will
the pathogen. Moreover, the plant or the crop may compensate for the infection.
At the plant level, there have been relatively few studies in which the level of
disease at different growth stages has been weighted (Gaunt, 1995). Shtienberg
and co-workers (1990), however, have developed such a model in order to
generalize yield assessments for potatoes attacked by early and late blight, caused

by *Alternaria solani* and *Phytophthora infestans*, respectively. Loss was described as a function of the AUDPC relative to a reference crop (RAUDPC: relative area under the disease progress curve) multiplied by the effect of the disease on bulking rate. RAUDPC was obtained by integrating disease severity over the period of yield accumulation and weighting each chronological day by its relative contribution to host growth. When the model was evaluated with data from 53 epidemics, which were independent of those used for its development, differences between predicted and observed losses ranged from −11.2 per cent to 8.2 per cent and were less than 5 per cent in 80 per cent of the observations.

Compensation for disease may take place at the crop level when mixtures of resistant and susceptible plants are grown, the resistant plants compensating for the reduced productivity of the susceptible plants in the blend (Wilcox and St. Martin, 1998). Other factors are also involved in contributing to the superior yields of mixtures of genotypes as a means of controlling disease as opposed to planting a single genotype (see Section 3.4.1).

4.5.4 Coupling disease predictions with economic loss

The effect of disease on the economic return for growing a crop is difficult to gauge and is particularly pertinent to whether a farmer initiates control measures. There are, nevertheless, a number of computer programs available that give such advice, such as Blitecast and Simcast in the case of potato and *Phytophthora infestans*. Unfortunately, at lease in this case, the only palliative the farmer can apply is a relatively broad-spectrum biocide. More environmentally friendly solutions are desperately needed (see Chapters 5 and 12).

5 Inoculum Control

Summary

Controlling the inoculum of pathogens is an important way of protecting plants which is achieved with varying degrees of success by the application of a variety of techniques. Eradication may be possible from seed and propagative material and from the growing crop by roguing and pruning. Inoculum from crop residues may be reduced by ploughing and from the soil by solarization, the addition of amendments and, in some instances, by flooding. In the greenhouse screening out certain wavelengths of light that allow sporulation of pathogenic fungi may be an option but in the field when an attack is imminent there is little alternative, at the time of writing, to the application of pesticides. However, forecasting may reduce the number of applications and the amount of pesticide used so that they are only applied when conditions are favourable to the pathogen. Biological control is an alternative to the application of pesticides in some circumstances and there are about 80 products on the market. The screening, development and application of these are major endeavours which, in some cases, have resulted in products that give control which is at least as good as the application of chemicals. A measure of control of wind-borne pathogens may be achieved by intercropping and planting mixtures of varieties or multilines. Water-borne pathogens may be controlled by careful control of irrigation, and disinfecting farm equipment may help to limit the spread of soil-borne pathogens. One of the best methods of controlling plant disease is to use clean propagating material and here quarantine measures have an important role to play. Diseases caused by such pathogens as mycoplasmas, viruses and viroids are transmitted by vectors; control of their movements by reflective mulches and the use of plants that contain anti-feedants or interfere with ingestion or phloem finding may therefore reduce infection by these agents.

5.1 Introduction

In the previous three chapters methods were discussed for identifying plant pathogenic organisms, for describing the epidemics they cause and for quantifying the diseases and the extent to which they result in the failure of the crop to attain its maximum yield. In the present chapter it will be assumed that the causal

Introduction to Plant Pathology by Richard Strange
© 2003 John Wiley & Sons, Ltd   ISBN 0 470 84972 X (cased)   ISBN 0 470 84973 8 (pbk)

organism has been identified and that the losses it causes have been measured and are unacceptable. How then can it be controlled? Returning to Fry's equation, it is clear that disease is a product of pathogen, host and environmental factors acting over time (Fry, 1982; Section 3.1). Moreover, these factors interact (Figure 5.1).

Plant pathogens outside their hosts are often very sensitive to environmental conditions (arrow A). For example, a period of low humidity may be sufficient to abort a severe attack by fungal or Oomycete pathogens. The host, too, can contribute to environmental effects (arrow B) and so influence the pathogen indirectly. For example, the closure of the canopy in bean fields increases the humidity of the underlying atmosphere to the extent that mycelial bridges of the aerial blight fungus, *Rhizoctonia solani*, form, allowing the spread of the fungus from plant to plant (Yang, Berggren and Snow, 1990). Environmental factors such as drought also profoundly affect the host (arrow C) and these, in turn, affect the interaction of host and pathogen (arrows D and E). Plant pathogens may also affect the environment directly (arrow F) but they chiefly bring about environmental change indirectly through their parasitism of plants. For example, most of the elm trees in UK and the USA have been destroyed by Dutch elm disease and similar fates have befallen chestnut trees in North America owing to blight caused by *Cryphonectria parasitica* (Anagnostakis, 1987). Also infection of the jarrah forests of Australia by *Phytophthora cinnamomi* has affected the water budgets of catchment areas and led to increased salinity, a potential threat to the water supply of Perth (Weste and Marks, 1987). In this chapter control of plant pathogens by manipulation of the environment (arrow A) will be our primary concern.

Plant pathogens may be classified as monocyclic, polycyclic or polyetic, as discussed in Chapter 3. Disease intensity for monocyclic pathogens is directly related to the amount of inoculum present at the beginning of the season but for polycyclic pathogens disease intensity is exponentially related to the amount of inoculum. It follows that, when only one season is under consideration monocyclic diseases may be controlled by reduction of the initial inoculum, whereas polycyclic diseases may be controlled by reduction of both initial inoculum and inoculum multiplication. However, if more than one season is considered, this distinction disappears as monocyclic pathogens are usually polyetic, their inocu-

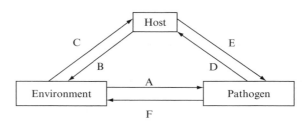

Figure 5.1 Diagram indicating the interactions of host, pathogen and environment (for explanation see text)

lum increasing from season to season. Besides reducing the initial inoculum and its multiplication, plant disease may also be controlled by impairing the effectiveness of inoculum and inhibiting its spread. In this chapter, we shall therefore be considering methods of limiting inoculum by reducing its sources, multiplication, effectiveness and spread.

5.2   Reducing the sources of inoculum

Infected plant material is normally the source of inoculum although for some pathogens vectors are important (Section 3.3.2). Infected seed and propagative material are particularly damaging since they may be transported to new areas and the inoculum they harbour will be present at the beginning of the season and consequently in a position to inflict maximum harm. On the other hand, providing that the pathogen can be detected efficiently (see Chapter 2), diseases arising from such sources can often be effectively controlled either by exclusion (Section 5.6.4), or by the treatments described below. Infected plants, crop residues and soil are also important sources of infection and control measures should be aimed at reducing the generation of inoculum from these sources as well.

5.2.1   Eradicating pathogens from seed and propagative material

Fungal, bacterial and viral pathogens of crop plants are often transmitted in seed or propagative material (Section 3.3.1). Treatment of seed with antimicrobial compounds represents one of their more benign uses and may result in substantial improvements in yield. For example, Trutmann, Paul and Cishabayo (1992) reported yield increases of between 17 and 35 percent from bean mixtures planted in the central African highlands when seed was treated with fungicide and insecticide. However, in rural communities with low literacy rates, care must be taken that treated seed is not mistakenly used as food. Hot-water treatment is one alternative to the use of biocides. For example, Fatmi, Schaad and Bolkan (1991) investigated a range of treatments for the eradication from tomato seed of *Clavibacter michiganense* subsp. *michiganensis*, a pathogen that causes a serious disease of tomato world-wide and for which there is no satisfactory means of control once it is established in the field. They found that hot-water treatment at 52°C for 20 min was totally effective in eliminating the bacterium.

Plant pathogens may also be eliminated from propagative material by hot-water treatment. Doornik (1992) found that treatment of Anemone corms infected by *Colletotrichum acutatum* with water at 47.5°C for 1.5 h or at 50°C for 1 h almost completely suppressed the symptoms of leaf curl and leaf necrosis caused by this fungus. Also, the root-knot nematodes, *Meloidogyne incognita* and *M. javanica*, could be successfully eradicated from grapevine rootstocks by hot-water treatment at 48°C for 20 min followed by further hot-water treatment at

50°C for 10 min. Although treatment at 50°C for 20 min or 51°C for 10 min eliminated the nematodes and could be used on dormant rootstocks, these treatments were lethal to active rootstocks (Gokte and Mathur, 1995).

Viruses, mycoplasmas and viroids may be eradicated from propagative material by thermotherapy or tissue culture or a combination of the two. For example, Monette (1986) cultured shoot tips of grapevine which were infected by grapevine fan-leaf virus (GFLV) and arabis mosaic virus (ArMV) at alternating temperatures of 6 h at 39°C followed by 18 h at 22°C with a 16 h photoperiod. After 40 days of this treatment both viruses were eliminated. Similarly, Möllers and Sarkar (1989) found that three mycoplasma-like organisms could be eliminated from *Catharanthus roseus* by tissue culture and regeneration of plants. In contrast to the successful use of thermotherapy to eliminate viruses, viroids may be eliminated by prolonged exposure to low temperature. For example, potato spindle tuber viroid (PSTVd) was eliminated from potato plants by a long period at low temperature followed by meristem culture (Griffiths, Slack and Dodds, 1990).

The disadvantages of meristem culture are that the survival rate may be low and the regeneration time for explants may be long, about e.g. 8 months for potato. In consequence, some workers have attempted to eliminate viruses from less exacting material. For example, Griffiths, Slack and Dodds (1990) worked with potato plantlets established from nodal cuttings. By using a combination of heat treatment and ribavirin (1-β-D-ribofuranosyl-1,2,4-triazole-3-carboxamide) in the medium, titres of potato viruses M, S and X were reduced by 10- to 60-fold and some virus-free plants were obtained within 6 weeks.

5.2.2 Eradicating sources of inoculum by roguing and pruning

Removing infected plants may be an effective means of controlling disease but it is labour intensive. For example, some success has been claimed in controlling African cassava mosaic virus (ACMV) by this means (Fauquet and Fargette, 1990). African farmers also frequently remove *Striga* from infected plants in an attempt to reduce its disastrous effects on yields. However, since the host plant is already seriously compromised before the pathogen is visible above ground the effect on the current crop is not great, the benefit being more apparent in subsequent seasons owing to the reduction in the pathogen's seed bank (Parker, 1991).

Self-seeded volunteer plants often provide a rich source of inoculum for the subsequent crop and are an important means by which pathogens survive between seasons. This 'green-bridge effect' is significant in a number of diseases such as take all of wheat (Section 3.3.1) in temperate countries but in tropical countries, where there is usually no natural seasonal break between successive crops, the carry over of inoculum poses a particularly acute problem. This is exacerbated in the case of pathogens with wide host ranges such as *Colletotrichum gloeosporioides*, which may pass from one crop to another (Sections 3.4.1 and 5.6.1).

Some pathogens survive from season to season on alternate hosts and reservoir plants (see Chapter 3) and therefore the elimination of these may drastically reduce the availability of inoculum for the cultivated host. The barberry eradication programme in the USA was a classic example of this approach, the rationale for the operation being Craigie's discovery in 1926 that barberry was the alternate host of the wheat stem-rust fungus, *Puccinia graminis* f. sp. *tritici* (Figure 5.2). As may be appreciated from the diagram, the life cycle of this pathogen is complicated, involving not only two hosts but also five different types of spore. One of these is the uredospore which is not only produced on wheat but is also infective for this plant rather than barberry. Since, in a susceptible cultivar, the latent period between the arrival of a uredospore on the plant and sporulation is only 7–8 days, epidemics of the polycyclic pathogen may develop rapidly. Towards the end of the season uredosori cease to give rise to uredospores and teleutospores are produced instead, committing the fungus to completing its cycle through the barberry plant. At first sight, elimination of the barberry plant would seem to be a perfect control measure but, unfortunately, in the south of the USA the fungus is able to overwinter in the uredial form and from this base it spreads progressively northwards as the summer advances (Figure 5.3). Nevertheless, it is probable that losses have been reduced by the eradication of barberry for two reasons. First, the absence of the alternate host makes overwintering impossible in the North of the country and there is therefore a delay of several weeks before inoculum reaches these areas from the South. Second, the sexual phase of the fungus occurs on barberry (Figure 5.2); elimination of this host therefore reduces the opportunity for genetic recombination and thus the evolution of races with new virulence characteristics. Some evidence for this view has been provided by Groth and Roelfs (1987) who found that there was a general trend of decreasing genetic diversity of North American populations of *P. graminis* f. sp. *tritici* during the period 1918–82. If such decreases in variation involved decreases in the recombination of virulence factors (or, perhaps more likely, as we shall see in Chapter 10, the segregation of genotypes lacking avirulence genes) the effect would have been to prolong the life of some wheat cultivars.

Many plant viruses have wide host ranges involving non-cultivated plants that often provide reservoirs from which infection may spread to economic hosts (Section 3.3.1). The importance of such sources of inoculum can hardly be over-emphasized when the cultivation of plants in new areas is being contemplated since they may acquire lethal infections from indigenous plants. Both African cassava mosaic virus and swollen shoot of cocoa are thought to have arisen by this means. Therefore the identification and removal of such reservoir plants is an important aspect of plant-disease control.

Pruning is often a practical method for reducing inoculum levels of perennial crop plants. For example, *Colletotrichum gloeosporioides*, the cause of ripe rot of muscadine grapes, can survive in mummified fruit that remains attached to the vine as well as in infected pedicels and fruit spurs. Removal of these by severe pruning in the dormant season is one of the best methods of control (Milholland,

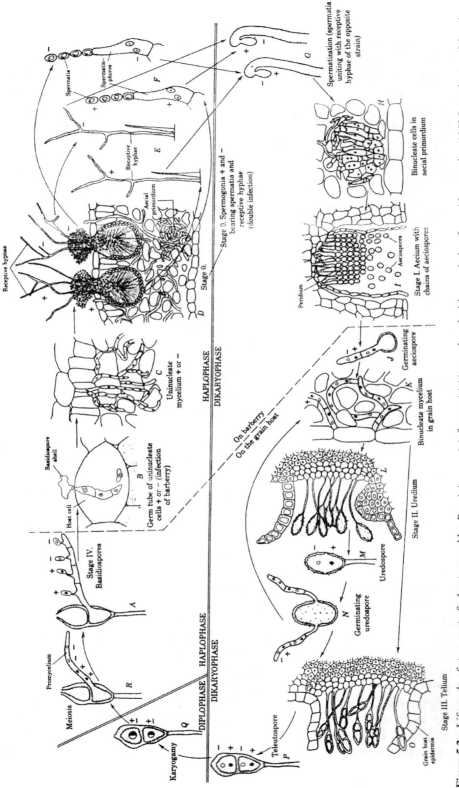

Figure 5.2 Life cycle of stem rust of wheat caused by *Puccinia graminis* f. sp. *tritici* (reproduced with permission from Alexopoulos, 1952 Introductory Mycology, John Wiley, New York)

Figure 5.3   The seasonal spread of *P. graminis* f. sp. *tritici* across the USA

1991). Another aspect of pruning is that it may be used to render the microclimate around the plant less conducive for infection. For example, leaf removal has been used with some success to control both Botrytis bunch rot and summer bunch rot of grapes, the latter being caused by a complex of organisms (English *et al.*, 1993; Stapleton and Grant, 1992).

Burning has long been used as a means of removing debris after harvesting a crop and there is little doubt that it is an effective means of reducing inoculum when the debris is infected (Hardison, 1976). For example, pruning lowbush blueberry by burning was subsequently found to have been an effective means of controlling mummy-berry disease, caused by *Monilinia vacinii-corymbosi*. Plants treated in this was way grew vegetatively during the first summer after pruning and produced fruit the next. However, damage caused to the upper layers of the soil by the high temperature of the fire as well as increased fuel costs led to the discontinuation of the method and the substitution of flail mowing as an alternative. In subsequent years, an increase in mummy-berry disease was noted which rose over a 12-year period to a level that was 90-fold that of plots pruned by burning (Lambert, 1990). Burning has several other environmental disadvantages besides the potential damage to the upper layers of the soil already noted; these include the generation of smoke and carbon dioxide and, in dry countries, the possibility of the fire getting out of control.

5.2.3   Reducing the role of crop residues as sources of inoculum

Debris of annual crop plants, as noted previously, is often an important source of inoculum for the subsequent season's crop and, where a pathogen's ability to survive in the soil is poor, inoculum production may be reduced by ploughing in such material. For example, the survival time of *Xanthomonas campestris* pv. *campestris* in crop debris could be reduced if fields were disked or ploughed. Similarly, Summerell and Burgess (1989) found that the recovery of *Pyrenophora tritici-repentis* from naturally infected wheat-stubble samples which had been buried was 6 per cent, 2 per cent and 0 per cent after 26, 52 and 104 weeks, respectively, whereas the figure was 46 per cent from samples which had been retained on the soil surface for 104 weeks. Dillard and Cobb (1993) studied the survival of *Colletotrichum lindemuthianum*, the cause of anthracnose of bean, on debris of the crop in New York State. No colonies of the fungus could be recovered on agar plates after overwintering for 5 months but when the debris was shaken with water and the liquid dispensed on the underside of the first trifoliate leaf of the host, lesions occurred from which the fungus could be recovered. In further experiments using the plant assay, mild symptoms developed from pod tissue that had been buried at a depth of 10 cm for 593 days. Thus the host plant functioned as a highly selective medium for the pathogen, enabling its detection at low but epidemiologically significant levels.

As mentioned in the previous section, burning when used as a pruning technique for lowbush blueberry was a highly effective means of controlling mummyberry disease. It is also thought to be an effective strategy for limiting inoculum emanating from infected debris but this view is seldom backed with quantitative data (Hardison, 1976).

5.2.4   Eradicating inoculum from the soil

It has always been difficult to eradicate infectious agents of plants from the soil and with the banning of methyl bromide as a fumigant it is now even more imperative to discover satisfactory methods. In the greenhouse it may be feasible to use steam sterilization but in open fields other more practicable techniques must be found. During the last two to three decades a simple procedure, termed soil solarization, has been developed which is particularly appropriate to countries with warm sunny climates. It involves placing a plastic sheet on the soil surface which, when illuminated by the sun, causes temperatures in the soil beneath to rise sufficiently to kill soil-inhabiting plant pathogens (Katan, 1981; Figure 5.4).

There are now many reports of the successful use of this technique. For example, Antoniou, Tjamas and Panagopoulos (1995), working in Greece, found that bacterial canker of tomato, caused by *Clavibacter michiganensis* subsp. *michiganensis*, was controlled by mulching soil for 6 weeks with transparent polyethylene whereas methyl bromide applied at a rate of $70\,g\,m^{-2}$ was ineffective. Similarly, López-Herrera, Verduvaliente and Melerovaras, *et al.* (1994) found, in Spain, that primary inoculum of *Botrytis cinerea*, consisting of mycelium and sclerotia was eradicated by means of a film of transparent polyethylene. Loss of viability was remarkably rapid, mycelium buried at depths of up to 25 cm requiring treatment for only 2 days but sclerotia at depths greater than 5 cm requiring longer.

The heat generated by solarization may not kill a pathogen outright, but the organism may be weakened, resulting in a reduction of aggressiveness for its host and greater susceptibility to attack by other components of the soil microflora. For example, colonies arising from microsclerotia of *Verticillium dahliae* that had been heated were slower growing and slower to melanize than those of controls, rendering them more susceptible to microbial attack. In consequence, sublethal heating combined with exposure to the thermotolerant biocontrol agent, *Talaromyces flavus*, acted additively to suppress Verticillium wilt of egg plant (Tjamos and Fravel, 1995). The authors conclude that there is considerable scope for combining sublethal solarization with thermotolerant biocontrol agents. Moreover, the lower temperatures would allow solarization of land that was already planted, obviating the need for a fallow period for the solarization treatment.

This last point may be of particular relevance to farmers in the San Joaquin valley of California where few have adopted solarization to control Verticillium wilt of tomato and eggplant as it would involve the loss of a season's production. However, Morgan and co-workers (1991) found that plastic mulches were as effective in controlling wilt of cherry tomatoes when applied for the entire growing season as those applied to fallow soil. This success was attributed to the practice of staking cherry tomato vines, leaving the soil surface relatively unshaded.

Another approach to soil solarization is to use the technique in combination with amendments. Cabbage residues have proved particularly effective since, on heating, they generate volatile compounds that are antimicrobial. Gamliel and Stapleton (1993) have characterized several of these as alcohols, aldehydes, sulphides and isothiocyanates.

5.2.5 Control of soil-borne plant disease through the addition of minerals and amendments and the alteration of pH

Soil is a complex medium consisting of solid, aqueous and gaseous phases. Within this a multitude of interactions of soil nutrients, pH, plants, plant pathogenic

(a)

(b)

microorganisms and the general soil microflora occurs. A number of experiments have been performed in which the concentration of a single macro- or micro-element has been increased but these have only infrequently given good disease control. One successful example is the excellent control of club root of brassicas which was obtained in the Salinas Valley of Central California. Here applications of $CaCO_3$ at about 4 t/ha, sufficient to raise the pH to at least 6.9, were made annually. Care, however, must be taken in making such heavy applications as they can result in the unwanted side effects of deficiencies in micronutrients such as boron, iron or manganese. However, although these may be deleterious to the plant they may also be unfavourable to the pathogen. Jones, Engelhard and Woltz, (1989), for example, found that liming the soil controlled wilt of tomato caused by *Fusarium oxysporum* f. sp. *lycopersici* owing to the reduction of availability of Mn, Fe and Zn which caused the suppression of growth and sporulation of the fungus.

Meyer, Shew and Harrison, (1994) have pointed out that soils which are suppressive to *Thielaviopsis basicola* usually have a low pH and high levels of exchangeable aluminium and these observations were mirrored by results of experiments in which they showed that the germination of endoconidia and chlamydospores as well as the radial growth of the fungus on agar medium was suppressed by aluminium at about 100 p.p.m., particularly at pH values less than 5.6. In other experiments, Benson (1995) showed that damping-off in bedding plants of snapdragon, periwinkle and petunia, grown in peat-vermiculite mixes, could be controlled by the addition of 10–50 meq (90–450 p.p.m.) aluminium.

Hydroponically grown plants offer ideal experimental conditions for testing the effect of nutrients on plant disease. For example, crook root of watercress caused by the Plasmodiophoromycete, *Spongospora subterranea* subsp. *nasturtii*, which is also the vector of two watercress diseases, watercress yellow spot (WCYSV) and watercress chlorotic leaf spot agent can be controlled by the addition of zinc sulphate at 0.3–0.5 μg/ml to the water (Tomlinson, 1960; Tomlinson and Hunt, 1987). In another hydroponic system, significant control of crown and root rot of cucumber, caused by *Pythium ultimum*, was obtained by the addition of 100 p.p.m. silica as potassium silicate to recirculating nutrient solutions (Chérif and Bélanger, 1992).

Other workers have supplemented chemical amendments with less well-defined biological materials. Huang and Kuhlman (1991) tested the effect of 17 chemicals supplemented with milled pine bark on damping-off of slash pine seedlings in soils infested with *Rhizoctonia solani*, a binucleate *Rhizoctonia* sp. and *Pythium aphanidermatum*. The *Rhizoctonia* species were inhibited by several of the compounds as was zoospore production by *P. aphanidermatum*. Some of these were therefore mixed together with milled pine bark to give a formulation termed SF-21 which was partially effective in controlling damping-off caused by these organisms.

Figure 5.4 (a) An aerial photograph of a wheat field – the dark transverse strips are the solarized plots, compared with the light transverse non-solarized plots (courtesy of J. Katan); (b) control of *Orobanche* sp. in a field of carrots by solarization – the whole field was solarized except for one plot (control) in which the plants are yellow and stunted (courtesy of J. Katan). A colour reproduction of this figure appears in the colour section

More recently, with the term organic being perceived favourably by the general public, there has been increased interest in the use of organic amendments for the control of soil-borne plant pathogens. In a review, Lazarovits (2001) documents the control of a wide spectrum of soil-borne plant pathogens with organic amendments containing high nitrogen, such as poultry manure, soymeal and meat and bone meal. Control was mediated by the ammonia and/or nitrous acid generated from these amendments and the amounts of these were modulated by pH, organic matter content, soil buffering capacity, and nitrification rate of the soil. Volatile fatty acids also made an important contribution to control in several liquid manures. In the cases of other products such as ammonium lignosulfonate, a byproduct of the pulp and paper industry, the mechanism of disease control was not clear, although displacement of pathogens may have played a role since organic amendments led to increases in soil micro-organism populations by up to 1000-fold.

5.2.6 Control of soil-borne plant pathogens by flooding

Flooding of banana groves to control Panama wilt of bananas, caused by *F. oxysporum* f. sp. *cubense* was an early application of the technique. Unfortunately, it proved to be a palliative rather than a completely effective control measure owing to the difficulty in eradicating the fungus completely. Although survival in flooded soil was less than 90 days, the fungus remained viable for over 120 days on the lake floor or in the water. When bananas are planted on virgin land economic production is possible for an average of 26 years but after the first flooding this was reduced to 6 years (Wardlaw, 1972). Further reduction of the period for economic production occurred after subsequent floodings. In contrast, the practice of growing rice in paddy fields (i.e. flooding the soil) destroys many soil-borne pathogens (Thurston, 1990).

5.3 Reducing inoculum multiplication

A number of fungal pathogens of plants, such as *Botrytis cinerea*, have quite closely defined light requirements for sporulation giving an opportunity, where a susceptible crop is grown under glass, to screen out these wavelengths.

Early work by Reuveni, Raviv and Bar (1989) with *B. cinerea* showed that simply excluding ultraviolet irradiation (310 nm) did not prevent sporulation if blue light (480 nm) was also at a low level. However, reducing ultraviolet irradiation and increasing the amount of blue light was effective and this was achieved with polyethylene containing hydroxybenzophenone and hydrated cobalt oxide co-polymerized with ethylene vinyl acetate. Spore production by *B. cinerea* grown *in vitro* under this material was reduced 500-fold compared with cultures grown under clear polyethylene sheeting and, in greenhouse tests with similar material, the number of infection sites on tomatoes and cucumbers was halved.

Köhl and co-workers (1995a) have also investigated the suppression of sporulation of *Botrytis* spp. but as a means of control of onion leaf spot. They pointed out that such a strategy is only feasible if the development of an epidemic results mainly from inoculum from within the crop rather than arrival from outside. In order to simulate a hypothetical biocontrol agent that would compete for necrotic plant tissue, which the pathogen had killed and on which it normally sporulates, they removed such tissue. This resulted in a reduction of the number of lesions on green leaves at the end of the season from 1.1 to $0.6\,cm^{-2}$. In further experiments Köhl and co-workers (1995b) tested a number of potential antagonists for their ability to suppress sporulation of *Botrytis allii* and *B. cinerea* and their ability to survive periods of dryness such as might occur in the field. They found that *Alternaria alternata, Chaetomium globosum, Ulocladium atrum* and *U. chartarum* suppressed sporulation of the two fungi under continuous wet conditions and when these were interrupted with a drying period of 9 h imposed 16, 40 and 64 h after application of the antagonists. In contrast, *Gliocladium roseum, G. catenulatum* and *Sesquicillium candelabrum*, which efficiently suppressed sporulation of the fungi under continuous wet conditions, were of only low or moderate efficiency if wetness was interrupted with a period of dryness 16 h after application.

Soil-borne pathogens such as species of *Phytophthora* usually require high moisture levels for the production of zoosporangia and the release of zoospores. For example, wet conditions favoured the production and release of zoospores of *P. citrophthora* and led to severe rot of citrus, (Feld, Menge and Stolzy, 1990; Section 3.5.1). Adequate drainage would be expected to limit both zoospore production and spread (Section 5.6.2). The author experienced a rather similar situation in Iran where cantaloupe melons were grown on ridges, irrigation water being supplied by the furrows between the ridges which were fed from bore holes. Where the ridges were high there was comparatively little 'green-death' disease, caused by *Phytophthora drechsleri*. In contrast, where the ridges were lower and the furrows correspondingly shallower, most plants died. The explanation lay in the acute susceptibility of the crown of the plant and the comparative resistance of the roots (Alavi, Strange and Wright, 1982). Where water in the furrows never reached the crown, the plants remained healthy. In contrast, where the crown was flooded the plant became infected by zoospores in the irrigation water. These were rapidly carried from plant to plant, causing devastation (Figure 5.5).

5.4   Reducing the effectiveness of inoculum

If it has proved impossible to reduce the sources of inoculum (Section 5.2) and its multiplication (Section 5.3) consideration may be given to the reduction of its

Figure 5.5 Cantaloupe melons devastated by *Phytophthora drechsleri*; water from a bore hole had been allowed to flood the irrigation furrows (see top left of picture), carrying with it zoospores of the fungus which were able to infect the highly susceptible crown of the plant (courtesy of Dr Ahmad Alavi, Plant Pests and Diseases Research Institute, Tehran, Iran). A colour reproduction of this figure can be seen in the colour section

effectiveness. Here forecasting has an important role to play but, even if a farmer is warned that an attack by a plant pathogen is imminent, spraying his crop with a pesticide is usually the only palliative procedure he can adopt. Although such compounds can be very effective fears have been expressed on several bases. These include the possible chemical pollution of foods and feeds, the hazard to those who work with the compounds and the broad spectrum of their activity which may result in unwelcome side effects. Therefore, in this text, such biocides are not treated but rather the potential of an alternative, namely biological control. This is widely seen as a 'green' and more environmentally-friendly method for controlling plant pests and diseases with the result that, at the time of writing, there are over 80 biological control products on the market. However, it should be recognized that these too may be hazardous, as a brief perusal of Table 5.1 will demonstrate – for example, HCN, a product of *Pseudomonas fluorescens*, is hardly a benign compound! In fact, some biological control agents are described as plant growth promoters, plant strengtheners or soil conditioners in order to avoid stringent toxicological testing (Paulitz and Bélanger, 2001). Nevertheless, plants do support a considerable microflora around both their aerial and subterranean organs and it would therefore seem to make sense to exploit this in plant protection.

5.4.1 The role of forecasting

Weather conditions often play a crucial role in the multiplication, spread and effectiveness of inoculum. It is not surprising therefore that weather forecasts have long been used to predict the outbreak of a number of plant diseases. However, these rely on a causal connection between specific weather conditions and the outbreak of disease. Obtaining such data is time-consuming. For example, Coakley (1988) has suggested that a minimum of 8–12 years of disease data obtained from fields infected with a natural source of inoculum is necessary to define the climatic factors crucial in determining disease occurrence, although shorter periods may be adequate if data are available from several sites. During these shorter periods, however, it is less likely that the full range of weather conditions will be represented.

Many fungal pathogens require wetness for several hours to gain entry into their hosts. For example, infection of lettuce by downy mildew, *Bremia lactucae*, oc-curred mainly when leaves of the plant remained wet after 10 a.m. However, the wet period need not necessarily occur at or near the time of inoculation. Lacy (1994) found that a continuous period of 24 h wetness or one interrupted by a 12-h dry period allowed the development of significant numbers of lesions of *Septoria apiicola* on celery leaves provided that it occurred within 15 days of inoculation.

The potato blight organism, *Phytophthora infestans*, requires high humidity to sporulate (Section 3.5.2) and consequently the occurrence of humid conditions is a good predictor of infection, allowing several forecast models to be developed. Two of these, Blitecast and Simcast, have a daily time-step and use rainfall, temperature and leaf-wetness duration (i.e. times during which relative humidity exceeds 90 per cent) to predict spraying schedules. Blitecast predicts the timing of the initial and subsequent sprays. Simcast also takes into account the decrease over time of the protectant biocide owing to its removal by rain or chemical degradation. Hijmans, Forbes and Walker (2000) have linked these models to a global climate data base (global information system, GIS) to predict how many sprays would be required to control late blight and thus, indirectly, to estimate the global severity of the disease. They showed that the optimum number of sprays was especially high in the tropical highlands of Latin America, Africa and Asia, southeastern Brazil, central-southern China and in many coastal areas (Figure 5.6). Comparison of these data with actual amounts of biocide used demonstrated that these were insufficient to control the disease in many African countries but excessive in countries such as Ecuador, Costa Rica and Indonesia.

5.4.2 Biological control

Pathogens of non-cultivated plants have been kept in check over millions of years by natural means and pathogens of cultivated plants controlled over hundreds, if not thousands, of years by traditional agricultural methods (Thurston, 1990).

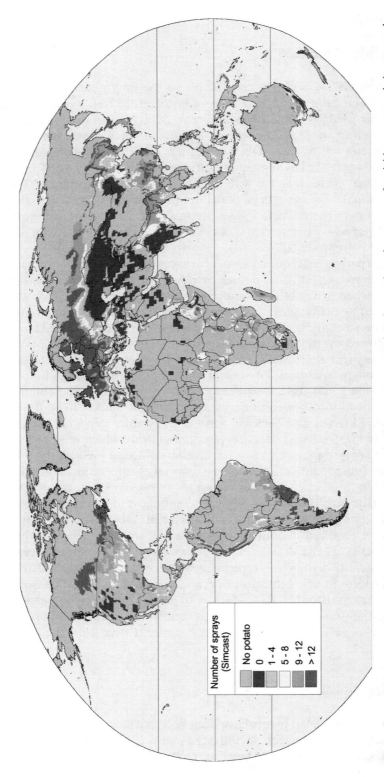

Figure 5.6 Predicted global late blight severity for potato production zones expressed as the number of protectant fungicide sprays needed to control late blight of potatoes caused by *Phytophthora infestans*; predictions were based on Simcast, a late blight forecasting model, linked to global climate surfaces in a geographic information system (courtesy of R J Hijmans and Plant Pathology)

Some of these may depend upon increasing the populations of other members of the biota which reduce disease of the target crop, usually by one or more of five means: antibiosis, competition, parasitism, interference with virulence mechanisms and the induction of systemic resistance. The first four of these would be expected to act directly on the pathogen by reducing inoculum, its multiplication or its effectiveness and will be discussed in the succeeding sections. The indirect action of acquired and induced systemic resistance will be discussed in Chapter 11 (Section 11.3.7 and 11.3.8) in the context of active defence mechanisms.

Antibiosis

Antibiotic compounds are implicated in many reports of the successful control of plant pathogens by microorganisms. The compounds range from HCN to enzymes and the microorganisms involved are often species of *Trichoderma* and *Gliocladium* among fungi and *Bacillus* and *Pseudomonas* among bacteria (Table 5.1; Figure 5.7).

Some of the earliest work on the isolation and identification of antibiotic compounds from biocontrol organisms concerned those synthesized by fungi. For example, Ghisalberti and co-workers (1990) compared the ability of isolates of *T. harzianum* to reduce take-all of wheat, caused by the fungus, *Gaeumannomyces graminis* f. sp. *tritici*. The most effective isolate, designated 71, produced two pyrone antibiotics which suppressed the disease, one of which was also shown previously to be synthesized by *T. koningii* and *T. hamatum*.

Gliocladium virens has been known as an antagonist of several important soil-borne plant pathogens for a number of years. These include *Rhizoctonia solani*,

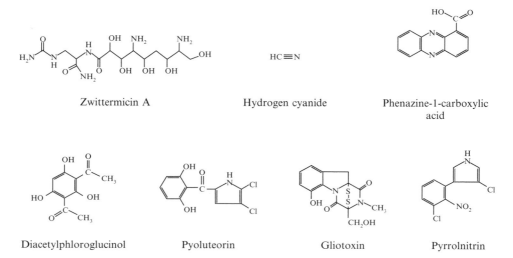

Zwittermicin A	Hydrogen cyanide	Phenazine-1-carboxylic acid
Diacetylphloroglucinol	Pyoluteorin	Gliotoxin	Pyrrolnitrin

Figure 5.7 Chemical structures of some antibiotics produced by bacterial and fungal biocontrol agents; (for further details see Table 5.1 and text)

Sclerotinia sclerotiorum, Sclerotium rolfsii and *Pythium ultimum*. Although culture filtrates of the fungus were inhibitory to *P. ultimum* and contained several enzyme activities, the filtrates remained effective after their inactivation by heat or size fractionation (Roberts and Lumsden, 1990). The active compound was identified as gliotoxin (Figure 5.7).

Laetisaria arvalis, a basidiomycete, inhibited the growth of the plant pathogenic fungi, *Rhizoctonia solani* and *Phoma betae*, as well as the Oomycete, *Pythium ultimum* in both laboratory and field trials. The active factor, which was isolated and characterized as a long chain fatty acid (8-hydroxyocta deca-9(Z),12-(Z)-dienoic acid), was given the trivial name laetisaric acid (Bowers *et al.*, 1986). A systematic study of analogues of the compound, using *Pythium ultimum* as a test organism, revealed that a 12-carbon *a*-hydroxy fatty acid was the most active (Bowers, Evans and Katayama, 1987). The authors therefore suggested that, unlike higher plants and animals, *P. ultimum* was unable to metabolize laetisaric acid via lipid β-oxidation beyond this compound.

At the time of writing, bacteria are the microorganisms that are known to produce the most diverse range of antimicrobial compounds. The first commerical biological-control agent was probably strain K84 of *Agrobacterium* which has been used successfully to control crown gall disease caused by *Agrobacterium tumefaciens* in many countries of the world (Penyalver, Vicedo and Lopez, 2000). This strain harbours a plasmid, pAGK84, which encodes the production of a bacteriocin and immunity to it. One fear is that the plasmid could be transferred to pathogenic strains of *Agrobacterium tumefaciens*. In order to counter this a strain, K1026, has been genetically engineered to remove the transfer region of the plasmid (Penyalver, 2000).

Loeffler and co-workers (1986) found that *Bacillus subtilis* gave good control of *Rhizoctonia solani* in many crops. The bacterium produces bacilysin and fengymycin A and B which are composed of a C_{15}-C_{18} lipid moiety and a peptide moiety of eight amino acid residues. Substitution of a D-valine for a D-alanine differentiates fengymycin B from fengymycin A. Bacilysin inhibits yeasts and bacteria and fengymycin inhibits filamentous fungi. More recently, He and co-workers (1994) described the novel aminopolyol, zwittermicin A as the antibiotic produced by *Bacillus cereus* UW85, another successful biocontrol agent which is effective against damping-off and root rot of soybean caused by *Phytophthora sojae* (Figure 5.7; Osburn *et al.*, 1995).

In some instances, an interest in biological control has been fired by the observation that, under continuous cropping to a particular plant, the soil becomes suppressive to the soil-borne pathogens of the crop. The best known example is that of take-all decline. Take-all is caused by the fungus, *Gaeumannomyces graminis* var. *tritici*. The disease often becomes progressively more serious when monoculture is practised but in succeeding seasons spontaneously declines (Section 3.5.1; Cook *et al.*, 1995). Moreover, the suppressive nature of the soil may be transmitted to soil that is conducive to the disease by the addition of as little as 1 per cent of the suppressive soil. Although siderophores have been implicated as contributors to disease suppression in some instances (see next section) in most cases one or more antibiotics are involved (Cook *et al.*, 1995).

It is often difficult to determine directly if an antibiotic is present in sufficient concentrations at the right place and the right time to explain the control of a pathogen. An indirect but powerful alternative is to generate mutants deficient in antibiotic production. For example, Poplawsky and Ellingboe (1989), using take-all of wheat as their model, investigated two bacterial strains, designated 111 and NRRL B-15135 which were effective in suppressing the disease and 10 mutants of these affected in antibiotic production by the insertion of the transposon Tn5. All four of the mutants from one strain, which were antibiosis-negative, were less effective suppressors of disease than the parent strain as were three of the four mutants of the other strain. The remaining antibiosis-negative mutant gave control which was as effective as that of the parent strain while two mutants with increased antibiosis towards the pathogen did not give higher levels of disease suppression.

Strains of *Pseudomonas* produce several antibiotics. These include phenazine-1-carboxylic acid, phenazine-1-carboxamide (PCA), anthranilic acid, diacetylphloroglucinol, pyoluteorin, pyrrolnitrin and viscosinamide (Figure 5.7; Table 5.1). Thomashow and Weller (1990) reviewed the importance of PCA, produced by strain 2–79 of *Pseudomonas fluorescens* (Figure 5.7). The compound was present in washings of roots and rhizosphere soil of seedlings colonized by the bacterium but strains of the bacterium which contained Tn5 insertions and were unable to produce the compound (*phe⁻* strains) were less suppressive. However, some residual suppressive activity of these *phe⁻* strains was detected and attributed to anthranilic acid as well as a siderophore (see next section). Reciprocally, Timms-Wilson and co-workers (2000) have shown that insertion of a single copy of the genes required for PCA production into the chromosome of a wild-type *P. fluorescens* enhanced its ability to control damping-off disease of pea seedlings.

Variation in sensitivity of strains of a pathogen to antibiotics produced by biocontrol organisms may be encountered. For example, Mazzola and co-workers (1995) found that two of three isolates of *G. graminis* f. sp. *tritici* that were insensitive to PCA at 1.0 mg/ml were not suppressed or were suppressed to a lesser extent by the PCA producing bacteria, *Pseudomonas fluorescens* 2–79 and *P. chlororaphis* 30–84. Similarly, isolates of the fungus that were insensitive to diacetylphloroglucinol at 3.0 mg/ml were not suppressed by *P. fluorescens* Q2–87. In consequence, they suggested that mixtures of organisms differing in their mechanisms of pathogen suppression should be used to manage the disease.

Woo and co-workers (2002) have described experiments with just such mixtures of organisms and mechanisms in the postharvest control of infection of apple fruit by *Botrytis cinerea*. They used *Pseudomonas syringae* pv. *syringae* (strain B359) in combination with *Trichoderma atroviride* (strain P1) which separately reduced the percent infected wounds in the fruit from about 90 per cent for controls to about 58 per cent and 62 per cent, respectively. Together they reduced the number of infected wounds to 10 per cent. The mechanisms

Table 5.1 Some products of biocontrol agents with activity against plant pathogens

Class of compound	Example	Producing organism	Target organism	Reference
Amino acid derived	Bacilysin	*Bacillus subtilis*	Yeasts Bacteria	Loeffler *et al.*, 1986
Aminopolyol	Zwittermicin A	*Bacillus cereus*	Fungi	He *et al.*, 1994
Cyanide	HCN	*Pseudomonas fluorescens*	Fungi	Voisard *et al.*, 1989
Enzymes	Chitinase	*Serratia marcescens*	Fungi	Jones *et al.*, 1986
Fatty acid	Laetisaric acid	*Laetisaria arvalis*	Fungi	Bowers *et al.*, 1986, 1987
Furanone	3-(1-hexenyl)-5-methyl-2-(hH)-furanone	*Pseudomonas aureofaciens*	*Pythium ultimum* *Rhizoctonia solani* *Fusarium* spp. *Thielaviopsis basicola*	Paulitz *et al.*, 2000
Lipopeptide	Iturin	*Bacillus subtilis*	Fungi and bacteria	Hiraoka *et al.*, 1992
	Surfactin	*Bacillus subtilis*	Fungi and bacteria	
	Viscosinamide	*Pseudomonas fluorescens* DR54	*Pythium ultimum*	Thrane *et al.*, 2000
Nicotinic acid derived	AFC-BC11	*Burkholderia cepacia*	*Rhizoctonia solani*	Kang *et al.*, 1998
	2-methylheptyl isonicotinate	*Streptomyces* sp.	*Fusarium* spp.	Bordoloi *et al.*, 2002
Nucleotide	Agrocin 84	*Agrobacterium* strain K84	*Rhizoctonia solani* *Agrobacterium tumefaciens*	Kerr and Tate, 1984
		Agrobacterium strain K1026		Penyalver, Vicedo and Lopez, 2000

Class	Compound	Producing organism	Target	Reference
Peptide	Dimerum (a siderophore)	Trichoderma virens	Fungi	Wilhite, Lunsden and Straney 2001
	Gramicidin S	Brevibacillus brevis	Botrytis cinerea	Edwards and Seddon, 2001
Phenazine	Phenazine-1-carboxylic acid	Pseudomonas fluorescens	Gaeumannomyces graminis f. sp. tritici	Thomashow and Weller, 1990
	Phenazine-1-carboxamide	Pseudomonas chlororaphis PCL 1391	Fusarium oxysporum f. sp. radicis-lycopersici	Chin et al., 2001
Phenol	Anthranilic acid	Pseudomonas fluorescens	Gaeumannomyces graminis f. sp. tritici	Thomashow and Weller, 1990
	Diacetyl-phloroglucinol	Pseudomonas fluorescens		Keel et al., 1992, Landa et al., 2002
	Flavipin	Epicoccum nigrum	Monilinia laxa	Madrigal, Tadeo and Melgarejo, 1991
Piperazine	Pyoluteorin	Pseudomonas fluorescens	Pythium ultimum	Maurhofer et al., 1994
	Gliotoxin	Gliocladium virens	Fungi	Roberts and Lumsden, 1990
Pyrone	6-pentyl-α-pyrone	Trichoderma spp.	Gaeumannomyces graminis f. sp. tritici	Ghisalberti et al., 1990
Phenyl-pyrrole	Pyrrolnitrin	Pseudomonas spp.	Fungi	Thomashow and Weller, 1990

responsible were the toxicity to fungi of the lipodepsipeptides, syringomycin E and syringopeptin 25A produced by the bacterium and the chitinase activity of *T. atroviride*. Antifungal synergism against several fungi was also recorded *in vitro*. The authors suggest that the synergism was brought about again by the lipodepsipeptides of the bacterium and the cell wall degrading enzymes, which included chitinase, of the fungus and postulated that these allowed better access of the toxins to the plasmamembrane, their site of action.

Competition

An ability to compete successfully with a pathogen is an important property of biological control organisms. Often, successful competition occurs at the infection court, preventing the ingress of the pathogen, although in some instances, an ability of the biocontrol agent to limit reproduction of the pathogen can also be important.

The fungus, *Idriella bolleyi*, controls take-all of wheat, caused by *Gaeumannomyces graminis* var. *tritici*, by competition for both nutrients and infection sites. It does this by exploiting senescing cortical cells of the plant, which occur naturally early in its growth, and rapidly producing spores. These are carried down the root by water and continue its colonization (Allan, Thorpe and Deacon, 1992; Douglas and Deacon, 1994; Lascaris and Deacon, 1994)

Some bacteria, particularly *Pseudomonads*, are aggressive root colonizers and promote plant growth. This has led to the use of the term Plant Growth Promoting Rhizobacteria (PGPR) which, while descriptive of the phenomenon, does not indicate the mechanism by which it occurs. Although suppression of major or minor pathogens of the plant may be one mechanism by which growth is promoted, increased growth may also occur in some instances in the absence of these. Here it is thought that one or several of a variety of mechanisms may be involved such as the associative fixation of nitrogen, solubilization of minerals such as phosphorus, promotion of mycorrhizal function, regulation of ethylene production, the release of phytohormones and the decrease of heavy metal toxicity (Whipps, 2001). Where suppression of disease is important, competition for iron may be a decisive factor.

Iron has an exceedingly low solubility in water and is therefore often limiting for both plants and microorganisms. Both plants and microorganisms capture iron by the production of iron-binding compounds known as siderophores (Figure 5.8). These have been defined as 'low molecular weight, virtually Fe(III)-specific ligands produced as scavenging agents in order to combat low iron stress' (Neilands and Leong, 1986). A role for them in disease suppression may be inferred if the addition of ferric iron nullifies the effect. For example, Becker and Cook (1988) found that 7 per cent of nearly 5000 strains of bacteria isolated from wheat roots produced a zone of inhibition against the fungal pathogen, *Pythium ultimum* var. *sporangiferum*, when assayed *in vitro*. When they were tested by applying them at a rate of 10^7–10^8 colony-forming units/

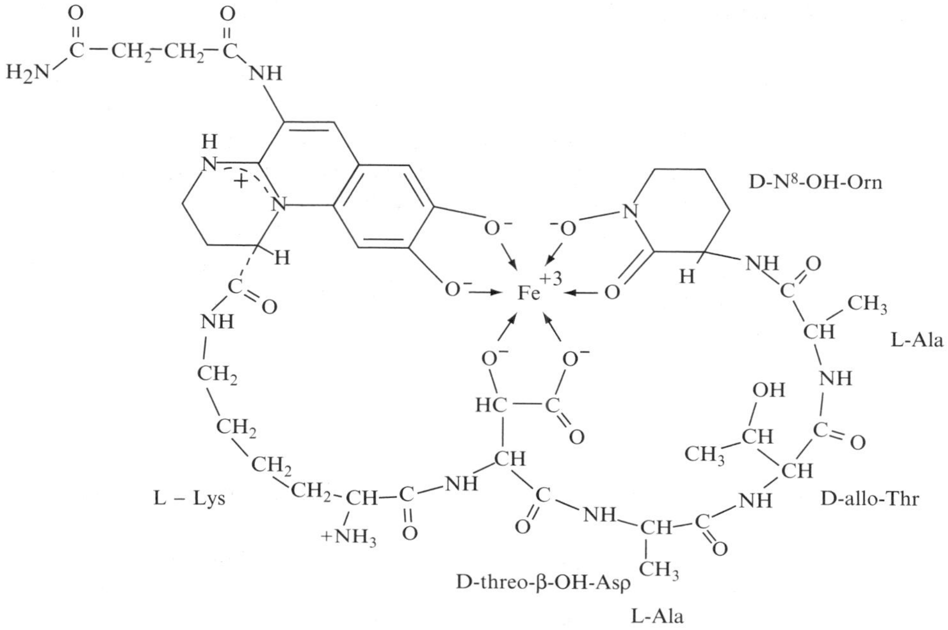

Figure 5.8 Ferric pseudobactin, a siderophore; siderophores are iron scavengers that are used by both plants and microorganisms in order to combat conditions of low iron concentration

seed and the seed sown in a loam naturally infested with the pathogen, a third of them increased the height of wheat seedlings by 10–30 per cent. With some isolates, both the inhibition of the pathogen *in vitro* and the growth promoting effect on the plant *in vivo* were nullified by the addition of iron as ferric chloride, implying the involvement of siderophores in both phenomena, although abolition of both phenomena by gene disruption would be stronger evidence.

Lim and Kim (1997) isolated a strain, GL20, of *Pseudomonas fluorescens* from the rhizosphere of ginseng which promoted growth of the plant. The bacterium produced a hydroxamate siderophore in an iron deficient medium which inhibited spore germination and germ tube elongation of *Fusarium solani*. In pot trials of GL20, using kidney bean as the test plant, incidence of disease caused by *F. solani* was reduced by 68 per cent and plant growth was increased nearly 1.6-fold.

One requirement of root colonizers is that they should be rhizosphere competent but the factors controlling this are not clear (Whipps, 2001). Katsuwon and Anderson (1990) pointed out they must be able to survive the defence mechanisms of the host such as the production of hydrogen peroxide (H_2O_2) and the superoxide anion (O_2^-) and for these they would probably need to produce catalase and superoxide dismutase, respectively, to nullify their effects.

Van Dijk and Nelson (2000) found that competition for fatty acids produced by seeds and roots of plants explained the suppression by *Enterobacter cloacae* of the germination of sporangia of *Pythium ultimum* and damping-off caused by the

Oomycete. Germination of *P. ultimum* is elicited by linoleic acid and cottonseed exudates. Mutants of the bacterium which were reduced in β-oxidation and fatty acid uptake and were therefore unable to metabolize linoleic acid did not suppress germination of sporangia of the pathogen exposed to linoleic acid or cottonseed exudate. They also failed to suppress rot of cotton seedlings. Complementation by clones of the wild type loci restored both the ability to inhibit sporangial germination and suppress disease. Inhibition of spore germination of *P. ultimum* was also invoked as the most significant factor in the control by *Pseudomonas fluorescens* of damping-off of sugar beet (Ellis, R.J.*et al.*, 1999).

Suppression of sporulation as a means of control has already been mentioned (see Section 5.3).

Parasitism

Parasitism of plant pathogens is an important mechanism by which many biocontrol agents suppress plant disease and the association of the agent and the pathogen may range from simple attachment and subsequent plundering of the pathogen's resources to complete lysis (Whipps, 2001). Fungal pathogens are also susceptible to virus infection leading to hypovirulence of the pathogen (i.e. reduced virulence) and presenting another opportunity for biocontrol.

Pythium oligandrum is a ubiquitous Oomycete with biocontrol properties that are of potential use against a wide range of plant pathogenic fungi (Picard, Tirilly and Benhamou, 2000). For example, in experiments with pepper and soil infested with *Verticillium dahliae*, a serious pathogen of several crops, shoot and fruit weights were significantly higher when *P. oligandrum* was present. In dual culture, *P. oligandrum* parasitized *V. dahliae*, reducing its growth and formation of microsclerotia. However, when pepper was grown in soil infested with *P. oligandrum* alone, fresh weights of shoots and fruits were 40–50 per cent higher than in its absence, demonstrating that factors other than control of the pathogen were important in promoting plant vigour (Al Rawahi and Hancock, 1998).

Sporidesmium sclerotiorum is an obligate pathogen of the sclerotia of five important plant pathogens, *Sclerotinia sclerotiorum, S. minor, S. trifoliorum, Sclerotium cepivorum* and *Botrytis cinerea*. After infection, glucanase activity of the sclerotia increases, resulting in the production of glucose, which is readily assimilated by the mycopathogen and allows it to grow out of the sclerotium and into the soil for a distance of up to 3 cm. Other sclerotia within this radius are infected. Exploitation of this property by disking in inoculum of spores of the fungus resulted in 53 per cent control of lettuce drop caused by *S. minor* with inoculum levels of the biocontrol agent as low as 2 spores/cm^2 of the soil surface (200 g/ha). This contrasted with a previous test in which a sand-sclerotia-*Sporidesmium* mixture was added to a field at the rate of 2300–23 000 kg/ha (Adams, 1990).

In many cases, breaching the plant pathogen's cell wall would seem to be a necessity for successful biocontrol activity. In consequence, considerable effort has been expended on identifying cell wall degrading enzymes that might be involved, such as β-1,3 and β-1,4 glucanases, chitinases, cellulases and proteases. For example, Picard and co-workers (2000) investigated the effects of cellulases of *Pythium oligandrum* on the parasitism of *Phytophthora parasitica* and found they were effective in degrading both carboxymethylcellulose and wall-bound cellulose from the Oomycete. However, the best evidence for their involvement consists of genetic modification of the production of the enzymes and demonstration of a corresponding effect on control. Much of this type of work has been concentrated on species of *Trichoderma*. Migheli and co-workers (1998) transformed *T. longibrachiatum* with extra copies of the *egl1* gene encoding the production of β-1,4 endoglucanase. Transformants with constitutive or inducible expression of the enzyme were generally more suppressive than the wild type strain when applied to seed of cucumber which was planted in soil infested with *Pythium ultimum*.

As chitin is an important component of the cell walls of true fungi, it is not surprising that considerable effort has been expended in implicating chitinases as important weapons in the armoury of biocontrol agents. Numerous papers report the expression of chitinases during mycoparasitism (cited in Whipps, 2001). For example, expression of the endochitinase gene, *ech42* of *T. harzianum* occurred before contact with the host, *Rhizoctonia solani*, whereas that of the *N*-acetylhexosaminidase gene, *nag1* did not occur until after penetration. Studies with mutants, however, suggest that the interaction of mycopathogen and host is complex and specific to the particular species concerned. For example, *ech42* minus mutants of *T. harzianum* were impaired in regard to the biocontrol of *Botrytis cinerea* but not that of *Pythium ultimum* or *Sclerotium rolfsii*. Transformation of plants with genes encoding cell wall degrading enzymes of their fungal pathogens are more resistant and this has been taken as evidence of their role in biocontrol (Whipps, 2001). However, the environments of the interaction of a biocontrol agent and pathogen are likely to be very different from those of a transgenic plant and its pathogen.

Kerry (1990, 2000) has reviewed progress towards the control of plant parasitic nematodes by microbial agents. Several soil bacteria produce nematocidal compounds and some have indirect effects which make roots less attractive to nematodes, such as altering root exudates or inducing resistance, but greater attention has been paid to parasitic fungi. In particular nematode trapping fungi have continued to attract attention since Dopf first described the activities of *Arthrobotrys oligospora* in 1888. This organism produces adhesive traps, whereas *A. dactyloides* has constricting ring traps. However, products for the control of cyst and root-knot nematodes are being developed from two other, egg-infecting, species, *Verticillium chlamydosporium* and *Paecilomyces lilacinus* (Bioact Corporation Pty Ltd., Sydney, Australia). The processes of control can be complex. For example, *V. chlamydosporium* affects *Heterodera schachtii* in several ways: by reducing the numbers of females on the host plant, reducing the numbers of eggs per female, causing infection of a high percentage of the eggs and reducing the dimensions of the female. Other considerations in developing a viable biocontrol agent for nematodes include its establishment and maintenance in the soil and its efficacy under a variety of conditions.

Hypovirulence is a well-established phenomenon in a number of fungal pathogens and has been associated with their infection by double-stranded RNA (dsRNA; Elliston, 1985; Nuss and Koltin, 1990). *Cryphonectria parasitica* and *Ceratocystis ulmi*, the causal agents of chestnut blight and Dutch elm disease, respectively, both harbour dsRNA. Studies of the organization and expression of the dsRNA from *C. parasitica* as well as its replication suggest that it is viral in origin and therefore it has been referred to as a hypovirus (Heiniger and Rigling, 1994; Liu and Milgroom, 1996).

Because of the hypovirulent phenotype conferred on fungal pathogens by mycoviruses, there has been considerable interest in their use as biocontrol agents (Heiniger and Rigling, 1994; Bertrand, 2000). However, success presupposes that the hypovirus should be infectious and readily transmissible to non-infected strains. Transmission normally occurs by anastomosis but this is often controlled by several vegetative incompatibility (*vic*) genes. Liu and Milgroom (1996) found that hypovirus transmission in *C. parasitica* readily occurred between isolates of the same vegetative compatibility group but was negatively correlated to the number of *vic* genes by which isolates of the fungus differed, transmission falling to 3 or 4 per cent when the isolates differed by more than two genes.

Hypovirulence may also be conferred by dsRNAs which are confined to mitochondria, mutant mitochondrial DNAs and mitochondrial plasmids (Bertrand, 2000). The recovery of chestnut trees in a small isolated grove of the Kellog Forest in Michigan has been attributed to a 973 bp insert in the first exon of the mitochondrial small-subunit rRNA gene of the pathogen (Baidyaroy *et al.*, 2000).

Inhibition of virulence mechanisms as a means of biological control

The production of toxins is an important virulence factor for many plant pathogens and, for some, essential for their pathogenicity (see Chapter 8). The degradation of such toxins could be a target for a successful biological control agent. Zhang and Birch (1997) have shown that *Pantoea dispersa* (synonym *Erwinia herbicola*) gave almost complete control of sugarcane leaf scald caused by *Xanthomonas albilineans* even when co-inoculated with a 10-fold excess of the pathogen. The pathogen produces a family of toxins and antibiotics known as the albicidins which selectively block DNA replication in bacteria and chloroplasts. In consequence, sugarcane infected by the bacterium is chlorotic. Zhang and Birch (1997) screened isolates of bacteria from sugarcane infected with the pathogen for albicidin resistance. They found that one albicidin resistant isolate, identified as *Pantoea dispersa*, was highly effective in detoxifying albicidin and they speculated that this attribute not only allowed *P. dispersa* to compete with *X albilineans* at wounds, which are the sites of infection, but also protected the plant by degradation of the pathogen's toxins.

5.5   Screening, development and application of biological control agents

An effective biological control agent giving good protection against disease caused by plant pathogenic organisms must have the following characteristics.

(1)   It must be able to control the pathogen by inhibiting its development, making it vulnerable to other members of the prevailing microflora or killing it. Mechanisms by which it may do this include competition for nutrients or potential sites of penetration of the host, production of antibiotics or lytic enzymes, parasitism or inhibition of the means by which the pathogen is able to attack its host, e.g. inhibiting enzymes necessary for penetration or destroying toxins.

(2)   It must be able to establish itself at the appropriate location and at a sufficient density to give effective control. For air-borne pathogens, the biocontrol agent must be able to compete with the naturally occurring microflora and withstand fluctuations in the microclimate normally associated with the crop. These may include high temperatures and high light intensities as well as wash-off by rainfall. For soil-borne pathogens the biocontrol agent must be able to compete with the soil microflora and to grow in the rhizosphere environment, i.e. it must be rhizosphere competent.

In aggregate, these requirements are formidable but in 2001 they were fulfilled by at least 80 commercially available preparations (Paulitz and Bélanger, 2001). These are the successful results of various approaches to the screening, development and application of such agents.

5.5.1   Screening for biological control agents

The environment contains a multitude of potential biocontrol agents. These may form part of the resident microflora of aerial and root surfaces of plants as well as existing independently in the soil. In order to exploit these resources efficiently it is necessary to select promising sources of potential biocontrol agents and to use appropriate screening procedures to single out the microorganisms responsible. Soils with a history of suppressiveness towards a given pathogen are obvious sources of potential biocontrol agents but opinions diverge as to how screening for the effective microorganisms should be done. Handelsman and co-workers (1990), in their search for biocontrol agents of *Phytophthora megasperma* f. sp. *medicaginis*, the causal agent of damping-off of alfalfa seedlings, developed a technique which was used to screen 700 isolates of root-associated bacteria. Alfalfa seedlings were grown in small culture tubes filled with vermiculite and inoculated with the isolates. Two days later, 10^3 zoospores of the fungus were added and, after incubation for a further 5–7 days, the seedlings were scored for viability. Promising candidate organisms were screened again but under more stringent conditions by adding the bacterial isolates to 3-day-old seedlings at the same time as a greater number of zoospores (10^4). Only one isolate, identified as *Bacillus cereus* and designated UW85, reduced mortality from 100 per cent in controls to 0 per cent in three screens. When seed was treated with this isolate and planted in soil known to be infested with the pathogen, emergence was significantly greater than controls. The bacterium is now known to produce a novel antibiotic, zwittermicin A (Table 5.1 and Figure 5.7; He *et al.*, 1994), as well as a compound designated antibiotic B, and the organism was also effective in suppressing diseases of cucumber, groundnuts, soybean, tobacco and tomato (Osburn *et al.*, 1995).

Schisler and Slininger (1997) have developed a novel method for selecting antagonists of Fusarium dry rot of potatoes, which mimics the natural situation closely, and the technique has also been adapted in the author's laboratory to select for potential biocontrol agents of *Phytophthora infestans* (Figure 5.9). In brief, loam sterilized by autoclaving, live test soil and potato periderm were combined in the proportions 93:5:2 by weight to form an antagonistic soil mix (ASM). This was adjusted to a moisture level of $0.2\,m^3\,m^{-3}$ and incubated in a sealed beaker for 7 days at 18°C. The ASM was mixed with zoospore suspension and the mixture was adjusted to a moisture level of $0.35\,m^3 m^{-3}$ before applying to wounded tubers of the blight susceptible cultivar King Edward 2 days later. Microorganisms from soils that were reproducibly suppressive were isolated on a variety of media and checked individually in the assay for their ability to suppress lesion development. Fifteen bacteria were identified by routine bacteriological tests and sequencing of ribosomal DNA that reproducibly inhibited colonization of potato tubers by *P. infestans* (Hollywood and Strange, in preparation).

Some workers have first screened for potential control agents *in vitro* but caution must be exercised here as the methods adopted may have profound effects on the results. For example, Dickie and Bell (1995) analysed nine factors which influenced the outcome of screens for endophytes of grapevines that were antagonistic to *Agrobacterium vitis*. All nine factors had significant effects on the diameter of the zones of inhibition but the most important were the strain of *A. vitis*, its growth history and the medium, defined media giving smaller zones than complex media. Moreover, Handelsman and co-workers (1990) point out that their effective isolate of *B. cereus* (UW85) would have been missed had such a procedure been adopted as it was not effective against *P. megasperma* f. sp. *medicaginis in vitro*. Nevertheless, *in vitro* preliminary screens have proved useful in a number of instances. For example, Smith, Wilcox and Harman (1990) developed a medium selective for *Trichoderma* and *Gliocladium*, two fungi well known as antagonists of soil-borne pathogens, and tested the ability of the isolates obtained to overgrow or inhibit the apple pathogen, *Phytophthora cactorum, in vitro*. Their experiments gave a high rate of success: 11 of 14 isolates collected from a soil which was suppressive to *Aphanomyces* root rot of pea being antibiosis-positive and seven giving control *in vivo*, suggest-

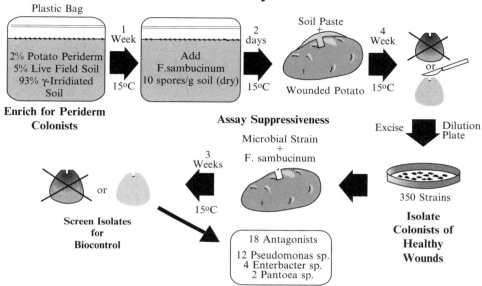

Isolation of Antagonists of Fusarium Dry Rot

Figure 5.9 Screening technique for antagonists of potato pathogens (courtesy of David Schisler) A sample of field soil is enriched for periderm colonists by incubation for 1 week with periderm and γ-irradiated soil before inoculation with the test pathogen, *Fusarium sambucinum*. After incubation for a further 2 days the mixture is applied as a paste to wounded potato and incubated for 4 weeks. Samples of potato tissue that remained healthy were excised and plated. The assay was repeated with the isolated organisms in order to determine those that were antagonistic to *F. sambucinum*

ing that the initial *in vitro* screen was useful in selecting candidates for the *in vivo* screen.

As pointed out in Section 5.3, suppression of sporulation of a pathogen by a biocontrol agent may be an effective means of control. For example, *Ulocladium atrum* reduced the sporulation of *Botrytis cinerea* on dead lily leaves under a wide range of microclimatic conditions and also competed successfully with naturally occurring saprophytes (Köhl *et al.*, 1995a,b,c). Where the biocontrol agent is applied to live tissue the induction of host resistance may be another reason for suppression of the pathogen (see Chapter 12, Section 12.4.3).

5.5.2 The development of biological control agents

Biocontrol agents normally possess several of the following characteristics:

(1) ability to associate sufficiently closely with the plant to exert an effect on the pathogen, i.e. to be phyllosphere or rhizosphere competent,

(2)   ability to compete with the pathogen for nutrients or niches, e.g. infection courts,

(3)   production of antibiotic compounds,

(4)   production of lytic enzymes effective against the pathogen,

(5)   ability to parasitize the pathogen,

(6)   ability to interfere with the reproduction of the pathogen,

(7)   ability to interfere with the virulence mechanisms of the pathogen,

(8)   the induction of host defence mechanisms.

No one pathogen is likely to have all these properties so it is often advantageous to combine biocontrol agents which exert control by different mechanisms. For example, Paulitz, Ahmad and Baker (1990) found that a combination of *Pythium nunn* and a rhizosphere competent mutant of *Trichoderma harzianum* gave greater control of *Pythium* damping-off of cucumber in some treatments than either biocontrol fungus alone. Similarly, as noted previously (Section 5.4.2), variation in sensitivity of the take-all pathogen, *Gaeumannomyces graminis* f. sp. *tritici*, to antibiotics produced by biocontrol agents may necessitate mixtures of these organisms which differ qualitatively in antibiotic production in order to obtain adequate control. In work with this pathogen, Duffy, Simon and Weller (1996) found combinations of *Trichoderma koningii* with bacteria such as *Pseudomonas chlororphis* 30–84 and *P. fluorescens* Q2-87 gave more effective disease control than either the fungus or the bacteria alone. Also, the experiments of Woo and co-workers (2002), previously mentioned (Section 5.4.2), showed that a combination of *Trichoderma atroviride*, which produced chitinase, and *Pseudomonas syringae* pv. *syringae*, which produced lipodepsipeptides, gave better post-harvest control of infection of apple fruit by *Botrytis cinerea* than either separately.

Considerable effort has been expended on transforming fungi and bacteria with genes that improve the biocontrol properties possessed by the wild type or add to them. For example, Delany and co-workers (2001) genetically modified strain F113 of *Pseudomonas fluorescens* by introducing two plasmids in separate experiments. One genetically modified strain contained the multicopy plasmid, pCU8.3 and the other pCUP9. Both plasmids contained the genes for the synthesis of the antimicrobial compound 2,4-diacetylphloroglucinol (*phlA, C, B and D*) as well as a putative permease gene (*phlE*). The modified strains produced significant and substantial increases in the antibiotic in the early logarithmic phase of growth and in the stationary phase compared with the wild-type strain and they were as effective as a proprietary fungicide in controlling damping-off of sugarbeet seedlings caused by *Pythium ultimum*.

5.5.3 The application and establishment of biological-control agents

Often the major difficulty in the use of biocontrol agents is getting them to the right place at the right time in sufficient density to be effective and then maintaining them there. Application as seed dressings is very attractive but it is necessary for the shelf-life of the control agent to be sufficiently long to be practicable and preferably to match that of the seed itself. Formulation of the agent has an important role to play here. For example, in initial field trials of *B. cereus* UW85, the bacterium was added to alfalfa seeds in 1 per cent methylcellulose (Handelsman *et al.*, 1990) but later, Osburn and co-workers (1995) found that clay-granule formulations applied infurrow gave the most consistent results. Nagtzaam and Bollen (1994) tested seeds of chinese aster and tomato which had been coated with a soil-oatmeal culture of *Talaromyces flavus* as 40 per cent of a mixture together with quartz flour and polymer binder and found that spores of the biocontrol agent could be recovered after storage for 17 years at room temperature.

Nelson and Craft (1991) added strains of *Enterobacter cloacae* to golf-course turf for the control of dollar spot caused by *Sclerotinia homoeocarpa*. Providing that the applications of the bacterium in a corn-meal – sand mix at a rate of $465 \, cm^3$ to $1.4 \, m^2$ plots were made monthly, the control obtained was as effective as the application of fungicides. Given the context of the frequent top dressings of sand and peat often applied to golf-course greens, it would seem feasible to include an appropriate inoculum of *E. cloacae* with these.

The successful trials with *Sporidesmium sclerotiorum* as a control agent for several sclerotia producing fungi have already been mentioned (Section 5.4.2).

One further aspect that should be addressed before the widespread application of any biological control agent is its environmental impact. Gullino, Migheli and Mezzalana (1995) evaluated the biological, ecological and genetic risks presented by the application of strains of *Fusarium oxysporum* isolated from rhizosphere of carnations grown in a *Fusarium*-suppressive soil, since such saprophytic strains are potentially excellent antagonists of formae speciales of the fungus that cause wilts in many plant species. They considered persistence and survival of the biocontrol strains, their effects on microbial communities, their genetic stability, their capacity for transfer of genetic material to pathogenic strains and their pathogenicity and toxicology. Populations of introduced strains generally decreased from 5×10^5 colony forming units (cfus)/g soil to 10^3–10^4 cfus/g soil after 6–10 weeks and then remained stable. The introduction did not affect population dynamics of other microorganisms and no transfer of genetic material was obtained in the experiments reported since, as the authors point out, these biocontrol agents lack a sexual cycle. They are therefore more acceptable than sexually reproducing agents, although the formation of anastomoses with other fungi cannot be ruled out. None of the 10 plants used to detect pathogenicity was affected by the antagonistic *Fusarium* spp. and the toxigenicity tests with 10 strains, apart from one exception, proved negative.

Another aspect that should be considered is the possibility that a biocontrol organism may be pathogenic to humans or other members of the biota. For example, bacteria belonging to the complex *Burkholderia cepacia* are used as biocontrol agents and in bioremediation but some strains are plant pathogens and others are opportunistic pathogens of humans with cystic fibrosis (Parke and Gurian-Sherman, 2001).

5.6   Reducing the spread of inoculum

Inoculum may be spread by wind, water and soil and by infected plant material and vectors. Reduction of the dissemination of disease by these means is an important endeavour of plant pathologists.

5.6.1   Controlling the spread of wind-borne inoculum

The spread of wind-borne pathogens is influenced by host distribution, as pointed out in Chapter 3 (Sections 3.2.2 and 3.4.1). Low densities of susceptible plants reduce the opportunity of infection, particularly if susceptible plants are mixed with genotypes which are resistant. Possibly this accounts for the survival of rubber trees in the Amazon basin where they grow wild among the other members of the flora despite the presence of the leaf-blight fungus, *Microcyclus ulei*. In contrast, when attempts were made to grow rubber trees in pure stands in South America, the venture was a failure owing to the leaf-blight pathogen.

With the requirement of today's intensive farming methods, driven by the necessity of feeding the world's population of over six billion, planting a crop at low density is normally not an option although intercropping different species is often practised by subsistence farmers. This practice, however, is risky where pathogens with wide host ranges such as *Colletotrichum gloeosporioides* are encountered (Section 3.4.1). Whether plants are grown in monoculture or inter-cropped, appropriate exploitation of the diversity of resistance genes in the population is crucial. There are three ways in which this may be done:

(1)   planting multilines – these are cultivars that are made up of a mixture of isolines differing in their genes for resistance to a particular pathogen

(2)   planting mixtures – these are heterogeneous crops of a single species;

(3)   amalgamating several resistance genes in the same genetic background.

Procedures (1) and (3) may be viewed as breeding approaches to reducing crop losses caused by disease while procedure (2) may be regarded as a deployment approach.

Multilines and mixtures are essentially similar with regard to their pathology but the great advantage of mixtures is that their components may be varied at will. For example, they may be varied to confer resistance against multiple pathogens and to take advantage of genotypes with superior agronomic characters as well as to exploit local environmental conditions. Not surprisingly multilines and mixtures have received considerable attention (Wolfe, 1985; Garrett and Mundt, 1999; Zhu et al., 2000).

Zhu and co-workers (2000) reported some very successful experiments in the control of rice blast by planting mixtures of resistant and susceptible rice varieties over a large area in Yunnan Province, China. The susceptible variety was glutinous or 'sticky' rice, which is used for confectionery and speciality dishes, and the resistant plants were four different mixtures of hybrid varieties. The 'sticky' rice was planted as single rows interspersed between four or sometimes six rows of hybrid rice. Grain production of 'sticky' rice was 89 per cent higher and blast caused by *Magnaporthe grisea* 94 per cent less when planted in mixtures compared with monocultures.

Several factors are thought to contribute to the decreased levels of disease in genetically diverse plant populations. The increased distance between susceptible plants is regarded as the most important since inoculum levels fall off rapidly with distance from the infected plants (Section 3.2.2). A second reason is the barrier provided by the resistant plants to spread as inoculum may alight on resistant plants rather than susceptible ones. Third, the induced resistant responses caused by incompatible inoculum are essentially non-specific, allowing the plant to resist challenge by compatible inoculum which may be of a completely different pathogen (see Chapter 11, Section 11.3.7), a factor which has been demonstrated to be of considerable importance in small grain cereals (Chin, Wolfe and Minchin, 1984; Calonnec, Goyeau and deVallavielle Pope, 1996). In plants which are heterogeneous with regard to architecture, as in the case of mixtures of 'sticky' and hybrid rice which are of different heights, differences in microclimate such as temperature, light and humidity may also play a role (Zhu et al., 2000).

Whether to supplant low-yielding, genetically heterogeneous crops with high-yielding but genetically more uniform plants is a particularly cruel dilemma in Developing Countries where food security is tenuous. One approach has been to introduce high-yielding varieties as a component of a mixture. For example, Trutmann and Pyndji (1994) reported experiments in Zaire in which local mixtures of bean (*Phaseolus vulgaris*) were mixed with 25 or 50 per cent of high-yielding varieties that were resistant to angular leaf spot caused by *Phaeoisariopsis griseola*. Increased yields were attributed to reduction in angular leaf spot as well as the higher-yielding component of the mixture.

Another philosophy in distributing plants with resistance genes is to amalgamate the resistance genes themselves into a single cultivar, a process which is becoming ever more feasible with the growing list of cloned resistance genes (see Chapter 10, Section 10.5.3). Schafer and Roelfs (1985) have estimated that a single cultivar containing four or five effective resistance genes would remain resistant for many years, possibly centuries. However, this assumes that each gene conferring resistance is independent of all other resistance genes. As we shall see (Chapter 10, Section 10.4), this is not necessarily so.

Surveys of the races of important pathogens in a given area are invaluable in the early detection of new virulences and may be crucial in preventing the occurrence of widespread epidemics. For example, in 1976 the Physiologic Race Survey of Cereal Pathogens (now the United Kingdom Cereal Pathogen Virulence Survey, UKCPVS) was instituted as a direct result of an unexpected epidemic of yellow rust caused by *Puccinia striiformis* on the previously resistant wheat cultivar Rothwell Perdix. The surveys have had significant effects on the use of resistance genes by breeders and the deployment of varieties with these genes by farmers (Bayles, Clarkson and Slater, 1997). For breeding, knowledge of resistance genes that are effective is essential and, for growing the crop, selection of varieties by farmers is strongly influenced by official evaluations of disease resistance.

In the industrialized world, mankind survives on about 20 predominant species of cultivated plants but during history around 3000 plants have been used as food (Vietmeyer, 1986). There is therefore a great diversity of plant material that can potentially be exploited although few non-traditional species are likely to have yields which can compete with modern crops. Traditional farmers in some parts of the world still plant a great diversity of species. For example, as many as 250 plant species are used in the village gardens of West Java (Thurston, 1990). The same author also visited a traditional Mexican farm in 1980 where he found 17 different types of bean on the 1.5 ha plot but the farmer did not mention disease. Thurston concluded that the diversity of the many varieties probably provided protection against any pathogens. Clearly such practices are labour intensive and it would also be important to know if there were a penalty to pay in terms of yield.

5.6.2 Controlling the spread of water-borne inoculum

Plants are particularly at risk from pathogens such as Oomycetes, fungi and Plasmodiophoromycetes that produce zoospores when free water is available. This may occur as a result of heavy rainfall or as a consequence of irrigation. The example of the infection of cantaloupe melons in Iran has already been mentioned (Section 5.3). In California farmers recycle water as there are controls on irrigation water run-off. Such water may be a source of zoospore inoculum of several species of *Pythium* and *Phytophthora*. Café and Duniway (1995) have studied the dispersal of *Phytophthora capsici* and *P. parasitica* from point sources in irrigation furrows. Dispersal of the fungi was monitored in tomato, pepper and squash irrigated on a 14-day cycle. Repeated irrigations dispersed *P. capsici* and *P. parasitica* up to 70 m from the source and fruit infection increased with increasing distances downstream, suggesting an accumulation of secondary inoculum with the repeated flow of water. In contrast, disease severity was highest on roots of tomato and pepper at the source of inoculum and decreased rapidly with distance from this.

Well-drained soil is a first line of defence in the control of such organisms but even the best-drained soil is likely to be flooded by heavy rain and, in the case of blight of pepper, caused by *Phytophthora capsici*, rainfall had the largest direct effect on the rate of disease progress (Bowers, Sonoda and Mitchell, 1990). Where plants are irrigated rather than rainfed, control of inoculum in the irrigation water may be an option. For example, Californian farmers using recycled irrigation water employ filters, clarifiers and chlorine injectors to prevent the spread of disease. As previously noted (Section 5.3) in Iran, infection of cantaloupe plants by *Phytophthora drechsleri* can be reduced if irrigation furrows are dug deeply and irrigation water is not allowed to make direct contact with the crown of the plant.

Rain splash is an important means by which plant pathogens are disseminated (Section 3.5.2). Control in these instances may be obtained by treatments that reduce the number and size of the droplets and their velocity. One way of doing this is to increase the spacing between plants. For example, good control of bacterial pathogens such as wildfire of tobacco caused by *Pseudomonas tabaci* was obtained by this means. The use of appropriate mulches can also be effective. For example, as noted in Section 3.5.2, strawberry plants mulched with straw had only 15 per cent infection with *Phytophthora cactorum* compared with those mulched with plastic, ca. 82 per cent infection (Madden and Ellis, 1990).

5.6.3 Controlling the spread of soil-borne inoculum

Soil-borne pathogens generally spread slowly but the speed of spread may be increased by cultivation methods. Kite-shaped patches of disease may be one indication of this, the pathogen being spread more effectively in the direction taken by farm equipment than at right angles to it (Figure 5.10). In order to prevent spread to new areas it may be of value to disinfect equipment such as ploughs, harrows and discs.

Armillaria is a virulent pathogen of woody plants which spreads through the soil, commonly from stumps of old trees by means of rhizomorphs and by mycelium at sites of root contact. At present there are only two methods used to combat the pathogen, the use of less susceptible species and the removal of tree roots (Chapman and Xiao, 2000). Both have drawbacks, resistant species may not be available and removal of tree roots is expensive and causes major disturbance of the soil. Previously recommended methods had included soil fumigation around infected hosts and direct injection of fumigants into the hosts but these are expensive and may have deleterious effects on other members of the soil microflora. An alternative, which has been used successfully with other wood

Figure 5.10 Spread of disease by farm equipment: kite-shaped patches in a sugarbeet field caused by *Rhizoctonia solani* (courtesy of Mike Asher, Broom's Barn Research Station, Higham, Bury St Edmunds, Suffolk 1P28 6NPR)

rotting fungi, is to inoculate stumps with fungi that compete with the pathogen. For example, Pearce and Malajczuk (1990) found that *Coriolus versicolor, Stereum hirsutum* and *Xylaria hypoxylon* significantly reduced stump colonization by *A. luteobubaliana* and *C. versicolor* inhibited rhizomorph growth strongly or completely (Pearce, 1990). Stumps were also naturally colonized below ground by a *Hypholoma* species which partially or completely excluded *A. luteobubaliana*. In other studies, Chapman and Xiao (2000) found that *Hypholoma fasciculare* overran colonies of *Armillaria ostoyae* in culture and colonized discs of tree roots where the pathogen was well established. When roots were inoculated with both fungi simultaneously, *H. fasciculare* competed successfully with *A. ostoyae*. *H. fasciculare* therefore appears to be a promising biocontrol agent and the authors describe a simple method for inoculating stumps which may prove to be at least a partial solution to the *Armillaria* problem.

5.6.4   Controlling the spread of inoculum in infected propagative material: the role of quarantine

As remarked by Abdallah and Black (1998), the import of vegetative propagating material whether for research, development and breeding purposes or for direct commercial use, has undoubtedly been responsible for the introduction and spread of many important plant pathogens throughout the world. Such material may harbour shoot-infecting and vascular-infecting fungi and bacteria, phloem- and xylem-limited bacteria, viruses and viroids and nematodes, especially those of the endoparasitic and shoot-feeding types (Black, 2001). In order to limit the dissemination of these important pathogens, quarantine measures have been established. Details of these may be found in the text of the International Plant Protection Convention which was approved at an FAO conference in Rome in 1997 and may be accessed from the CABI Crop Protection Compendium.

Quarantine measures for specific pathogens generally involve the inspection of the plant material to be exported and certification to the effect that stocks are free of a given pathogen or that its incidence does not exceed a certain threshold. However, pathogens such as phytoplasmas, viruses and viroids are often not easily detected owing to their extremely irregular distribution in host plants, some being confined to vascular tissue. Here, molecular biology techniques are invaluable and promise to be increasingly used in the future as their specificity, sensitivity and reliability become better established. For example, Heinrich and co-workers (2001) have developed a sensitive diagnostic test which is quick and specific for phytoplasmas. They constructed primers for PCR based on rDNA sequence information from an Austrian isolate of European stone fruit yellows (ESFY). Since these were effective at high annealing temperatures their specificity is high and there is a consequent decrease in the risk of false positives. The primers allowed the amplification of rDNA sequences from the phytoplasmas responsible for apple proliferation, European stone fruit yellows and pear decline. Strains of phytoplasmas were identified by RFLP of the amplicons.

Two difficulties commonly encountered in quarantine matters are the sheer scale of the requirement and the cost (Martin, James and Levesque, 2000). One way of reducing cost is to detect several target pathogens simultaneously, a process known as multiplexing. These are often PCR-based and involve primer pairs that are specific for each organism or for a group of organisms, such as a virus family if sequences are conserved among them, enabling all members to be detected. Alternatively, where conservation at the primer sites is not complete, degenerate primers may be used (Martin, James and Levesque, 2000). In some instances immuno-capture-PCR has been demonstrated to be more sensitive than alternative techniques. For example, in the detection of plum pox virus, immunocapture PCR was more sensitive than RT-PCR and 2000 times more sensitive than polyclonal antibody ELISA (Candresse et al., 1994). More recently,

Heinrich and co-workers (2001) have developed an immuno-capture-PCR procedure for apple proliferation phytoplasma that was both sensitive and suitable for large-scale testing of apple material *in vivo* and *in vitro*.

During the 1980s a Federal plant quarantine facility was built at Beltsville, Maryland, USA, in an area that was remote from agricultural crops. There the greatest effort is expended on detection of pathogens that may be latent in germplasm. The tests include simple observation, sap transmission, electron microscopy and serology. *Prunus* germplasm provides a considerable challenge since many of the uncharacterized infectious agents can only be detected by grafting and the subsequent appearance of symptoms in the fruit. As a result, such germplasm may take several years to clear quarantine (Waterworth, 1993).

5.6.5   Controlling the spread of inoculum by vectors

Jones (1987) has reviewed the role of plant resistance to vectors in controlling the diseases they transmit but according to a literature survey there has been surprisingly little published work since. The light absorption properties of the plant may critically affect the alighting behaviour of vectors. For example, there was 75 per cent less infestation of non-glaucous compared with glaucous wheat by *Sitobion avenae*. The sensitivity of vectors to light has also found application in the use of mulches. Greenough, Black and Bend (1990) found that plastic mulch with an aluminium surface reduced the number of thrips in fields of solanaceous crops by up to 68 per cent and the incidence of tomato spotted wilt virus by up to 78 per cent. More recently, Riley and Pappu (2000) evaluated management practices for reducing the same vector and virus in tomato. Host plant resistance and reflective mulch significantly reduced thrips and virus infection but early planting on black plastic with an intensive insecticide treatment resulted in the highest yields. In KwaZulu-Natal Province, South Africa, potato virus Y is a major cause of crop losses in pepper. Budnik, Laing and daGraca (1996) compared the efficacy of five treatments to reduce transmission of the virus by aphids. They found that white plastic mulch was the most effective, resulting in a 32 per cent yield increase over the untreated control.

Both physical factors such as leaf hairs and chemical factors such as pheromones may be important in determining the initial settling and feeding behaviour of vectors (Chapter 12, Section 12.2.1). For example, as mentioned in Chapter 3, *Myzus persicae* was repelled by the alarm pheromone E-β-farnesene from *Solanum berthaultii* (Section 3.3.2). Wang and co-workers (2001) have genetically modified a cytochrome P450 hydroxylase gene specific to the trichome gland in tobacco using both antisense and sense co-suppression strategies to investigate its function. These resulted in a 19-fold increase in cembratriene-ol (CBT-ol), giving a concentration that was equivalent to 4.3 per cent of the dry weight of the leaf. This compound is a precursor of cembratriene-diol (CBT-diol), which showed a reduction of 41 per cent or greater. Transgenic plants were less colonized by aphids and were more aphicidal than the wild type.

Some plants interfere with ingestion or phloem finding and it appears that this may be the mechanism involved in resistance to *Macrosiphum euphorbiae* conferred by the *Mi* gene (Kaloshian, 2000). Work in this area has been aided by employing electrical devices to monitor the probing activity of aphids. Using this technique, Annan and co-workers (2000) showed that a variety of cowpea resistant to *Aphis craccivora* caused severe disruption of penetration activity as demonstrated by frequent penetration attempts that were abruptly terminated and in shorter penetration times than those on a susceptible variety. These phenomena would be expected to reduce the opportunities for acquisition and transmission of viruses.

Some attention has been given to the possibility of protecting plants from viral infection by physically excluding vectors. For example, Harrewijn, Denouden and Piron (1991) found that the material Lutrasil LS 10 gave 100 per cent protection against transmission of potato virus Y and potato leaf-roll virus by aphids. Webb and Linda (1992) evaluated a light-weight floating row cover, either used alone or in combination with a white-on-black polyethylene mulch as a method of controlling aphid-borne mosaic viruses, whitefly-induced leaf silvering and insects directly damaging zucchini squash. Dramatic increases in yield were recorded, especially if covers were left in place for at least 1 week after plants began to bloom. Once plants were uncovered, colonization was slow except by *Bemisia tabaci*.

As discussed above (Sections 5.6.3 and 5.6.4) man may also act as a vector for plant pathogens by transporting either them – or plant parts containing them – to other localities. However, he can act in more direct ways during crop husbandry practices such as pruning and thinning. For example, if tomato plants were thinned in the morning, when they were still laden with dew, the incidence of bacterial spot caused by *Xanthomonas vesicatoria* pv. *vesicatoria* was 87 per cent but this was reduced to 44 per cent if thinning were delayed to the afternoon when the foliage was dry. Since farm workers preferred not to thin the plants during hot afternoons, two treatments for hand disinfection were tested, povidone-iodine and 70 per cent ethanol, so that thinning could be done in the morning. Use of these disinfectants reduced infection by 81 per cent and 65 per cent, respectively but unfortunately, there were objections to both: povidone-iodine has a sticky texture and it was feared that it might induce allergenic dermatitis if used extensively. With the alcohol procedure the fear was different and was succinctly described as field consumption (Pohronezny *et al.*, 1990)!

6 Locating, Penetrating and Colonizing the Host

Summary

Since many pathogens have to breach the barriers of plant waxes, cutin and suberin that cover plants as well as plant cell walls before establishing a parasitic relation with their hosts, the physical and chemical characteristics of these are first discussed. Some soil-borne pathogens locate their hosts through chemical signals and these are also important in subsequent events such as the germination of propagules, chemotropism of germ tubes and the differentiation of infection structures, the last of these also being influenced by physical features of the host. Adhesion is often required for successful penetration, particularly where this is achieved by the exertion of mechanical force. However, enzymes that degrade the surface layers of plants, such as waxes, cutin and suberin are also critical for entry by many pathogens. Once past these surface layers the pathogen usually has to breach the cell wall and for this a range of pectolytic enzymes, cellulases and xylanases as well as enzymes involved in the degradation of lignin are required. In some instances, other enzymes are inferred to have important roles to play in pathogenicity or virulence such as proteases and membranlytic enzymes. The degradative enzymes of bacterial pathogens may be globally regulated and in some cases the autoinducer N-3-(oxohexanoyl)-L-homoserine lactone is involved. Products of degradative enzymes acting on host tissues are sources of nutrition for necrotrophic pathogens but the subtler biotrophic pathogens feed through specialized structures called haustoria. In rust fungi evidence is becoming available which shows that specific genes are induced in these structures, some of which have transport functions. Viruses and viroids lack motility and therefore face a particular problem in colonizing their hosts after entry. In tobacco mosaic virus (TMV) newly synthesized viral RNA complexes with movement protein which is encoded by the virus and is transported via microtubules and microfilaments to plasmodesmata. These are enlarged under the influence of the movement protein, allowing movement of the virus to the adjacent cell. Long distance spread is by passive movement in the xylem or phloem.

Introduction to Plant Pathology by Richard Strange
© 2003 John Wiley & Sons, Ltd   ISBN 0 470 84972 X (cased)   ISBN 0 470 84973 8 (pbk)

6.1   Introduction

Once a pathogen has arrived in the vicinity of a potential host plant or, as may happen in the case of soil-borne pathogens, a plant root has arrived in the vicinity of a pathogen, subsequent events depend on the production and perception of signals by both partners. First we shall be concerned with those factors external to the plant that influence the behaviour of propagules of the pathogen and then the host and pathogen factors that are involved in the mechanisms of penetration. Since some of these, such as degradative enzymes, are also involved in further colonization of the host after penetration, this aspect of parasitism will also be considered.

In soil, pathogens may be influenced by compounds exuded from the host root. Motile stages may be attracted or repelled and the germination of sessile propagules stimulated or inhibited. Air-borne pathogens generally rely upon large populations of propagules to ensure that at least some of them alight on a suitable host. At this point adhesion is a necessity to prevent the propagule being washed off the plant and, for at least one fungal pathogen, adhesion has been established as a prerequisite for germination. Following adhesion, germination, which may be under the control of topological or chemical signals from the host, occurs and in some instances such signals lead to the differentiation of infection structures. These, too, require firm anchoring to the surface of the plant if any mechanical force is to be exerted.

Some pathogens enter their hosts via natural openings such as stomata, nectaries, hydathodes or lenticels while others require wounds, which may be made by physical phenomena such as wind or hail or by biological agents such as vectors or herbivores. They therefore require neither mechanical force nor enzymes to establish their initial beachhead in the host. Others use mechanical force to effect entry but often these rely upon the secretion of enzymes as well. Such enzymes include lipases and cutinases for breaching the wax and cuticle of aerial parts of the plant as well as enzymes for degrading cell-wall constituents such as pectic substances, cellulose and lignin.

A significant number of pathogens live extra-cellularly in their hosts and some are confined to conducting tissue. These would seem to have only limited requirements for degradative enzymes. Biotrophs often feed through haustoria, which penetrate the host cell wall, almost certainly through the agency of degradative enzymes, and invaginate but do not penetrate the host plasma membrane. For necrotrophs the role of degradative enzymes seems clear. They are required not only for penetration and colonization of plant tissue but also to reduce the high molecular weight components of these tissues to products which they can metabolize. In the case of soft rotting organisms this often results in the 'mushy' symptoms that give these diseases their name. Despite the seemingly obvious necessity of enzymes for pathogenicity or virulence, unequivocal demonstration of such roles has been difficult to achieve owing to the numbers and diversity of isozymes produced by pathogens and subtleties in their regulation.

Colonization of the host by viruses is a special case as they lack enzymes with which to degrade plant cell walls but spread from cell to cell through plasmodesmata and then, in the case of systemic infections, by way of the phloem or xylem.

6.2 The physical and chemical characteristics of materials that cover plants

Since many pathogens need to penetrate the outer coverings of plants and plant cell walls in order to establish a parasitic relationship with their hosts, the characteristic of these will be described first.

Aerial parts of plants are generally covered with a cuticle consisting of an outer, thin layer of wax and an inner layer of cutin. The outer waxy layers are usually complex and a stereoscan electron microscopy survey of 13 000 species representing all the major groups of seed plants has classified them into 23 types. This physical diversity is also paralleled by their chemical diversity. For example, Shepherd and co-workers (1999) found that the epicuticular waxes of red raspberry (*Rubus idaeus*) consisted of a complex mixture of free primary alcohols and their acetates, secondary alcohols, ketones, terpenoids including squalene, phytosterols, tocopherol, amyrins, alkanes and long chain alkyl and terpenyl esters. Similarly, Jetter and Riederer (2000) examined the rodlet-shaped wax crystals of the fern, *Osmunda regalis*, and found 139 compounds belonging to 14 homologous series. These included alkanes alkyl esters, primary alcohols, secondary alcohols, ketones, aldehydes, fatty acids and β-sitosterol. Clearly surface waxes are far from simple and potentially contain an abundance of physical and chemical cues, which may influence the prepenetration stages of organisms seeking entry into plants.

Cutin, which is found under the wax surface, is an amorphous substance composed of hydroxylated and epoxylated fatty acids with either 16 or 18 carbon atoms and these are linked to each other by means of ester bonds (Figure 6.1). Underground parts of plants are generally covered with suberin. This is more complex than cutin and consists of alcohols and monobasic acids with 18 to 30 carbon atoms as well as hydroxylated or dibasic acids with 14 to 20 carbon atoms together with phenolic components (Kolattukudy, 1980). The phenolics are unique and distinct from lignin, consisting primarily of hydroxycinnamates such as feruloyltyramine (Bernards and Lewis, 1998; Figure 6.1). Comparatively recently, evidence has been presented for the old hypothesis that glycerol is an important monomer of suberin (Moire *et al.*, 1999). Suberization is also an active mechanism by which plants defend themselves from attack by pathogens (see Chapter 11, Section 11.3.4).

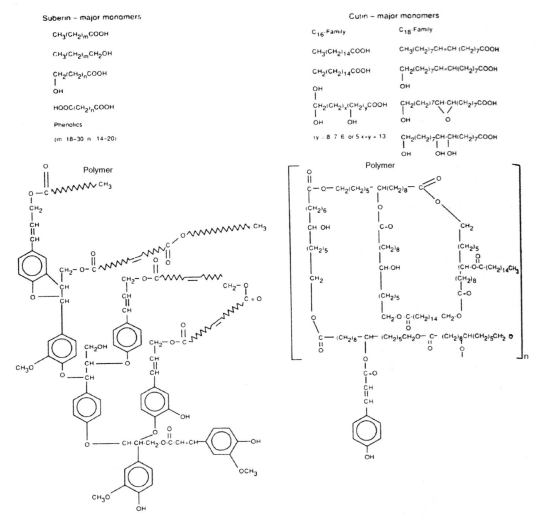

Figure 6.1 The chemical structures of suberin and cutin monomers and their proposed polymers (from Kolattukudy, 1980 and reproduced with permission from the author and Science)

6.3   The physical and chemical characteristics of plant cell walls

The cell walls underlying cutin or suberin consists of two parts, the outer primary wall and the inner secondary wall. Both are composed of a crystalline microfibrillar phase of cellulose, a β-1,4 linked glucan, embedded in an amorphous phase. The microfibrils are about 30 nm in diameter and consist of aggregates of 20–70 linear chains of cellulose, which are held together by hydrogen bonds between the sugar OH groups. The incidence of microfibrils in primary walls is sparse and this comparative scarcity allows the cell to grow. Once growth has ceased, the secondary wall is laid down which contains an orderly arrangement of cellulose microfibrils as well as hemicelluloses. The latter are a heterogeneous group of saccharide polymers which may be extracted with alkali and, on hydrolysis to their monomers, yield glucose, xylose and arabinose (Figure 6.2). In monocotyledonous plants arabinoxylan is not only the predominant hemicellulose but may account for up to 60 per cent of the total wall carbohydrate. It is composed of a β-1,4-linked D-xylopyranosyl backbone which is frequently substituted at the O-2 and O-3 positions by various mono- or oligo-saccharide units, consisting mainly of arabinosyl-, xylosyl- and/or glucosyluronic acids.

The middle lamella lies between cells and is composed largely of pectic substances. The predominant motif is a linear chain of galacturonic acid residues which are α-1,4 linked. Where this is the major constituent the substrate is referred to as polygalacturonate or pectate but often the carboxyl groups of the galacturonate residues are esterified with methanol and then the term pectin is used. The carboxyl groups of polygalacturonate form salts with the divalent cation, calcium, thus linking neighbouring chains. This gives rise to the gel properties of pectin and provides the cohesive element that cements cells together. The 'smooth' regions of pectate or pectin may be interspersed by the occasional rhamnose residue and these may be substituted with oligosaccharides composed of neutral sugars such as arabinose, galactose and xylose giving rise to so-called 'hairy' regions! In addition, some of the galacturonate residues are acetylated at the C-2 or C-3 position (Figure 6.2). Arabinose and galactose are also prominent constituents of the middle lamella, forming the polymers, arabinan and galactan, respectively and are linked to the rhamnogalacturonan backbone (Figure 6.2).

Lignin is another important component and constitutes between 15 and 35 per cent of the dry weight of woody tissues. It is a heterogeneous polymer formed by the polymerization of up to three components – coumaryl, coniferyl and sinapyl alcohols (Figure 6.3).

Proteins, which may have structural or enzymatic functions, can account for as much as 15 per cent of the cell wall (Showalter, 1993). Most of those that have been characterized are from dicotyledonous species. They consist of extensins, which are hydroxyproline-rich glycoproteins (HRGPs), glycine-rich proteins (GRPs), proline-rich proteins (PRPs, an abbreviation not used to describe such proteins in this text in order to avoid confusion with pathogenesis-related proteins – see Chapter 11, Section 11.3.6), arabino-galactan proteins (AGPs) and solanaceous lectins. Variants of HRGPs are found in monocotyledonous species, which contain threonine and histidine hydroxyproline-rich glycoproteins (THRGPs and HHRGPs, respectively).

Although once the secondary cell wall has been laid down its apparently rigid structure would appear to preclude any further change it is, nevertheless, dynamic and responds to stimuli such as challenge with pathogenic organisms. These responses will be considered in Chapter 11 (Sections 11.3.3–11.3.5).

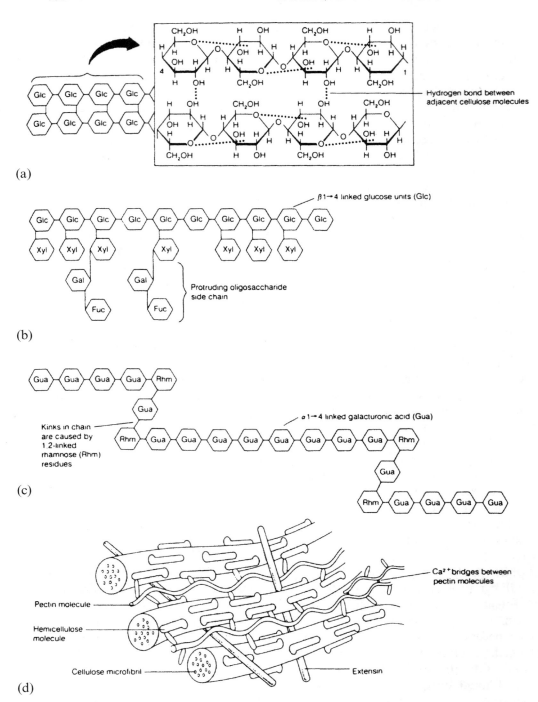

(a)

(b)

(c)

(d)

Figure 6.2   The chemical structures of (a) cellulose, (b) hemicellulose, (c) pectin, and (d) a model of the interconnections among these components (reproduced from Taiz and Zieger, 1991, with permission from the authors and the Benjamin/Cummings Publishing Company Inc.)

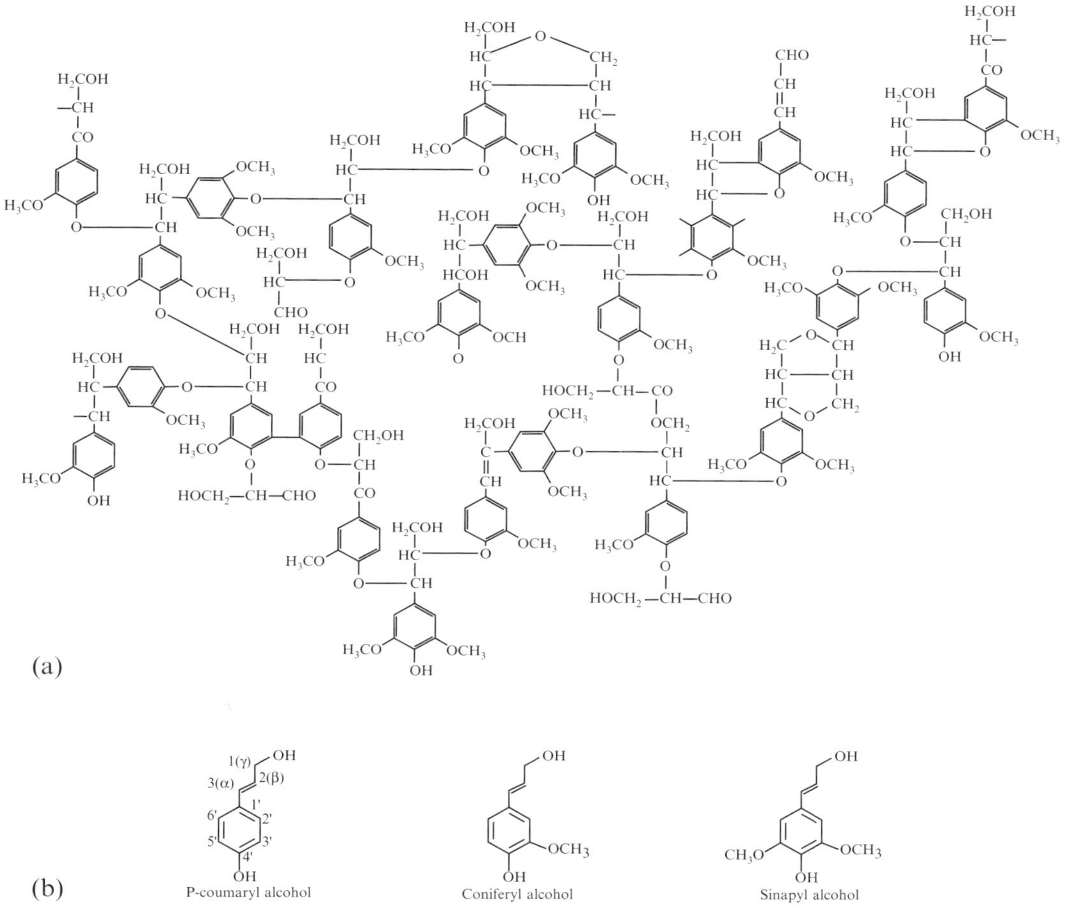

(a)

(b)

P-coumaryl alcohol   Coniferyl alcohol   Sinapyl alcohol

Figure 6.3   The structure of lignin, (a) partial structure of a hypothetical lignin molecule from *Fagus sylvatica*, (b) the structures of lignin monomers ((a) reproduced from Taiz and Zieger, 1991, with permission from the authors and the Benjamin/Cummings Publishing Company Inc.)

6.4 Chemotaxis, encystment and chemotropism

Several taxa of soil-borne pathogens have motile stages. These include Chytridiomycetes (Section 1.3.2), nematodes (Section 1.3.3), Oomycetes (Section 1.3.5), Plasmodiophoromycetes (Section 1.3.6) and bacteria (Section 1.3.8). Since plants release as much as 20 per cent of their photosynthate into the rhizosphere it is not surprising that some of the compounds found in their exudates influence the motility of these pathogens. Furthermore, evidence is accumulating that root cap border cells have important roles to play in determining whether or not the roots themselves become infected by such organisms.

Border cells originate from the root cap meristem and are attached to the root periphery by a water-soluble polysaccharide matrix, but their middle lamellae are solubilized by pectolytic enzymes in the cell wall. As a result, the cells immediately disperse when root tips are placed in water (Hawes *et al.*, 1998). Border cells affect both bacterial and fungal plant pathogens by chemo-attraction or repulsion and there is some evidence that they may act as decoys of pathogenic fungi, thus protecting the root from infection. For example, Hawes and co-workers (1998) found that when pea roots were uniformly infected with spores of a virulent strain of the fungus, *Nectria haematococca*, hyphal growth was confined to the apex forming a mantle, which, on placing in water, fell off as a unit. The mantle supported copious growth of the fungus but at the surface of the root only newly formed border cells and not fungal hyphae were apparent (Figure 6.4).

Zoospores of Oomycetes frequently accumulate at the zone of elongation of roots and it was initially thought that there were generally no appreciable differences in accumulation between plants that are resistant and those that are susceptible (Wynn, 1981). However, more recent work has refuted this view for a number of plant and pathogen combinations. For example, Mitchell and Deacon (1986) investigated four species of *Pythium, P. graminicola* and *P. arrhenomanes*, which infect only grass species and *P. aphanidermatum* and *P. ultimum* which have broad host ranges. When grass roots or roots of dicotyledonous plants were placed in zoospore suspensions, larger numbers of the two graminicolous species accumulated on grass roots than on roots of dicotyledonous plants whereas there was no differential effect on the species with broad host ranges. More specifically, zoospores of *Phytophthora megasperma* f. sp. *sojae* which infect soybean were strongly attracted to two flavonoids produced by the plant, daidzein and genistein – concentrations as low as $0.01\ \mu M$ being effective (Figure 6.5). In contrast, strains of *P. megasperma* that were infective for lucerne and Douglas fir did not respond to these compounds (Morris and Ward, 1992). In further work, Tyler and co-workers (1996) investigated 59 compounds which were structurally related to genistein and daidzein and found that 43 elicited some response. In particular, the possession of a hydroxyl group at the 4′ and 7 positions of the isoflavone molecule or comparable positions of related molecules was necessary for high activity. Wide differences in sensitivity of strains of the fungus were noted and also some analogues were repellent. Using a different Oomycete, *Aphanomyces cochlioides*, Horio and co-workers (1992) found that another flavonoid, cochliophilin A, from spinach is an effective attractant of zoospores at $0.001\ \mu M$. (Figure 6.5).

The volatile compound, isovaleraldehyde, is effective as a chemoattractant for zoospores of *Phytophthora palmivora* (Cameron and Carlile, 1978) and *P. cinnamomi* (Cahill and Hardham, 1994a) and the latter Oomycete was also responsive to several amino acids, alcohols and phenolics. Additionally, Cahill and Hardham (1994b) demonstrated that zoospores of *P. cinnamomi* showed strong electrotaxis towards a positively charged nylon membrane. Exploiting both chemotaxis and

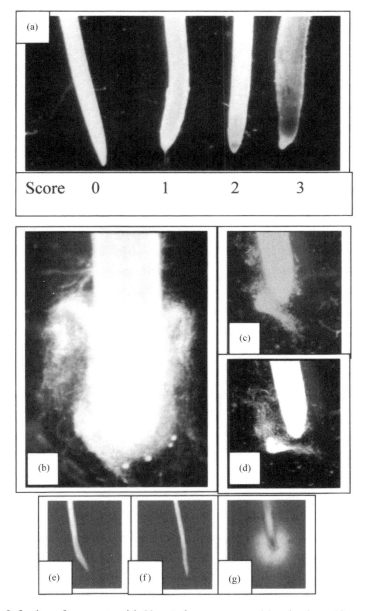

Figure 6.4   Infection of pea roots with *Nectria haematococca*: (a) only about 10 per cent of pea seedlings developed lesions at the root tip, comprising a 1–2 mm region consisting of the root cap and apical meristem – these were scored on a 0–3 scale; (b) some root tips scored as 0 were found, on microscopic examination, to be covered in a mantle of the fungus; (c) and (d) in some cases the mantle detached spontaneously on immersion in water; (e) when root tips from which mantles had become detached were surface sterilized and plated onto culture medium, no fungus grew as in controls (f); in contrast, where lesions were scored, the surface sterilized roots yielded copious fungal growth (g) (courtesy of Martha Hawes and the American Society of Phytopathology). A colour reproduction of this figure can be seen in the colour section

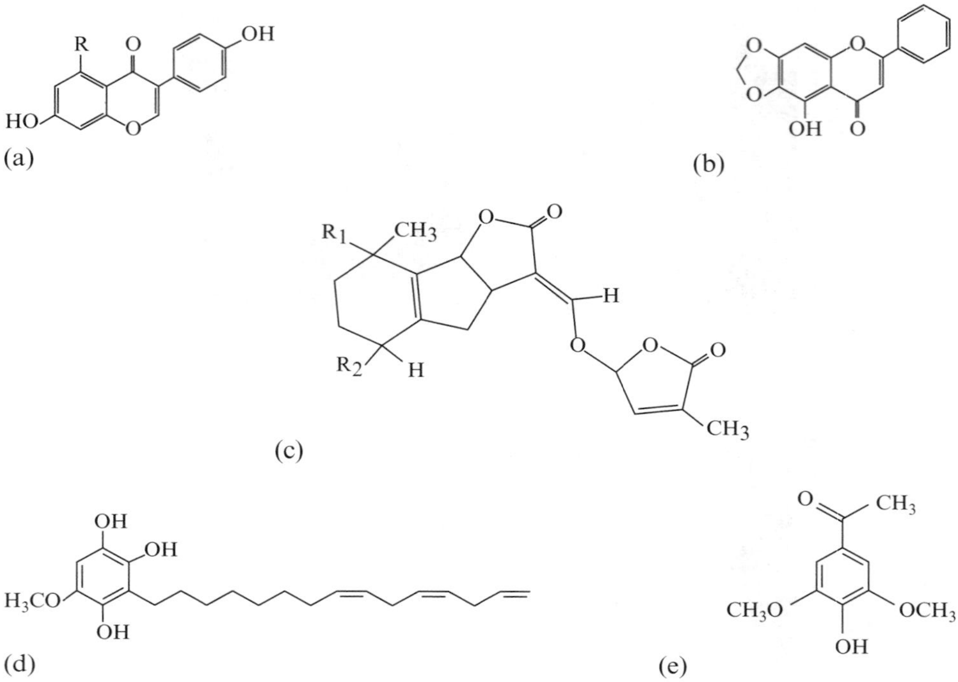

(a)

(b)

(c)

(d)

(e)

Figure 6.5 Structures of attractants and stimulants of soil-borne plant pathogens: (a) R = H, daidzein, R = OH, genistein, two compounds found in soybean that attract the zoospores of *Phytophthora megasperma* f. sp. *sojae* at a concentration of $0.01\,\mu M$; (b) cochliophilin A from spinach, an effective attractant of zoospores of *Aphanomyces cochlioides* at $0.001\,\mu M$; (c) $R_1 = CH_3$, $R_2 = OH$, strigol, a germination stimulant of species of *Striga* found in the root exudates of cotton, a non-host, and the root exudates of the host species maize, proso millet and sorghum; $R_1 = R_2 = H$, sorgolactone; (d) sorgoleone, stimulant of species of *Striga* found in the root exudates of sorghum; (e) acetosyringone, a plant wound product that causes chemotaxis of *Agrobacterium* towards roots and induces virulence genes involved in the transfer of DNA from the bacterium to the host

electrotaxis as well as a specific monoclonal antibody they were able to detect as few as 40 zoospores/ml in an assay time of less than 40 min (see Section 2.5.4).

Zoospores of *Pythium dissotocum* accumulate on border cells of cotton and penetrate them within minutes, although the chemoattractant has not been identified (Hawes, personal communication).

Acetosyringone, which is secreted by tobacco and other plants in response to wounding, attracts *Agrobacterium tumefaciens*, the cause of crown gall (Ashby, Watson and Shaw, 1987; Figure 1.19). The same compound also activates plasmid-encoded virulence genes, which are involved in the transfer of DNA from the bacterium to the host (Ashby *et al.*, 1988; Chapter 7, Section 7.4).

After a zoospore has arrived at the surface of the host plant it normally encysts, the cysts then germinate and the resulting germ tubes usually grow towards the root, i.e. they show tropism. Jones, Donaldson and Deacon (1991) found that only glutamic and aspartic acids elicited zoospore taxis, encystment, cyst germination and tropism of germ tubes of *Pythium aphanidermatum* although a number of other compounds elicited one or more of these four responses. The concentrations of the two acids used was quite high (25 mM), but it does not appear to be known whether such concentrations occur in the rhizo-spheres of susceptible plants or whether a synergistic mixture of these and other compounds is responsible for the sequence of the four events leading to infection. Donaldson and Deacon (1993), investigating encystment of three species of *Pythium*, found that only *P. catenulatum* responded to fucosylated xyloglucan, fucoidan and a methylglucuroxylan, only *P. aphanidermatum* responded to arabi-noxylan and only *P. dissotocum* responded to a non-fucosylated xyloglucan. However, all three species encysted in response to gum arabic (100 μg/ml), sodium alginate (250–500 μg/ml) or polygalacturonate (500–1000 μg/ml). Morris, Bone and Tyler (1998) extended their work with *Phytophthora sojae* and flavonoids and found that daidzein and genistein (Figure 6.5), beside acting as chemo attractants of zoospores of the fungus, also exerted a chemotropic effect on hyphae from geminating cysts similar to that of roots of soybean.

Clearly, our knowledge of host factors that are interpreted as signals by pathogens and cause them to aggregate, encyst, germinate and grow towards the host is fragmentary. For a review of these topics, readers are referred to the paper by Islam and Tahara (2001). Signals from the pathogen that elicit re-sponses in the host will be reserved for Chapter 11.

6.5 Passive entry through natural openings

Some fungi and bacteria enter their hosts passively through stomata, nectaries, hydathodes or lenticels. In particular, as in the case of wildfire of tobacco, bacteria notoriously enter plants through stomata by the agency of wind-driven rain. Similarly, nectaries provide the point of entry of the fireblight pathogen *Erwinia amylovora* (Sections 1.3.8, 3.3.1, 3.5.2 and Figure 1.21). Although stomata or trichomes provide a means of entry into tomato leaves by *Clavibacter michiganensis* subsp. *michiganensis*, it was not clear that infection by these routes could cause the marginal necrosis of leaves, sometimes known as 'firing', so often seen. Carlton, Braun and Gleason (1998) found that hydathodes were the route by which the bacterium entered and caused such lesions. Under high humidity, hydathodes exude droplets of guttation fluid but these are withdrawn when conditions become dryer. In these circumstances, should the guttation droplets become contaminated with the bacterium, infection leading to 'firing' occurs. Pathogens, which enter through lenticels, are generally those that require wounds, lenticels being of secondary importance as a means of entry (Agrios, 1997).

6.6 The roles of physical and chemical signals in the germination of propagules of plant pathogens and the differentiation of infection structures

Many fungi, on encountering their host or some other solid substrate germinate, producing germ tubes which may differentiate into infection structures. These vary from being simple appressoria to complex structures such as 'cushions'. Similarly, parasitic angiosperms such as *Striga* elaborate haustoria (Figure 6.6). The stimuli provided by the host for germination, growth and the differentiation of infection structures are hydrophobicity, hardness, chemical components and topographical features (Mendg, Hahn and Deising, 1996).

6.6.1 Hydrophobicity

Lee and Dean (1994) found a correlation between hydrophobicity of the contact surface and the formation of appressoria by the fungus, *Magnaporthe grisea* but with *Phytophthora palmivora* the situation was more complex (Bircher and Hohl, 1997). Germlings of this organism that were free floating or were in contact with

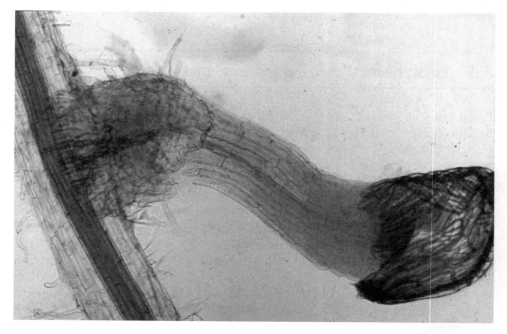

Figure 6.6 Germinating seed of *Striga hermonthica* giving rise to a haustorium which has attached to a host root (reproduced courtesy of Dr Chris Parker and CAB International). A colour reproduction of this figure can be seen in the colour section

smooth substrates under high nutrient conditions (20 per cent pea broth) did not form appressoria regardless of the hydrophobicity of the contact surface. In contrast, under low nutrient conditions (5 per cent pea broth), appressoria were formed on smooth substrates that were hydrophobic but not hydrophilic substrates. However, if these same surfaces were scratched, appressoria formed on the scratches, irrespective of the levels of nutrients or substrate hydrophobicity (Figure 6.7).

6.6.2 Hardness

Xiao and co-workers (1994) reported that although conidia of *M. grisea* started to germinate whether they contacted a liquid or solid surface, appressoria only formed on solid surfaces and not on liquid or agar surfaces. On freshly prepared agar surfaces, germ tubes of the fungus penetrated directly without the differentiation of appressoria but if the agar were allowed to dry partially appressoria were formed.

6.6.3 Chemical signals

Several chemical components of host plants have been implicated in the germination of propagules of plant pathogens and the differentiation of infection structures. In particular, the wax on the surface of aerial parts of the plant is a rich source of diverse compounds, which may play these roles (Section 6.2). For example, wax from rice leaves relieved the self-inhibition of conidia of the rice-blast pathogen, *Magnaporthe grisea*, and stimulated appressorium production (Hegde and Kolattukudy, 1997; Uchiyama and Okuyama, 1990). In isolates of *Colletotrichum gloeosporioides* that infect avocado, specific components of waxes were involved since the surface wax of this host but not of other plants triggered both conidial germination and the development of infection structures (Podila, Rogers and Kolattukudy, 1993). The fatty alcohol fraction of the wax, which comprised 5 per cent of the total, was the most effective and synthetic aliphatic *n*-fatty alcohols with 24 or more carbons were also active. Reciprocally, wax of avocado was ineffective in inducing differentiation of other species of *Colletotrichum* which are pathogenic to other plants.

Species of *Colletotrichum* which infect ripe climacteric fruit, such as tomato, banana and avocado are responsive to the ripening hormone ethylene (see Chapter 7) forming multiple appressoria on glass surfaces in the presence of micromolar concentrations of the gas. Confirmation of the role of ethylene in these processes was obtained in experiments with transgenic tomatoes that did not produce ethylene. On these the fungus was unable to germinate or form appressoria unless exogenous ethylene was supplied (Flaishman and Kolattukudy, 1994). Thus, these fungal pathogens of ripe climacteric fruit recognize the hosts' ripening

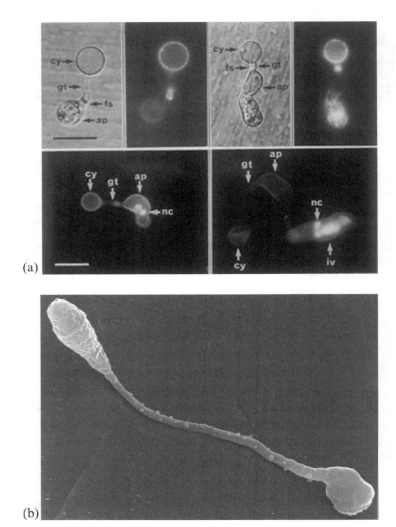

Figure 6.7 Simple appressoria formed by (a) *Phytophthora palmivora* (courtesy of Elmon Schmelzer and European Journal of Plant Pathology); cy = cyst, gt = germ tube, fs = fals sept, ap = appressorium and (b) *Magnaporthe grisea* – here the germinated spore (upper left) has given rise to a germ tube and an appressorium (lower right) (courtesy of Nick Talbot)

signal and this allows them to form the structures necessary for the launch of a successful attack at the appropriate time.

Flavonoids exuded from the roots of legumes stimulate the spore germination of several soil-borne fungi. For example Bagga and Straney (2000) found that naringenin was a powerful stimulant of the germination of macroconidia of *Nectria haematococca*, stimulation occurring in the low μM range. Other flavonoids, such as luteolin and hesperitin, were also effective but at higher concentrations. Effectiveness of the flavonoids as stimulants of spore germination was related to their ability to inhibit cAMP phosphodiesterase activity and thus increase cAMP levels.

Signalling compounds, passed between parasitic plants and their hosts, have been reviewed by Estabrook and Yoder (1998). Seeds of *Striga* normally germinate only when in close proximity to roots of host plants. However, some non-host plants also stimulate germination and, in fact, the first germination stimulant to be isolated, strigol, was from a non-host plant, cotton (Cook *et al.*, 1966; Figure 6.8). SXSg from *Sorghum bicolor* was the first stimulant to be isolated from a host plant but it is quickly auto-oxidized to the inactive quinone sorgoleone (Figure 6.5). Subsequently, Siame *et al.* (1993) reported that strigol was the major germination stimulant of *S. asiatica* from roots of maize (*Zea mays*) and proso millet (*Panicum milaceum*) and a minor component of the stimulant activity of sorghum. The major stimulant from sorghum was sorgolactone, a compound closely related to strigol (Hauck, Muller and Schildknecht, 1992; Figure 6.5).

When in proximity to a host, germinating seed of *Striga asiatica* differentiate an attachment organ, the haustorium (Figure 6.6). A number of compounds from host plants induce haustorium differentiation and these include the flavonoids, xenogosins A and B, ferulic acid and 2,6-dimethoxy-*p*-benzoquinone, (DMBQ) this last being oxidatively released from the host root by an enzyme of the pathogen (Chang and Lynn, 1986; Figure 6.8).

White rot caused by *Sclerotium cepivorum* is one of the most important diseases of onion. Work summarized by Coley-Smith (1990) and carried out in his laboratory over a number of years has demonstrated that the germination of microsclerotia of the fungus is stimulated by alk(en)yl-1-cysteine sulphoxides from the host.

In most agricultural soils, nutrients are limiting and, as a result, many soil-inhabiting organisms, including pathogenic fungi, are essentially dormant. Dormancy may be broken by the addition of nutrients such as those supplied by germinating seeds and plant roots. Despite the widespread occurrence of this phenomenon, relatively few studies have been made of the compounds responsible (Nelson, 1990).

Stimulants may also play an important role in the establishment of infection by aerial organisms. In particular, pollen and intact anthers are a rich source of nutrient and may enhance the virulence of facultative pathogens. For example, *Fusarium graminearum* causes head blight of wheat but plants are only susceptible when warm, moist weather coincides with anthesis. When test plants, from which

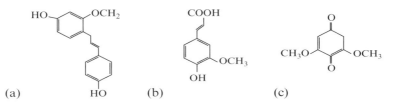

Figure 6.8 Structures of compounds that induce haustorium formation in *Striga* (a) xenogosin A, (b) ferulic acid, and (c) 2,6-dimethoxy-*p*-benzoquinone (DMBQ)

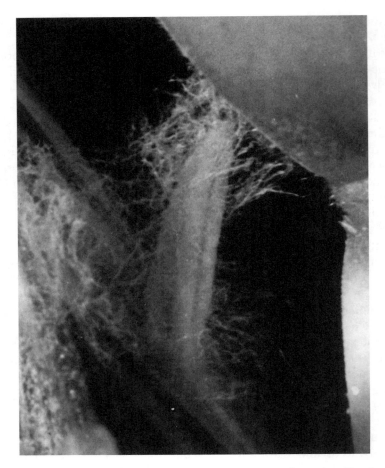

Figure 6.9   Luxuriant growth of *Fusarium graminearum* on an anther protruding from a wheat floret

anthers had been removed, were inoculated with an aerosol of macroconidia of the fungus they remained essentially free of infection, whereas the fungus grew prolifically on the extruded anthers of control plants and caused heavy infection (Strange and Smith, 1971; Figure 6.9).

In a cup plate assay of anther extracts, hyphae from germinating macroconidia of *F. graminearum* were 70 per cent longer than controls. The stimulants responsible were isolated and identified as choline and glycinebetaine with EC_{50} concentrations of 10 and 30 parts per billion, respectively (Strange, Majer and Smith, 1974). Two other *Fusarium* species were stimulated at similar concentrations but not *F. nivale* which has now been reclassified as *Microdochium nivale* (Strange, Deramo and Smith, 1978). When the compounds were applied to slivers of filter paper as substitutes for anthers they promoted infection (Strange, Deramo and Smith, 1978).

The effect of choline and glycinebetaine on *F. graminearum* has been reinvestigated (Wiebe, Robson and Trinci, 1989; Robson *et al.*, 1991, 1992, 1994, 1995). They found that the compounds are transported into hyphae of the fungus by different permeases and are converted into a common active component which increases the hyphal growth unit but not fungal biomass. (The hyphal growth unit is defined as the total length of the hyphae of a colony divided by the number of hyphal tips). In other words, choline and glycinebetaine reduce branching relative to that in control colonies, allowing growth to be concentrated in extension. This could be advantageous to the pathogen since it would increase the probability of rapid entry into the host, thus avoiding desiccation should the weather become dry.

6.6.4   Topographical features

The topologies of plant surfaces provide signals to many fungal pathogens. For example, rust fungi usually enter their hosts through stomata, their topology triggering the development of infection structures. Elegant work by Staples' group (Hoch *et al.*, 1987; Staples and Hoch, 1987) has shown that a simple ridge, $0.5\,\mu\text{m}$ high on a flat surface, was optimal for the differentiation of infection structures of *Uromyces appendiculatus*, the bean rust fungus. A ridge of almost exactly this height ($0.487 \pm 0.07\,\mu\text{m}$) was found on guard cells of the host, *Phaseolus vulgaris*. However, when the germ tube of the rust grew over the ridge it appeared to be flattened, suggesting the application of force (Terhune, Bojko and Hoch, 1993). Moreover, experiments in which hyphae were perturbed with micropipettes demonstrated that the area that was most responsive for appressorium formation was $0-10\,\mu\text{m}$ behind the apex of the germ tube, perturbation more than $40\,\mu\text{m}$ from the apex being entirely ineffective (Correa and Hoch, 1995).

Similar work has been done with 26 other rust species in which four types of behaviour were recognized (Allen *et al.*, 1991). In group 1, which contained the important cereal rusts *Puccinia graminis* f. sp. *tritici* and *P. recondita*, no appressoria were formed on any membrane whether they were smooth or possessed a single ridge, apart from one isolate of *P. recondita* which formed a few appressoria erratically. However, more recent work by Read and co-workers (1997) has shown that *P. graminis* f. sp. *tritici* and *P. hordei* do differentiate infection structures at high frequency (83–86 per cent) over multiple ridges that were closely spaced (optimally $1.5\,\mu\text{m}$ and $2.0\,\mu\text{m}$ high), such topology reflecting the close spacing of cell junctions associated with guard cells of cereals. Moreover, the frequency was increased by the presence in aqueous solution of the leaf alcohol *trans*–2-hexen-1-ol to 88 per cent (Collins, Moerschbacher and Read, 2001). In Allen and co-worker's group 2, which contained *U. appendiculatus*, there was a relatively sharp optimum height for the ridge of about $0.5\,\mu\text{m}$ at which more than 80 per cent of germ tubes differentiated infection structures. A third group consisted of fungi which responded with increased appressorium differentiation to increased height of ridges but there was no decline once the maximum had been reached. Allen and co-worker's final group of rust fungi, exemplified by *Phakopsora pachyrhizi*, a pathogen of soybean, formed appressoria on ridges of all heights as well as smooth polystyrene membranes and silane-treated glass.

For rust fungi which enter via stomata, locating a stoma may be facilitated by responding to other topological signals. For example, germ tubes of *P. graminis* f. sp. *tritici* orient themselves at right angles to leaf veins which, owing to the manner of their distribution, maximizes the chance of the tube encountering a stoma (Figure 6.10).

Fungi that penetrate plants directly also respond to the topography of their hosts. For example, *Rhizoctonia solani* produces infection cushions both on hypocotyls of cotton seedlings and on artificial replicas of hypocotyls (Armentrout and Downer, 1987; Armentrout *et al.*, 1987).

6.7 Adhesion

Fungi which enter their host plants directly without the aid of a wound or a vector, must adhere firmly to them and this is particularly true of those that use mechanical force. Jones and Epstein (1989) found that macroconidia of the squash pathogen, *Nectria haematococca* (*Fusarium solani* f. sp. *cucurbitae*), adhered to polystyrene Petri dishes if they were harvested at 24°C but not if they were harvested at 1°C although the attachment process itself appeared to be temperature insensitive. Adhesion was also inhibited by sodium azide and cycloheximide, suggesting a requirement for respiration and protein synthesis. In

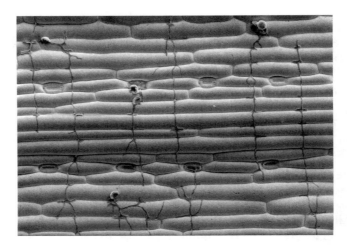

Figure 6.10 Germ tubes of *P. graminis* f. sp. *hordei* orienting themselves at 90° to the grooves on the host surface corresponding to the epidermal cell junctions (courtesy of Nick Read)

experiments with another fungus, *Uromyces appendiculatus*, adhesion of germlings was directly related to the hydrophobicity of the substrate (Terhune and Hoch, 1993).

Kuo and Hoch (1996) have investigated adhesion of the pycnidiospores of *Phyllosticta ampelicida*, the causal agent of black rot of grape. In this species, adhesion was a prerequisite for germination but, in contrast to the work with macroconidia of *Nectria haematococca*, the attachment of spores to poorly wettable surfaces such as polystyrene was not inhibited by sodium azide. Moreover, adhesion occurred in water within a few minutes and in less than 0.03 s if the water was acidified. Both of these observations suggest that, unlike macroconidia of *Nectria haematococca*, attachment is not dependent on metabolic activities of the spores. On wettable surfaces such as nutrient- and water-agars or heat-treated glass the pycnidiospores did not attach firmly and did not germinate.

Adhesion is also a necessity for fungi that employ appressoria for the direct penetration of plants. The chemical nature of the glue that sticks the appressoria of *Magnaporthe grisea* to surfaces has been reported to consist of a mixture of lipids, proteins, sugars and water but unidentified substances made up nearly a quarter of the material investigated (Ebata, Yamamoto and Uchiyama, 1998).

Adhesion is also crucial to the successful parasitism of plants by bacteria. Many bacteria produce fimbriae and they play a role, for example, in the attachment of *Pseudomonas syringae* pv. *phaseolicola* to bean leaves (Romantschuk and Bamford, 1986) as well as *Pseudomonas solancearum* to walls of tobacco leaf cells (Young and Sequeira, 1986). Mutation of the *chvB* locus of *Agrobacterium tumefaciens* gave rise to phenotypes that were non-motile, avirulent and could not attach to host roots. However, motility, virulence and attachment were restored if the mutants were grown in a medium of high osmolarity containing Ca^{++} ions (Swart *et al.*, 1994a, b). Under these conditions, the mutants were able to produce an active 14-kDA outer membrane protein, termed rhicadhesin, which is also important in the attachment of *Rhizobium leguminosarum* bv. *viciae* to pea roots (Broek and Vanderleyden, 1995).

The haustoria of parasitic angiosperms such as *Striga* must also adhere firmly to the roots of their hosts in order to penetrate them and experiments with monoclonal antibodies suggested that pectin, which accumulated between the pathogen and the host root, fulfilled this function (Neumann *et al.*, 1999).

6.8 Breaching the cell wall by mechanical force

Although for many plant pathogens a capacity to breach the cell walls of their hosts is not required for entry since they rely on wounds, natural openings or vectors, many fungal pathogens achieve entry by mechanical force or enzyme activity or a combination of both.

One method fungi have developed for applying considerable pressure on a restricted area is to produce melanized appressoria which adhere tightly to surfaces and within which massive turgor pressures are developed. In the rice-blast pathogen, *Magnaporthe grisea*, germinating conidia differentiate melanized appressoria in response to the physical cues of hydrophobicity and hardness (see Sections 6.6.1 and 6.6.2) as well as chemical signals (Section 6.6.3; Figure 6.11). Synthesis of melanin is crucial since buff mutants, which are defective in melanin production, are non-infective. Melanin is permeable to water but is impermeable to solutes, including glycerol. De Jong and coworkers (1997) found that this sugar alcohol accumulated to molar concentrations in appressoria allowing the generation of osmotic pressures as high as 8.0 MPa (80 bar – roughly 40 times the pressure of a car tyre)!

Restriction enzyme mediated integration (REMI) is a mutagenesis technique that has been used to identify a number of genes involved in pathogenesis, including penetration. Clergeot and co-workers (2001) have described a non-pathogenic mutant of *M. grisea*, named *punchless*, obtained by this technique. The gene

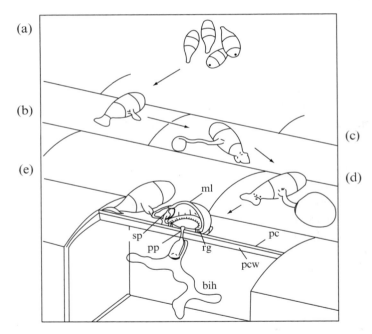

Figure 6.11 Penetration of host leaves by the rice blast fungus, *Magnaporthe grisea*: (a) conidia are disseminated in water drops splashing from a sporulating lesion; (b) a strong glue sticks conidia to the host surface; (c) appressorium differentiates rapidly after conidium germination; (d) mature appressorium is a melanized, dome-shaped and thick-walled cell; (e) penetration peg breaches the plant cuticle and cell wall by mechanical force – after penetration, a bulbous infection hypha invades the epidermal cell (abbreviations: ml, melanin layer; sp, septum; pp, penetration peg; rg, ring of conidial glue; bih, bulbous infection hypha; pc, plant cuticle; pcw, plant cell wall). Reproduced with permission from the Proceedings of the National Academy of Sciences USA

disrupted, *PLS1*, encodes a putative integral membrane protein of 225 amino acids with homology to the tetraspanin family. The authors suggest that *PLS1* is essential for the differentiation of the penetration peg (Figure 6.11).

6.9 Breaching the cell wall by the production of degradative enzymes

A considerable literature has accumulated that implicates degradative enzymes in pathogenesis or virulence. Early work was particularly concerned with pectic enzymes, which are likely to be important not only directly in ingress and destruction of structural materials, but also indirectly as a source of nutrient for the pathogen, since the depolymerization of pectic substances to monomers or oligomers of a low degree of polymerization would be readily assimilated. However, partial depolymerization may give rise to oligomers that function as elicitors of defence reactions (see Chapter 11, Section 11.4.2). More recently, other enzymes such as lipases, cutinases and proteases have been investigated, in some instances with particular reference to the ability of an organism to penetrate its host. A further point for consideration is that some enzymes are able to kill cells.

Before the routine application of molecular biology techniques to plant pathology, the case for the involvement of degradative enzymes in pathogenesis or virulence was usually made on the basis of six criteria set out by Cooper (1983), to which two others have been added more recently:

(1) the ability to produce enzymes *in vitro*,

(2) detection of the enzymes in infected tissue,

(3) depletion of plant material such as the middle lamella,

(4) correlation of enzyme production with virulence or pathogenicity,

(5) reproduction of symptoms of the disease with purified enzymes,

(6) reduction of symptoms *in vivo* when enzyme activity is inhibited (Cooper, 1983).

To these must now be added the following:

(7) fusing the promoter of the gene specifying the enzyme of interest to a reporter gene such as *GUS* (encoding glucuronidase),

(8) genetically engineering alterations in enzyme production and demonstrating corresponding alterations in pathogenicity or virulence. Such alterations

may take the form of gene complementation of a deficient mutant, heterologous gene expression, antisense gene expression and, most directly, gene
disruption.

The first of these points is the weakest. Most fungal and bacterial pathogens
of plants produce enzymes that degrade plant materials *in vitro*, but so do a great
many saprophytic species since they are requirements for metabolizing the dead or
dying vegetation that constitutes most of the carbonaceous material which is
deposited on the soil surface. Furthermore, enzymes produced by an organism in
culture may differ considerably from those produced in the plant. Thus, the simple
demonstration of the production enzymes *in vitro* must be regarded as only an
indication that they may have a role to play *in vivo* or pathogenesis. In the case of
necrotrophs, for example, some enzymes may be required purely for the saprophytic phase of growth. Indeed, the ability of many plant pathogens to secrete a
multiplicity of degradative enzymes has often thwarted their unequivocal demonstration as pathogenicity or virulence factors (see below).

6.9.1 Enzymes that degrade the surface layers of plants: waxes, cutin and suberin

Tariq and Jeffries (1987), on the basis of cytochemistry, invoked the presence of
lipolytic enzymes in the penetration of bean leaf tissues by *Sclerotinia sclerotiorum* and Kunoh and co-workers (1988) showed that spores of *Erysiphe* =
Blumeria graminis germinating on the surface of barley leaves eroded the surface
wax. Nicholson and co-workers (1988) found that the contact stimulus provided
by the leaf or by a cellophane surface caused the release of proteins with esterase
activity which could be resolved into three bands by electrophoresis. More
recently, Hoshino and co-workers (1997) reported the purification and characterization of a lipolytic enzyme that was active at low temperatures from *Typhula
ishikariensis*, a fungus that causes disease of winter grasses and cereals in
Norway, although the role in disease of the enzyme was not established. Some
evidence that lipases may be required for infection of tomato leaves by *Botrytis
cinerea* was obtained by Comménil, Belingheri and Dehorter (1998). They found
that the fungus produced both a cutinase of 18 kDa and a lipase of 60 kDa when
grown on a medium with apple cutin as sole carbon source. Polyclonal antibodies
raised against the lipase did not prevent germination of conidia of the fungus on
leaves of the plant, but the germ tubes failed to penetrate the cuticle.

Similar work with *Alternaria brassicicola* demonstrated the presence of an
80-kDa lipase in the water washings of ungerminated spores which cross-reacted
with anti-lipase antibodies from *Botrytis cinerea*. When the anti-lipase antibodies
were added to a conidial suspension of *A. brassicicola* prior to inoculation,
blackspot lesions were reduced by 90 per cent on intact cauliflower leaves, but
not on leaves from which surface wax had been removed (Berto *et al.*, 1999).

Cutinases are widespread in plant pathogenic fungi and may be constitutive or inducible. For example, Gindro and Pezet (1997) showed that a cutinase was constitutive in the cell walls and cytoplasm of ungerminated spores of *Botrytis cinerea*, whereas Tenhaken *et al.* (1997) found that *Ascochyta rabiei* secreted a cutinase into the culture medium only when cutin or hydroxylated fatty acids were present. Derepression of cutinase in *Fusarium solani* f. sp. *pisi* occurred *in vivo* in response to cutin monomers of the host released by small amounts of the constitutive enzyme in the spore. Secretion of the induced enzyme was then targeted to the infection structure (reviewed in Kolatukudy *et al.*, 1995). When cutinase from this fungus was inhibited either by diisopropyl fluorophosphate or antibody, there was a marked reduction of infection of intact pea stems (Maiti and Kolattukudy, 1979).

Francis, Dewey and Gurr (1996) have demonstrated the importance of cutinases in the pathogenicity of *Erysiphe = Blumeria graminis* for barley. They found that the formation of appressorial germ tubes by this fungus was increased *in vitro* by cutin monomers, the products of cutinases acting on cutin, but inhibited on barley leaves by the esterase inhibitors ebelactone A and ebelactone B. Moreover, application of the ebelactones to barley leaves prevented infection of the host and the development of sporulating lesions.

Genetic studies have been useful in determining the requirement of cutinases for penetration of plants, although here, as with other cell-wall degrading enzymes, the issue is complicated by the ability of plant pathogens to produce multiple isozymes. Dickman and Patil (1986) mutagenized a strain of *Colletotrichum gloeosporioides* pathogenic to papaya (*Papaya carica*), and selected for cutinase⁻ phenotypes in a plate assay. Two cutinase⁻ mutants were obtained which were unable to penetrate intact papaya fruit, but did so if the fruit were wounded or treated with partially purified cutinase. Conversely, Dickman, Padila and Kolattukudy (1989) cloned and characterized a cutinase gene from *Fusarium oxysporum* f.sp. *pisi* and transferred it to a *Mycosphaerella* sp. which caused disease in papaya only if the fruit were wounded. The transformants were able to infect intact papaya fruit but infection was prevented by antibodies raised against the *Fusarium* cutinase (Figure 6.12).

In contrast, disruption of a cutinase gene of *Alternaria brassicicola* reduced its saprophytic growth on cutin but not its pathogenicity (Köller *et al.*, 1995). Apparently, the gene in the wild type was not expressed on leaf surfaces but two other cutinase genes, whose role in pathogenicity has not been determined, were. In a different approach, van Kan and co-workers (1997) obtained evidence for high levels of expression of cutinase A of *Botrytis cinerea* during spore germination and penetration of gerbera flowers and tomato fruits by fusing the promoter of the fungal enzyme to the reporter gene, glucuronidase (GUS) and transforming the fungus with the construct. However, transformants of *Botrytis cinerea* in which the cutinase A gene had been disrupted were unimpaired with regard to penetration and symptom expression in both plants. Clearly, it would be advantageous to use pathogens which only have single copies of cutinase genes for disruption experiments such as strain 77-2-3 of *Fusarium solani* f. sp. *pisi*.

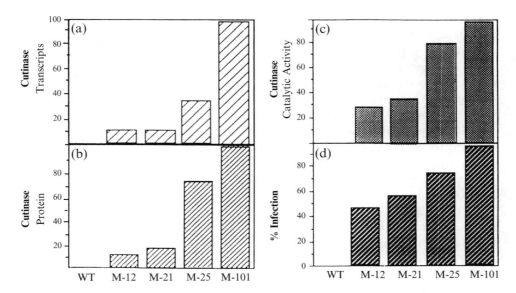

Figure 6.12   Correlation of cutinase transcripts (a), protein (b), and enzymatic activity (c), with infectivity (d) of strains of *Mycosphaerella* sp. transformed with a cutinase gene; WT = wild type and M-12, M-21, M-25 and M-101 are the four transformants. Reproduced by permission of Nature

Unfortunately, there have been conflicting results with this organism. Stahl and Schäfer (1992) disrupted the gene and showed that the mutant retained virulence, suggesting that cutinase did not play a significant role in pathogenesis. However, Rogers, Flaishman and Kolattukudy (1994) using the same strain and a mutant in which the cutinase gene had been disrupted, found that there was a significant decrease in pathogenicity.

Although the degradation of cutin has been extensively studied, less attention has been paid to the more complicated substrate, suberin. Zimmermann and Seemuller (1984) found that *Fusarium solani* and *Armillaria mellea* released enzymes when grown on suberin *in vitro* which degraded it to aliphatic monomers and also cleaved the model substrate, *p*-nitrophenyl butyrate.

6.9.2   Pectolytic enzymes

Pectolytic enzymes are classified according to their catalytic activity and this is complicated by the fact that the substrate is heterogeneous. There are two mechanisms by which the linear chains of pectate or pectin are broken, hydrolysis and β-elimination, the latter giving rise to oligomers which are 4,5 unsaturated at the non-reducing end (Figure 6.13). For pectate and pectin the hydrolytic enzymes are referred to as polygalacturonases and pectinases, respectively whereas for the enzymes that cleave the β-1,4 bonds by β elimination the terms are pectate lyases and pectin lyases, respectively. These names are often prefixed by exo- or endo- according to whether the enzymes attack the polymer chains at the ends giving rise to dimers or in the middle, producing mixtures of oligomers (Figure 6.13). Pectin is de-esterified by the action of pectin methyl esterase (Figure 6.13).

Studies of two pectin lyases from *Aspergillus niger* have shown that they share only 17 per cent sequence homology with pectate lyases but, despite this, some of their structural features, such as amino acid stacks and the asparagine ladder are remarkably similar (Mayans *et al.*, 1997). However, the two enzymes diverge markedly in their substrate-binding clefts and by the fact that calcium is required for pectate lyase activity but not for pectin lyase activity.

In addition and perhaps because of their enzymic activities, endo-polygalacturonases, endo-pectate lyases and endo-pectin lyases kill cells. The simplest explanation for this is that the osmotically sensitive protoplast is rendered more vulnerable by the digestion of the enzymes' substrates (Collmer and Keen, 1986).

Pectic enzymes are regulated by diverse factors. These include the presence of pectin, growth phase, catabolite repression, plant extracts, temperature, anaero-biosis, iron limitation and nitrogen starvation (Hugouvieux-Cotte-Pattat *et al.*, 1996). For example, in soft-rot *Erwinias*, pectate lyase production is increased by plant cell walls, polygalacturonate and oligogalacturonate but repressed by glucose and 4,5 unsaturated digalacturonate concentrations that support bacter-ial growth. Evidence has also been obtained in *Erwinia chrysanthemi* for the role of a cyclic AMP receptor protein in activation of some pectinolysis genes but repression of one of these, the *pelA* gene, important for pathogenicity (Reverchon *et al.*, 1997). Scott-Craig and co-workers (1990) found that a gene from the maize pathogen, *Cochliobolus carbonum*, encoding an endopolygalacturonase was ex-pressed when the fungus was grown on pectin as carbon source but not on sucrose. In contrast, Leone (1990) found that 11 isolates of *Botrytis cinerea* required phosphate for production of the polygalacturonase *in vitro* and gave rise to spreading lesions on French bean and tomato leaves when phosphate and glucose were included with the inoculum droplets.

The induction of pectic enzymes during cell-wall penetration by bacteria is thought to involve the following sequence of events: in response to a basal level of secretion of pectic enzymes acting on the cell walls of the host, oligogalacturonate (particularly dimers) are produced. The oligogalacturonates are taken up by the cell and metabolized via 5-keto-4-deoxyuronate (DKI), 2,5-diketo-3-deoxgluco-nate (DKII) and 2-keto-3-deoxygluconate (KDG) to pyruvate and 3-phospho-glyceraldehyde. In *Erwinia chrysanthemi*, DKI, DKII and KDG interact with the repressor of a regulator gene, *kdgR* which controls the expression of 11 genes involved in pectinolysis. Thus the presence of DKI, DKII and KDG and their interaction with this repressor explains the coordinated expression of these genes (Hugouvieux-Cotte-Pattat *et al.*, 1996).

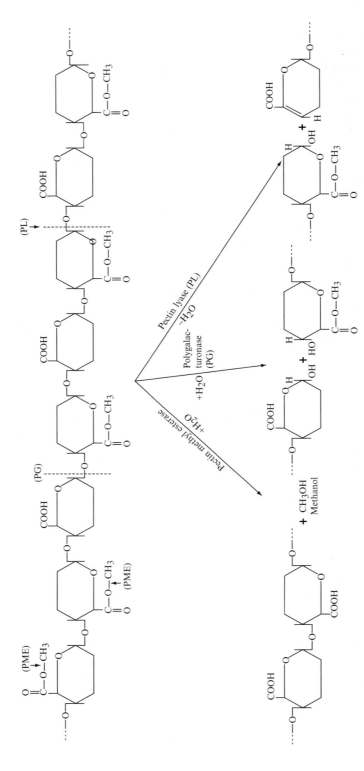

Figure 6.13 Mode of action of pectic enzymes, PME = pectin methyl esterase PG – polygalacturonase, PL = pectate lyase (courtesy of George N Agrios and Academic Press)

Polygalacturonases

Agrobacterium vitis was formerly classified as biovar 3 of *A. tumefaciens*. Unlike most strains of *A. tumefaciens*, which have wide host ranges, *A. vitis* is specific for grapevines. Both tumorigenic and non-tumorigenic strains produce polygalacturonase in culture showing that the gene is chromosomal rather than plasmid borne (McGuire *et al.*, 1991; see Section 7.4) and the enzyme has been isolated from necrotic lesions on grape roots. A strain of the bacterium carrying a Tn5 insertion in the polygalacturonase gene *pehA* multiplied at a reduced rate, produced smaller tumours and did not form necrotic lesions. An explanation of the effect on tumorigenicity may lie in the release of galacturonic acid by the enzyme since this monomer enhances the expression of virulence genes (*vir* genes) in the related *A. tumefaciens*. These data strongly implicate the polygalacturonase as a virulence factor (Rodriguez-Palenzuela, Burr and Collmer, 1991).

Flego and co-workers (1997) have shown that the *pehA* polygalacturonase gene of another plant pathogenic bacterium, *Erwinia carotovora*, is modulated by calcium. Western analysis showed that production of the enzyme was repressed in tobacco plants grown in the presence of 10 mM or 30 mM $CaCl_2$ and that such plants were more resistant to the bacterium. Moreover, when *pehA* expression was put under the control of a calcium insensitive promoter the resistance engendered by elevated calcium levels was lost. Another aspect of these experiments is that pectate lyases (see below) of the same organism are promoted by calcium. The authors explain this paradox by suggesting that polygalacturonase is required early in infection but that once invasion has occurred, its action releases calcium from the plant cell walls. This would repress *pehA* expression but stimulate the activity of pectate lyases which may be required for the next phase of host colonization.

Cryphonectria parasitica, the fungus that destroyed the American chestnut trees (Section 5.1), produced polygalacturonase in culture and caused browning of the inner bark of the tree. A hypovirulent strain produced less polygalacturonase *in vivo* and smaller cankers on American chestnut than the wild type. Further evidence implicating the polygalacturonase as a virulence factor came from comparisons of the susceptible American chestnut and the more resistant Chinese chestnut. The latter developed smaller cankers and lower levels of polygalacturonase were detected in them than in the larger cankers on American chestnut. Possibly a polygalacturonase inhibiting protein (see Chapter 9, Section 9.3.2) was responsible for this difference since a protein extract from Chinese chestnuts was 15 times more inhibitory to the fungal polygalacturonase than a similar extract from American chestnuts (Gao and Shain, 1995).

Yao, Conway and Sams (1996) purified a polygalacturonase from rotted cortical tissue of apples infected with *Penicillium expansum*. Using degenerate primers based on the amino acid sequence of the protein, they amplified a DNA sequence from the fungus, which gave a predicted amino acid sequence that exactly matched that of the purified enzyme. A cloned 212-bp PCR product hybridized with 1.5-kb RNA molecules extracted from apples rotted by *P. expansum* but no transcripts were detected in uninfected apples or in fungal mycelium grown on apple pectin as the sole carbon source. These results demonstrated that the polygalacturonase gene was expressed only in the invaded fruit.

Gene disruption experiments have provided more direct evidence of the role of polygalacturonase enzymes in virulence. Shieh and co-workers (1997) showed that disruption of the gene in *Aspergillus flavus* decreased invasion of cotton bolls by the fungus. Similarly, ten Have and co-workers (1998) eliminated an endo-polygalacturonase (*Bcpg*1) from *Botrytis cinerea* by transformation-mediated gene replacement. Although the mutants were still able to infect tomato leaves and fruits as well as apple fruits there was a significant decrease in growth of the lesions beyond the point of inoculation in the three host tissues. Isshiki and co-workers (2001) demonstrated that a polygalacturonase, *Acpg1*, was probably the only polygalacturonase in *Alternaria citri* and that disruption by integration of an internal fragment of the gene reduced virulence for citrus and reduced its ability to macerate potato tissue. In contrast disruption of the homologous gene in *Alternaria alternata* rough lemon pathotype which was 99.6 per cent similar in nucleotide sequence had no effect on pathogenicity, possibly owing to the fact that this organism produces a host selective toxin (see Chapter 8, Section 8.4.1).

Pectate lyases

The danger of extrapolating results for enzyme activity obtained *in vitro* to those obtaining in the plant is dramatically demonstrated by work with the pectate lyases of *Erwinia chrysanthemi*. A mutant of strain EC16 was produced by directed deletions or insertions in four pectate lyase genes as well as an exopoly-galacturonate lyase and an exo-poly-α-D-polygacturonosidase (products of these last two enzymes are dimers rather than mixtures of oligomers). The mutant did not cause pitting of semi-solid pectate agar medium, a standard test of pectolytic activity in bacteria, but still macerated leaves of chrysanthemum, although less actively than the parent strain. This result was explained by the finding that the mutant produced a second set of enzymes *in planta* and also in minimal media containing chrysanthemum extracts or cell walls as the sole carbon source but not in minimal media containing pectate. Sterile preparations of the enzymes macerated chrysanthemum leaf tissue (Kelemu and Collmer, 1993).

Greater success has attended experiments designed to assess the role of pectate lyase in the fungus, *Colletotrichum gloeosporioides*. Yakoby and co-workers (2001) disrupted the *pel*B gene by homologous recombination and found that decay of avocado fruits was reduced by 36–45 per cent. Also, Rogers and co-workers (2000) showed that disruption of both *pel*A and *pel*D of *Nectria haematococca* drastically reduced virulence of the fungus for pea epicotyls but that disruption of one or other of the genes had no effect. Virulence of the double mutant could be restored by complementation with the *pel*D gene or addition of either of the purified enzymes PLA or PLD.

6.9.3 Cellulases and xylanases

Both cellulose and arabinoxylan are depolymerized by β-1,4-glycanases, usually termed cellulases and xylanases, respectively. Genes encoding these enzymes have conserved regions, which are common to both and occur in discrete domains. It is probable that both types of enzyme arose from progenitor sequences by gene duplication, mutation and domain shuffling. Some show mixed specificity, hydrolysing not only the β-1,4 bonds of cellulose but also those of xylan, chitin and related substrates.

Before the cellulose chains that make up microfibrils can be depolymerized, the microfibrils must be rendered amorphous and it is thought that this is achieved by free-radical attack (Veness and Evans, 1989). After amorphogenesis the cellulose chains are readily depolymerized by cellulases which may attack the β-1,4 linkages within the chains, endocellulases, or at the ends, exocellulases. These normally yield the dimer, cellobiose, which in turn is split by exocellobiohydrolases and β-glucosidases.

Walker, Reeves and Salmond (1994) obtained evidence for the role of a cellulase in the pathogenicity *Erwinia carotovora* subsp. *carotovora* for potato. Using ethylmethylsulfonate (EMS) as a mutagen, they obtained a mutant which was incapable of degrading carboxymethylcellulose and which was significantly impaired in its ability to macerate potato tissue. Western analysis with polyclonal antibodies against the cellulase, CelV, showed that neither the enzyme nor a truncated version of it was produced by the mutant. Complementation of the mutant by cosmids containing *celV* restored synthesis and secretion of the enzyme as well as the ability to macerate potato tuber tissue.

A similar, clear-cut story for the role of cellulases in fungal infections of plants does not seem to have been established. Sposat, Ahn and Walton (1995) disrupted the gene *CEL1* in the maize pathogen, *Cochliobolus carbonum* but pathogenicity was not affected, presumably owing to the ability of the fungus to secrete other cellulases and, indeed, these were detected in culture filtrates of the fungus grown on cellulose or cell walls of maize.

Since arabinoxylans are such a dominant feature of the cell walls of monocotyledonous plants, it is not surprising that production of xylanases is a prominent feature of their pathogens. For example, *Septoria nodorum*, a pathogen of wheat produces more xylanase-degrading enzymes than pectinases and strains of *Erwinia chrysanthemi* which infect maize secrete more xylanase activity than those that are pathogens of dicotyledonous plants. Once again, it has been difficult to prove their involvement in pathogenicity or virulence. Wu, S.C. and co-workers (1997) deleted two xylanase genes of the rice blast fungus, *Magnaporthe grisea*. The double mutant only accumulated half the mycelial mass of the wild type when grown on rice cell walls or xylan as sole carbon source but strains carrying mutations at one or other or both loci were as virulent as the parent. Analysis of the culture filtrate of the double mutant revealed four further xylanases.

As with enzymes that depolymerize pectin, xylanases are also capable of killing plant cells (Ishii, 1988).

6.9.4   Enzymes involved in the degradation of lignin

The complexity of lignin (Figure 6.3) has hampered elucidation of its degradation. Among the best-studied systems are those from two white rot fungi, *Phanerochaete chrysosporium* and *Trametes versicolor*. Both produce lignin peroxidase, a heme peroxidase which causes the oxidative depolymerization of lignin, but other components are also important such as manganese-dependent peroxidases, laccases, H_2O_2 generating enzymes, veratryl alcohol, lignin and manganese.

The regulation of enzymes that degrade lignin and their role in pathogenicity does not seem to have attracted a great deal of attention. Work *in vitro* has shown that the heme peroxidase of *Phanerochaete chrysosporium* is triggered by nitrogen starvation (Tien and Kirk, 1983) and lignolytic activity is enhanced by the lignin degradation product, veratryl alcohol (Leesola *et al.*, 1984). More recently, tryptophan was found to be highly stimulatory to lignin peroxidase activity in cultures of four white rot fungi (Collins *et al.*, 1997). The authors propose that tryptophan stabilizes lignin peroxidase by acting as a reductant for the enzyme. Dutton and Evans (1996) have suggested that in white rot fungi, oxalate acts as a potential electron donor for lignin-peroxidase reduction and chelates manganese. It is proposed that the latter activity would allow dissolution of the manganese cation from a manganese-enzyme complex causing the stimulation of extracellular manganese peroxidase activity.

Schultz and Nicholas (2000) have pointed out that a satisfactory explanation as to why angiosperm sapwood but not heartwood or gymnosperm wood is usually degraded by white-rot fungi has never been given. They suggest that the phenolics present in lignin and heartwood extractives act as free radical scavengers and may retard white-rot fungi, whereas these fungi may rapidly colonize angiosperm sapwood as it has a relatively low free-phenolic content.

6.9.5   Proteases

Despite the fact that proteins may make up to 15 per cent of the cell wall of some plants (Section 6.3), proteases produced by pathogens and their role as pathogenicity or virulence factors has received little attention. Movahedi and Heale (1990a,b) found an aspartic protease secreted by *Botrytis cinerea* both in culture and in infected carrots. When spores of *B. cinerea* were treated with pepstatin, a specific inhibitor of aspartic protease, there was a marked reduction in symptoms, not only when carrot was challenged but also strawberry, raspberry, cabbage and broad bean. Pagel and Heitefuss (1990) showed that several degradative enzymes appeared sequentially in potato tubers infected by *Erwinia carotovora* subsp. *atroseptica* including a protease which was detected 19 h after inoculation.

Dow and co-workers (1990) demonstrated that *Xanthomonas campestris* pv. *campestris* produced two proteases in culture, a serine protease and a zinc-requiring protease and these accounted for almost all the proteolytic activity of the wild-type organism. A mutant that lacked both proteases was less virulent than the wild type when introduced into the cut vein endings of turnip leaves.

Ball and co-workers (1991) obtained genetic evidence for the requirement of an extracellular protease in the pathogenicity of the fungus *Pyrenopeziza brassicae* for oilseed rape. An ultraviolet-induced mutant which was non-pathogenic and also deficient in extracellular protease production was transformed with clones from a genomic library of *P. brassicae*. Both pathogenicity and protease activity were restored by a transformant with a single cosmid insert.

More recently, Carlile and co-workers (2000) have reviewed the production of proteases by plant-pathogenic fungi and bacteria and have reported a cell-wall degrading trypsin, SNP1, produced during infection by *Stagnospora nodorum* (=*Septoria nodorum*). In this thorough piece of work, these workers showed that the fungus produced a protease when grown on wheat cell walls as the sole source of carbon and nitrogen, purified it and demonstrated that the pure protein degraded wheat cell walls, releasing hydroxyproline-rich proteins. This activity was inhibited by the trypsin inhibitors, aprotinin and leupeptin. Sequencing of the N-terminal region of the protein showed that it shared 89.5 per cent identity with a trypsin protease from *Fusarium oxysporum* and 84.2 per cent identity with ALP1 from *Cochliobolus carbonum*. The presence of the protease *in planta* was demonstrated by the presence of protease activity which co-eluted from a cation exchange resin with SNP1, by expression of the encoding gene in Northern blots and by the expression of green fluorescent protein when this was fused to the SNP1 promoter. It will be interesting to see what effect the deletion of this gene, as intended by the authors, has on the virulence of the fungus.

6.9.6 Membranlytic enzymes

In animal pathology, enzymes that catabolize lecithin are recognized as toxins but there have been few comparable studies in plant pathology. Tariq and Jeffries (1987), on the basis of cytochemistry, invoked the presence of lipolytic enzymes in the penetration of bean leaf tissues by *Sclerotinia sclerotiorum*. However, more attention has been paid to the alteration of membranes in connection with plant defence as part of the hypersensitive response (Chapter 11, Section 11.3.1) and to membrane dysfunction caused by toxins (Chapter 8, Section 8.3).

6.10 Global regulation of degradative enzymes

Several regulatory genes in plant pathogenic bacteria have been discovered by random insertional mutagenesis. For example, *aep* genes which *a*ctivate *e*xtracellular *p*rotein production activate the production of pectate lyase,

polygalacturonase, cellulase and protease in *Erwinia*. A regulatory mutant of
E. carotovora subsp. *carotovora*, created by the insertion of the sequence Mud1,
was less virulent than the wild type and was deficient in the production of
cellulase and protease but produced normal amounts of pectate lyase and polygalacturonase (Frederick *et al.*, 1997). The regulatory sequence was designated
rpfA for regulator of pathogenicity factors and sequencing showed high identity
of the predicted protein, a putative kinase, with two-component sensor-regulator
proteins from six other bacterial species. Similar work with *Xanthomonas campestris* pv. *campestris*, a pathogen of crucifers, was found to contain eight such
regulatory sequences designated *rpfA-H*. In mutants at two of these loci, *rpfB* and
rpfF, protease, endoglucanase and pectate lyase was restored by a small but as yet
unidentified diffusible compound (Barber *et al.*, 1997). The authors postulate
that the diffusible compound is synthesized under conditions of starvation such
as might occur in the xylem sap, the main habitat of the bacterium, after the
bacterial population had reached a certain density. This would then allow the
production of enzymes which would degrade plant macromolecules to substrates
that could be used by the bacterium.

In other gram negative bacteria such as *Pseudomonas aeruginosa* and *Erwinia
carotovora* (pathogens of animals and plants, respectively) mutants were also
found that were deficient in the production of enzymes required for pathogenicity, elastase for *P. aeruginosa* and pectinase, cellulase and protease for
E. carotovora. One class of mutants in each organism was found to be deficient
in production of the autoinducer N-3-(oxohexanoyl)-L-homoserine lactone
(HSL; Jones *et al.*, 1993). Addition of HSL to the mutants of *E. carotovora*
restored enzyme production and virulence.

Factors governing secretion of cell-wall degrading enzymes are important for
pathogenicity. Proteases are secreted by a one step, Sec-independent pathway,
whereas pectinases and cellulases are secreted by a two-step mechanism. The
enzymes cross the inner membrane by a Sec-dependent pathway but their passage
across the outer membrane is effected by proteins encoded in the *out* cluster
(Reeves *et al.*, 1993).

6.11   Nutrition of the pathogen

Although cell-wall degrading enzymes have been described above mainly in
relation to their ability to allow ingress of pathogens into plants, the products
of their activity are likely to be significant in terms of nutrition of the pathogen,
particularly when these are necrotrophs. Biotrophs, on the other hand, obtain
their nutrition by more subtle means. The enhancement of cytokinin levels in
their vicinity and the consequent mobilization of nutrients towards these sinks
are discussed in Chapter 7. Here we shall be concerned with the role of a
specialized organ, the haustorium, in obtaining nutrients from the host.

Haustoria in biotrophic fungi differ from the rather gross structures of the same name in angiosperm pathogens such as *Striga* (Figure 6.6). In the rusts they are knob shaped and are found penetrating through the cell wall of the host, invaginating the plasmamembrane but not penetrating it. They have long been suspected to be the organ by which the pathogen obtains its nutrients (Staples, 2001). Excellent evidence for this view has been provided by Mendgen's group. They found that more than 30 plant-induced genes were expressed preferentially in haustoria of the bean rust fungus (Hahn and Mendgen, 1997). One of these was identified as a putative amino acid transporter (Hahn *et al.*, 1997) and another as a hexose transporter (Voegele *et al.*, 2001). The hexose transporter is a single copy gene and was shown by immunocytochemisry to be expressed almost exclusively in the haustorium.

6.12 Movement of viruses through the plant

Viruses lack any means of self-locomotion, yet they move within and between plant cells and may become systemic throughout the plant. The mechanisms by which these movements are achieved are still largely speculative but a fundamental necessity is the possession by the virus of a functional movement protein. Such proteins have been identified in most families of plant viruses and complementation of a virus or a virus strain which is movement defective in a given host can occur with an unrelated virus which is movement competent (Carrington *et al.*, 1996). For example, potato leafroll luteovirus is normally restricted to phloem cells but when the plant is coinfected with potato virus Y it is able to move into mesophyll cells (Barker, 1987).

Viruses move from cell to cell through plasmodesmata but they may be replicated at sites that are some distance away. For tobacco mosaic virus (TMV) it is proposed that newly synthesized viral RNA complexes with the movement protein and is transported by microtubules and microfilaments to plasmodesmata. The size exclusion limit of normal plasmodesmata for the passive diffusion of molecules is less than 1 kD. This has to be increased to allow passage of virus particles and the increase is thought to be effected by the movement protein. Evidence supporting this notion was obtained from transgenic plants expressing the movement protein of TMV in which the size exclusion limit of plasmodesmata was about 10-fold higher than controls. There is now evidence that TMV movement protein associates with pectin methyl esterase and it is thought that this association facilitates viral movement, although the mechanisms remains speculative (Tzfira *et al.*, 2000).

Long-distance movement of viruses occurs through the phloem or xylem and normally requires an intact capsid protein. Once in the conducting tissues of the plant, movement of the virus and unloading follows that of solutes (Roberts *et al.*, 1997) but the mechanisms remain obscure (Tzfira *et al.*, 2000). Opalka and co-workers (1998) have investigated rice mottle yellow virus, an important pathogen of rice which causes considerable economic loss in East and West Africa. They showed that the virus becomes systemic by partial digestion of pit membranes and suggest that this results from programmed cell death and concomitant autolysis.

7 Subverting the Metabolism of the Host

Summary

Many of the symptoms of plant disease can be attributed to degradative enzymes or toxins but others are more indicative of hormone imbalance and these are usually associated with biotrophic rather than necrotrophic pathogens. The five 'classical hormones' and brassinosteroids exert multiple effects on the physiology and growth of plants and these are briefly described in order to provide a background to a discussion of their roles in disease. Enhanced concentrations of auxins and cytokinins have been demonstrated in several instances in which hypertrophy is a symptom. In addition, local increases in cytokinin concentrations caused by pathogens may give rise to green islands and result in the redirection of nutrients. Stunting in some instances can be attributed to reduced concentrations of gibberellins while in others enhanced concentrations of abscisic acid may be responsible. Ethylene has many effects on plants which have also been associated with infection such as epinasty, abscission of organs, chlorosis and necrosis as well as the promotion of certain defence responses. The brassinosteroids also cause diverse responses and some, such as ethylene production, proton pump activation and increased rate of stem elongation, may be involved in symptom expression. Peptide hormones have been discovered in plants relatively recently, but apart from systemin, an 18-amino acid peptide which is produced in response to wounding and induces the systemic production of protease inhibitors, little is known of their role in pathology.

Agrobacterium tumefaciens causes tumorous growths on a wide range of dicotyledonous plants. During infection, part of a plasmid is transferred to the plant and is stably incorporated into the plant genome. The integrated DNA (T-DNA) contains genes for auxin and cytokinin synthesis and their presence explains the ability of explants of tumours to grow in tissue culture without auxin or cytokinin supplements. T-DNA also contains genes for the synthesis of arginine derivatives, known as opines, which may be catabolized by the bacterium but are unavailable to the plant.

Introduction to Plant Pathology by Richard Strange
© 2003 John Wiley & Sons, Ltd   ISBN 0 470 84972 X (cased)   ISBN 0 470 84973 8 (pbk)

7.1 Introduction

The necrotic lesions caused by many pathogens may often be attributed to the secretion of enzymes or toxins by the pathogen. Enzymes have already been discussed (Chapter 6) and toxins will be reviewed in Chapter 8. In contrast, biotrophic pathogens cause little necrosis and initially this is true also of hemi-biotrophs – pathogens that have a period of 'peaceful co-existence' with their hosts immediately after infection. Although in some instances no symptoms are obvious, in others there may be quite gross effects such as abnormal growth, redirection of nutrients, stunting, epinasty, premature senescence and premature abscission of leaves and other organs. Many of these symptoms are mediated by altered concentrations of the five 'classical' plant hormones: auxins, cytokinins, gibberellins, ethylene and abscisic acid (Kende and Zeevaart, 1997). However, the extent to which they contribute to parasitism is debated although it does seem probable that the diversion of nutrients towards the lesions of some pathogens by enhanced concentrations of cytokinins has a role to play (Section 7.3.2).

More recently, oligosaccharins, brassinosteroids and jasmonates have been recognized as non-traditional regulators of plant growth, development and gene expression (Creelman and Mullet, 1997). Apart from jasmonates, which are involved in signal transduction in defence (Chapter 11, Sections 11.6.4 and 11.6.6), the role of these in symptom expression has been little studied. Polypeptides are now also known to be involved in signalling in plants. Initial work was concentrated on the 18-amino acid peptide systemin which was identified in 1991 as the signal for the systemic activation of defence genes in wounded tomato plants. A systemin receptor was isolated from tomato in 2002 and shown to be a member of the leucine rich repeat receptor kinase family (Scheer and Ryan, 2002). Other peptides include RALF which inhibits root growth and caused the rapid alkalinization of tobacco suspension cultures, a feature of the action of the biotic elicitor ergosterol (Chapter 11, Section 11.4.2; Pearce et al., 2001). With the exception of systemin it is too early to comment on the possible roles of these peptides in the pathogen-challenged plant but it would seem that exciting times are around the corner!

In this chapter, the biosynthesis and mechanisms of action of the 'classical' plant hormones as well as jasmonates and brassinosteroids in the healthy plant, as far as they are known, are first briefly described. Disease symptoms, which may be explained in terms of altered concentrations of these hormones, are then discussed and some mention is made of the action of systemin. Finally, crown gall and hairy root diseases are considered. In both diseases altered levels of auxins and cytokinins are well established since genes for their biosynthesis are transferred from their respective pathogens to the genomes of their hosts.

7.2   Biochemistry and mechanism of action of hormones in the healthy plant

Plant hormones have usually been discovered by investigating the reasons for the normal or abnormal stimulation of growth of whole plants or the failure of tissue explants to grow in culture without their inclusion in the media. For example, investigation of the abnormal tallness of rice plants suffering from the 'bakanae' or foolish disease led to the discovery of the gibberellins, first as secondary products of the infecting fungus, *Gibberella fujikuroi*, and subsequently as a large family of compounds found in uninfected plants (Section 7.2.3). Discovery of cytokinins resulted from the fractionation of undefined supplements added to defined media that allowed tissue such as tobacco explants to undergo cell division and differentiation (Section 7.2.2). Isolation of hormones is difficult as the compounds are generally present in low concentration and the bioassays used to detect them often long and laborious.

With the 'classical' hormones, research has moved on to the investigation of the means by which they bring about their effects. There has been some success in the search for receptors but, in general, the signal transduction pathway from hormone binding to effect is obscure. Part of the reason for this is that the hormones often interact with each other making the disentangling of cause and effect fraught.

7.2.1   Auxins

Indole acetic acid (IAA) is the principal auxin of higher plants. It is synthesized from tryptophan but three pathways have been proposed: via indole-3-pyruvic acid, tryptamine or indole-3-acetonitrile, the last precursor being found mainly in the Cruciferae and may be derived from indoleglucosinolates (Figure 7.1.). Recently, a tryptophan independent pathway has been described in plants (Taiz and Zeiger, 2002). Plant pathogenic bacteria such as *Agrobacterium tumefaciens*, however, use a further pathway in which tryptophan is converted to indole-3-acetamide from which IAA is released by the action of indole-3-acetamide hydrolase (Figure 7.1; Bartel, 1997). IAA may also be degraded by at least three pathways: decarboxylation, which is thought to play a minor role, and conversion to oxindole-3-acetic acid, either directly or via an aspartate conjugate (Taiz and Zeiger, 2002; Figure 7.2). In addition, the concentration of IAA in a plant may be regulated by the formation and hydrolysis of conjugates, transport and compartmentation, possibly in the vacuole. IAA that is synthesized in mature leaves may be transported in a non-polar fashion by the phloem. In contrast, in coleoptiles and vegetative shoots IAA transport is polar and basipetal (away from the shoot tip) whereas in roots it is acropetal (towards the root tip).

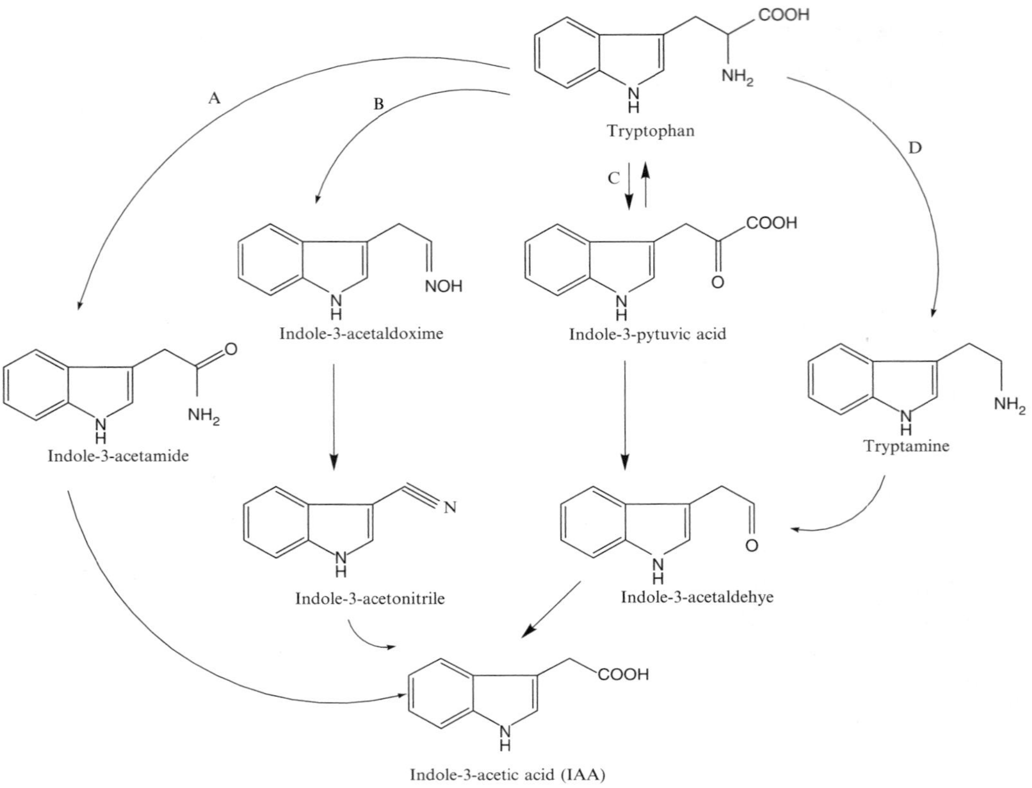

Figure 7.1   Biosynthesis of indole acetic acid (IAA): (a) bacterial pathway; (b), (c) and (d) alternative plant pathways

Auxins have a multiplicity of effects on plants: they induce cell elongation in stems and coleoptiles, increase the extensibility of the cell wall, mediate phototropism and gravitropism, promote the formation of lateral roots, delay leaf abscission and regulate fruit development. How they achieve all these effects, however, is by no means clear. One promising approach to solving this problem is to use the technique of differential display of mRNA. By comparing the expression of mRNA in auxin-treated tissue with that in controls, mRNA species that are up- or down-regulated may be identified. For example, Roux and co-workers (1998) found that four mRNAs were up-regulated in tobacco seedlings. However, only one of them showed homology with a previously described sequence. More recently microarray technology has been used and has shown a vast array of genes is regulated by application of IAA to *Arabidopsis thaliana*.

Several auxin-binding proteins (ABPs) have been identified. For example, a putative receptor protein was purified by affinity chromatography (Löbler and Klämbt, 1985a, b). The receptor was antigenic, allowing antibodies to be

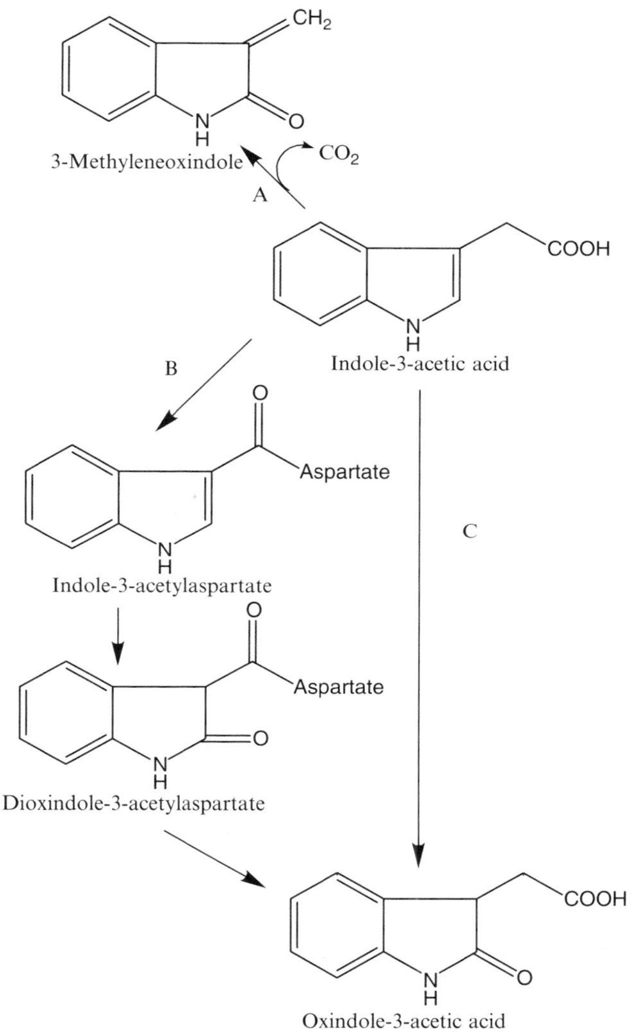

Figure 7.2 Biodegradation of indole acetic acid (IAA): pathways (A) in which IAA is degraded by the peroxidase route is a minor route; pathways (B) and (C) are the most common

prepared for immunocytochemistry and physiological studies. These showed that the protein was located on the outer epidermis of maize coleoptiles and auxin-induced growth was inhibited by the antibody. ABP1 of maize has been expressed in insect cells, purified and crystallized (Woo *et al.*, 2000). It is a high-affinity auxin receptor and is involved with cell expansion. However, cell division is stimulated by high auxin concentrations and this may be mediated by another ABP with a lower affinity (Chen, 2001).

7.2.2 Cytokinins

Plant tissue cultures were initially established on rich and undefined media. Eventually, as a result of work in Skoog's laboratory, a defined medium was discovered which contained macro- and micro-nutrients and vitamins to which IAA and cytokinin had to be added in order to obtain tissue proliferation. An artificial cytokinin, kinetin (from autoclaved herring sperm DNA!) was the first to be discovered and it was not until several years later that a natural cytokinin, zeatin, was isolated from immature endosperm of maize (Miller *et al.*, 1955; Letham, 1973; Figure 7.3.). The compound has since been found to be the most prevalent cytokinin in higher plants.

Cytokinins have been reviewed frequently, e.g. Mok and Mok (2001). They have been defined as compounds 'which, in the presence of optimal auxin, induce cell division in tobacco pith or similar tissue cultures' (Letham and Palni, 1983). Chemically, they are adenine derivatives that are substituted at the N^6 position. An isopentenyl transferase, which catalyses the addition of an isopentenyl group to adenosine monophosphate (AMP) is therefore a key enzyme in cytokinin synthesis (Figure 7.4.). Other cytokinins such as zeatin and dihydrozeatin are formed by modification of the isopentenyl group.

Although low concentrations of cytokinins exist free or as ribonucleotides or glucosides, they are also often found in tRNA as the base adjacent to the anticodon. The prenyl transferase that catalyses their synthesis in this position is thought to differ from the one catalysing the synthesis of free cytokinins.

Cytokinins are synthesized in the root and they are transported as their ribonucleotides to the shoot via the xylem. Their concentration may be regulated by formation of N-conjugates with glucose at the 7- or 9-positions, with alanine at the N-9 position or by cytokinin oxidases which remove the isopentenyl side chains from a number of cytokinins.

As with other plant growth regulators, cytokinins affect many facets of plant metabolism. Those of particular relevance to symptom expression are the delay of senescence, redirection of nutrients and the proliferation of plant organs. They also promote the maturation of chloroplasts and stimulate cell enlargement. Their mechanism of action is still unknown although they affect

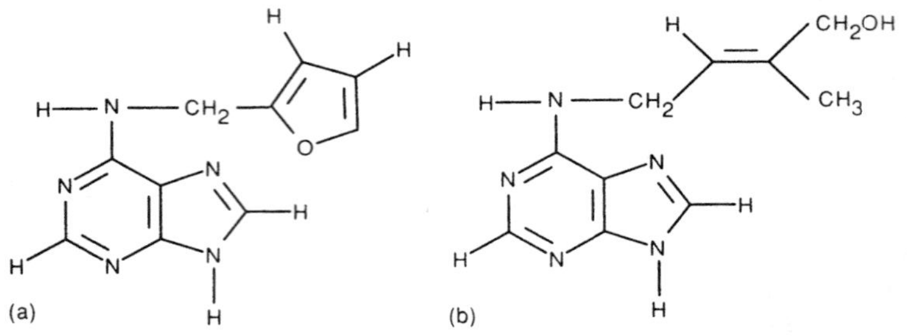

Figure 7.3 (a) Kinetin, and (b) *trans*-zeatin, two compounds with cytokinin activity

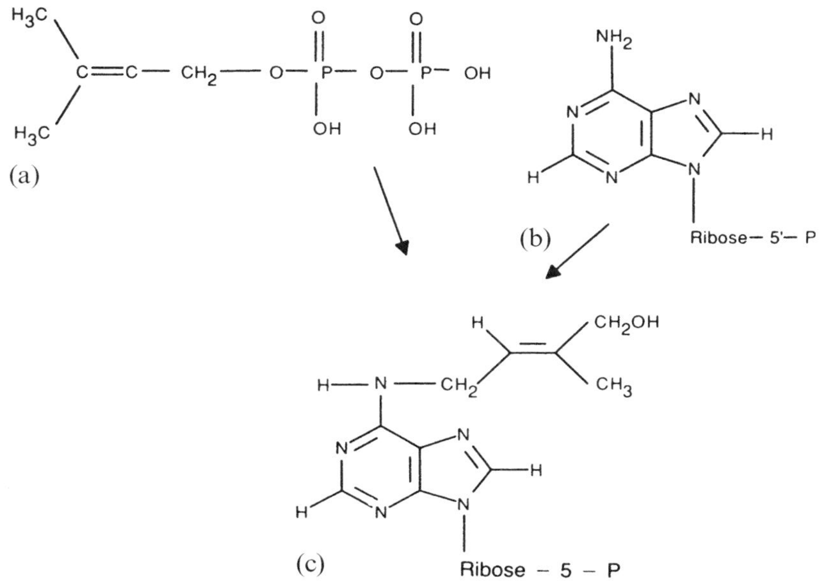

Figure 7.4 The biosynthesis of cytokinins: the key step in cytokinin synthesis is the transfer of an isopentenyl group from isopentenyl pyrophosphate (a) to the N^6 atom of an adenosine moiety and is catalysed by an isopentenyl transferase gene, *ipt*; in bacteria the adenosine moiety is adenosine monophosphate (AMP) shown in (b) but in plants it may be adenosine diphosphate (ADP) or adenosine triphosphate (ATP) to give (c) – this is readily converted to zeatin (Figure 7.3(b))

protein synthesis, possibly by stabilizing specific mRNAs. Cytokinin-binding proteins have been identified and these are likely to represent the start of one or more signal transduction pathways.

7.2.3 Gibberellins

Gibberellins are a large family of tetracylic diterpenoid growth factors currently numbering well over 100 (Richards *et al.*, 2001 and see http:// www.plant-hormones.bbsrc.ac.uk). However, only a few of these are biologically active, the presence of a 3β-hydroxyl group being necessary for activity. The others are precursors or deactivated products. Gibberellins are synthesized by the mevalonic acid pathway, the first committed step being *ent*-kaurene (Figure 7.5.). They are deactivated by 2β-hydroxylation. When this reaction is blocked, as it is in the pea mutant *slender*, GA_{20} accumulates in maturing seed and the plant has a spindly phenotype (Ross *et al.*, 1995).

As with other growth regulators, gibberellins, have multiple effects on plants. However, the one that first attracted attention was the promotion of stem

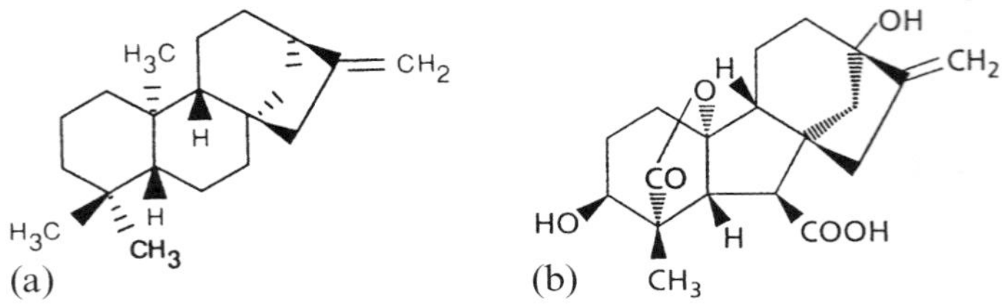

Figure 7.5 (a) *ent*-kaurene, the first committed step in gibberellin biosynthesis; (b) gibberellic acid (GA₁), the biologically active gibberellin that regulates stem elongation

elongation. Rice plants infected by *Gibberella fujikuroi* grew tall and often lodged. They were therefore said to be suffering from the bakanae or 'foolish' disease. The fungus produced the growth factor in culture and Japanese scientists obtained impure crystals of the compound in the 1930s but it was not until 20 years later that plants were found to have their own endogenous gibberellins. In addition to promoting stem elongation, which results from both cell elongation and increases in cell division, they also induce bolting in long day plants, cause some plants to enter the reproductive phase early or conversely cause others to revert to a juvenile stage. Other effects are the induction of maleness in flowers, promotion of fruit set and growth and stimulation of hydrolases. One hydrolase in particular is well documented, the α-amylase in the aleurone layers of germinating cereals, and this is commonly used as an assay in gibberellin work (Taiz and Zeiger, 1991). Gibberellins also promote trichome formation in *Arabidopsis* (Perazza, Vachon and Herzog, 1998).

Although a gibberellin receptor has not been isolated, pharmacological and mutant data support the view that the early events in signalling may involve G-protein-coupled receptors and/or α-subunits of G-proteins (Richards *et al.*, 2001).

7.2.4 Ethylene

The first effect of ethylene on plants was noticed by the 17-year old graduate student Dimitry Neljubow in 1886 as a result of experiments with etiolated pea seedlings. He showed that a component of the gas used for illumination was responsible for the seedlings growing horizontally in the laboratory air but vertically outside and identified it as ethylene. However, it was not until 1959 that it was recognized as an endogenous plant growth regulator (Burg and Thimann, 1959). Plants synthesize ethylene from methionine via S-adenosylmethionine (SAM) and 1-aminocyclopropane-1-carboxylic acid (ACC, Figure 7.6.). ACC synthase is encoded by a multigene family whose members are differentially expressed in response to developmental, environmental and hormonal factors (Kende and Zeevart, 1997). ACC is converted to ethylene by ACC oxidase, an enzyme that is also encoded by a multigene family whose members are differentially expressed according to the organ of the plant and the

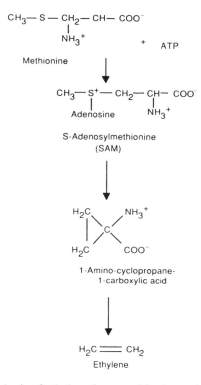

Figure 7.6 The biosynthesis of ethylene from methionine and adenosine triphosphate

stage of development. An alternative pathway for ethylene synthesis which is independent of ACC and is mediated by free radicals has been found in some plants (Chen, Tabner and Wellburn, 1990).

Concentrations of ethylene in the plant may be down-regulated by irreversible conjugation of its precursor, ACC, with malonate to form *N*-malonyl-ACC and by the conversion of ethylene itself to ethylene oxide and ethylene glycol. However, since ethylene is a gas its concentration in the plant is constantly reduced by diffusion and therefore metabolism is not necessary for its removal.

Ethylene receptors are histidine protein kinases. They are related to bacterial two-component regulators and in both *Arabidopsis* and tomato are encoded by a family of genes. Alteration in levels of the receptors is reflected in modification of sensitivity of a plant or tissue to ethylene (Ciardi and Klee, 2001). For example, Tieman and co-workers (2000) found that the ethylene receptor LeETR4 in tomato was a negative regulator of signal transduction since antisense plants for the gene encoding the receptor were more sensitive to ethylene treatment, displaying severe epinasty, enhanced flower senescence, and accelerated fruit ripening.

Plants are affected by ethylene in numerous ways. These include fruit ripening, the abscission of fruits, leaves and flowers, epinasty, alteration of seedling growth patterns, breaking of dormancy of seeds and buds, promotion or inhibition of stem elongation depending on the test species, induction of roots, inhibition of flowering (although this is promoted in pineapple and mango) and hastening of the onset of senescence in flowers and leaves. How these multiple effects are brought about is far from fully understood. The issue is further complicated by the interaction of ethylene with other hormones. For example, ethylene causes a rapid reduction in concentration of abscisic acid (ABA), while ABA is a potent antagonist of GA. Thus, the growth rate of the plant is regulated by the balance of ABA and GA, the balance itself being regulated by ethylene. Another significant observation concerns the effect of ethylene on the peel of citrus fruit. Ethylene is often used commercially to 'degreen' citrus fruit and it appears to do so by the induction of the synthesis of an enzyme, chlorophyllase, which destroys chlorophyll (Trebitsh, Goldschmidt and Riov, 1993). However, an unfortunate side-effect of ethylene treatment is that it enhances citrus stem-end rot caused by *Diplodia natalensis*, possibly by a direct stimulatory effect on the pathogen (Brown and Lee, 1993).

7.2.5 Abscisic acid (ABA)

ABA was discovered as a result of two independent investigations, one into a substance from immature cotton fruit which promoted abscission (Ohkuma *et al.*, 1963) and the other into a compound from sycamore leaves which sent buds into dormancy. Both phenomena were caused by the same compound which was named abscisic acid. Ironically, it is now recognized that ABA does not induce abscission, nor is it likely to be responsible for dormancy in resting buds of deciduous trees or potato sprouts. ABA is derived from carotenoids via xanthoxal and its biosynthesis increases in response to water stress (Figure 7.7). It is deactivated by oxidation to phaseic acid and by the formation of a glucose conjugate (Figure 7.7).

ABA has been described as an essential mediator in the triggering of plant responses to drought stress. One of the best documented of these is the closure of stomata which occurs within minutes of exogenous application of ABA. Conversely, enhanced water loss and an increased tendency to wilt are characteristics of ABA-biosynthetic mutants and are indicative of impaired stomatal regulation (Merlot and Giraudat, 1997). ABA also down-regulates pathogenesis-related β-1,3-glucanases (Rezzonico *et al.*, 1998). This finding is particularly intriguing since it provides a rationale for the frequent observation that drought-stressed plants are more susceptible to disease than controls. Other physiological responses include inhibition of auxin-induced growth and premature senescence. A start has been made in determining how all these diverse effects are mediated, principally using the model plant *Arabidopsis*, but much remains to be done (Merlot and Giraudat, 1997).

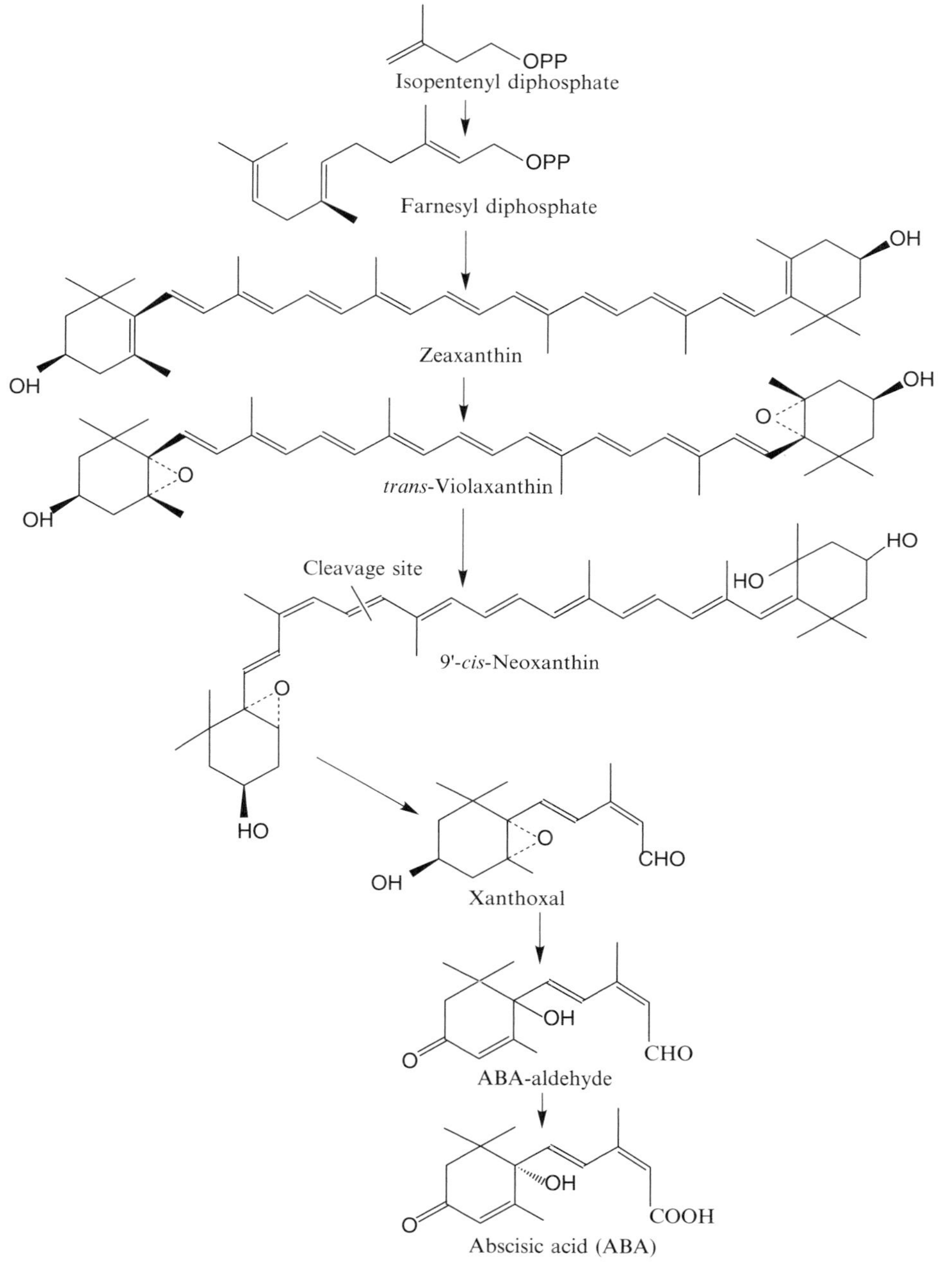

Figure 7.7   Absisic acid synthesis and metabolism

Figure 7.7   *continued*

7.2.6 Jasmonates

Jasmonic acid (JA) was first isolated in 1971 as a plant growth inhibitor from culture filtrates of the fungus, *Lasiodiplodia theobromae*, a plant pathogen with a wide host range. Since then it has become recognized as only one of a large group of biosynthetically and structurally related compounds known as the jasmonates or, since they are derived from the octadecanoid fatty acid, α-linolenic acid, the octadecanoids (Weiler, 1997). Increases in jasmonates may result from the release of linolenic acid from membranes by phospholipases. This is converted to 13-hydroperoxyl linolenic acid by lipoxygenase. Further reactions result in the formation of 12-oxophytodienoic acid, which is converted to jasmonic acid by a reduction step and three rounds of β-oxidation (Figure 7.8).

As with other regulators of plant growth and metabolism the effects of jasmonates are many and include the inhibition of germination of non-dormant seeds and root growth and the stimulation of the ripening of tomato and apple fruit.

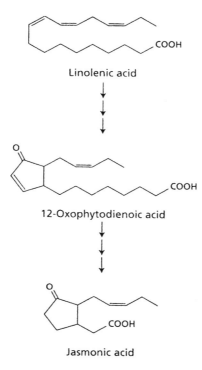

Figure 7.8 Synthesis of jasmonic acid from linolenic acid

They are also thought to be important in the formation of flowers, fruit and seed as their concentrations are relatively high in these organs (Creelman and Mullet, 1997). However, direct evidence is accumulating for fundamental roles of jasmonates in the signalling of plant defence (Chapter 11, Section 11.6.4). Also, indirect evidence has emerged from studies with the toxin coronatine, a structural analogue of jasmonic acid, which interferes with defence (Chapter 8, Section 8.5.1).

7.2.7 Brassinosteroids

The rapid germination of pollen grains and extension of their pollen tubes suggested that a growth promoter might be involved. This was confirmed when a crude extract of pollen from *Brassica napus* was shown to induce rapid elongation of internodes of pinto bean. The first brassinosteroid to be isolated was brassinolide and it still ranks as the most biologically active although, since the 1980s, a family of more that 40 related compounds has been identified. Brassinosteroids cause diverse responses when applied to plants; these include ethylene production, proton pump activation, reorientation of cellulose microfibrils, xylogenesis, pollen tube growth, increased rate of stem elongation and unrolling and bending at the sheath joint of grass leaf blades.

A gene encoding the brassinosteroid receptor BRI1 has been cloned from *Arabidopsis thaliana*. It is a transmembrane serine/threonine kinase with an extracellular domain consisting mainly of leucine-rich repeats. Interaction of the hormone and its receptor leads to the regulation of multiple enzymes which may be involved in cell elongation and division (Friedrichsen and Chory, 2001).

Roth, Friebe and Schnabl (2000) reported that an extract of seeds of *Lychnis viscaria*, which contained brassinosteroids, enhanced the resistance of tobacco, cucumber and tomato to viral and fungal pathogens. In cucumber, this was related to the greater expression of the PR-proteins, chitinase and beta-1,3-glucanase as well as peroxidase after challenge with powdery mildew.

7.2.8 Peptide hormones

Systemin was the first peptide hormone to be characterized. It was isolated from tomato and consists of 18 amino acids but is actually synthesized as a 200-amino acid precursor from which the much shorter systemin is released. Systemin is synthesized in response to wounding and acts systemically causing the upregulation of protease inhibitors. These inhibit insects that attempt to feed on the unwounded parts of the plant and may therefore be of relevance in preventing the spread of viruses that are vectored by insects. A systemin receptor has been identified as a member of the leucine-rich repeat receptor kinase family and, intriguingly, is similar to the brassinolide receptor BRI1 (Scheer and Ryan, 2002, and see Section 7.2.7). Other peptide hormones clearly have important roles to play in the physiology and development of plants such as RALF which, besides causing the rapid alkalinization of the medium of tobacco suspension cultures also causes the inhibition of root growth of tobacco and *Arabidopsis thaliana*.

7.3 The role of altered hormone levels in symptom expression

Since hormones play such vital roles in normal plant development it is not surprising that alterations of their concentrations by infectious agents is liable to have far-reaching consequences. These alterations may be brought about by synthesis or degradation of hormones by the pathogen, alterations of the host's metabolism of hormones and interference with the normal response of plant tissues to them.

7.3.1 Abnormal growth

Hypertrophy, hyperplasia and abnormal differentiation of organs are common symptoms of plant disease. The increased growth may take the form of general and sometimes gross swelling of plant parts, while in other instances lateral buds are released from apical dominance giving rise to witches' brooms or the normal differentiation of the plant is upset, resulting in the production of misplaced organs (Figure 1.19). Qualitative and quantitative alterations in one or more of three 'classical hormones' auxins, cytokinins and gibberellins have been implicated in several of these diseases.

Club-root is a well-known disease of brassicas caused by *Plasmodiophora brassicae*. Dekhuijzen and Overeem (1971) found that swollen roots contained 10–100 times as much cytokinin as healthy roots (Figure 1.14). However, it was not established how the increase was achieved. Here there are several possibilities such as synthesis by the pathogen, induction of enhanced synthesis in the host or inhibited metabolism.

More recently, Ludwig-Muller, Epstein and Hilgenberg (1996) measured free, ester- and amide-bound IAA in 24- and 30-day-old leaves and roots of Chinese cabbage that were healthy or infected by *Plasmodiophora brassicae*. Hydrolase activity was also measured in the same tissues. The amide conjugates, indole-3-acetic acid-alanine (IAAla), IAA-phenylalanine (IAPhe), but not IAA-aspartate (IAAsp) were dramatically enhanced in infected roots, whereas free IAA was only slightly increased compared to control tissue. Hydrolase activity was also enhanced in clubbed roots, but the substrate specificity differed from that found in seedlings with IAAsp hydrolysis being particularly strongly induced by infection with *P. brassicae*. The authors concluded that the specific auxin conjugate hydrolysed was influenced by the developmental stage and stress caused by infection with *P. brassicae*.

Lin and Lin (1990) have investigated the galls caused by the smut fungus *Ustilago latifolia* in *Zizania*, which are used as a summer vegetable in Taiwan. They found that *cis*-zeatin riboside concentrations in tRNA from galls were three to five times that of flowering *Zizania* and that the protein content of mature galls was about 30 times that of flowering tissue.

Gall formation by the bacterium *Erwinia herbicola* pv. *gypsophilae* on *Gypsophila paniculata* is conferred by the pathogenicity associated plasmid pATH and a similar plasmid is found in *Erwinia herbicola* pv. *betae* which confers pathogenicity for beet (*Beta vulgaris*) as well as *G. paniculata* (Figure 7.9). Many virulence genes are harboured by the plasmids including a cluster that specify IAA and cytokinin production. One of these, *etz*, is homologous to other cytokinin biosynthesis genes that have been described. Its role in cytokinin synthesis by *E. herbicola* pv. *gypsophilae* was demonstrated by Northern analysis in which it was shown that an *etz* transcript was present at high levels in the late logarithmic phase of growth of the bacterium in culture and very low in the stationary phase. Proof of the importance of the gene in gall formation was obtained by marker exchange of the gene with a mutant version. The mutant did not produce cytokinins, induced only small galls in *Gypsophila* cuttings and was almost completely symptomless in a whole plant assay (Lichter *et al.*, 1995). More recently, the importance of both IAA and cytokinins in gall formation was demonstrated by PCR using primer pairs that were based on IAA or cytokinin

Figure 7.9   Gall formation on *Gypsophila paniculata* by *Erwinia herbicola* pv. *gypsophilae* (left labelled *Ehg*) and by *Erwinia herbicola* pv. *betae* (right labelled *Ehb*: illustrations kindly provided by Isaac Barash and Shulamit Manulis). A colour reproduction of this figure can be seen in the colour section

biosynthetic genes. The primers were specific to gall-forming strains of *E. herbicola* and distinguished them from other gall-forming bacteria (Manulis *et al.*, 1998 and see Section 2.5.5). In infections by another bacterium, *Pseudomonas savastanoi*, auxin is the main virulence factor, stimulating division of cambium cells resulting in the proliferation of disorganized tissue (Iacobellis *et al.*, 1994).

Goethals and co-workers (2001) have reviewed the role of another bacterium, *Rhodococcus fascians*, in the formation of growth abnormalities (see Figure 1.19). These occur in a wide range of both monocotyledonous and dicotyledonous plants encompassing 39 families and 86 genera. The abnormalities include leaf deformation, witches' brooms and fasciation but leafy galls are the most severe symptom. They consist of centres of shoot amplification and shoot growth inhibition and remain green long after the mother plant has senesced. In intensive studies of one strain of the bacterium, an intact isopentenyl transferase gene (*ipt*, see Figure 7.4), which is responsible for catalysing the first dedicated step of cytokinin biosynthesis, was found to be an absolute requirement for leafy gall formation and cytokinins have been isolated from culture supernatants of the organism. However, the compounds responsible for gall formation do not appear to be simple cytokinins but rather N^6 substituted purine derivatives with unique activity. Molecular genetic analysis suggests that the substituents may be glutathione or an alternative peptide.

7.3.2 Redirection of nutrients

One of the properties of cytokinins is their ability to cause the accumulation of nutrients. When droplets of cytokinin solution are placed on detached leaves floated on water, the area under and surrounding the cytokinin droplet remains green while the remainder of the leaf senesces. This 'green island' effect is commonly encountered around the sporulating pustules of obligate pathogens such as rusts and mildews, suggesting that the poor yields associated with these infections is caused by the redirection of host nutrients to the infectious agent. Some more direct evidence for this view is now available. López-Carbonell, Moret and Nadal (1998) have measured the cytokinin, zeatin riboside, in leaves of four plants, *Hedera helix, Pelargonium zonale, Prunus avium* and *Rubus ulmifolius*, infected by fungi which gave rise to green islands; these were *Colletotrichum trichellum, Puccinia pelargonii-zonalis, Cercospora circumscissa* and *Phragmidium violaceum*, respectively. Zeatin riboside concentrations were always higher in the green islands than the surrounding senescing tissue and were almost the same as those found in healthy, green leaves.

Murphy and co-workers (1997), using a combination of High Performance Liquid Chromatograph (HPLC) and Enzyme-Linked Immunoassay (ELISA), have demonstrated that the hemibiotrophs, *Pyrenopeziza brassicae* and *Venturia inaequalis*, as well as the biotrophs, *Cladosporium fulvum* and *Erysiphe = Blumeria graminis*, produce cytokinins whereas the necrotrophs, *Botrytis cinerea* and *Penicillium expansum*, did not. The authors suggest, on the basis of this work and that of others, that there is a correlation between obligate parasitism and cytokinin production.

Evidence for the localization of cytokinins in fungal structures of *Puccinia recondita* f. sp. *tritici* was obtained by Hu and Rijkenberg (1998) using anti-cytokinin antibodies. Putting these lines of evidence together they amount to a strong case for the presence of elevated concentrations of cytokinins localized to areas in the vicinity of lesions caused by obligate pathogens. These act as powerful sinks and, as a result, nutrients destined for the economic part of the crop are redirected to the pathogen, consequently depressing yields. For example, a severe attack of stem rust of wheat caused by *Puccinia graminis* f. sp. *tritici* can cause total loss of grain yield (Strange, unpublished observations).

7.3.3 Stunting

Stunting is a common symptom of plant disease and in some cases has been attributed to reduced levels of hormones or an inability to respond to them.

Stunting is a symptom of rice tungro virus and this could be reversed by treating the plants with gibberellin (Thomas and John, 1981). It is also a symptom of sorghum infected by *Sporisorium reilianum*. Matheussen, Morgan and Frederiksen (1991) investigated the gibberellin content of infected plants.

Figure 7.10 Stunting of tobacco (right) caused by infection with tobacco mosaic virus compared with (left) uninfected control (courtesy of Alex Murphy). A colour reproduction of this figure can be seen in the colour section

They identified 11 gibberellins from diseased panicles and showed that the total of GA_{53}, GA_{44} and GA_{19} was reduced by about 90 per cent compared with controls.

Decreased growth may also result from abnormally high concentrations of abscisic acid, an antagonist of GA. For example, Whenham and Fraser (1981) found that abscisic acid concentrations in leaves of tobacco dwarfed by infection with tobacco mosaic virus were 6- to 18-fold higher than uninoculated controls depending on the strain used. They also showed that the reduction in leaf growth could be mimicked by spraying plants with ABA (Figure 7.10).

In contrast, stunting of tomatoes infected with citrus exocortis viroid was not relieved by the application of IAA, suggesting that this symptom may be caused by an inability to respond to plant hormones.

7.3.4 Chlorosis and necrosis

Premature chlorosis is a frequent symptom of disease and may often be mimicked by ethylene. For example, lethal yellowing of palms caused by a phytoplasma could be mimicked by application of the ethylene-releasing agent ethephon (Leon *et al.*, 1996). More directly, accumulation and increased activity of ACC synthase and ACC oxidase are localized in chlorotic tissue surrounding lesions caused by TMV in tobacco (de Laat and van Loon, 1983) and by *Phytophthora infestans* in tomato (Spanu and Boller, 1989). Further evidence for the role of ethylene in the development of chlorotic and necrotic symptoms has come from work with the tomato mutant *Never ripe* which is insensitive to ethylene. As a result fruit ripening in the homozygous plant (*Nr/Nr*) is delayed and incomplete. These plants also showed markedly reduced foliar chlorosis and necrosis when inoculated with either *Xanthomonas campestris* pv. *vesicatoria* or *Pseudomonas syringae* pv. *tomato* (Lund, Stall and Klee 1998). One particularly telling parameter in these experiments was that electrolyte leakage, an indicator of cell death (see Chapter 8, Section 8.3), was about four-fold higher in wild type plants infected with either pathogen than in *Nr/Nr* plants.

7.3.5 Epinasty

Epinasty, the downward curving of leaves, caused by the relatively greater growth of the upper compared with the lower side of the petiole, is also associated with the over-production of ethylene or increased sensitivity to the hormone (Tieman *et al.*, 2000). It is a symptom of many diseases. For example, tomato plants infected with *Fusarium oxysporum* f. sp. *lycopersici* or potato spindle tuber viroid (PSTVd) both demonstrate epinasty. In the latter a specific protein kinase gene, *pkv*, is activated by infection but not by the ethylene producing compound ethephon casting doubt on its involvement in the expression of this symptom.

7.3.6 Abscission

Defoliation is often a symptom of plant disease and has severe implications for plant growth. For example, an epidemic of Swiss needle cast, caused by the ascomycete *Phaeocryptopus gaeumannii*, is reported to be affecting over 100 000 ha of Douglas-fir forest plantations along the Oregon Coast, USA (Hansen *et al.*, 2000). In experimental plots defoliation was proportional to the number of stomata occluded by pseudothecia of the fungus. When this reached about 50 per cent, needles were shed. It will be interesting to learn if a build-up of ethylene within the needles is responsible since regulation of abscission by the gas is well documented. For example, a defoliating strain of *Verticillium dahliae* enhanced ethylene production in infected cotton plants (Tzeng and DeVay, 1985) and the capacity of leaf tissue to form ethylene as well as the concentration of abscisic acid increased in coconut palms infected with the phytoplasma causing lethal-yellowing disease (Leon *et al.*, 1996). Furthermore, treatment of symptomless palms with ethephon which releases ethylene resulted in symptoms that mimicked some of those of the disease. In contrast, decreases in IAA brought about by IAA oxidases may be the cause of premature abscission as in leaf drop of coffee, resulting from infection by *Omphalia flavida*, and infection of rose leaves by *Diplocarpon rosae*.

7.4 Crown gall

As long ago as 1897, Cavara in Italy reported the isolation of a bacterium from grape-vines which, when re-inoculated into the plant, gave rise to tumours (Figure 1.15). Subsequently the bacterium was named *Agrobacterium tumefaciens* and then *A. vitis*. As mentioned in Chapter 1, *Agrobacterium* is being retained as the generic name for plant pathogenic species of this group although on the basis of 16S rDNA analysis it should be *Rhizobium* (see Section 1.3.8).

Besides the presence of the bacterium, wounding was critical for the development of tumours. More specifically the size of tumour was dependent on the time allowed to elapse between wounding and inoculation, the optimum being 48 h (Braun, 1982). Tumour tissue was sterile showing that, although the bacterium was necessary for tumorigenesis, its continued presence was not required for tumour maintenance. Furthermore, tumour tissue may be kept in culture indefinitely on media devoid of auxin and cytokinin, which are normally absolute requirements for growth of plants in tissue culture. Had the bacterium been present it would have rapidly multiplied and swamped the cultures.

As a result of these experiments Braun and Mandle (1948) proposed that there was a tumour inducing principle (TIP). Many candidates were suggested for this role during a period of over 20 years including toxins, hormones, chromosomal DNA, bacteriophages and RNA. Eventually sufficient evidence accumulated to implicate DNA. This included the occurrence of certain arginine derivatives in galls, the opines, and the transfer of TIP between strains co-inoculated on the same plant.

Although there are several types of opines they are neither normal plant constituents nor specific to the plant in which they are found. Rather they are specific to the particular bacterial strain that induces the tumour. Moreover, the bacterium that has induced the production of a specific opine is also able to catabolize it. For example, octopine strains both induce and catabolize octopine and similarly nopaline strains induce and catabolize nopaline (Figure 7.11). These observations suggested the genes required for opine synthesis were transferred from the bacterium to the plant.

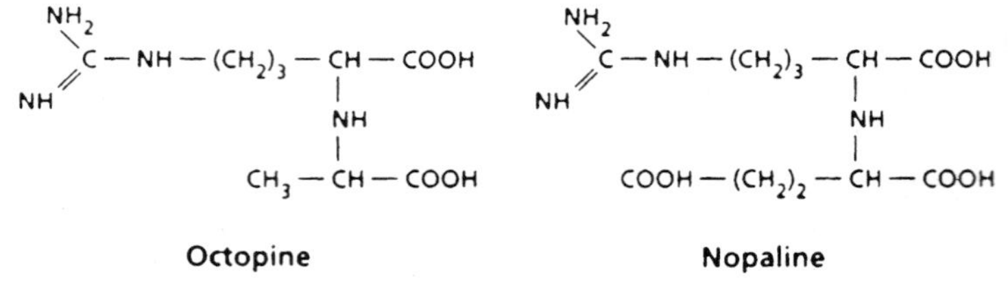

Octopine   **Nopaline**

Figure 7.11 Octopine and nopaline, compounds that are found in crown gall tumours; genes for their synthesis are present in the T-DNA that is transferred from *Agrobacterium tumefaciens* to the plant

Kerr (1969) showed that TIP could be transferred between strains of the bacterium *in planta* and it was not long before it was realized that all oncogenic (i.e. tumour inducing) strains carried a large plasmid of about 120×10^6 daltons. When such strains were cured of their plasmid by high temperature they were no longer oncogenic nor were they sensitive to the bacteriocin Agrocin 84. Transfer of the oncogenic or, as it is more commonly called the Ti (tumour inducing) plasmid among strains *in vitro* was achieved using the promiscuous plasmid Rp4 in the donor strain or by incorporation of antibiotic resistance in both donor and recipient, allowing the selection of the transconjugants on a medium containing the appropriate antibiotics. More simply, conjugation and transfer were promoted by incorporating an appropriate opine in the medium (Kerr and Ellis, 1982).

The transfer data provided proved that TIP was a plasmid but did not demonstrate how it induced the tumorous state. Nucleic acid hybridization experiments were helpful since they clearly showed that a portion of the Ti plasmid, the T-DNA, is incorporated into the host genome. For example, restriction enzyme digests of the Ti plasmid hybridized with DNA from tumour cells, and pulse labelled RNA from tumours hybridized with DNA from the Ti plasmid demonstrating that transcription of the transferred genes had occurred. Thus *Agrobacterium* may be thought of as the first genetic engineer!

In the past few years considerable progress has been achieved in elucidating the mechanism by which T-DNA is transferred to the plant host and incorporated into its genome as well as the prerequisites for this phenomenon (see Tinland, 1996, and Gelvin, 2000, for reviews). The principal events are as follows (Figure 7.12).

(1) *Induction:* the wounded root provides an acidic environment in which low molecular weight phenolic compounds such as acetosyringone (Figure 6.5) and sugars are released. The phenolic compounds induce the VirA protein which autophosphorylates and transphosphorylates the VirG protein.

(2) *Expression of vir proteins:* the active VirG protein stimulates other genes of the *vir* region of the plasmid such as those encoding the VirC1, VirC2, VirD1 and VirD2 proteins.

(3) *T-DNA processing:* the VirD1 and VirD2 proteins excise single-stranded T-DNA from the plasmid. VirD2 stays attached to the 5'-end of the released DNA.

(4) *Transport to host:* pili are assembled from proteins coded by the *virB* operon and the protein VirD4.

(5) *Nuclear targeting:* once in the plant cell the T-DNA becomes associated with VirE2 proteins which are also synthesized by the bacterium and exported via the pili. The VirD2 and VirE2 proteins, which contain plant-active nuclear localization signal sequences, mediate the entry of the T-DNA into the nucleus of the plant cell through nuclear pore complexes.

202

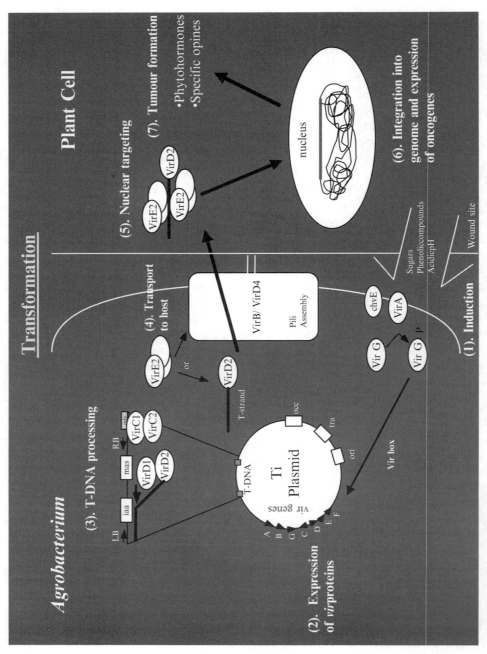

Figure 7.12 Diagram of the mechanism by which plants are transformed by *Agrobacterium* (courtesy of Derek Wood). For details see text

(6) *Integration into genome and expression of oncogenes:* once in the plant nucleus, the T-DNA integrates into the plant genome by illegitimate conjugation. Mutation studies suggest that VirD2 plays an important role in this process, as do some plant proteins, since mutants deficient in DNA repair and recombination have low rates of stable transformation.

(7) *Tumour formation:* tumours, containing opines specific to the *Agrobacterium* strain are formed as a result of the expression of the genes for opine, IAA and cytokinin synthesis.

T-DNA is delimited by two imperfect 25 bp direct repeats, known as the border sequences, but the DNA between these sequences is not required for transfer. This has important implications for genetic engineering since it is possible to excise the DNA and replace it with genes of choice (see Chapter 12, Section 12.6 for further details). The transferred DNA is not always stably incorporated as some revertants have been reported but Furner and co-workers (1986) obtained evidence of an *A. tumefaciens* transformation event in the evolution of *Nicotiana*.

Since tumour tissue may be propagated in tissue culture without the addition of IAA and cytokinins it was not surprising that the genes coding for these biosynthetic pathways were found to be present in T-DNA. Moreover, as might be expected from the undifferentiated tissue mass that comprises many tumours, IAA and cytokinin levels are elevated compared with normal tissue. Genes for the synthesis of the opines are also present on T-DNA. Mutation by integration of transposons such as Tn5 into T-DNA gave rise to tumours with altered morphology and capacity to synthesize opines. These mapped to four loci, two of which are concerned with hormone synthesis (Figure 7.13). Transposon mutagenesis of *tms* gave rise to avirulent phenotypes or shoot teratomas but virulence was restored by the addition of auxin. Auxin was also required for growth of these tumours *in vitro*. Similarly, transposon mutagenesis of *tmr* gave

Map of the T-DNA region of an octopine strain of *Agrobacterium tumefaciens*

T-DNA

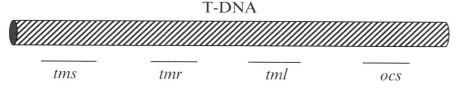

tms = locus for auxin biosynthesis: mutation here gives rise to a tumour with shoots.
tmr = locus for a cytokinin biosynthesis enzyme (isopentenyl transferase): mutation here gives rise to a tumour with roots.
tml = mutation here gives rise to larger tumours.
ocs = the octopine synthase gene: mutation here inhibits octopine synthesis.

Figure 7.13 Map of the T DNA region of an octopine strain of *Agrobacterium tumefaciens*

rise to tumour lines in which roots were formed. Normal tumours resulted when the tumours were supplemented with cytokinin. These tumours also required cytokinin for growth *in vitro*.

The route for the synthesis of IAA from tryptophan (Trp) adopted by bacteria is via indoleacetamide (IAM, Figure 7.1). Both the gene and the enzyme for tryptophan monoxygenase have been isolated and sequenced, the amino acid sequence of the enzyme corresponding exactly to that specified by the nucleotide sequence in the open reading frame of the *tms* gene. Similarly, the amino acid and nucleotide sequences of indoleacetamide hydrolase were found to correspond. When *E. coli* was transformed with both genes, the transformants secreted 58 μg IAA/mg cell dry weight, demonstrating that they could act coordinately to produce the growth regulator (Morris, 1986).

The *tmr* gene has also been sequenced. It codes for a 27 kDa protein which, when cloned and expressed in *E. coli*, gave rise to an isopentenyl transferase. As mentioned previously (Section 7.3), this enzyme catalyses the prenylation of adenine, a key step in the synthesis of cytokinins (Figure 7.4.).

T-DNA also contains genes that encode enzymes catalysing the synthesis of opines which are expressed when incorporated into the plant genome. *Agrobacterium tumefaciens* therefore appears to be a uniquely well-adapted pathogen. By integration of a segment of DNA into the genome of the host that codes for IAA, cytokinin and opine production it maintains the host's cells in a healthy condition and also provides itself with 'the pleasures of the table and the bed' since the opines are a source of nutrition exclusive to the bacterium, the host plant being unable to catabolize them and, as mentioned above, they promote conjugation, which may involve the transfer of the Ti plasmid between strains!

A. rhizogenes is a closely-related species which causes hairy-root disease. As with *A. tumefaciens* virulent strains contain a plasmid, termed the Ri plasmid, and part of this is integrated into the genome of the host, resulting in root proliferation rather than the unorganized tumours caused by *A. tumefaciens*.

A. tumefaciens and *A. rhizogenes* are both used as vectors in the transformation of plants since the hormonal genes involved in the abnormal outgrowths caused by these bacteria may be substituted by genes of choice (see Chapter 12, Section 12.6, for the use of these organisms in the genetic modification of plants).

8 Killing the Host – the Role of Toxins

Summary

Many plant pathogens elaborate phytotoxic compounds which produce a variety of symptoms on sensitive plants. They include wilting, water-soaking, chlorosis and necrosis. Bioassays, usually based on one of these, have been used to monitor the isolation of toxins from cultures of pathogens. Toxins have customarily been divided into host-selective (HSTs) and non-host-selective (NSTs). For the first class, production of the toxin is mandatory for pathogenicity whereas for the second it is not, although toxin production contributes to virulence. HSTs are found principally in species of *Alternaria* and *Helminthosporium*. In some instances considerable advances have been made in defining the genes necessary for toxin synthesis. Other work has shown the site of action in sensitive plants and the reason for insensitivity in plants that are not affected and are not hosts of the toxigenic organism. NSTs are found in many classes of plant pathogens and have a wide range of activities. A role for them in pathogenesis is more difficult to establish than for HSTs since they only contribute to virulence. Nevertheless, gene disruption experiments in which toxin synthesis has been negated have provided strong evidence for attributing a proportion of the symptoms caused by a pathogen to its toxin. In some instances specific compounds in the host regulate toxin synthesis. Where a crucial role for a toxin in a disease has been adduced there is a strong motive for biotechnological approaches to render the plant insensitive.

8.1 Introduction

It is difficult to formulate a generally acceptable definition of a toxin that has phytopathological significance. Some have used the term phytotoxin which is generally taken to mean a substance, usually secreted by a pathogen, that is toxic to plants. However, a number of compounds of fungal origin are not only toxic to plants but also to humans and animals. Since compounds synthesized by fungi that are toxic to man and his domesticated animals are defined as mycotoxins,

Introduction to Plant Pathology by Richard Strange
© 2003 John Wiley & Sons, Ltd   ISBN 0 470 84972 X (cased)   ISBN 0 470 84973 8 (pbk)

the term phytotoxin would logically carry the inference that they were produced by the plant rather than the pathogen. In order to avoid this difficulty the simple term toxin will be used in this text and, unless otherwise stated, it is to be taken that the compounds concerned are toxic to plants.

It was Anton de Bary in the 19th century who first suggested that one way in which plant pathogens damage their hosts was to secrete toxins into them. Gäumann (1954) went further and wrote that 'microorganisms are pathogenic only if they are toxigenic'. Today such an all-encompassing definition would not be accepted. For example, some plant pathogens damage their hosts, as we saw in Chapter 6, by secreting enzymes. Others do so by altering the concentrations of plant-growth regulators (see Chapter 7). Neither such enzymes nor plant-growth regulators are normally viewed as toxins.

Toxins affecting plants are divided into two classes: host-selective toxins (HSTs), affecting only plants that are hosts of the toxin-producing organism, and non-selective toxins (NSTs), causing symptoms not only on hosts of the pathogen but on other plants as well. HSTs are usually essential for pathogenicity, i.e. the ability to cause disease, mutants that have lost their ability to produce the toxin being non-pathogenic. In consequence the role of HSTs in disease is seldom disputed. In contrast, NSTs are not essential for pathogenicity but may contribute to virulence. Here there are often arguments as to their role in disease! The main ones are whether a toxin synthesized by a pathogen *in vitro* is also synthesized in the plant and whether the presence of the toxin in the plant is a prerequisite for assigning a role for it in disease. This last point is particularly contentious since it may be argued that absence of the toxin in the diseased plant may be indicative of binding or metabolism, either or both of which may be required for activity, but these cannot be distinguished from lack of production. As with attempts to deduce the role of enzymes in pathogenicity or virulence, molecular techniques have been invaluable in settling this point.

8.2   Macroscopic symptoms

Symptoms of toxic damage to a plant such as wilting, water-soaking, chlorosis and necrosis may be readily observed by the naked eye in many instances but the primary lesion is usually at the biochemical level. In some cases, however, wilting may be more a response to a physical rather than a biochemical lesion in that some pathogens produce materials that plug the water-conducting xylem of the plant or cause embolisms. For example, Van Alfen and co-workers (1987a, b) found that *Clavibacter michiganense* subsp. *insidiosum*, a pathogen of alfalfa, produced three types of extracellular polysaccharide, each of which was of the appropriate size to block one of the three known capillary pore sizes of the host's water-conducting pathway. Embolisms may result from the enlargement of the pore size of pit membranes and Sperry and Tyree (1988) have suggested that this is caused by the action of oxalic acid in combination with calcium.

Impairment of membrane function has long been thought to be the primary cause of wilting and certainly the common symptom of water-soaking, in which intercellular spaces become flooded with cell sap, may be attributed to this. In a greenhouse experiment, acute wilting was demonstrated in cuttings of chickpea exposed to low concentrations of solanapyrone A, the most toxic of the three solanapyrone toxins produced by *Ascochyta rabiei*. Here the upper leaves of the plant did not merely droop, but their petioles actually broke, a symptom which is typical of Ascochyta blight (Figure 8.1). The interpretation of this result was that the plant relies upon the turgor pressure of cells surrounding the stele for support and, when this is destroyed by the toxin (see Section 8.5.3), distal parts collapse under their own weight (Hamid and Strange, 2000). No doubt such collapse would be exacerbated by windy conditions in the field. The notion that membrane impairment was responsible for the symptoms in this case is given credence by the assay used to detect and isolate the toxins in the first place which relies upon disruption of the plasmamembrane (see Section 8.3)

Necrosis is usually the next stage in the demise of plant cells after water-soaking and chlorosis may result from perturbations of chlorophyll metabolism but its relation to toxin activity may be rather indirect. Other symptoms of toxin activity may be more subtle such as the impairment of defence responses.

Figure 8.1 A chickpea cutting after taking up 45 μg of solanapyrone A, the most potent of the solanapyrone toxins produced by the pathogen *Ascochyta rabiei*, and incubating for a further 96 h in water note the broken stem, a typical symptom of Ascochyta blight of chickpea. A colour reproduction of this figure can be seen in the colour section

8.3 Bioassay

Detection and analysis of an unknown toxin requires a suitable bioassay. The assay should be sensitive, give quantitative results and be simple and rapid to perform. In the first instance it is probably best if the assay is as non-specific as possible. Inhibition of root growth has frequently been used (Rasmussen and Scheffer, 1988a) and some toxins are sufficiently non-specific for microbiological assays to be employed (Barzic, 1999). Manulis, Netzer and Barash (1986) found that the inhibition of the incorporation of labelled amino acids into protein by plant cell suspensions was a rapid and non-specific means of measuring toxic activity.

Wilting is one of the most frequent and dramatic symptoms of toxin activity but it is, unfortunately, difficult to quantify. Assays usually involve placing cuttings in vials containing toxin solution but this is seldom taken up at uniform rates owing to differences in transpiration among the cuttings, thus subjecting tissues to differing doses. Other difficulties are the relatively large volumes of toxin required and the definition of a reliable scale for the severity of wilting.

Chlorosis is often a symptom of toxin action. Usually droplets of toxin preparation are placed on abraded leaves or injected into the mesophyll and the severity of chlorotic symptoms that develop measured on an arbitrary scale. For example, Fogliano and co-workers (1998) used a 0–6 scale, which ranged from no symptoms (0) through light and severe chlorosis to extended necrosis (5) and tissue drying (6). A more objective method is to measure chlorophyll content after extraction (Rasmussen and Scheffer, 1988a).

Many toxins affect membranes either directly or indirectly. These may be assayed by measuring electrolyte leakage from tissues such as discs cut from leaves or storage organs (Damaan, Gardner and Scheffer, 1974; Kwon et al., 1996). Another technique, which also relies upon membrane dysfunction and is used in the author's laboratory, is to incubate isolated cells or protoplasts with toxin and then add fluorescein diacetate (FDA). Cells or protoplasts with intact plasmamembranes take up the dye and, once inside the cell, the acetate groups are cleaved from the molecule by means of non-specific esterases. The fluorescein resulting from this reaction is unable to permeate intact membranes and therefore live cells with functional plasmamembranes fluoresce brightly when viewed under a fluorescence microscope. In contrast, cells with ruptured membranes do not retain fluorescein and remain dark (Figure 8.2). The assay can be performed in the wells of a microtitre plate using only 50 µl volumes of both toxin solution and cell or protoplast suspension. As with all assays of toxic compounds, it is important to use a dilution series in order to determine the LD_{50} dose. Usually a two-fold dilution series of the toxin is made directly in the micro-test plate and the cells or protoplasts added. The plate is incubated for 3 h before adding freshly prepared fluorescein diacetate and scoring under an inverted fluorescence

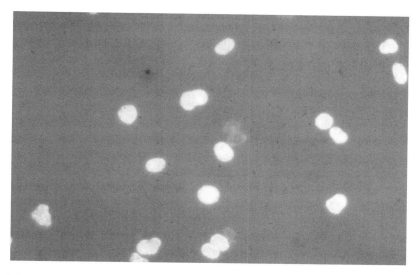

Figure 8.2 Assay of solanapyrone toxins from *Ascochyta rabiei* using cells from leaves of chickpea and the vital dye fluorescein diacetate. Live cells take up the dye and release fluorescein from the molecule by the action of non-specific esterases. Since intact plasmamembranes are not permeable to fluorescein they fluoresce yellow under a fluorescence microscope. Note by contrast the dead brown cells in the centre of the picture. A colour reproduction of this figure can be seen in the colour section

microscope. The percentage cell or protoplast death is recorded and corrected for control values. Transformation of these figures to probits (Finney, 1980) and plotting them against the $\log_2$ of the dilution factor allows a straight line to be drawn through the points and the dilution which causes 50 per cent cell death, the LD_{50} value, arbitrarily designated as 1 unit of activity, to be interpolated. Conversion to the arithmetic scale permits the number of units of activity in a preparation to be calculated (Figure 8.3). If cell-suspension cultures of the host are available, these may also be used in the assay.

Some toxins inhibit the dark fixation of carbon dioxide. For example, the percentage inhibition of this reaction by victorin (Section 8.4.2) was linear over at least two orders of magnitude in the ng/ml range.

As knowledge of the toxin's biological properties increases the assay may be refined. Phaseolotoxin was assayed initially by its ability to cause chlorotic lesions on bean leaves (Hoitink *et al.*, 1966). However, the accumulation of ornithine and the reversal of the chlorotic symptoms by arginine suggested that inhibition of ornithine carbamoyltransferase was the primary lesion of the toxin and therefore inhibition of this enzyme could be used as an assay (Mitchell, 1979; Turner and Mitchell, 1985; Section 8.5.1).

Once the chemistry of the toxin is known non-biological assays such as high performance liquid chromatography (HPLC) may be used (Hayashi *et al.*, 1990) or, if the toxin is immunogenic or can be rendered immunogenic, immuno-assays (Phelps *et al.*, 1990).

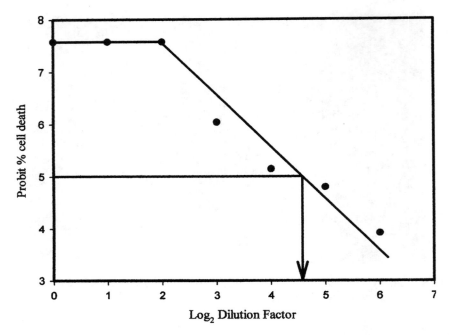

Figure 8.3 Graph of probit per cent cell death vs. $\log_2$ dilution factor of a toxin preparation; a probit value of 5 corresponds to 50 per cent cell death, i.e. the LD_{50} dose, and may be taken as representing 1 unit of activity – in the experiment presented this correspond to 4.6 on the $\log_2$ scale or a dilution factor of 24.25 on an arithmetic scale, and as only 50 μl of toxin are used this factor may be multiplied by 20 to give the number of units/ml

8.4   Host-selective toxins (HSTs)

Host-selective toxins (HSTs) are usually secondary metabolites and are normally produced as a family of compounds with one predominant type. Of the approximately 20 for which chemical structures are known, all are produced by fungi and most of them by members of the genera *Alternaria* and *Cochliobolus* (= *Helminthosporium* or *Bipolaris*). The exceptions are the toxins produced by *Phyllosticta maydis, Periconia circinata, Stemphylium vesicarium* and *Pyrenophora triticirepentis*, the last, being a protein, is the only known HST which is not a secondary metabolite (Tomas *et al.*, 1990; Walton, 1996).

8.4.1   Host-selective toxins from *Alternaria* species

Tanaka (1933) was the first to demonstrate the existence of an HST. He worked with a fungus he called *Alternaria kikuchiana*, which causes a leaf spot of certain varieties of pear, and found that 'the fungus free media showed striking virulence to fruits of a susceptible variety, while no effect was produced on those of a resistant (variety)... the appearance of black spots as the sign of this disease did not always require the penetration of the causal fungus, being caused by some toxic substance produced by the fungus'. The fungus is now generally considered to be a pathotype of *Alternaria alternata*, a poorly defined genus, which stands in much need of molecular taxonomic studies. An indication of this is that the pathotype infecting pear is one of six recorded from six different species of plant, each pathotype producing a toxin that selectively affects its host. The six pathotypes and their predominant toxins are the original one from Japanese pear, which produces AK-toxin, the others being from strawberry (AF-toxin), apple (AM-toxin), tangerine (two types producing different toxins – ACT-toxin and ACTG-toxin) and rough lemon (ACR-toxin; Figure 8.4). In addition, there are isolates which are selectively pathogenic to a specific genotype of tomato which have been given forma specialis status, i.e. *A. alternata* f. sp. *lycopersici*.

The structure of AK-toxin 1 was deduced from mass spectrometry and nuclear magnetic resonance, infra-red and ultraviolet spectroscopy in 1982, nearly 50 years after Tanaka's original observations of the selective toxicity of culture filtrates of the fungus. AK toxin exists as two homologues, AK toxins I and II, and these as well as AF toxins I, II and III and ACT toxin I and II all possess a common decatrienoic acid moiety (9,10-epoxy-8-hydroxy-9-methyl-decatrienoic acid; Figure 8.4).

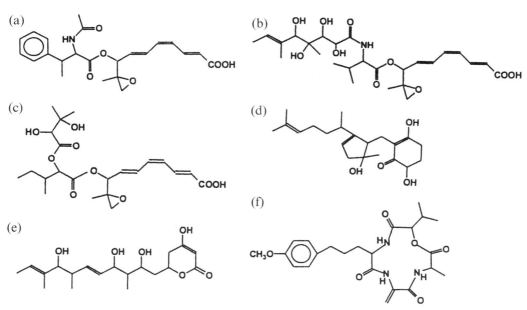

Figure 8.4 Structures of some host-selective toxins produced by pathotypes of *Alternaria alternata*: (a) AK-toxin, Japanese pear pathotype, (b) ACT-toxin, tangerine pathotype, (c) AF-toxin, strawberry pathotype, (d) ACTG-toxin, tangerine pathotype, (e) ACR-toxin, rough lemon pathotype, and (f) AM-toxin, apple pathotype; where more than one similar compound is synthesized by the pathogen the structure of the principal product is shown

AK-toxin 1 induced veinal necrosis in leaves of the sensitive cultivar Nijisseiki at 10^{-8} to 5×10^{-9} M. In contrast, no visible symptoms occurred on leaves of resistant cultivars or those of apple and strawberry at 1.2×10^{-4} M. The strawberry pathotype also affects cultivars of pear which are sensitive to AK-toxins. AF-toxin 1 at 0.1 µg/ml caused veinal necrosis on leaves and K^+ loss from leaf tissues of not only the susceptible strawberry cultivar Morioka-16 but also the pear cultivar, Nijisseiki (which is sensitive to AK-toxin 1). Despite the similarity of structures of all three AF-toxins, the activity of AF-toxins 2 and 3 differed from that of AF-toxin I. AF-toxin II caused necrosis on the pear cultivar Nijisseiki at 0.02 µg/ml as did AK-toxin but not on the strawberry cultivar Morioka-16 while AF-toxin III only affected cultivar Morioka-16 at 1 µg/ml (Nishimura and Kohmoto, 1983).

AK-toxin 1 and AF-toxin 1 were found after incubating spores of the pear and strawberry pathotypes of *A. alternata* on moist paper towels, concentrations reaching 0.02 pg/spore and 0.08 pg/spore, respectively (Hayashi *et al.*, 1990). Since 0.1 µg AF-toxin 1 is required to cause veinal necrosis and K^+ loss from leaf tissue of sensitive species it follows that 1250 spores would be required to cause the same effect. However, such high numbers of spores or high concentrations of toxin may not be required to allow the pathogen to attack the plant successfully. For example, inhibition of defence responses might occur at lower concentrations. Moreover, in the experiments of Hayashi and co-workers (1990) the spores were washed, raising the possibility that a proportion of the toxins may have been removed. Rather more recently than the original work, an outbreak of black spot of strawberry was attributed to the strawberry pathotype of *A. alternata*. Spores of the fungus released AF toxin-1 during germination and the isolates were pathogenic to the pear cultivarNijisseiki which, as noted above, is sensitive to this toxin (Wada *et al.*, 1996).

AM toxins caused veinal necrosis on susceptible apple cultivars at 10^{-8} M, on moderately resistant cultivars at 10^{-5} M and no symptoms on resistant cultivars at 10^{-4} M. The cyclic peptide synthetase gene involved in the synthesis of AM toxin has been cloned and sequenced (Johnson *et al.*, 2000). It is 13.1 kb in length and has no introns. Mutants obtained by transformation of a wild type toxigenic isolate with disruption vectors gave toxin-minus mutants which were unable to cause disease symptoms on susceptible apple cultivars, proving the essential involvement of the toxin in the disease.

Restriction enzyme mediated integration (REMI) has been used to produce toxin-minus mutants of the pear pathotype of *A. alternata* (Tanaka *et al.*, 1999). Recovery of genomic DNA flanking the integration site revealed two genes, *AKT1*, which encodes a carboxyl activating enzyme and *AKT2*, which encodes a protein of unknown function. Targeted gene disruption showed that both genes were necessary for pathogenicity and toxin production. Homologues of both genes were detected in tangerine and pear pathotypes of the fungus but not in other pathotypes or non-pathogenic isolates. They therefore are likely to be involved in the synthesis of the common decatrienoic acid moiety. Moreover, the authors suggest that this commonality of homologues and function may imply horizontal gene transfer.

Nine novel compounds closely related to ACTG-toxin and termed tricycloalternarenes were isolated from a strain of *Alternaria alternata*, obtained from *Brassica sinensis* (Nussbaum *et al.*, 1999) but it is not known at the time of writing if these toxins are selective for *Brassica* species. Liebermann, Nussbaum and Gunther (2000) also reported the isolation of nine new bicycloalternarenes as well as ACTG-toxins A and B but could not confirm the phytoxicity of these last two compounds. The specificity and toxicity of the ACTG toxins therefore seem to be issues that are in dispute.

Little is known about the mechanism of action of the host-selective toxins from these pathotypes of *A. alternata*, although it is thought that AF-, AC- and ACTG-toxins act at the plasma membrane whereas ACR- and AT-toxin affect mitochondria.

In contrast, information is available that goes far in explaining both the mechanism of action and the specificity of the toxins of *A. alternata* f. sp. *lycopersici*, a pathogen of tomato. The toxins, known collectively as AAL toxin, are mono-esters of propane-1,2,3-tricarboxylic acid and 2,4,5,13,14-pentahydroxyheptadecane and are structurally similar to the fumonisins (Figure 8.5). Both AAL toxin and fumonisin B_1 are sphinganine analogues (SAMs) which competitively inhibit *de novo* sphingolipid biosynthesis *in vitro*. As a result, sphingoid bases accumulate in both animal cells and plants. SAMs are toxic to sensitive tomato cultivars and induce apoptosis in them (Wang, H. *et al.*, 1996). Resistance to SAMs and to the pathogen is determined by the codominant *Asc* (**A**lternaria **s**tem **c**anker) locus, plants that are heterozygous (*Asc/asc*) having an intermediate reaction. AAL toxin is essential for pathogenicity as toxin-deficient mutants are unable to infect tomato plants with the *asc/asc* genotype. Brandwagt and co-workers (2000) have cloned the *Asc-1* gene and demonstrated by complementation that it conferred toxin insensitivity to plants with the *asc/asc* genotype. The latter were found to have several deletions in the gene, the most important of which appeared to be a 2-bp in one of the exons giving rise to a stop codon. Sequence data showed that the *Asc-1* gene is homologous to *LAG1* of *Saccharomyces cerevisiae*, which has been associated with longevity and facilitates the transport of glycosylphosphatdylinositol (GPI)-anchored proteins from the endoplasmic reticulum to the Golgi apparatus. Homologues of the gene are widely transcribed as indicated by records of expression sequence tags and full length cDNAs from fish, mammals, insects, higher fungi and monocotyledonous and dicotyledonous plants. In order to explain their results, Brandwagt and co-workers (2000) propose that *Asc-1* is involved in the production of sphingolipids by a SAM insensitive pathway, thus negating the catastrophic effect of inhibition of synthesis of these vital components of membranes. In a later paper Brandwagt and co-workers (2002) presented further evidence for this view by showing that plants overexpressing the tomato *Asc-1* gene are highly insensitive to both AAL toxins and fumonisin B_1. Toxin insensitivity in this instance is therefore attributed to a

(a)

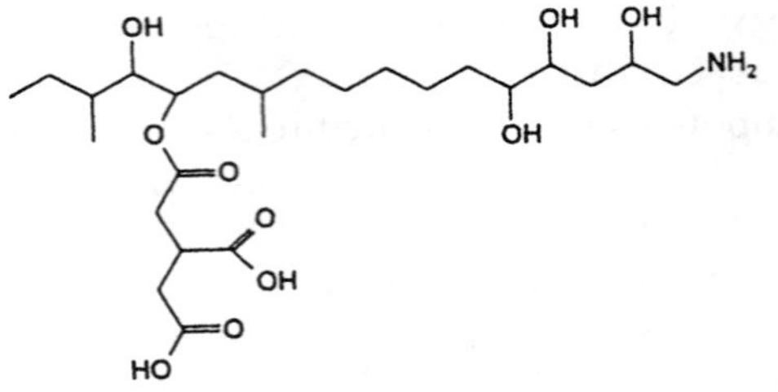

(b)

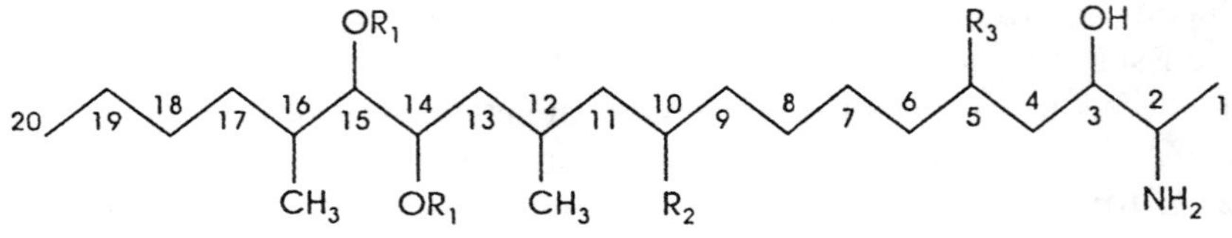

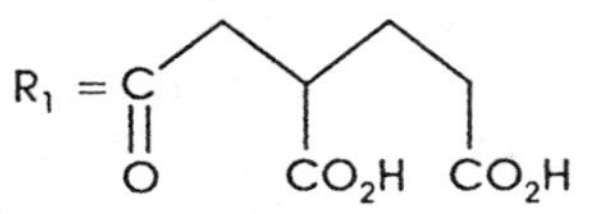

	R_2	R_3
Fumonisin B$_1$	OH	OH
Fumonisin B$_2$	H	OH
Fumonisin B$_3$	OH	H
Fumonisin B$_4$	H	H

Figure 8.5   (a) Structure of AAL toxin, a host-selective toxin from *Alternaria alternata* f. sp. *lycopersici*, which specifically affects tomatoes with the *asc/asc* genotype, and (b) structures of the fumonisin toxins from *Fusarium moniliforme*

completely different mechanism from those normally found in insensitive plants, i.e. toxin degradation or lack of recognition.

8.4.2   Host-selective toxins from *Helminthosporium* species and *Mycosphaerella zeae-maydis*

Victorin, the HST from Helminthosporium victoriae

In 1947, Meehan and Murphy demonstrated that *Helminthosporium victoriae* produced a toxin in culture that specifically affected oats derived from the cultivar Victoria (Meehan and Murphy, 1947). During the 1930s, crown rust of oats, so named because the crenulations around the uredospore of the fungus resemble a crown, caused by *Puccinia coronata* f. sp. *avenae* was a serious problem in the USA. However, the *Pc-2* gene from the cultivar Victoria, discovered in Uruguay, gave good resistance to the disease. It was therefore introgressed into the oat cultivars grown in the USA. Concomitantly, a new disease appeared which specifically attacked cultivars containing the *Pc-2* gene. The disease was named Victoria blight and the pathogen *Helminthosporium victoriae*, both names reflecting the parentage of susceptible cultivars. Subsequent experiments have failed to separate *Pc-2*, the gene conferring resistance to *P. coronata*

f. sp. *avenae* from *Vb*, the gene conferring susceptibility to Victoria blight. The genes are therefore either the same or very tightly linked (Walton, 1996).

Although Meehan and Murphy's original work on the specificity of the toxic effect of culture filtrates of *H. victoriae* for oats of Victoria parentage was published in 1947, the unusual chemical structures of the toxins were fully elucidated only 40 years later. They consist of a group of pentapeptides which are structurally related. The most abundant member of the group is victorin which contains three chlorine atoms and consists of covalently linked glyoxylic acid, 5,5-dichloroleucine, threo-β-hydroxylysine, erythro-β-hydroxyleucine, α-amino-β-chloroacrylic acid and 2-alanyl-3,5-dihydroxy-cyclopentenone-1 (Figure 8.6).

The genetics of victorin production have not been fully elucidated and will presumably be complex since the toxin is composed predominantly of non-protein amino acids, each of which requires at least one specific enzyme for its elaboration. The whole structure would then have to be assembled, presumably by a synthetase along the lines of HC-toxin (see below). Nevertheless, crosses between strains of *H. victoriae* which synthesize victorin and are infective for oats and strains of *H. carbonum* which synthesize HC-toxin and are infective for maize were informative. Four classes of progeny which segregated 1:1:1:1 were obtained. They were those that produced both toxins and were infective for both oats and maize, those that produced victorin and were infective for oats, those that produced HC-toxin and were infective for maize, and those that produced neither toxin and were not infective for either plant.

In susceptible cultivars the gross effect of the toxin is that leaves become a bronze colour but, at the biochemical level, the toxin causes many symptoms such as membrane depolarization, ion leakage and inhibition of protein synthesis. Wolpert and Macko (1989) found that an active ^{125}I-labelled derivative of the major toxin, victorin C, bound selectively *in vivo* to a 100 kDa protein only in susceptible genotypes and suggested that this protein was the toxin receptor. However, *in vitro* binding analysis showed that the labelled victorin derivative bound to a similar protein from resistant genotypes. Moreover, the proteins from both sources could not be distinguished immunologically (Wolpert and Macko, 1991). The protein was identified as the P protein subunit of glycine

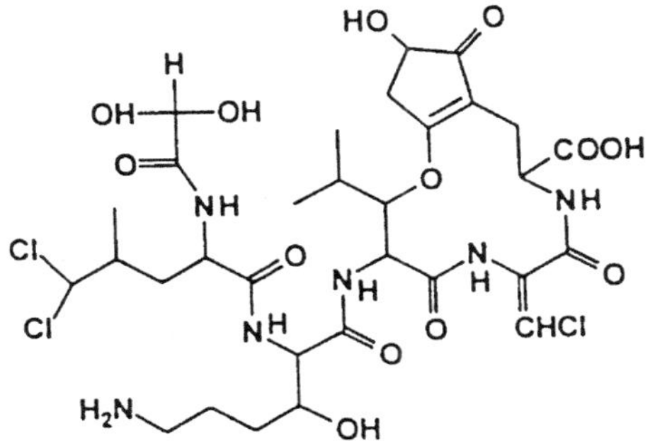

Figure 8.6 Structure of victorin, the major host-selective toxin from *Helminthosporium victoriae*

decarboxylase (GDC), a four component multienzyme complex (Wolpert *et al.*, 1994). GDC is located in the mitochondrial matrix and, together with serine hydroxymethyltransferase, catalyses the conversion of two glycine molecules to serine. Further work showed that victorin also bound to a 15-kDa protein which was identified on the basis of amino acid sequence data as the H protein component of GDC. Pre-treatment of leaf slices from susceptible plants with victorin at a concentration of 81 pM inhibited glycine decarboxylase activity by 50 per cent. Much higher concentrations (100 μg/ml) were required to inhibit GDC of resistant plants and the inhibition was only 26.5 per cent relative to control samples without toxin. Further evidence for GDC as a significant site of action of victorin has come from studies of light and carbon dioxide concentrations on symptoms (Navarre and Wolpert, 1999). GDC is required by plants to reclaim carbon and energy lost under the photorespiratory conditions of high light intensity and temperature and low carbon dioxide concentration. Furthermore, the extreme importance of the enzyme complex has been demonstrated by the fact that mutants of both *Arabidopsis* and barley compromised in GDC activity die under such conditions. Accordingly, the typical bronzing symptoms in leaves treated with toxin were moderated when they were kept in continuous dark, or in elevated CO_2 levels (1 per cent or 5 per cent) or in combinations of these treatments (Figure 8.7). Although these experiments provide powerful evidence for the inhibition of GDC as the mode of action of victorin C, the molecular basis of its specificity is still unresolved. Solution of this problem is keenly awaited since it is also likely to shed light on the mechanism of resistance of oats carrying the *PC-2* gene for resistance to crown rust.

HC-toxin from Helminthosporium carbonum

As in the case of victorin, HC toxin consists of a family of related compounds. Here four are known, but the most abundant is a cyclic peptide, consisting of D-proline, L-alanine, D-alanine and L-Aeo, where Aeo represents 2-amino-9,10-epoxi-8-oxodecanoic acid (Figure 8.8; Walton, Earle and Gibson, 1982; Rasmussen and Scheffer, 1988b).

Production of the toxin is controlled by several genes at a complex locus termed *Tox2*. This contains two copies of *HTS1* which encodes a large peptide synthetase of about 574 kDa. Disruption of both copies is required for the mutant to become non-toxigenic and avirulent for maize lines susceptible to the wild type (Panaccione *et al.*, 1992). *Tox2* is present on a 4-Mb chromosome in some toxigenic isolates and on a 2.3-Mb chromosome in others but in crosses of the two types the chromosomes in the progeny did not segregate independently. Had they done so 25 per cent of the progeny would have been expected to have inherited both chromosomes, 50 per cent one or other and 25 per cent neither. The last would be expected to be non-toxigenic. Instead, despite their difference in size, the chromosomes behave as homologues during meiosis, all progeny inheriting only one copy of the *Tox-2* locus. This suggests that the difference in

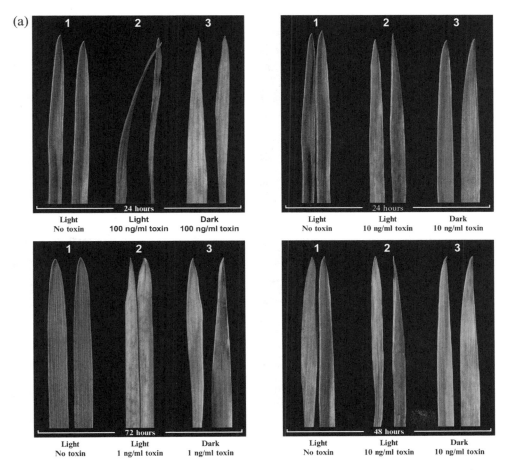

Figure 8.7   The effect of victorin on a susceptible cultivar of oat incubated for various times with 1, 10 or 100 ng of toxin; note the moderating influence of (a) darkness, and (b) high CO_2 concentration (courtesy of Tom Wolpert and Academic Press). A colour reproduction of this figure can be seen in the colour section

size of the *Tox-2* bearing chromosomes arose by translocation. Translocation could be an important means by which new pathogens or new races of pathogens could arise (Canada and Dunkle, 1997).

TOXC is another locus involved in production of HC toxin. It is present in most toxigenic strains as three copies on the same chromosome as the *Tox2* locus. When all three copies were mutated by targeted gene disruption the fungus grew and sporulated normally *in vitro* but was not pathogenic and did not produce HC toxin. Since *TOXC* has high homology to the beta subunit of fatty acid synthase from several eukaryotes it is plausible that it contributes to the synthesis of the decanoic backbone of Aeo and NMR data support this (Ahn and Walton, 1997; Cheng *et al.*, 1999).

(b)

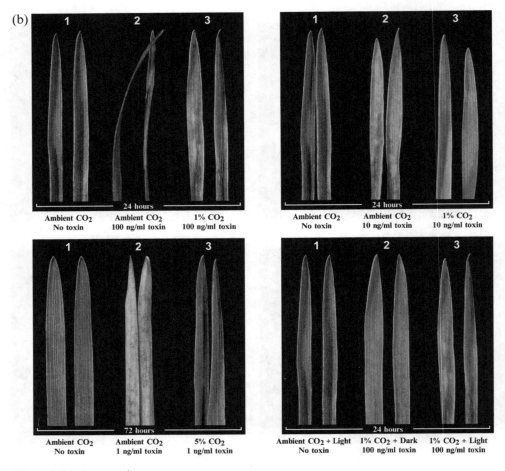

Figure 8.7   *Continued*

A universal problem facing all toxigenic organisms is the avoidance of poisoning themselves with their own toxin! Pitkin, Panaccione and Walton (1996) have found a gene in toxigenic strains of *H. carbonum* that may fulfil this role. The gene, named *TOXA*, is present as two linked copies and flanks the *HTS1* peptide synthetase gene. Sequence data predicted that the gene codes for a 58 kDa hydrophobic protein with 10–13 membrane spanning regions, a structure which is very similar to several members of a family of genes that confer antibiotic resistance. Targeted gene disruption experiments showed that it was possible to mutate one of the copies of the gene but not both. Taking all these data together, the authors proposed that *TOXA* codes for an HC-toxin efflux pump which protects the fungus from the toxin by secreting it.

Targeted gene disruption has shown that another locus in the *TOX2* complex, *TOXE*, is concerned with the regulation of toxin biosynthesis and pathogenicity. *TOXE* is required for the expression of *TOXA, TOXC* and

TOXD, the last being a gene that is unique to toxigenic isolates but whose involvement in toxin production has not been defined (Ahn and Walton, 1998).

The toxin is not stored in dormant spores of the fungus but is synthesised concomitantly with spore germination and this is matched by increasing amounts of transcripts of *HTS1* (Jones and Dunkle, 1995). Moreover, toxin production and appressorial formation appear to be coordinately regulated since spores which were incubated under conditions which did not induce appressoria also failed to synthesize toxin (Weiergang *et al.*, 1996).

Resistance of maize to *H. carbonum* and insensitivity to HC-toxin are dominant traits controlled by a single Mendelian locus, *Hm1*. Plants with the genotypes *Hm1/Hm1* or *Hm1/hm1* inactivate the toxin by reducing the carbonyl group on the side-chain of Aeo (Figure 8.8; Meeley and Walton, 1991; Meeley *et al.*, 1992). A duplicate gene, *Hm2*, has also been cloned but this gives only partial, adult resistance.

Sensitive cultivars with the genotype *hm1/hm1* are affected by doses of 100–1000 ng/ml and both the ketone and epoxide function of the 2-amino-9,10-epoxy-8-oxo-decanoic acid moiety are essential for activity (Kim, Knoche and Dunkle, 1987). Originally the primary site of action of the toxin was thought to involve the synthesis of chlorophyll. Rasmussen and Scheffer (1988a) hypothesized that this effect was achieved by the prevention of the formation of δ-aminolevulinic acid (ALA) and this was supported by experiments in which ALA reversed the effect of the toxin. Later evidence, however, suggests that histone deacetylase is the site of action. This enzyme reversibly deacetylates the core histones H3 and H4 in chromatin. In other systems alteration of the acetylation of these histones affects gene expression and in maize such genes might be involved in defence; for example, those encoding pathogenesis-related proteins or proteins that strengthen cell walls (Brosch *et al.*, 1995).

Multani and co-workers (1998) showed that resistance is the wild-type condition in maize and that susceptibility arose from a transposon insertion in *Hm1* and a deletion of *Hm2*. Moreover, homologues of these genes are present in other monocotyledonous plants but none of these is susceptible to *H. carbonum* race 1. In rice, the homologues were found in chromosomal regions that were syntenic with the maize loci, suggesting that they are related to the maize genes by vertical descent. Thus, the function of *Hm*-homologous sequences in plants may be specifically to defend them against HC toxin or related molecules. The authors further speculate that HC toxin or related compounds may have been important

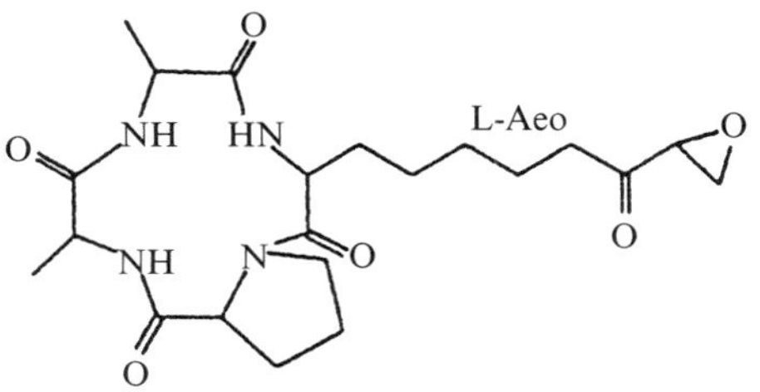

Figure 8.8 Structure of HC-toxin, the major host-selective toxin from *Helminthosporium carbonum*

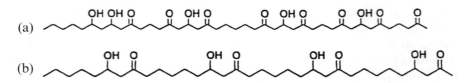

Figure 8.9 Structures of (a) the major toxin of *Helminthosporium maydis* race T, and (b) the toxin of *Mycosphaerella zeae-maydis*

in plant disease before the existence of the HC toxin reductase function encoded by the *Hm* homologues.

Polyketide toxins from *Helminthosporium maydis* race T and *Mycosphaerella zeae-maydis*

Helminthosporium maydis race T and *Mycosphaerella zeae-maydis* (anamorph, *Phyllosticta maydis*) both produce families of long chain polyketides (Figure 8.9) and both specifically affect maize with the T cytoplasmic factor for male sterility, *H. maydis* race T being responsible for the disastrous epidemic of Southern Corn Leaf Blight (Section 1.3.2). The role of the toxins in each disease differ in that T-toxin contributes to virulence, the non-toxin producing race O also causing disease, whereas toxin-minus mutants of *M. zeae-maydis*, developed by restriction enzyme mediated integration (REMI), are non-pathogenic (Yun, Turgeon and Yoder, 1998).

Two genes, involved in toxin production by *H. maydis* race T, have been identified: *Tox1A*, which is predicted to encode a polyketide synthase, and *Tox1B*, which is located on another chromosome. Since T-toxin is a mixture of neutral compounds with odd numbers of carbon atoms and condensation of acetate in polyketide synthesis gives rise to aliphatic acids with an even number of carbons, it is probable that *Tox1B* encodes a decarboxylase. Examination of the REMI mutants did not reveal integration in an ORF and therefore the genes involved in synthesis of the toxins in this species remain to be discovered (Kodama *et al.*, 1999).

The reason for specificity of the polyketide toxins from the two fungi for *Tcms* maize lies in their altered mitochondrial genome, giving rise to a chimeric open reading frame, t-URF13. This encodes a 13 kDa protein, URF-13, which is located on the inner mitochondrial membrane. When the gene is expressed in *E. coli*, yeast mitochondria or tobacco it confers sensitivity to T-toxin, probably owing to the formation of pores in membranes (Walton, 1996).

HS-toxin from *Helminthosporium sacchari*

The symptoms of eyespot of sugarcane, caused by *H. sacchari* consist of small eye-shaped lesions from which long red runners emanate and which may travel

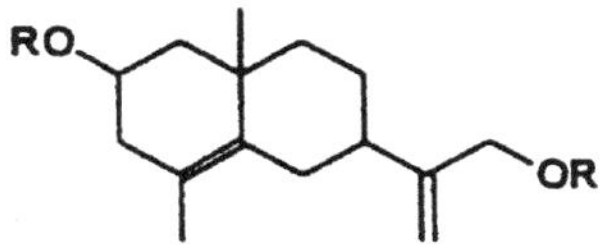

Figure 8.10   Structure of the toxin from *Helminthosporium sacchari* R = 5-O-(β-galactofuranosyl)-β-galactofuranoside

up the whole length of the leaf. When the fungus was cultured on a medium containing an extract of the host the filtrate induced similar lesions. The toxin is a terpenoid with two furano-digalactoside groups (Figure 8.10). Other toxin analogues have been found in cultural filtrates of the fungus but these are generally less active and may compete with the toxin (Nakajima and Scheffer, 1987). At temperatures above 32°C sugarcane is insensitive to the toxin and this is thought to reflect changes in a toxin receptor.

8.4.3   Host-selective toxins from other species

Proteinaceous toxins from Pyrenophora tritici-repentis

Tan spot caused by *Pyrenophora tritici-repentis* (anamorph, *Drechslera tritici-repentis*) is an important disease of wheat world-wide (for a review see Ciuffetti and Tuori, 1999). Two types of symptoms are recorded, tan necrosis and extensive chlorosis. Five races of the fungus have been described based on the phenotype on four differential wheat lines (*Triticum aestivum*). Races 1 and 2 cause both necrosis and chlorosis, race 2 only necrosis, races 3 and 5 producing chlorosis but on different cultivars and race 4 which produces neither symptom (Lamari *et al.*, 1995). Necrosis is caused by Ptr toxA, a 13.2 kDa protein, and chlorosis by Ptr toxB, a 6.61 kDa protein (Strelkov, Lamari and Ballance, 1998). Ptr toxA was shown unequivocally to be an HST by transforming a non-pathogenic, nontoxin-producing strain to a pathogenic state with the gene responsible for toxin production (Ciuffetti, Tuori and Gaventa 1997). Crosses of plants resistant to the fungus with those that were susceptible showed that the development of both types of symptom was controlled by four dominant and independently inherited genes (Gamba and Lamari, 1998). Martinez and co-workers (2001) have cloned the *ToxB* gene from race 5. The gene contains a 261 bp open reading frame that encodes a 23-amino acid signal peptide and the 64-amino acid toxin. Both the *ToxA* and the *ToxB* gene have been expressed heterologously, the former in *E. coli* and the latter in the yeast, *Pichia pastoris.* Another HST, Ptr ToxC, has been shown to be involved in the chlorosis response (Ciuffetti and Tuori, 1999). It is a polar, non-ionic, compound of low molecular weight (Effertz *et al.*, 2002). In the same paper the authors reported that insensitivity to the crude and partially purified toxin was located within 5.7 cM of the marker *Xglil* on the short arm of chromosome 1A.

Peritoxins from Periconia circinata

Periconia circinata is the cause of milo disease of *Sorghum bicolor*. Symptoms consist of root and crown rot. Only pathogenic isolates of the fungus produce the toxins which mimic the disease symptoms. These include inhibition of growth of primary roots and mitosis, induction of electrolyte leakage and up-regulation of a group of 16 kDa proteins. Culture filtrates of the fungus were fractionated by Macko and co-workers (1992) who described two toxic compounds, peritoxins A and B as well as two biologically inactive congeners, periconins A and B (Figure 8.11). Resistance to the toxins and the pathogen is recessive and is controlled by the *Pc* locus, i.e. only genotypes that are *pc/pc* are resistant. These results suggest that sensitive plants (*Pc/Pc* or *Pc/pc*) synthesize a receptor of the toxins but this does not seem to have been identified. However, the effects of the toxins are prevented if cells or tissues are pre-treated with inhibitors of RNA or protein synthesis, suggesting that protein synthesis or a protein with a high turnover rate is a prerequisite for sensitivity to the toxins (Dunkle and Macko, 1995).

SV toxins from Stemphylium vesicarium

Stemphylium vesicarium causes brown spot of European pear (*Pyrus communis*). The disease is of considerable economic importance and its control necessitates 15–25 applications of fungicide from fruit set to preharvest. Culture filtrates of the fungus were selectively toxic to cultivars of pear susceptible to the fungus. Singh and co-workers (1999) isolated two active compounds, designated SV-toxin I and SV-toxin II, but at the time of writing no structures appear to be available. The toxins cause veinal necrosis in susceptible cultivars at 10 and 100 ng/ml, respectively, but resistant cultivars were insensitive at 1 mg/ml. Electron microscope studies revealed that the toxins caused invaginations of the plasmalemma at both ends of plasmodesmata 3 h after treatment, suggesting that this is the site of action (Singh *et al.*, 2000).

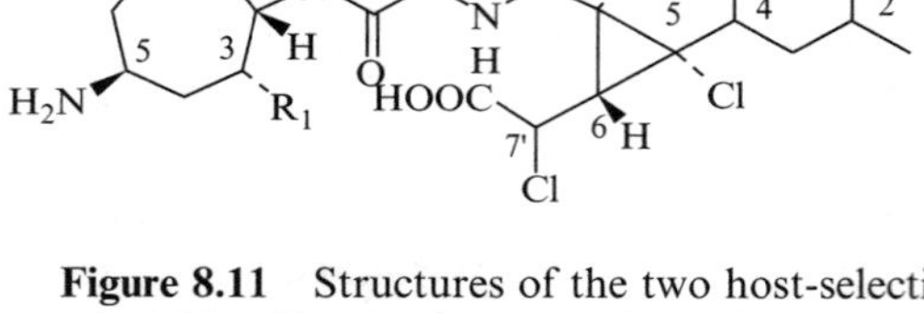

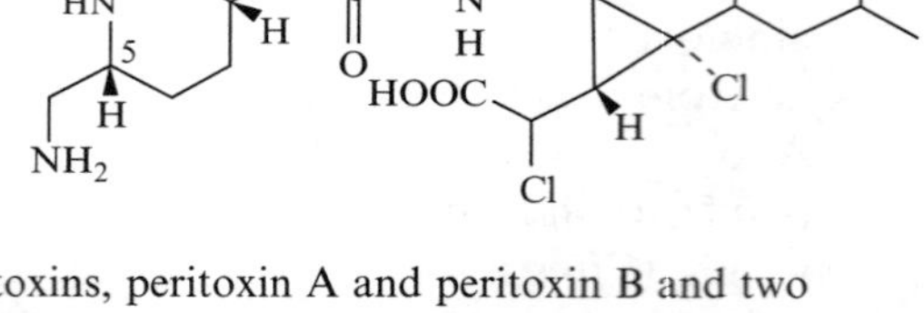

Figure 8.11 Structures of the two host-selective toxins, peritoxin A and peritoxin B and two related but inactive compounds, periconin A and periconin B from *Periconia circinata*, (a) R_1 = R_2 = OH peritoxin A, R_1 = OH, R_2 = H peritoxin B; R_1 = H, R_2 = OH periconin A (b) perconin B

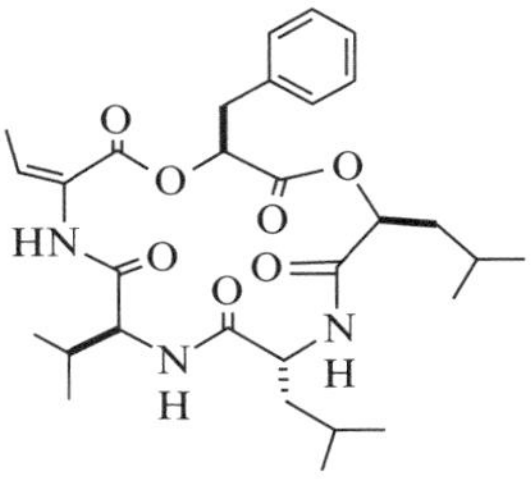

Figure 8.12 Structure of phomalide, a host-selective toxin from *Phoma lingam*

Phomalide from Phoma lingam

Phoma lingam is a pathogen of *Brassica* species and is well-known for the production of non-host specific toxins known as the sirodesmins (see Section 8.5.3). Pedras (1997), however, has shown that the fungus produces the HST phomalide (Figure 8.12). Originally the compound was missed as it is only produced in liquid cultures over a period of 24–60 h and its production is inhibited by sirodesmin PL. Moreover, production is also inhibited by the cruciferous phytoalexin, brassinin. Phomalide caused lesions on canola (*B. napus*) that closely resembled those caused by the pathogen, but mustard, which is resistant to the fungus, was only slightly sensitive to the toxin.

8.5   Non-host-selective toxins (NSTs)

Many non-host-selective toxins (NSTs) have been characterized, some organisms producing more than one class of compound and the same compound being produced by more than one organism. In the succeeding sections a selection of these phytotoxic compounds will be discussed according to the classification of the organisms rather than that of the compounds as, in some instances, this allows their role in pathogenesis to be discussed more easily. Moreover, there is evidence in a few cases, particularly in bacteria, of horizontal transfer of genes encoding enzymes required for toxin synthesis

The importance of NSTs in disease is more difficult to establish than that of HSTs since the former usually only contribute to virulence whereas the latter are necessary for pathogenicity. Evidence for the role of an NST usually relies upon one or more of the following principles, which will be referred to where relevant in discussions of individual toxins:

(1)   production of the toxin by the pathogen *in vitro*,

(2)   production of the toxin *in vivo*,

(3)   reproduction of disease symptoms with pure toxin,

(4)   correlation of virulence of the pathogen with toxigenicity,

(5)　correlation of susceptibility of the host with toxin sensitivity,

(6)　attenuation of virulence when genes involved in toxin synthesis are disrupted,

(7)　restoration of the virulence phenotype in toxin minus mutants by transformation with the genes required for toxigenicity.

The last two points are the strongest but, as in the case of cell-wall degrading enzymes, difficulties may arise in effecting gene disruption if there is more than one copy of the genes encoding toxin synthesis (cf. HC toxin Section 8.4.2).

8.5.1　NSTs from bacteria

Thaxtomins

Scab of potatoes is caused by four species of *Streptomyces* of which *S. scabies* and *S. acidiscabies* are the most commonly reported. Symptoms consist of tuber lesions with erumpent borders, resulting from hypertrophy, which may surround a central area of necrosis. Early work showed that the symptoms could be attributed to two piperazine toxins, given the trivial names thaxtomin A and thaxtomin B (Lawrence, Clark and King, 1990). More recently, a study of 37 isolates of *Streptomyces scabies*, which were drawn from eastern and central Canada and included five from the American Type Culture Collection (ATCC), showed that pathogenicity for potato was perfectly correlated with the production of thaxtomin (Loria *et al.*, 1997). Furthermore, Kinkel and co-workers (1998) in a study of 78 Streptomyces isolates, showed that thaxtomin A production in culture was significantly and positively correlated with the percentage of tuber surface infected. Thaxtomin A was the predominant compound produced but minor amounts of 10 other compounds have also been detected (Figure 8.13).

Two peptide synthetase genes, *txtA* and *txtB*, have been cloned from *S. acidiscabies*. Disruption of *txtA* abolished the production of thaxtomin A and these mutants were avirulent for potato. Both virulence and thaxtomin A production were restored by introduction of a thaxtomin synthetase cosmid

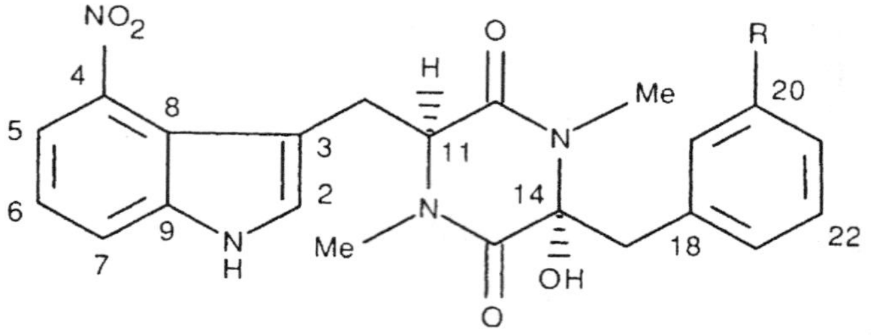

Figure 8.13　Structures of thaxtomins A (R = OH) and B (R = H), two piperazine toxins produced by species of *Streptomyces* that are important in the development of scab symptoms on potatoes

into the mutant, demonstrating the importance of the toxin in the disease syndrome (Healy *et al.*, 2000).

Thaxtomin A is derived from the combination and methylation of two modified amino acids, α-hydroxy-*m*-tyrosine and 4-nitrotryptophan whereas in thaxtomin B, α-hydroxyphenylalanine is substituted for α-hydroxy-*m*-tyrosine. Chemical studies have defined the attributes of the molecules required for toxicity: elimination of the nitro group, movement of the 4-nitro group to the 5, 6 or 7 positions of the indole ring, replacement of the phenyl side-chain, or conversion to a D,L configuration completely destroyed toxicity to potato tubers. Therefore the 4-nitrotryptophan and phenylalanine groups linked in an L,L-configured cyclodipeptide are requirements for phytotoxicity (King, Lawrence and Calhoun, 1992).

Both monocotyledonous and dicotyledonous plants are affected by thaxtomin A. At concentrations in the nanomolar range, the toxin causes stunting and hypertrophy of radish seedlings. This plant-growth regulator effect is also obvious in the erumpent lesions on potatoes that are typical of the disease. At higher concentrations the toxin causes cell death resulting in pitting of the tuber, also a symptom of the disease (Loria *et al.*, 1997).

Potato cultivars differ in their resistance to scab and some evidence points to the possibility that resistant cultivars are more able to detoxify thaxtomin A by glucosylation. The glucose derivative of the toxin is six times less toxic than the native compound to potato tuber slices and greater amounts of glucose transferase, the enzyme responsible for glucosylation, were present in resistant than in susceptible potato cultivars and selections (Acuna *et al.*, 2001).

Bukhalid, Chung and Loria (1998) have cloned a gene, *nec1*, from an isolate of *S. scabies* which conferred on a non-pathogen, *Streptomyces lividans*, the ability to colonize and cause necrosis of potato discs and produce scab-like lesions on immature potato tubers. The transformant did not produce thaxtomin A but did produce an unidentified water-soluble compound that caused necrosis of potato discs. Moreover, the gene was found in two other species of *Streptomyces*, *S. acidiscabies* and *S. turgidiscabies*. The authors suggest that the gene was acquired by horizontal transfer, possibly mediated by the gene ORF*ntp*, to which it is linked. The latter gene has a high level of identity to transposases found in members of the *Staphylococcus aureus* IS256 family (Bukhalid et al., 2002).

NSTs from pathovars and isolates of Pseudomonas syringae

Bender (1998, 1999) and Bender, Alarcon-Chaidez and Gross (1999) have reviewed NSTs from pathovars and isolates of *Pseudomonas syringae*.

Tabtoxin Tabtoxin is a toxin precursor produced by several pathovars and isolates of *Pseudomonas syringae*. It is a dipeptide of either threonine or serine linked to tabtoxinine-β-lactam, which is the active moiety (Figure 8.14). The

Figure 8.14 (a) Tabtoxin (R = CH₃) and (2-serine) tabtoxin (R = H), toxin precursors produced by several pathovars and isolates of *Pseudomonas syringae*; the compounds are rather unstable and rapidly convert to (b) the corresponding inactive δ-lactams–the active moiety is tabtoxinine-β-lactam

compound is rather unstable and rapidly converts to the inactive δ-lactam. Since tabtoxinine-β-lactam is structurally related to lysine, it is not surprising that labelling and genetic studies have shown some commonality of the pathways leading to the two compounds (Liu and Shaw, 1997).

Tabtoxin causes chlorosis and at one time the site of action was believed to be ribulose-1,5-bisphosphate carboxylase. This was found to be incorrect (Turner, 1986) and the mode of action of the toxin has been established as the irreversible inhibition of glutamine synthetase. As a result ammonia builds up insufficient amounts to cause chlorosis. One pathovar of *P. syringae*, designated *P. syringae* pv. *tabaci*, causes wildfire disease of tobacco (so-named because the disease spread like wildfire if the plants were not sufficiently widely spaced see Sections 3.2.1 and 5.6.2). A related species, previously called *P. angulata*, causes angular leaf spots in tobacco (the name referring to the fact that the lesions are bounded by the veins of the leaf which are arranged in an angular fashion). However, the two organisms could not be separated from each other, morphologically, serologically or physio-logically. Spontaneous tox⁻ mutants of *P. syringae* pv. *tabaci* arise *in vitro* at the high rate of 1 in 10^{-3} owing to the high frequency of excision of the tabtoxin gene cluster from the chromosome and these mutants gave lesions that were indistin-guishable from *P. angulata*, thus providing an explanation of the similarity of the organisms but the difference in symptoms they generate.

Lydon and Patterson (2001) have developed two sets of PCR primers that gave amplification products of 829 bp and 1020 bp corresponding to sequences in the *tblA* and *tabA* genes, which are required for tabtoxin production. When tested on 32 strains of *Pseudomonas syringae*, amplicons were only obtained with tabtoxin-producing pathovars, making this a useful test for toxigenic strains of the bacterium.

Protection of the bacterium's own glutamine synthetase from the toxin may be by adenylation, which renders the enzyme less sensitive to inactivation by tabtox-inine-β-lactam. Production of β-lactamases which hydrolyse the β-lactam ring to the non-toxic tabtoxinine is a second mechanism by which the bacterium may protect itself from its own toxin.

Phaseolotoxin *Pseudomonas syringae* pv. *phaseolicola*, the causal agent of halo blight of bean, also produces a toxin which induces chlorosis in its host and which is necessary for systemic invasion of the plant by the pathogen (Patil, Hayward and Emmons, 1974). The 'native' toxin is a tripeptide composed of homoargi-nine, alanine and N^6-(N^1-sulpho-diaminophospinyl)-L-ornithine and has been given the trivial name phaseolotoxin (Figure 8.15). The toxin is also produced by *P. syringae* pv. *actinidiae*, the causal agent of bacterial canker of kiwifruit (Sawada, Takeuchi and Matsuda, 1997). It is a potent and reversible inhibitor of ornithine transcarbamoylase (OCTase), an enzyme that catalyses the synthesis of citrulline from ornithine and carbamoyl phosphate. In the plant the 'native' toxin is attacked by peptidases to release N^6-(N^1-sulpho-diaminophospinyl)-L-ornithine (PSorn) and it is this compound that is the ultimate inhibitor of OCTase. As a result, the synthesis of arginine is blocked since production of its precursor, citrulline, is inhibited. This in turn leads to inhibition of protein accumulation and chlorosis is thought to be a consequence of the lower protein levels. Significantly, tissue that had become chlorotic as a result of toxin treatment, regreened on the addition of arginine (Turner and Mitchell, 1985).

The pathogen does not inhibit the synthesis of its own citrulline as it has two OCT genes one of which, *argK*, encodes an enzyme that is resistant to phaseolotoxin. Expression of this gene is coordinately regulated with phaseolotoxin production, both being produced at 18–20°C but not at 28–30°C, which is the optimum for growth of the bacterium. Hatziloukas and Panopoulos (1992) sequenced the gene and produced transgenic tobacco plants with enzyme activity. Some of these were insensitive to the toxin, suggesting that it may be possible to incorporate this pathogen derived resistance into bean, the natural host of the bacterium (see Chapter 12, Section 12.7.2).

Coronatine Coronatine is produced by *P. syringae* pv. *atropurpurea* and some other pathovars of this species. Bender, Malvick and Mitchell (1989) have shown that the genes for synthesis of this toxin by *P. syringae* pv. *tomato* are located on a plasmid and this is also true of *P. syringae* pv. *syringae* from cherry but in *P. syringae* pv. *morsprunorum* they were found in genomic DNA (Liang *et al.*, 1994). Detailed work with *P. syringae* pv. *glycinea* strain PG4180 has shown that the toxin is encoded by a single 18.8-kb operon. The toxin consists of two components, coronafacic acid and coronamic acid, linked by an amide bond (Figure 8.16). Coronafacic acid is synthesized from a branched polyketide and this is

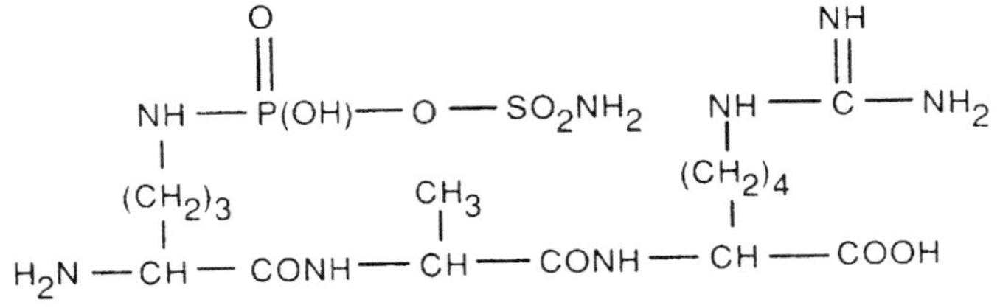

Figure 8.15 Phaeolotoxin, a toxin synthesized by *Pseudomonas syringae* pv. *phaseolicola* and *P. syringae* pv. *actinidiae* that reversibly inhibits the enzyme ornithine transcarbamoylase

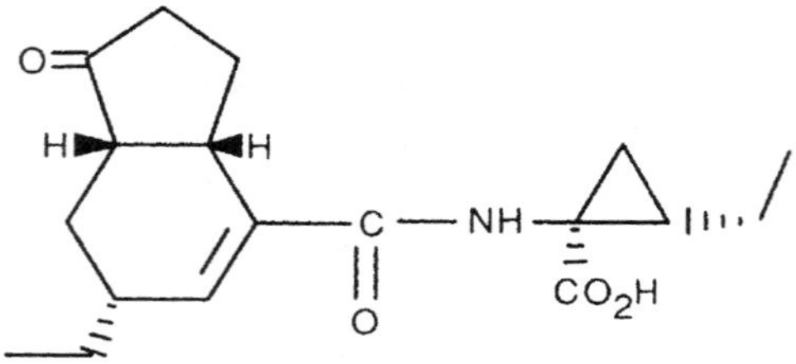

Figure 8.16   Structure of coronatine, a toxin synthesized by several pathovars of *Pseudomonas syringae* which may act by inhibiting active defence in the host

coupled to coronamic acid by the enzyme coronafacate ligase. The gene for this enzyme has been mapped to the 5′ end of the 18.8-kb operon and has been expressed in both *E. coli* and *P. syringae* as a translational fusion to the maltose-binding protein. These experiments showed that expression of the gene was temperature sensitive and also demonstrated that it contained an ATP-binding site which may be involved in activating coronamic acid before ligation by adenylation (Rangaswamy *et al.*, 1997).

Mitchell (1989) found that there was no rigid specificity of the coupling of neutral amino acids to coronafacic acid and that the toxin produced *in vitro* was predicated by the amino acid present in the medium. By growing the bacterium in media containing the appropriate amino acid, he obtained N-coronafacoyl-L-valine, N-coronafacoyl-L-alanine, N-coronafacoyl-L-norvaline, N-coronafacoyl-L-norleucine, N-coronafacoyl-L-isoleucine and N-coronafacoyl-L-allo-isoleucine.

Coronatine causes chlorosis in a number of plants and also hypertrophy of potato-tuber tissue but, at the time of writing, no specific receptor has been identified. Hypertrophy suggested that the toxin might mimic the action of one of the plant-growth substances such as auxins, cytokinins, gibberellins, abscisic acid or ethylene (see Chapter 7). Kenyon and Turner (1992) showed that some of the symptoms could be induced by indole acetic acid but that greater mimicry was achieved by application of the ethylene precursor 1-amino cyclopropane carboxylic acid (ACC). For example, the hypertrophy in potato tubers and necrosis in tobacco leaves caused by both coronatine and ACC were indistinguishable. Treatment of the hypocotyls of mung beans with coronatine caused a massive increase in the activity of the enzyme ACC synthase and its product ACC. The authors therefore concluded that the stimulation of ACC synthase leading to ethylene formation was the mode of action of the toxin.

Several authors have drawn attention to the similarities between the structures of coronatine and the plant-growth regulator, methyl jasmonate, which is derived from the octadecanoid signalling pathway and is elicited by stress (Weiler, 1997). As discussed in Chapter 11, Section 11.6.4, activation of this pathway leads to the expression of several defence responses. Therefore the possibility arises that coronatine might interfere with such defence reactions. Some evidence for this conjecture has been obtained by Mittal and Davis (1995). They found, in a comparison of two strains of *P. syringae* pv. *tomato*, one of which produced coronatine and the other which did not, that the toxin-producing strain multi-plied 10^4 to 10^6 times in leaves of *Arabidopsis thaliana* that had been dipped in a suspension of the bacterium and caused severe disease symptoms. In contrast, the tox⁻ strain only multiplied 10- to 70-fold and did not produce symptoms. Similar results were obtained in tomato, the natural host of the bacterium. When leaves of *A. thaliana* were inoculated by the harsher method of infiltration the tox⁻ strain also multiplied by 10^6 times over the experimental period of 4 days but only caused mild symptoms. Analysis of mRNAs of plants inoculated by infiltration with both strains showed that the tox⁻ strain induced higher levels of transcripts of three defence-related genes, phenylalanine ammonia lyase, glutathione-S transferase and ELI3 (see Chapter 11) than the tox⁺ strain. Thus coronatine may play a critical role in the early stages of infection by suppressing the active defence of the plant.

Further evidence for this view has been established by the study of mutants of *Arabidopsis thaliana* that are insensitive to coronatine. For example, Feys and co-workers (1994) found that such a mutant, *coil*, was more resistant to a *Pseudomonas syringae* pv. *atropurpurea* and was also male sterile and insensitive to methyl jasmonate. An allele of *coil* designated *coil-20* was isolated in a screen for mutants of *A. thaliana* with increased resistance to *Pseudomonas syringae* (Kloek *et al.*, 2001). Resistance was correlated with the accumulation of elevated levels of salicylic acid and the hyperactivation of *PR-1* expression, a gene encoding a pathogenesis-related protein involved in defence (see Chapter 11, Section 11.3.6) again leading to the conclusion that coronatine promotes virulence by inhibiting defence responses.

Lipodepsinonapeptides   The liopodepsinonapeptide toxins, syringomycin and syringopeptin, produced by *Pseudomonas syringae* pv. *syringae* cause necrotic symptoms in plants infected by the bacterium. The toxins consist of a polar peptide head and a hydrophobic 3-hydroxy fatty acid tail, which may vary in length. For example, three forms of syringomycin are produced in which the fatty acid component is decanoic, dodecanoic or tetradecanoic. The fatty acid is at-tached to an N-terminal serine residue of the peptide by an amide bond and to the C-terminus by an ester linkage to 4-chlorothreonine to form a macrocyclic ring. Chlorination of the compound is necessary for toxic activity and besides 4-chlor-othreonine the toxin contains several other non-protein amino acids. Both toxins form multimers which insert into membranes and at nanomolar concentrations cause the formation of pores of about 1 nm diameter. As a result, ion transport across the plasmamembrane is disrupted resulting in lysis (Bender, Alarcon-Chaidez and Gross, 1999). The contribution of the toxins to virulence was shown by mutation analysis, those mutants that were defective in the synthesis of both toxins being reduced in virulence by 60–70 per cent (Scholz-Schroeder *et al.*, 2001a).

The synthesis of syringomycin and syringopeptin occurs by a nonribosomal thiotemplate mechanism and is controlled by two gene clusters, *syr and syp* which, in strain B301D, are adjacent to each other (Scholz-Schroeder *et al.*, 2001a). Several genes are found in each cluster and their individual contributions

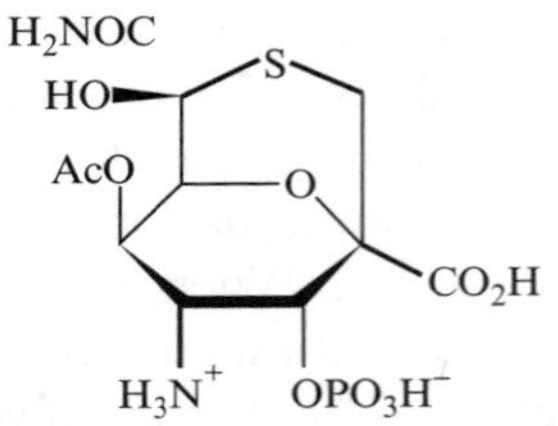

Figure 8.17 The structure of tagetitoxin from *Pseudomonas syringae* pv. *tagetis* the toxin is an inhibitor of chloroplast RNA polymerase

to toxin synthesis and its regulation are the subject of active experimentation at the time of writing (Lu, Scholz-Schroeder and Gross, 2002). The most striking gene in these clusters is *syrE* which is predicted to encode a 1039 kDa nonribosomal peptide synthetase containing eight amino acid activation modules.

Tagetitoxin   This toxin is produced by *Pseudomonas syringae* pv. *tagetis* (Figure 8.17). The bacterium is a pathogen of several species of plants in the Compositae family including African marigold (*Tagetes erecta*) from which the pathovar name of the organism is taken, sunflower (*Helianthus annuus*) and dandelion (*Taraxacum officinale*). Apical necrosis is one of the most striking symptoms of infection and the bacterium also causes necrotic leaf spots which are sometimes accompanied by chlorotic halos. The toxin acts by inhibiting chloroplast RNA polymerase (Matthews and Durbin, 1990; Lukens, Mathews and Durbin, 1987).

Albicidin

Leaf scald, caused by *Xanthomonas albilineans*, is a serious disease of sugarcane. The xylem-invading bacterium produces several toxins of which the major one is albicidin which has only been partially characterized. It is a low molecular weight compound with several aromatic rings and is bactericidal to Gram-positive and Gram-negative bacteria at concentrations as low as 1 ng/ml. Symptoms of leaf scald include chlorosis, necrosis, wilting and sometimes sudden death of the plant after a long latent period. The chlorotic symptoms are caused by inhibition of DNA synthesis in chloroplasts caused by the toxin. Tox⁻ mutants are unable to cause disease symptoms. Recently a gene has been cloned from *Pantoea dispersa* that codes for a peptide which detoxifies albicidin. Transformation of sugarcane with the gene confers resistance against the bacterium (Zhang, Xu and Birch, 1999, and see Chapter 12, Section 12.7.2).

8.5.2   NSTs from Oomycetes

Production of NSTs has been studied intensively in only a few Oomycetes despite the massive necrosis engendered by many organisms belonging to this group of plant pathogens. Several 10 kDa proteins, termed elicitins, have been isolated from species of *Phytophthora* and *Pythium*. Although their most significant property in a plant pathology context is their ability to elicit active defence responses (see Chapter 11, Section 11.4.2), in some cases they appear to perform a dual role as both toxins and elicitors. Perez and co-workers (1997) for example, have mapped the elicitor and necrosis inducing sites of *Phytophthora* elicitins and Boissy and co-workers (1996) have suggested that it may prove feasible to engineer a non-toxic elicitin that functions purely as an elicitor rather than a toxin.

Phytophthora nicotianae is a serious pathogen of tomato, causing crown and root rot. It secretes a peptide, phytophorin, of 3.3 kDa consisting of 25 amino acids, which is considerably smaller than the elicitins. The peptide caused wilting and chlorosis in tomato leaves at 7.5 and 4.5 μM concentrations, respectively. A feature of the toxin is that it forms aggregates with non-toxic components with probable molecular weights ranging from 3.5 to 30 kDa which are heat and pH stable and cause electrolyte loss from tomato tissue (Capasso *et al.*, 1997).

8.5.3 NSTs from fungi

NSTs from Deuteromycetes and Ascomycetes

Many toxins have been described from these two groups of fungi, which are considered together as a number of Deuteromycetes are known to have Ascomycete perfect stages, e.g. *Fusarium moniliforme* (anamorph) = *Gibberella fujikuroi* (teleomorph) and *Leptosphaeria maculans* (anamorph) = *Phoma lingam* (teleomorph). In particular, a large number of phytotoxic compounds are produced by species of the genera *Alternaria* and *Fusarium*.

Toxins from Leptosphaeria maculans (anamorph Phoma lingam)   *L. maculans* is a devastating pathogen of cruciferous crop plants, causing losses in Canada alone in excess of $30 million (Pedras *et al.*, 1996). Virulent isolates produce epipolythiodioxopiperazine toxins such as sirodesmin PL (Figure 8.18) and the depsipeptide HST, phomalide (Figure 8.12 and see Section 8.4.3). The toxicity of sirodesmin PL has been attributed to the reactivity of its disulphide bridge and may be reversed by zinc ions (Rouxel, Kollmann and Bousquet, 1990).

Trichothecene toxins from Gibberella species   Species of the genus *Gibberella* (anamorph *Fusarium*) have long been known as the producers of several mycotoxins. One class of these toxins, the trichothecenes, has received particular attention since they are potent inhibitors of protein synthesis in eukaryotic cells (Figure 8.19). Some evidence points to a role of trichothecenes in virulence of plant pathogenic species for their hosts. For example, the virulence of *Gibberella zeae* (anamorph *Fusarium graminearum*) for wheat has been correlated with the

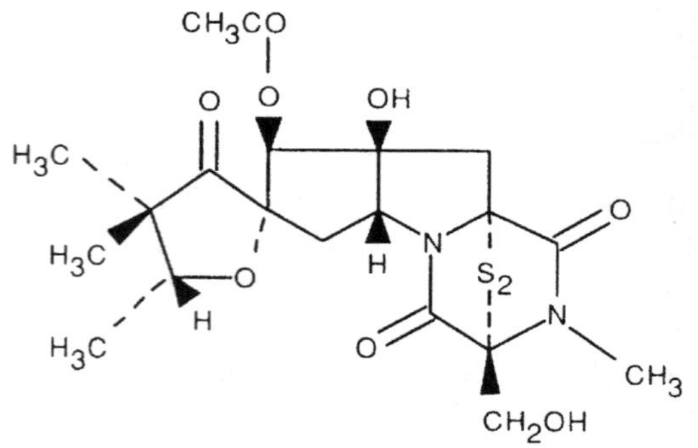

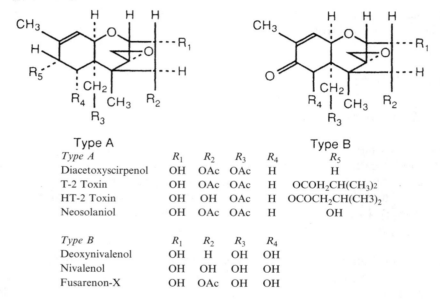

Figure 8.18   The structure of sirodesmin PL, a toxin from *Leptosphaeria maculans* containing a disulphide bridge

Type A				
Type A	R_1	R_2	R_3	R_4	R_5
Diacetoxyscirpenol	OH	OAc	OAc	H	H
T-2 Toxin	OH	OAc	OAc	H	OCOH$_2$CH(CH$_3$)$_2$
HT-2 Toxin	OH	OH	OAc	H	OCOCH$_2$CH(CH3)$_2$
Neosolaniol	OH	OAc	OAc	H	OH

Type B	R_1	R_2	R_3	R_4
Deoxynivalenol	OH	H	OH	OH
Nivalenol	OH	OH	OH	OH
Fusarenon-X	OH	OAc	OH	OH

Figure 8.19   The structures of some important trichothecene toxins from the fungal genus *Fusarium*

ability to synthesize trichothecenes *in vitro* and resistant cultivars were more proficient in degrading the compounds than susceptible ones (Miller and Arnison, 1986). Other experiments which support a role for trichothecenes in virulence of *Gibberella* species for their plant hosts have involved disruption of genes required for their biosynthesis.

Desjardins, Gardner and Weltring (1992) showed that disruption of the gene *Tri5* which is required for trichothecene production reduced the virulence of *G. pulicularis* (anamorph *Fusarium sambucinum*) for parsnip roots. Similarly, Proctor, Hohn and McCormick (1995) disrupted the same gene in *G. zeae*. The transformants were screened for inability to produce trichothecenes and by Southern blot to confirm that disruption of the gene had occurred. *Tri5*⁻ mutants were less virulent on seedlings of the wheat cultivar Wheaton and plants of this cultivar developed head-blight symptoms more slowly when inoculated at anthesis in growth chambers. Virulence on seedlings of common rye was also reduced but was maintained at wild-type levels in Golden Bantam maize. Confirmation of reduced virulence of $Tri5^-$ mutants on wheat in the field was obtained when the cultivars Wheaton and Butte 86 were inoculated at anthesis (Desjardins et al., 1995). Reciprocally, Nicholson and co-workers (1998) found, using a competitive PCR assay for fungal DNA, that colonization of wheat grain by trichothecene producing strains of F. graminearum was greater than non-producing strains. In further experiments with maize, Harris and co-workers (1999) showed that $Tri5^-$ mutants were still pathogenic for maize but were less virulent than the wild-type strain from which they were derived when inoculated through the silks. They conclude that the trichothecenes may enhance the spread of the fungus in this host.

Alexander, McCormick and Hohn (1999) have isolated a gene, Tri12, which appears to act as a trichothecene pump, protecting the producing fungus, Fusarium sporotrichioides from the effects of these toxins. When the gene was disrupted, growth on complex media and levels of trichothecene were reduced; moreover, incorporation of trichothecene in the medium inhibited growth of the mutant.

Introduction of an O-acetyl group at the C-3 position is one means by which trichothecenes may be rendered less toxic. Kimura and co-workers (1998) cloned a gene responsible for this activity, i.e. trichothecene 3-O-acetyltransferase from F. graminearum in the fission yeast Schizosaccharomyces pombe, transformants being selected by their ability to grow in the presence of T-2 toxin. Thus the possession of this gene provides another means, besides the efflux system, for protecting fungi that produce trichothecenes from their effects.

The fumonisins   The fumonisin toxins were discovered as a result of an investigation into the high level of oesophogeal cancer in the Transkei region of South Africa. They were isolated from cultures of Gibberella fujikuroi (anamorph Fusarium moniliforme) grown on maize and the most active compound, which has structural features similar to that of a toxin from Alternaria alternata f. sp. lycopersici (AAL toxin, see Section 8.4.1 and Figure 8.5) was designated fumonisin B_1 (FB_1). In a quantitative assay, Abbas and co-workers (1999) found that the concentration of FB_1 that caused half the maximal toxicity for Arabidopsis thaliana was less than 1 ppm. As stated previously (Section 8.4.1), FB_1 is a sphinganine analogue which, in both plant and animal cells, competitively inhibits sphingolipid biosynthesis, causing sphingoid bases to accumulate.

When a strain of Gibberella fujikuroi lacking the fum1 gene, which is necessary for fumonisin production, was crossed with a $fum1^+$ the progeny exhibited a range of virulence. However, high levels of virulence were always associated with production of fumonisins. Nevertheless, a survey of field strains did identify a fumonisin non-producer which was 'quite high in virulence' (Desjardins et al., 1995).

The solanapyrones   The solanapyrones were first isolated from culture filtrates of *Alternaria solani*, the causal agent of early blight of potato, by Ichihara, Tazaki and Sakamura (1983, Figure 8.20). Subsequently, Alam and co-workers (1989), using cells isolated from leaves of chickpea in a live/dead cell assay, found solanapyrones A and C in culture filtrates of an entirely different fungus, *Ascochyta rabiei* (teleomorph, *Didymella rabiei*), a pathogen that causes a devastating blight of chickpea. Formulation of a defined growth medium for the fungus allowed the production of solanapyrone B as well as solanapyrones A and C (Chen and Strange, 1991) and a mobile phase was developed which enabled all three compounds to be separated isocratically by reversed phase HPLC (Chen, Peh and Strange, 1991). Although the bioassay (see Section 8.3) and the structures of the compounds suggested that membranes might be the site of attack, Mizushina and co-workers (2002) have demonstrated that solanapyrone A is an inhibitor of DNA repair polymerase α and β.

Direct evidence for the role of the solanapyrone toxins in either early blight of potatoes or Ascochyta blight of chickpea is lacking but several lines of indirect evidence suggest that they may be pathogenicity or virulence factors for chickpea. First, all reliably identified isolates of the fungus produce at least one of the toxins in culture, including 20 isolates from Turkey whose identity was confirmed by the sequences of the ITS1 and ITS2 regions of their rDNA (Dolar and Strange, unpublished results). Since the fungus reproduces sexually on debris remaining on the ground after harvest, loss of toxigenicity would be likely to occur were it not necessary for pathogenicity or virulence. Second, symptoms of the disease – epinasty, chlorosis, necrosis and breakage of stems and petioles – are mimicked by the toxins. The last in particular is characteristic of the disease and, as previously suggested (Section 8.2), may be explained by the loss of turgor of parenchyma cells surrounding the stele as a consequence of the attack by the toxin on their plasmamembranes. Such stems would have only the inadequate support of their steles and not the additional strength imparted by the turgor of the surrounding cells (Figure 8.1). Use of the fluorescein diacetate assay to detect and monitor the isolation of the solanapyrone toxins from *A. rabiei* clearly demonstrates that the toxins (Section 8.3) attack the integrity of the plasma membrane of chickpea cells. Comparison of the order of sensitivity of cells isolated from several cultivars with literature records of disease scores on the unreliable 1–9 scale showed that there was a positive correlation between sensitivity to

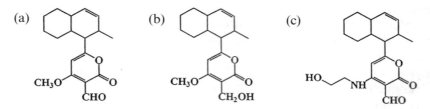

Figure 8.20 Structures of the solanapyrones, toxins found in the culture filtrates of *Alternaria solani* and *Ascochyta rabiei* (a) solanapyrone A, (b) solanapyrone B, and (c) solanapyrone C

solanapyrones A and B and severity of disease symptoms, although this was not significant (Hamid and Strange, 2000). Whether a reliable and standard method for scoring disease and extension of the experiment to other cultivars differing in resistance to the pathogen will show a significant correlation remains to be seen.

One reason for differences in sensitivity of the cultivars of chickpea to solana-pyrone A may lie in the glutathione/glutathione-S-transferase system by which plants detoxify xenobiotics. Inclusion of glutathione in the assay reduced toxin titres and solanapyrone A formed an adduct with glutathione spontaneously. Moreover, there was an inverse and significant correlation of solanapyrone A sensitivity of cultivars with both glutathione content and glutathione-S-transfer-ase activity. One way of boosting the glutathione concentration and glutathione-S-transferase activity of plants is to use compounds termed safeners, which are normally used to protect crops from herbicides. Cells from chickpea plants treated with the safener, dichormid, were less sensitive to solanapyrone A in the bioassay (Hamid and Strange, 2000).

Toxins from Alternaria species   Species from this genus produce a wide array of compounds with biological activity, some of which are phytotoxic and may be significant virulence factors. Many of them may be classified chemically as lactones, perylenequinones, anthraquinones and cyclic peptides.

Brefeldin A is a lactone toxin produced by *Alternaria carthami*, an important pathogen of safflower (*Carthamus tinctorius*) in which it causes a leaf-spot disease and wilting (Figure 8.21).

Several perylenequinones have been described from species of *Alternaria* such as the altertoxins (Figure 8.22) which have phytotoxic activity but most interest

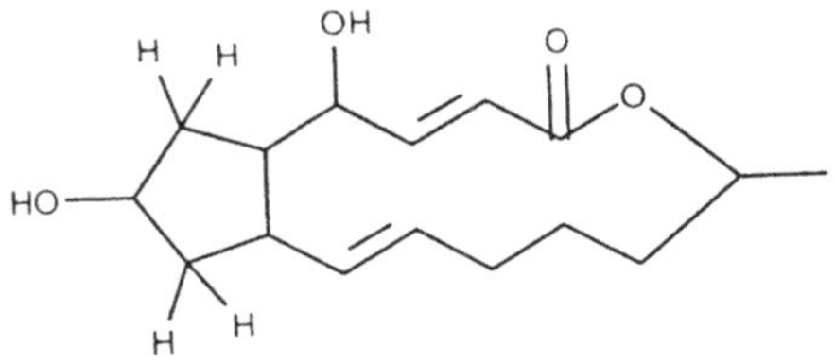

Figure 8.21   Brefeldin A, a toxin produced by the safflower pathogen, *Alternaria carthami*

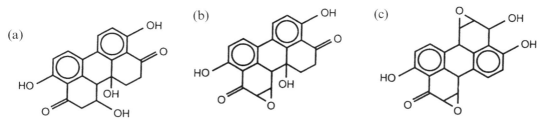

Figure 8.22   The altertoxins which are not only phytotoxic but mycotoxic (i.e. toxic to animals) and mutagenic: (a) altertoxin I, (b) altertoxin II, and (c) altertoxin III

has centred on their activity as mycotoxins and mutagens. For example, although altertoxin II is toxic to tomato leaves, it is also mutagenic in the Ames test at 3.4 µg/plate with *Salmonella typhimurium* as test organism and toxic to lung fibroblast cells of Chinese hamster at 0.02 µg/ml.

The cyclic peptide tentoxin from *A. alternata* is perhaps the most intensively studied toxin from a species of *Alternaria* (Figure 8.23). It is synthesized by a multienzyme of about 400 kDa which produces dihydrotentoxin and this is converted by another enzyme to tentoxin. The toxin causes chlorosis in germinating seedlings of some dicotyledonous plants but not maize, tomato and members of the Cruciferae and Graminae (Durbin and Uchytil, 1977). In isolated thylakoids it inhibits ATP synthesis at micromolar concentrations by its action on coupling factor 1 of the chloroplast. Substitution of aspartate for glutamate at position 83 in the β-subunit of this ATPase confers tentoxin sensitivity on plants that are normally resistant (Avni *et al.*, 1992). Holland and co-workers (1997) found that chlorosis of tobacco seedlings did not correlate with inhibition of ATP synthesis but rather with overenergization of thylakoids. They suggest that this leads to the generation of oxygen radicals and support this view by three experiments: the protonophore, gramicidin D, relieved both overenergization and chlorosis; ascorbate, which quenches free radicals, protected against chlorosis; transgenic plants overexpressing superoxide dismutase were partially resistant to tentoxin.

Toxins from Fusarium   Several toxigenic *Fusarium* species have been described for which no perfect stage is known. For example, fumonisins are produced by *F. proliferatum* and *F. nygami* and a number of *Fusarium* species produce enniatins. These are cyclohexadepsipeptides (Figure 8.24). Hermann, Zocher and Haese (1996) have shown the importance of these compounds in the virulence of *F. avenaceum* for potato tuber tissue by gene disruption. Enniatins are synthesized by a multienzyme, enniatin synthetase, encoded by the gene *esyn 1*. The gene was disrupted by homologous recombination using a fragment of *esyn 1* from *Fusarium scirpi*. Four independent mutants were obtained that did not express enniatin synthetase and were significantly reduced in virulence.

Naphthazarin toxins are produced by *Fusarium solani* and these compounds are present in the xylem of diseased citrus trees infected with the fungus (Nemec,

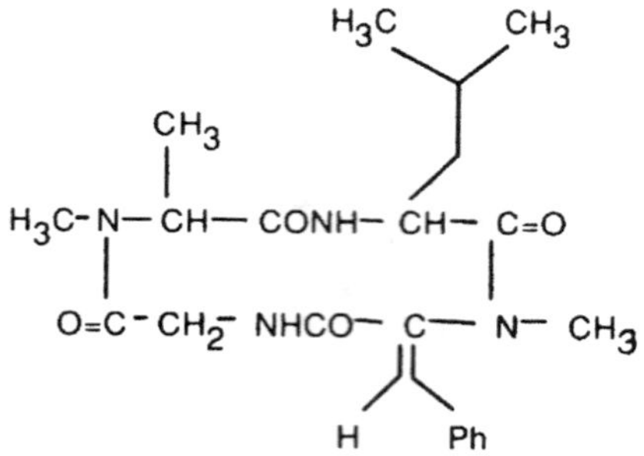

Figure 8.23   Tentoxin, a toxin produced by some isolates of *Alternaria alternata* that causes chlorosis in some dicotyledonous plants by overenergizing thylakoids

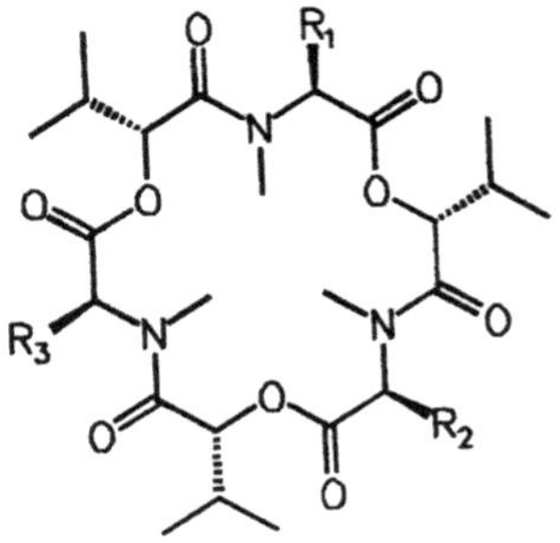

Figure 8.24 Structures of enniatin toxins and beauvericin: Enniatin A, $R_1 = R_2 = R_3 = $ sec-butyl; Enniatin A, $R_1 = $ isopropyl, $R_2 = R_3 = $ sec-butyl; Enniatin B, $R_1 = R_2 = R_3 = $ isopropyl; Enniatin B1, $R_1 = R_2 = $ isopropyl, $R_3 = $ sec-butyl; Beauvericin, $R_1 = R_2 = R_3 = $ benzyl

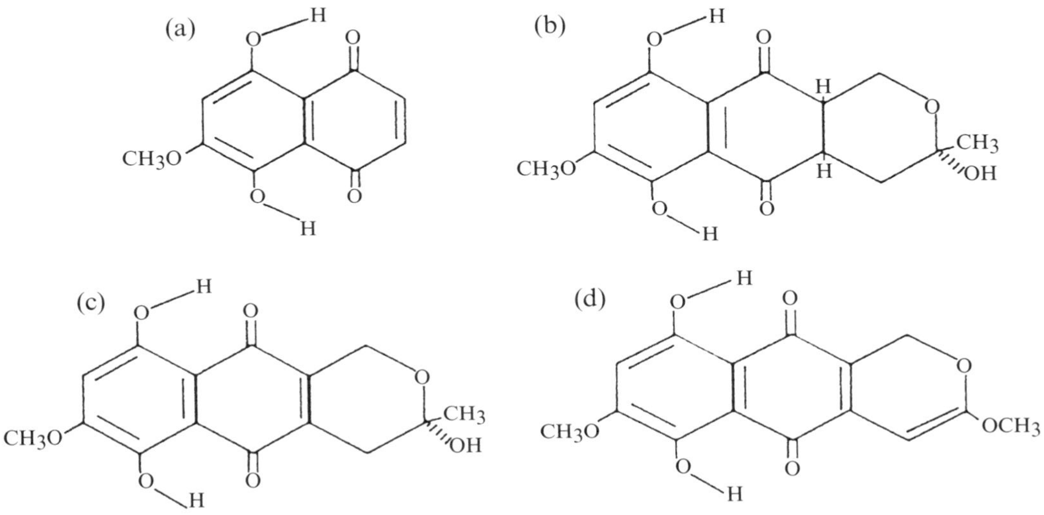

Figure 8.25 Structures of naphthazarin toxins produced by *Fusarium solani* and present in citrus trees infected by the fungus (a) methoxynaphthazarin, (b) dihydrojusarubin, (c) fusarubin, and (d) anhydrofusarubin

1995; Figure 8.25). The author suggests that since the compounds enhance membrane permeability this might explain the disruption of the hydraulic conductivity in roots of blighted trees and its eventual breakdown.

NSTs from Basidiomycetes

The Basidiomycetes comprise a large number of plant pathogenic fungi but many of these are biotrophs such as the rusts and smuts which, since they rely upon the living host, would be unlikely to produce highly phytotoxic compounds. Nevertheless, there are a number of necrotrophic pathogens of trees. One of these, *Heterobasidion annosum*, produces several low molecular weight toxins (Zweimüller *et al.*, 1997). Of these, fomannoxin appears to be the most toxic, inhibiting

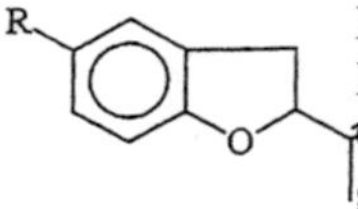

Figure 8.26 Structures of fomannoxin and its detoxification products: R = CHO, fomannoxin; R = CH$_2$OH, fomannoxin alcohol; R = COOH, fomannoxin acid; R = COO-β-glucosyl, fomannoxin acid-β-glucoside

the growth of several organisms as well as that of plant cells and protein synthesis in protoplasts of *Picea abies* and *Nicotiana tabacum* (Sonnenbichler *et al.*, 1989; Figure 8.26). Callus and suspension cultures of *Pinus sylvestris* were also inhibited by fomannoxin but the suspension cultures were able to detoxify fomannoxin by reducing the aldehydic group to the corresponding alcohol and by oxidation and conjugation to give fomannoxin carboxylic-β-glucoside.

8.6   Control of toxin biosynthesis

In some plant pathogens, toxin biosynthesis *in vitro* appears to require factors found in the host. For example, *Helminthosporium sacchari* toxin was only synthesized when an extract of sugar cane was included in the medium. The active factor was identified as serinol (2-amino-1,3-propanediol, Pinkerton and Strobel, 1976). Similarly, arbutin, a phenolic glucoside, and D-fructose were identified as signal molecules which were either necessary for the induction of syringomycin in *Pseudomonas syringae* pv. *syringae* or enhanced its production (Quigley and Gross, 1994). Mo and co-workers (1995) analysed the leaves of sweet cherry for compounds that activated the *syrB* gene which is required for the synthesis of syringomycin (Section 8.5.2). Using a *syrB–lacZ* fusion to monitor fractionation of the leaf extracts they identified two flavonol glycosides, quercetin 3-rutinosyl-4′-glucoside and kaempferol 3-rutinosyl-4′-glucoside and a flavanone glucoside, dihydrowogonin 7-glucoside. The flavonoid glycosides were similar in activity to arbutin and their activity was enhanced by D-fructose which itself had low activity.

Production of the solanapyrone toxins by *Ascochyta rabiei* was originally also thought to require an unidentified component from the host. Although the fungus grew on Czapek Dox liquid medium it did not produce the toxins unless the medium was supplemented with a hot-water extract of chickpea seed. Substitution of the chickpea extract with a complex mixture of amino acids, vitamins and metal cations gave toxin titres that were higher than the standard hot water extract and all three solanapyrones (A, B and C) were routinely obtained (Figure 8.20). Systematic elimination of components of the supplement showed that the cation fraction stimulated fungal growth and was necessary for toxin production. Zinc was essential for toxin production and omission of manganese or calcium or cobalt reduced toxin production by 74 per cent, 59 per cent and 35 per cent, respectively (Chen and Strange, 1991).

Several other reports in the literature suggest that host factors may be of importance in the production of toxins but the compounds concerned have not been identified. For example, Beausejour and co-workers (1999) attempted to identify the compounds required for the synthesis of the thaxtomins. They found that oat bran was the best medium for production of thaxtomin A and suberin was the only plant polymer that allowed the production of the toxin in minimal medium. They suggest that extracellular esterases allow the release of toxin inducing molecules from lipid precursors.

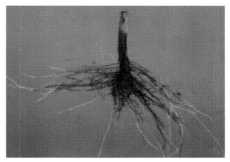

Figure 3.4 Symptoms of *Stagnospora nodorum* (= *Septoria nodorum*) on wheat leaves and ears (copyright Peter Scott)

Figure 3.6 Pepper showing roots infected by *Phytophthora capsici* (courtesy of Dr Jean Beagle Ristaino, Department of Plant Pathology, North Carolina State University)

Figure 5.4 (a) An aerial photograph of a wheat field – the dark transverse strips are the solarized plots, compared with the light transverse non-solarized plots (courtesy of J. Katan); (b) control of *Orobanche* sp. in a field of carrots by solarization – the whole field was solarized except for one plot (control) in which the plants are yellow and stunted (courtesy of J. Katan)

Figure 5.5 Cantaloupe melons devastated by *Phytophthora drechsleri*; water from a bore hole had been allowed to flood the irrigation furrows (see top left of picture), carrying with it zoospores of the fungus which were able to infect the highly susceptible crown of the plant (courtesy of Dr Ahmad Alavi, Plant Pests and Diseases Research Institute, Tehran, Iran)

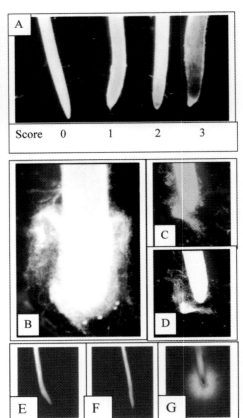

Figure 6.4 Infection of pea roots with *Nectria haematococca*: (a) only about 10 per cent of pea seedlings developed lesions at the root tip, comprising a 1–2 mm region consisting of the root cap and apical meristem – these were scored on a 0–3 scale; (b) some root tips scored as 0 were found, on microscopic examination, to be covered in a mantle of the fungus; (c) and (d) in some cases the mantle detached spontaneously on immersion in water; (e) when root tips from which mantles had become detached were surface sterilized and plated onto culture medium, no fungus grew as in controls (f); in contrast, where lesions were scored, the surface sterilized roots yielded copious fungal growth (g) (courtesy of Martha Hawes and the American Society of Phytopathology)

Figure 6.6 Germinating seed of *Striga hermonthica* giving rise to a haustorium which has attached to a host root (reproduced courtesy of Dr Chris Parker and CAB International)

Figure 7.9 Gall formation on *Gypsophila paniculata* by *Erwinia herbicola* pv. *gypsophilae* (left labelled *Ehg*) and by *Erwinia herbicola* pv. *betae* (right labelled *Ehb*: illustrations kindly provided by Isaac Barash and Shulamit Manulis)

Figure 7.10 Stunting of tobacco (right) caused by infection with tobacco mosaic virus compared with (left) uninfected control (courtesy of Alex Murphy)

Figure 8.1 A chickpea cutting after taking up 45 μg of solanapyrone A, the most potent of the solanapyrone toxins produced by the pathogen *Ascochyta rabiei*, and incubating for a further 96 h in water. Note the broken stem, a typical symptom of Ascochyta blight of chickpea

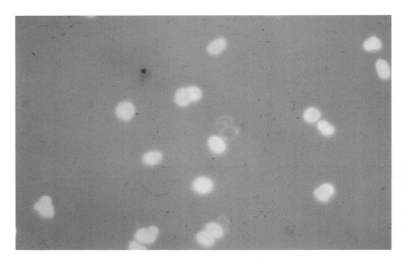

Figure 8.2 Assay of solanapyrone toxins from *Ascochyta rabiei* using cells from leaves of chickpea and the vital dye fluorescein diacetate. Live cells take up the dye and release fluorescein from the molecule by the action of non-specific esterases. Since intact plasmamembranes are not permeable to fluorescein they fluoresce yellow under a fluorescence microscope. Note by contrast the dead brown cells in the centre of the picture.

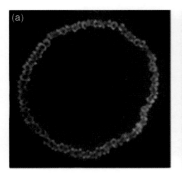 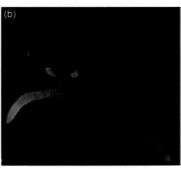

Figure 9.8 (a) Fluorescence associated with avenacin A-1 in the epidermal cell layer of an oat root; (b) oat seedlings viewed on an ultraviolet transilluminator, wild type on the left and a saponin-deficient mutant on the right (courtesy of Anne Osbourn of the John Innes Centre, Norwich)

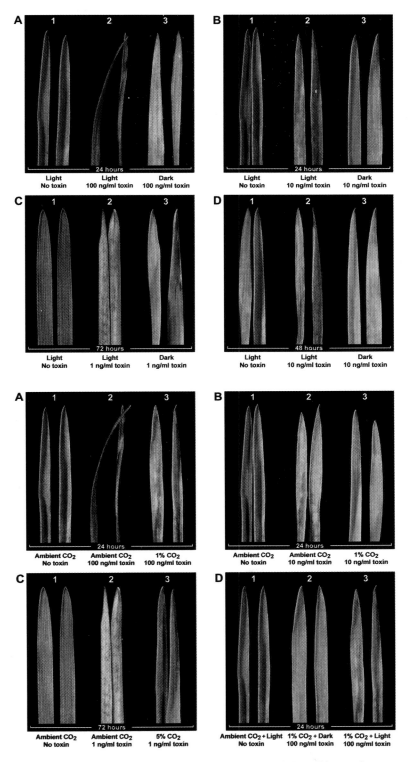

Figure 8.7 The effect of victorin on a susceptible cultivar of oat incubated for various times with 1, 10 or 100 ng of toxin; note the moderating influence of (a) darkness, and (b) high CO_2 concentration (courtesy of Tom Wolpert and Academic Press)

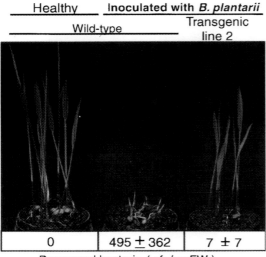

Healthy	Inoculated with *B. plantarii*	
Wild-type		Transgenic line 2
0	495 ± 362	7 ± 7

Recovered bacteria (cfu/mgFW)

Figure 9.15 The effectiveness of transforming rice with a thionin gene from oats in protecting the plant from infection by *Burkholderia plantarii*: uninoculated controls on the left, inoculated wild-type plants in the centre, and inoculated transformants on the right; numbers below the pots are colony-forming units of the bacterium/mg fresh weight of plant material recovered from the plants (courtesy of Yuko Ohashi and the American Society of Phytopathology)

Figure 10.1 Rust of flax caused by *Melampsora lini* in a segregating F_2 family of *Linum marginale*: plant on right, a fully susceptible reaction; plants at left and in centre are resistant (courtesy of Greg Lawrence)

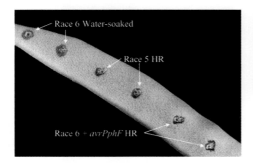

Figure 10.2 Cloning of the avirulence gene *avrPphF* from *Pseudomonas syringae* pv. *phaseolicola* race 5. Race 6 gives a susceptible water-soaked reaction whereas race 5 gives a hypersensitive response (HR). The HR was conferred on race 6 when transformed with a clone containing *avrPphF* (courtesy of John Mansfield)

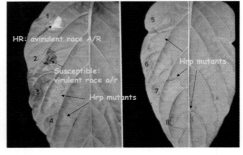

Figure 11.2 Reactions of pepper to *Xanthomonas campestris* pv *vesicatoria*: virulence, avirulence and the null reaction to *Hrp* mutants; note the water-soaked lesions of the areas of the leaf inoculated with the virulent isolate and compare with the hypersensitive cell death caused by the avirulent race – the *Hrp* mutants are deficient in the type III secretory system and therefore cannot export effector molecules that cause either the susceptible reaction or the hypersensitive response (courtesy of John Mansfield)

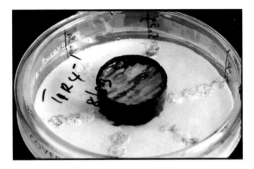

Figure 12.1 Test for stimulation of germination of *Striga* by ethylene-producing microorganisms: cultures of organisms were placed in the centre of the Petri dish and the seed of *Striga* were distributed on glass fibre filter discs radiating from the centre; germination was scored after 72 h (courtesy of Dana Berner)

Figure 12.3 Control of anthracnose of avocado (cv. Fuerte) caused by *Colletotrichum gloeosporioides* by subjecting freshly harvested fruit to 30 per cent CO_2 in air for 24 h (courtesy of Dov Prusky)

Figure 12.9 Comparison of genotypes of rice resistant and susceptible to *Xanthomouas oryzae* pv. *oryzae* and a susceptible genotype transformed with the resistance gene *Xa21* (all plants were challenged with the pathogen); from left to right, IRBB21 (resistant), IR24 (susceptible), TP 309 (susceptible), and TP309 (transformed with *Xa21*) (courtesy of Pamela Ronald, University of California, Davis, USA)

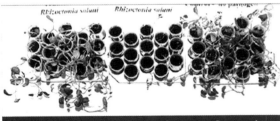

Figure 12.5 Potatoes transformed with an endochitinase gene from *Trichoderma harzianum* showing resistance to *Rhizoctonia solani* (left) compared with untransformed plants (centre) and uninoculated and untransformed plants (right) courtesy of Dr Matteo Lorito

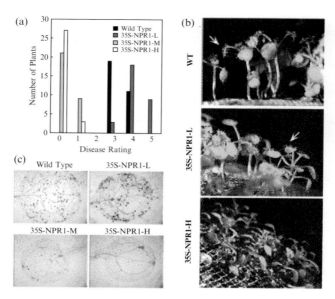

Figure 12.6 The effect of transforming *Arabidopsis thaliana* with the gene *NPR1* on disease development by *Perenospora parasitica*. (a) Disease ratings of wild-type and *NPR1* cDNA transgenic plants after infection with *P. parasitica* Noco (ratings were defined as follows: 0, no conidiophores on the plant; 1, no more than 5 conidiophores per infected plant; 2, 6–20 conidiophores on a few infected leaves; 3, 6–20 conidiophores on most of the infected leaves; 4, 5 or more conidiophores on all infected leaves; 5, 20 or more conidiophores on all infected leaves). (b) Conidiophores observed in wild-type and *NPR1* cDNA transgenic plants 7 days after inoculation with *P. parasitica* Noco. (c) Trypan blue staining of *P. parasitica*-infected leaves of wild-type and *NPR1* cDNA transgenic plants 7 days after infection; seedlings of wild-type and transgenic plants were stained with trypan blue and observed under a compound microscope (Cao and Dong, 1998 reproduced courtesy of the Proceedings of the National Academy of Sciences USA)

Figure 12.8 Control of crown gall by silencing of the genes used by the bacterium for the synthesis of IAA and cytokinins: (a) tomato transgene; (b) tomato control; (c) *Arabidopsis* transgene; (d) *Arabidopsis* control (courtesy of Escobar *et al.* and Proceedings of the National Academy of Sciences USA)

9 The Plant Fights Back – 1. Constitutive Defence Mechanisms

Summary

Plants possess many classes of compounds with anti-microbial properties which may confer at least a measure of resistance. Since they are present before microbial challenge they have been termed phytoanticipins. Two types of evidence have been put forward to support their role as resistance factors. The first is correlative. The better the correlation between possession of high concentrations of the compound and resistance, the more likely it is that the inhibitor is the cause of the resistance. This evidence is strengthened if it can be shown that varying the concentration of inhibitor in the plant leads to variation in resistance. The second type of evidence is genetic and relies upon obtaining mutants that are less sensitive to the inhibitor or insensitive and showing that they are more virulent than the wild type. Such mutants may lack the target of the antimicrobial compound and, in these instance, crossing the mutants with the wild type and showing that virulence and insensitivity are always inherited together constitutes strong evidence that the compound is a resistance factor. Alternatively, wild-type pathogens may be transformed with genes which encode enzymes that render the compound non-inhibitory. Here, increased virulence of the transformants would be evidence for the antimicrobial compound playing a role in resistance.

9.1 Introduction

As stated in Chapter 1 (Section 1.1), plants are primary producers, providing a rich source of nutrients not only for the human population but also for all other organisms that require fixed carbon compounds for sustenance. It is not surprising, therefore, that some of these have developed the ability to parasitize plants,

Introduction to Plant Pathology by Richard Strange
© 2003 John Wiley & Sons, Ltd   ISBN 0 470 84972 X (cased)   ISBN 0 470 84973 8 (pbk)

employing such weapons as degradative enzymes and toxins (Chapters 6 and 8) and subverting the metabolism of the host to their own ends (Chapter 7). Despite this onslaught plants have survived. They therefore must have adequate mechanisms of defending themselves. Some of these are constitutive and are considered in this chapter while others are elaborated in response to attack and will be reserved for Chapter 11 (Section 11.3.3).

9.2   Physical barriers

Unlike animals, plant cells have walls, which present a formidable barrier to any invading organism. Often these are thick, representing a considerable investment by the plant and where they occur on the outside of the plant they are usually covered with cutin or suberin (Section 6.2). It is difficult to prove that such structures have a role in resistance directly. Instead, the best evidence is indirect and comes from mutants of pathogens deficient in one or more cell-wall-degrading enzymes and their complementation with genes cloned from the wild type (Section 6.9). Lignin is often a component of secondary cell walls and confers considerable resistance to microbial decay but, since lignification is also an active response to microbial challenge, further consideration of this topic will be reserved for Chapter 11.

Bark undoubtedly provides physical protection against potential invaders, but it is probable that its chemical properties are also important. For example, Woodward and Pearce (1988) found the stilbene glucosides, astringin and rha-ponticin, were present at high concentrations (10–15 mg/g) in the bark of Sitka spruce (Section 9.3.1) and were inhibitory to two weakly pathogenic fungi at concentrations above 1 mg/ml.

In some ergots pollination competes with the invading fungus. For example, in ergot of pearl millet (*Pennisetum americanum*) caused by *Claviceps fusiformis*, infection takes place through stigmas, but 6 h after pollination a constriction occurs in these, owing to the collapse and desiccation of cells, rendering them impenetrable to the fungus. Resistance is therefore related to the length of time that elapses between stigma emergence and anthesis, when pollen becomes available for fertilization (protogyny). Cultivars in which protogyny of individual florets was 48 h or less were resistant but those in which the stigmas remained receptive for up to 6 days were highly susceptible (Willingale, Mantle and Thakur, 1986). Less directly, McLaren (1997) found that ergot of sorghum caused by *Claviceps africana* was more severe when the viability of pollen was reduced by temperatures of less than 16°C, 23–27 days before flowering.

9.3   Chemical barriers

9.3.1   Low molecular weight compounds

Plants usually contain low molecular weight compounds that are anti-microbial and have been termed phytoanticipins since they are present before microbial challenge. They may be extracted with suitable solvents but, once isolated and identified, the question arises as to whether they have a role to play in resistance, an important point that has been critically reviewed by Mansfield (2000). Initial evidence is normally correlative, high concentrations being found in resistant plants or tissues and low concentrations in susceptible ones. For example, Assef, Assari and Vincent (1986) working with *F. oxysporum* f. sp. *albedinis*, the cause of the devastating bayoud disease of date palms, reported an inhibitory factor from a single resistant type. The evidence is more convincing if resistance is correlated with concentration of the compound in several cultivars and still more so if varying the concentration of the compound, for example, by alteration of the environment or better still, genetically, leads to variation in the resistance of the plant.

An alternative approach to establishing the role of an antimicrobial compound in resistance of a plant is to determine if genotypes of the pathogen that are tolerant of the compound are more virulent than those that are sensitive. Tolerance may result from absence of target molecules of the anti-microbial compound, enzymes that detoxify it or the avoidance of its release from an inactive precursor. These mechanisms are susceptible to genetic analysis and as we shall see in the following sections this approach provides the strongest evidence.

Phenols and quinones

The work by Walker and associates on onion smudge caused by *Colletotrichum circinans* is a classic since it gave a satisfactory biochemical answer for the first time for differential resistance of cultivars to a plant disease (Link, Angell and Walker, 1929a, b; Link and Walker, 1933). Onion cultivars with outer scale leaves that were either red or yellow were resistant to the fungus but those with white scale leaves were susceptible. Aqueous extracts of the coloured scale leaves were anti-fungal whereas those of white scale leaves were not. Fractionation of the extracts from the coloured scale leaves led to the isolation of two compounds which were identified as catechol and protocatechuic acid (Figure 9.1). Further work showed that environmental conditions which led to the depletion of the compounds, such as exposure of the bulbs above ground and leaching, increased susceptibility, strengthening the case for a role of the compounds in resistance (Link, Angell and Walker, 1929a, b; Link and Walker, 1933).

Alkyl phenolic compounds have been implicated in latency and in the resistance of a number of plants to pathogens. For example, some isolates of

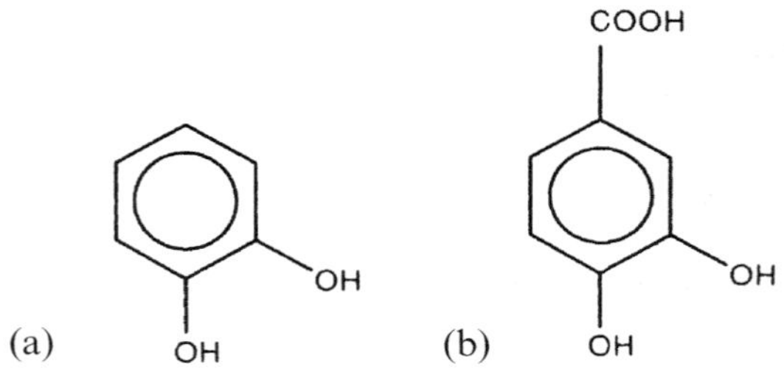

Figure 9.1 (a) Catechol, and (b) protocatechuic acid, two compounds found in the scale leaves of some onion cultivars but not others that explained the differential resistance to onion smudge disease caused by *Colletotrichum circinans*

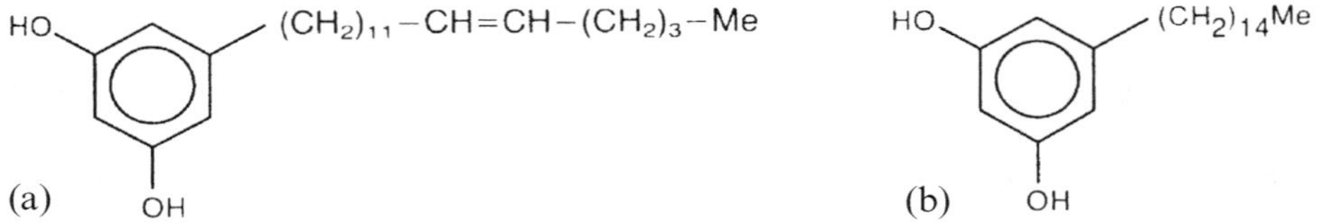

Figure 9.2 Two antifungal alkyl phenol compounds: (a) 5-(12-*cis*-heptadecenyl) resorcinol, and (b) 5-pentadecenyl resorcinol from the peel of mangoes that maintain the latency of *Alternaria alternata* until ripening when their concentrations fall precipitately

Alternaria alternata cause black-spot disease of mango (*Mangifera indica*) but the fungus remains latent until the fruit is ripe. Two antifungal compounds, 5-(12-*cis*-heptadecenyl)-resorcinol and 5-pentadecylresorcinol, were implicated as the latency factors as their initial high concentrations in the peel of unripe fruit dropped dramatically on ripening (Cojocaru *et al.*, 1986; Droby *et al.*, 1986; Figure 9.2)

More recently, Suzuki and co-workers (1996) and Bouillant and co-workers (1994) have identified five 5-alk(en)ylresorcinols as antifungal agents from etiolated rice seedlings. As they were present in high concentrations, it was suggested that they had an important role to play in the resistance of rice at the seedling stage. A similar role was suggested for an alkenyl benzoquinone from sorghum seedlings, which was present in appreciable amounts on the root surface (Suzuki *et al.*, 1998; Figure 9.2).

Long chain aliphatic and olefinic compounds

In a review of postharvest pathogen quiescence, Prusky (1996) has summarized his many years' work on the latency of *Colletotrichum gloeosporioides* in avocado. *C. gloeosporioides* is the most important cause of fruit rot of avocado and, although infecting the fruit at a very early stage during the growing season, remains latent until 7–15 days after harvest (Prusky *et al.*, 1990). Prusky discovered that the cause of latency was not lack of nutritional factors but rather the presence of two antifungal compounds, *cis,cis*-1-acetoxy-2-hydroxy-4-oxo-henei-cosa-12,15-diene and the less toxic monoene,1-acetoxy-2,4-dihydroxy-*n*-hepta-deca-16-ene (Figure 9.3). Spore germination of the fungus was inhibited by the diene at concentrations of 790 µg/ml but concentrations in unripe peel were ca 1600 µg/ml (1200 µg/g fresh weight of peel). During ripening, concentrations of the diene fell around 10 fold and cultivars that were more susceptible showed an earlier decrease than the more resistant.

The diene is a substrate for lipoxygenase. When the activity of this enzyme was inhibited by infiltration with α-tocopherol, the decrease in concentration of the diene as well as the onset of lesion development was delayed. Similarly, protec-tion of the diene by infiltration of the specific lipoxygenase inhibitor ETYA (5,8,11,14 eicosatetraynoic acid) delayed symptom development in avocado discs inoculated with the fungus but the fungus itself was not affected. An increase in specific activity of lipoxygenase during fruit ripening and increasing the activity of the enzyme by treatment with jasmonic acid, which decreased the concentration of the diene and enhanced fruit susceptibility, provided further evidence for the involvement of the diene in latency.

Prusky also showed that epicatechin was the natural inhibitor of lipoxygenase in the peel of unripe avocado. When the epicatechin concentrations of cultivars that differed in susceptibility were compared, concentrations of the compound decreased more rapidly in those in which symptoms appeared early (Ardi *et al.*, 1998). Moreover, epicatechin inhibits pectic enzymes such as pectate lyase and polygalacturonase which are likely to be important in fungal attack (Wattad, Dinoor and Prusky 1994).

In other experiments Prusky and co-workers (1990) found that inoculation of freshly harvested fruit increased the concentration of the diene suggesting that the compound could be regarded as a phytoalexin (Chapter 11, Section 11.3.2). Prusky, Wattad and Kobiler (1996) further corroborated the role of the diene in experiments involving exposure to the ripening hormone, ethylene. Although ethylene stimulated spore germination, appressorium formation and fungal proliferation of conidia placed on avocado wax or intact avocado (see Section 6.6.3.) it did not activate lesion development. Rather, lesion development occurred in parallel with decreasing concentrations of the diene. More recently, Leiken-Frenkel and Prusky (1998) showed that the avocado diene could be synthesized in idioblast cells and exported to the pericarp of fruit. Also a

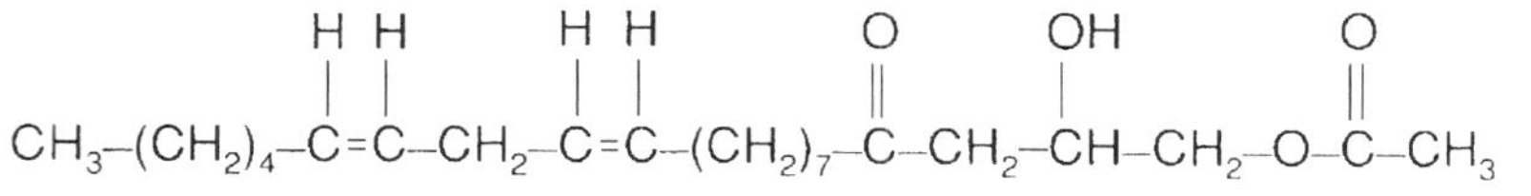

Figure 9.3 *Cis,cis*-1-acetoxy-2-hydroxy-4-oxo-heneicosa-12,15-diene, the major antifungal compound found in the peel of avocado pears that explains the latency of infections by *Colletotrichum gloeosporioides* in unripe fruit; concentrations decline on storage to non-inhibitory levels allowing proliferation of the fungus

compound with one further double bond (a triene) was isolated from these cells
and two other compounds (Domergue *et al.*, 2000).

Aldehydes

Penicillium digitatum and *P. italicum* are wound pathogens of citrus but seldom
cause disease in the orchard despite the presence of inoculum and wounds. It is
thought that the monoterpene aldehyde, citral, is responsible for the resistance
(Figure 9.4). The concentrations of this compound decrease after harvest concomitantly with greater susceptibility to the pathogens.

Spendley and co-workers (1982) isolated two long-chain aldehydes, α- and β-
triticene from homogenates of wheat which are thought to be formed from
inactive precursors by enzymatic action (Figure 9.4). They were active against
fungal pathogens of wheat grown *in vitro* at concentrations of 10–100 µg/ml but
their role in resistance does not seem to have been investigated (Spendley and
Ride, 1984).

Unsaturated lactones

Unsaturated lactones are usually found in plants as glucosides. For example, the
tuliposides are found in tulips and other species such as *Alstromeria*, with
particularly high concentrations being recorded in pistils (Figure 9.5). Their
role in the defence of tulip bulbs against *F. oxysporum* f. sp. *tulipae* has been
investigated. Although the pathogen may be continuously present in the soil, the
plant is only susceptible in the few weeks before harvest when the outer scale
leaves, which are normally white and high in tuliposide content, turn brown.
Concomitantly the concentrations of the tuliposides decrease precipitously but
rise again after a few days' storage when the bulbs become resistant once more.
There is therefore a good temporal correlation between the presence of the
inhibitors and resistance.

Although tulips are not normally susceptible to *Botrytis cinerea* under field
conditions, infection may take place after artificial inoculation and storage at
high humidity. Pistils, which contain high concentrations of the tuliposides, never

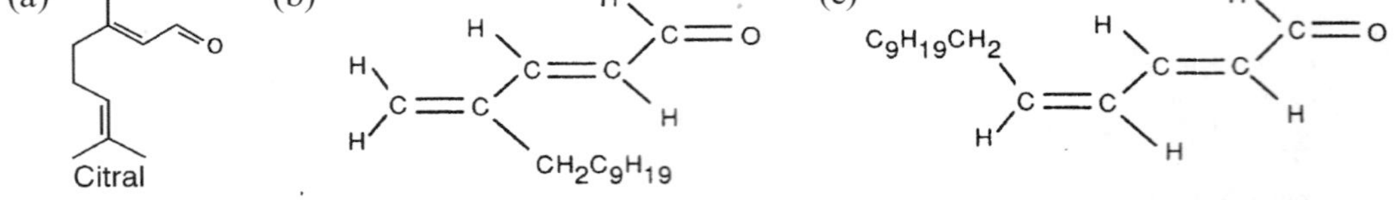

Figure 9.4 (a) Citral, an antifungal aldehyde from citrus thought to explain the resistance of
citrus fruit to infection by green and blue mould (*Penicillium digitatum* and *P. italicum*,
respectively; (b) and (c) antifungal aldehydes from wheat

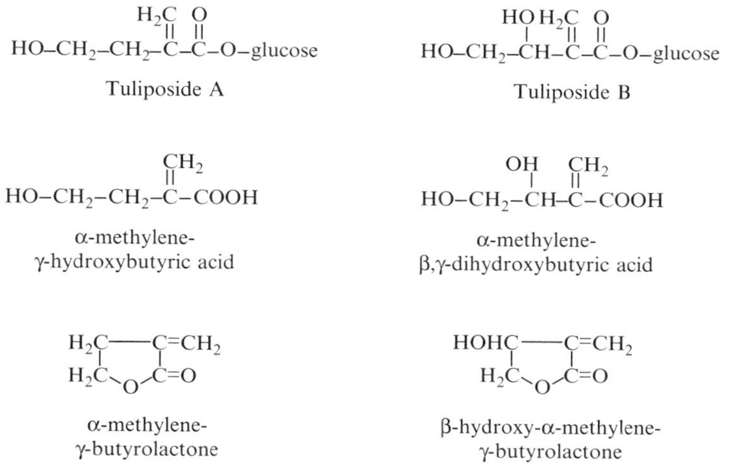

$$\underset{\text{Tuliposide A}}{\text{HO-CH}_2\text{-CH}_2\text{-}\overset{\displaystyle \overset{\text{H}_2\text{C}}{\|}}{\text{C}}\text{-}\overset{\displaystyle \overset{\text{O}}{\|}}{\text{C}}\text{-O-glucose}}$$

$$\underset{\text{Tuliposide B}}{\text{HO-CH}_2\text{-CH-}\overset{\displaystyle \overset{\text{HOH}_2\text{C}}{|}}{}\overset{\displaystyle \overset{\text{O}}{\|}}{\text{C}}\text{-O-glucose}}$$

$$\underset{\substack{\alpha\text{-methylene-}\\ \gamma\text{-hydroxybutyric acid}}}{\text{HO-CH}_2\text{-CH}_2\text{-}\overset{\displaystyle \overset{\text{CH}_2}{\|}}{\text{C}}\text{-COOH}}$$

$$\underset{\substack{\alpha\text{-methylene-}\\ \beta,\gamma\text{-dihydroxybutyric acid}}}{\text{HO-CH}_2\text{-CH-}\overset{\displaystyle \overset{\text{CH}_2}{\|}}{\text{C}}\text{-COOH}}$$

α-methylene-
γ-butyrolactone

β-hydroxy-α-methylene-
γ-butyrolactone

Figure 9.5 The tuliposides, antifungal unsaturated lactones found in tulips and other species and their metabolites; these compounds confer resistance to *Botrytis cinerea* since infection results in their conversion to the corresponding unsaturated lactones which are inhibitory whereas infection with *B. tulipae* leads to conversion to the corresponding carboxylic acids which are stimulatory to the fungus

become infected. In contrast, *B. tulipae* can infect all parts of the plant. The explanation of the difference in susceptibility to these two pathogens appears to lie in their effects on membranes and their metabolism of the lactones. *B. cinerea* damages membranes, causing the tuliposides, which are present in the vacuole, to leak out. The fungus then converts them to the active lactones (Figure 9.5). *B. tulipae* causes less membrane damage and converts the tuliposides to the corresponding hydroxycarboxylic acids which, rather than being inhibitory, actually stimulate the growth of the fungus (Figure 9.5).

Cyanogenic glucosides

Many plants contain high concentrations of cyanogenic glucosides. Some parts of the important cereal sorghum, for example, contain as much as 35 per cent of their dry weight as dhurrin (Figure 9.6). Selmar, Irandoost and Wray (1996) isolated a glycoside of dhurrin from the plant and suggested that it was a translocatable form of the compound. On infection or injury cyanogenic plants release HCN (Myers and Fry, 1978). Pathogens of cyanogenic plants evade the toxic effects of HCN by the production of the inducible enzyme formamide hydro-lyase which converts HCN to $HCONH_2$ and may also have cyanide insensitive respiratory pathways.

Although the production of HCN may confer resistance to cyanide-sensitive microorganisms, it may also inhibit normal defence reactions. Rubber clones that were highly cyanogenic were more susceptible to the leaf pathogen, *Microcyclus*

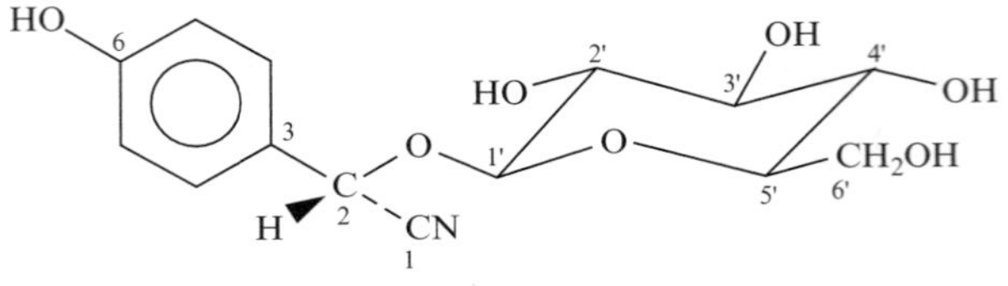

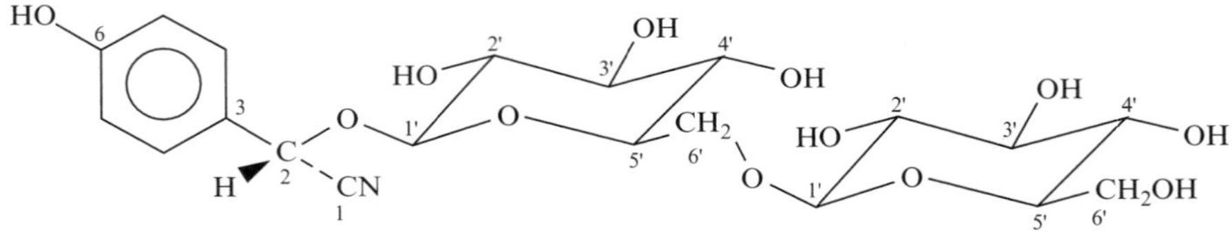

Figure 9.6 Dhurrin and its glycoside, cyanogenic compounds found in sorghum

ulei, than low-cyanogenic clones owing to the accumulation of lower concentrations of the phytoalexin, scopoletin (Lieberei *et al.*, 1989). When the HCN was removed by passing a stream of moistened air over infected leaves or by an alkaline trap, scopoletin accumulated and there was a decrease in the size and number of lesions.

Saponins

Saponins are glycosylated secondary metabolites that occur widely in the plant kingdom and lyse cells which contain sterols in their membranes. Indeed, they were originally recognized by their ability to cause lysis of red blood cells but have attracted the attention of plant pathologists owing to their potent antifungal activity.

In Tomatine is a saponin found in high concentrations in green tomatoes which affords protection against infection by *Fusarium solani* (Figure 9.7). As the tomato ripens and turns red the concentration declines rapidly and the tomato becomes susceptible to the fungus. Défago and Kern (1983) and Défago, Kern and Sedlar (1983) obtained mutants of the fungus which were insensitive to tomatine and could parasitize green tomatoes. Biochemical analysis showed that their membranes were deficient in sterols. Genetic studies revealed that deficiency in membrane sterols, insensitivity to tomatine and pathogenicity to green tomatoes were always inherited together, strongly supporting the role of tomatine in the resistance of tomatoes to the wild type pathogen.

In other work with tomato, Melton and co-workers (1998) transformed two races of *Cladosporium fulvum*, a pathogen that has a gene-for-gene relationship with this host (see Chapter 10, Sections 10.5.1 and 10.5.3) with a cDNA encoding tomatinase from the necrotroph *Septoria lycopersici* (see below). Transformants sporulated more profusely on susceptible tomato lines and also caused more

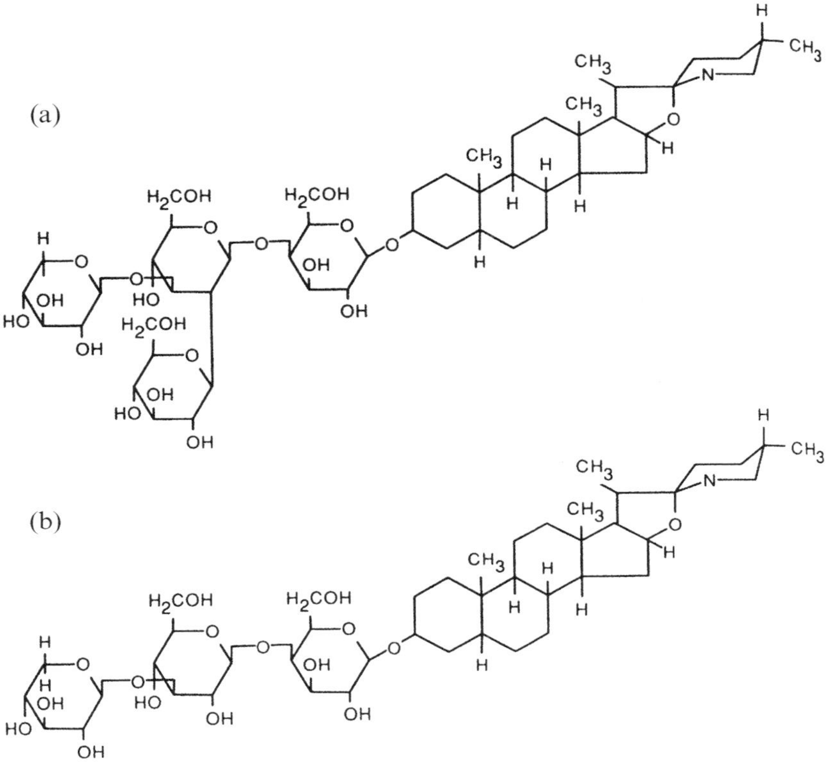

Figure 9.7 Structures of (a) α-tomatine, and (b) β₂-tomatine, a more weakly antifungal derivative, produced by the action of the enzyme tomatinase from two tomato pathogens, *Septoria lycopersici* and *Verticillium albo-atrum*

extensive infections of seedlings of resistant lines. Tomatine therefore appears to play a role in limiting the development of *C. fulvum* in tomato, whether the plant contains effective resistance genes or not.

Further recent work has centred on the avenacins, a group of structurally related triterpenoid saponins found in oats, and their role in resistance to the take-all fungus *Gaeumannomyces graminis* (Figure 9.8). *G. graminis* var. *tritici* is a major pathogen of wheat and barley but oats are resistant. However, oats are susceptible to *G. graminis* var. *avenae*, which is relatively insensitive *in vitro* to avenacin-1, the major saponin component of oat roots. The reason for this insensitivity is attributed to an enzyme, avenacinase, which detoxifies the saponin by removal of both the β,1–2 and the β,1–4 linked terminal D-glucose residues. The gene encoding avenacinase was cloned by Bowyer and co-workers (1995). They purified the enzyme from culture filtrates of the fungus and raised poly-clonal antibodies to it. This allowed them to screen a λ cDNA expression library of the fungus prepared from mRNA collected during active secretion of the enzyme. The gene was then isolated from a genomic DNA clone by homology to the cDNA

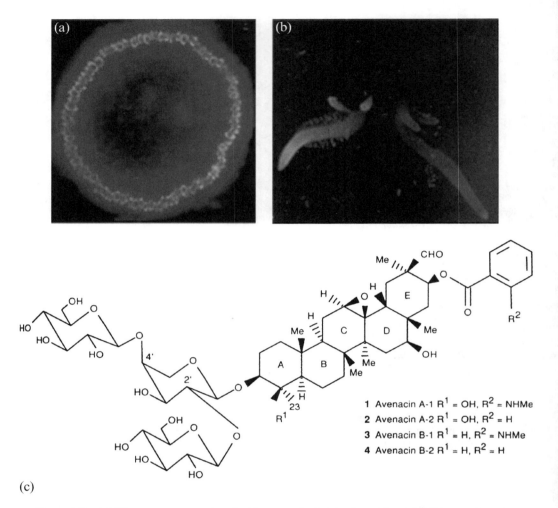

Figure 9.8 (a) Fluorescence associated with avenacin A-1 in the epidermal cell layer of an oat root; (b) oat seedlings viewed on an ultraviolet transilluminator, wild type on the left and a saponin-deficient mutant on the right (courtesy of Anne Osbourn of the John Innes Centre, Norwich); (c) chemical structures of the avenacins. A colour reproduction of this figure can be seen in the colour section

and the isolation confirmed by heterologous expression in *Neurospora crassa*. Disruption of the gene encoding the enzyme in *G. graminis* var. *avenae* by homologous recombination gave avenacinase⁻ mutants which were sensitive to avenacin-1 and were not pathogenic to oats. The possession of avenacinase therefore appears to be an absolute requirement for the pathogenicity of *G. graminis* var. *avenae* for oats. However, the avenacinase-minus mutant retained wild-type levels of pathogenicity for wheat which is not known to contain saponins (Osbourn, 1996). As pointed out by Mansfield (2000) this result shows that there is no hidden role for avenacinase in basic parasitic ability in the absence of its primary substrate, avenacin-1.

In further work on tomatine, mechanisms of detoxification have been studied. These differ according to the organisms concerned but the tomatinase enzymes of *Septoria lycopersici* and *Verticillium albo-atrum*, two fungal pathogens of tomato, act in a similar manner to avenacinase in that they remove the β,1–2 linked terminal D-glucose molecule by hydrolysis (Figure 9.7). Moreover, since tomatinase and avenacinase are closely related enzymes (Osbourn *et al.*, 1995), this similarity was exploited by using a cDNA clone of avenacinase gene as a probe to isolate the tomatinase gene from *S. lycopersici*.

Papadopolou and co-workers (1999) have obtained more direct evidence for the role of saponins in resistance. They mutagenized seed of *Avena strigosa* and detected deficiency of saponins in the roots by the absence of their characteristic blue fluorescence (Figure 9.8). Saponin deficiency was inherited as a single recessive allele and the mutants were susceptible to infection by *G. graminis* var. *tritici*.

Terpenoids

The terpenoids known as the duvatrienediols are found in the cuticular wax of tobacco leaves (Cruickshank, Perrin and Mandrym, 1997; Figure 9.9). When they were removed by dipping the leaves for 1 s in acetone, inoculation with the blue-mould pathogen, *Peronospora tabacina*, gave rise to disease severities which were three times greater than in undipped controls. Resistance was restored when the duvatrienediols were added back to the dipped leaves (Reuveni *et al.*, 1987). In addition, resistance and duvatrienediols increased as plants aged. Further work involved a quantitative assay of the duvatrienediols and other components of the cuticle of 25 *Nicotiana* species (Kennedy *et al.*, 1992). The compounds were dissolved in acetone and applied to leaf discs of a variety of *N. tabacum* that was susceptible to the pathogen. Values for the inhibition of sporangial germination were 3.0 and 2.9 μ/cm^2 for α- and β-duvatrienediols, respectively, 0.4 μg/cm^2 for labdenediol and 4.7 μg/cm^2 for sclareol and episclareol ([13-R]-labda-14-ene-8-α-,13-diol and [13-S]-labda-14-ene-8-α-,13-diol, respectively). In a similar study, Kono and co-workers (1991) have extracted novel diterpenes from healthy rice

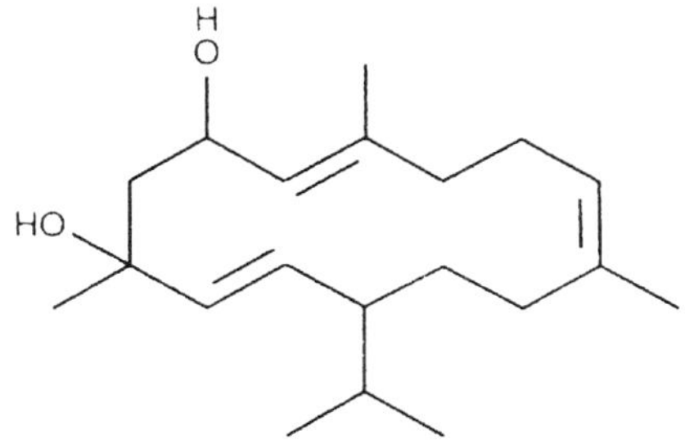

Figure 9.9 Duvatrienediol, an antifungal compound that exists as two isomers, α- and β-that are implicated in the resistance of tobacco to blue mould caused by *Perenospora tabacina*

leaves (Figure 9.10) which were active against the rice pathogen, *Xanthomonas campestris* pv. *oryzae*.

Stilbenes

The heartwood of trees is usually very resistant to decay. One reason for this is the relatively high concentrations of stilbenes such as pinosylvin and pinosylvin monomethylether often found in this tissue (Figure 9.11; Hart, 1981; Ceumene *et al.*, 1999, 2001). In surveys of the sensitivity of a range of fungal species from the Thelephoraceae, Polyporaceae and Agaricaceae to these compounds, most did not grow above a threshold of about 50 μg/ml. As mentioned above (Section 9.2), the stilbene glucosides astringin and rhaponticin occur in considerable concentrations in the bark of Sitka spruce and when excised bark discs were subjected to fungal challenge, they were converted to the corresponding aglycones, astringenin and isorhapontigenin, respectively (Figure 9.12). Tests *in vitro* showed that the aglycone, isorhapontigenin was more antifungal than the glucoside, rhaponticin (Woodward and Pearce, 1988).

Schultz and co-workers (1990) surveyed a number of synthetic stilbenes for activity against the white-rot fungus, *Coriolus versicolor* and the brown-rot fungi, *Gloeophyllum trabeum* and *Poria placenta* and showed that there was a linear relation between the activity of the compounds against the brown-rot fungi and hydrophobicity.

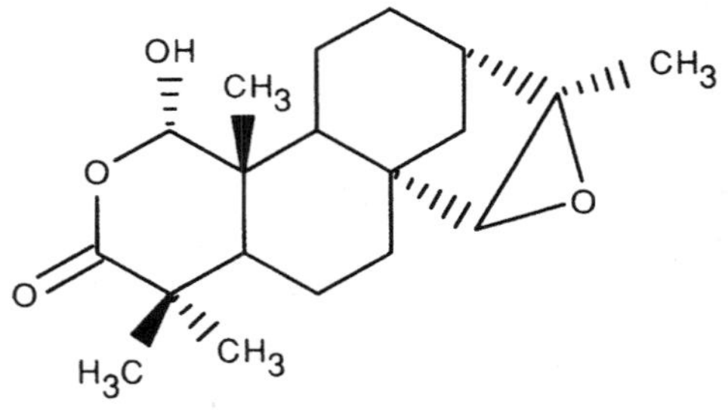

Figure 9.10 Oryzalide-A a novel diterpene from rice with activity against *Xanthomonas campestris* pv. *oryzae*

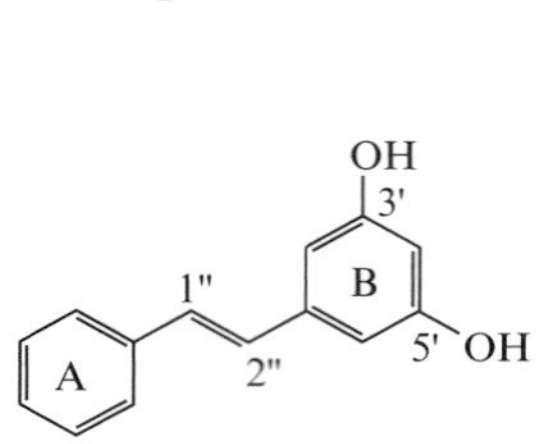

Figure 9.11 Pinosylvin an antifungal compound which, together with its monomethyl ether, is found in high concentrations in heartwood and explains the resistance of this tissue to decay

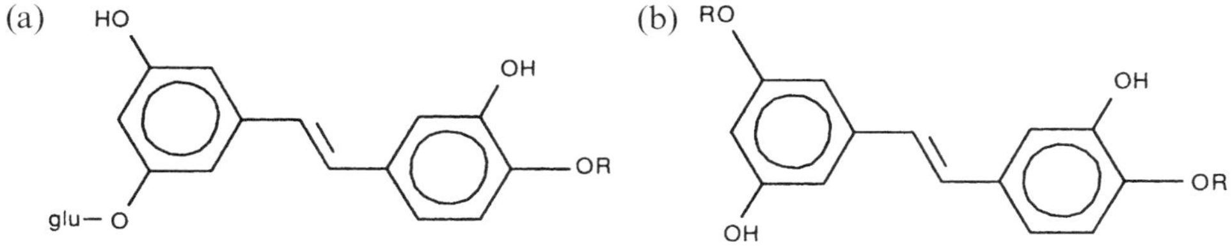

Figure 9.12 (a) Rhaponticin (R = CH3) and astringin (R = H), and (b) their corresponding aglycones isorhapontegenin and astringenin, antifungal compounds from Sitka spruce

Glucosinolates

Glucosinolates are a group of compounds, which consist of β-D-thioglucose, a sulphonated oxime and a variable side-chain. They are found mainly in the order Capparales (Bennett and Wallsgrove, 1994). Disruption of plant tissue leads to the hydrolysis of glucosinolates by a specific thioglucosidase termed 'myrosinase' giving a variety of products, collectively known as 'mustard oils' and which are responsible for the flavours of many food plants of the cabbage family (Figure 9.13). The glucosinolate/myrosinase system is thought to be effective as a general means of defence of plants against herbivores and pathogens but some pests and pathogens have evolved mechanisms of circumventing them. For example, Giamoustaris and Mithen (1997) investigated the relationship between glucosinolate concentration and resistance of 33 lines of oilseed rape to two fungal pathogens, *Leptosphaeria maculans* and *Alternaria* spp. under field conditions, both of which are inhibited *in vitro* by the concentrations of glucosinolates that occur in *Brassica* crops and wild species. They found that there was no correlation between the two variables for *L. maculans* and that the level of infection of both leaves and pods by *Alternaria* was positively correlated with glucosinolate concentration! One interpretation of this result is that while the glucosinolates may afford some degree of protection against non-specialized pathogens, such as *Botrytis*, specialized pathogens of plants rich in glucosinolates have become adapted to them. More recently, Li and co-workers (1999) found no correlation between glucosinolate concentrations and resistance of Chinese and European cultivars of oilseed rape to *Sclerotinia sclerotiorum*. However, in this system some lines showed good local and systemic induction of glucosinolates and this was correlated with resistance.

Cyclic hydroxamic acids and related benzoxazolinone compounds

Cyclic hydroxamic acids are found as glucosides in maize, rye and wheat but not in rice, barley or oats (Figure 9.14). On injury or pathogen attack, the toxic aglucones are released and these rapidly decompose to form the corresponding ring contracted benzoxazolinone compounds, MBOA and BOA (Figure 9.14). Friebe and co-workers (1998) have investigated the ability of three varieties of

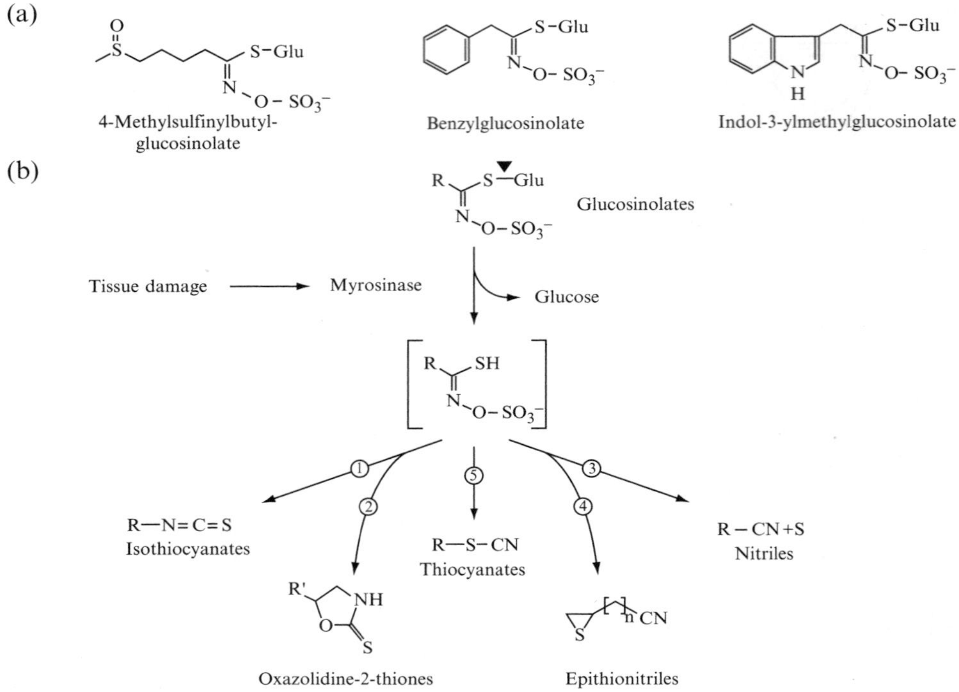

(a)

4-Methylsulfinylbutyl-
glucosinolate

Benzylglucosinolate

Indol-3-ylmethylglucosinolate

(b)

Figure 9.13 (a) Examples of alphatic, aromatic and indole glucosinolate structures; (b) upon tissue damage, glucosinolates are hydrolysed by myrosinases – at neutral pH, the unstable aglucones (bracketed) rearrange to form a variety of structures according to the nature of the side chain (R) and conditions such as pH (Reproduced by permission of Trends in Plant Science)

Gaeumannomyces graminis – vars. *tritici, graminis* and *avenae* – to detoxify these compounds. They found that the tolerance of *G. graminis* varieties was correlated with their ability to detoxify the compounds, forming *N*-(2-hydroxy-4-methoxyphenyl)-malonamic acid and *N*-(2-hydroxyphenyl)-malonamic acid, respectively. Moreover, the ability of the isolates to cause root-rot symptoms in wheat corresponded with their ability to detoxify them.

9.3.2   High molecular weight compounds

Tannins

A methanol extract of cocoa (*Theobroma cacao*) inhibited the growth of germ tubes of *Crinipellis perniciosa*, the cause of the highly destructive witches' broom disease of cocoa (Brownlee *et al.*, 1990). The anti-fungal agent did not diffuse through dialysis tubing and was the major constituent of the dialysate. It also bound to erythrocytes and protein. The active component was identified as

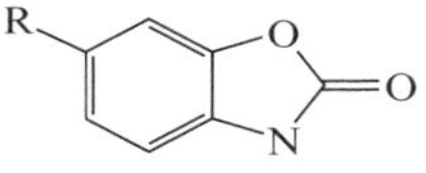

Figure 9.14   Aglucones of cyclic hydroxamic acids, antifungal compounds found as glucosides in some cereals (a) R = H, BOA; (b) OCH₃, MBOA

polymeric procyanidin, i.e. condensed tannin of variable structure. Further work by Andebrhan and co-workers (1995) showed that, for the inhibition of basidiospore germination of the pathogen, the degree of polymerization was important, mixtures of procyanidins with 5–8 monomeric units being more effective than those with 3–5 units.

Anti-microbial proteins and peptides

In addition to structural proteins such as hydroxyproline-rich glycoproteins (HRGPs, see Chapter 6), plants contain multi-gene families which code for proteins and small peptides which are anti-microbial. Some of these are constitutively expressed, typically in storage and reproductive organs, while in other organs, such as leaves, they are induced in response to pathogen attack – in some cases, possibly exclusively so. In this chapter we shall consider anti-microbial proteins and peptides that are constitutive and for which there is some evidence that they have a role to play in defence. They have been reviewed by Shewry and Lucas (1997), who divided them into 11 classes (Table 9.1), and by Broekaert and co-workers (1997).

Endohydrolases   These are enzymes with β-1,3-glucanase and chitinase activity and although they are commonly elicited by infection as members of a family of

Table 9.1   Classes of constitutive anti-microbial proteins in plants

Class	Probable reason for anti-microbial activity
Endohydrolases: β-1,3-glucanase; chitinase	Attack on structural components of pathogen wall
Chitin-binding proteins	Interference with synthesis of pathogen wall
Thionins	Destabilization of fungal membranes
Thaumatin	Alteration of permeability
Defensins	Destabilization of fungal membranes
Non-specific phospholipid transfer proteins (LTPs)	Possibly destabilization of fungal membranes
Albumins	Destabilization of fungal membranes
Ribosome-inactivating proteins	Inhibition of peptide elongation
Lysozyme	Digestion of bacterial cell-wall polymers
Proteinase inhibitors	Inhibition of digestive enzymes of invertebrate pests, vectors and pathogens
Polygalacturonase-inhibiting proteins	Inhibition of polygalacturonase

pathogenesis-related proteins (see Chapter 11) some are also constitutive. For example, Leah and co-workers (1991) found such proteins in the seeds of barley and Chye and Cheung (1995) showed that β-1,3-glucanases are highly expressed in the lactifers of the rubber tree (*Hevea brasiliensis*). Anguelova, van der Westhuizen and Pretorius (1999), investigating adult resistance of wheat to leaf rust (*Puccinia recondita* f. sp. *tritici*) conferred by the *Lr35* gene, found four intercellular proteins of 35, 33, 32 and 31 kDa, which were serologically related to β-1,3-glucanase. They were present in resistant and susceptible isolines during all stages of plant growth. However, resistance was associated with high constitutive levels of β-1,3-glucanase activity and possession of the *Lr35* gene.

In contrast to these results, Havana tobacco plants and *Nicotiana sylvestris* deficient in β-1,3-glucanase activity obtained by anti-sense transformation were more resistant to virus infection (Beffa *et al.*, 1996). The increased resistance was manifest as reduced lesion size, lesion number and virus yield in the local-lesion response of Havana 425 tobacco to tobacco mosaic virus (TMV) and of *Nicotiana sylvestris* to tobacco necrosis virus (TNV). Mutants also showed decreased severity of mosaic disease symptoms, delayed spread of symptoms and reduced yield of virus in the susceptible response of *N. sylvestris* to TMV. Symptoms in both plant species were positively correlated with β-1,3-glucanase content in a series of independent transformants, suggesting that this enzyme plays an important role in viral pathogenesis. It is likely to do this by degrading callose, a substrate for the enzyme, and deemed to be a physical barrier to the spread of virus.

Evidence for the constitutive expression of chitinases has been obtained for a number of plant species. For example, MunchGarthoff and co-workers (1997) found high levels of chitinase expression in a number of wheat cultivars where they may serve to release oligomers of the appropriate size to function as elicitors of lignification (see Chapter 11, Section 11.4.2). Chitinases are also highly expressed in ripening grape berries (Robinson, Jacobs and Dry, 1997).

The anti-fungal activity of these enzymes is presumably attributable to their activity on glucan and chitin, the predominant components of fungal cell walls. Experiments *in vitro* showed that a combination of both enzymes was inhibitory to more fungi than either enzyme alone (Mauch, Mauchmani and Boller, 1988). Chitinases may also play a role in the resistance of plants to insects and nematodes since chitin is also present in their exoskeletons.

Chitin-binding proteins   These proteins are characterized by a chitin-binding domain consisting of a sequence of 30–43 amino acids. Many are lectins which bind tightly to glycans of glycoproteins, glycolipids or polysaccharides (Goldstein and Hayes, 1977). Most lectins that have been characterized are secretory proteins which accumulate in vacuoles or the cell wall and intercellular spaces (Chrispeels and Raikhel, 1991). Some chitin-binding proteins such as one obtained from the intercellular washing fluids of the leaves of sugarbeet (*Beta vulgaris*, Nielsen *et al.*, 1997) and an agglutinin from stinging nettles (*Urtica dioica*, Van Parijs *et al.*, 1992), are constitutive but others, with which they have homology, are induced in response to wounding and pathogen attack. For example, hevein, found in the rubber tree (*Hevea brasiliensis*), is a small (4.7 kDa) cysteine-rich protein consisting essentially of the chitin-binding domain. Its expression in leaves, stem and latex is induced by wounding (Broekaert *et al.*, 1990).

Thionins Thionins are peptides of about 5 kDa which are rich in cysteine and basic amino acids. There are two types: type I thionins are constitutive whereas type II are pathogen induced and will be discussed in Chapter 11 (Section 11.3.6). They are very toxic not only to fungi but also to bacteria, plant and animal cells (Bohlmann and Apel, 1991). An investigation of β-pur-othionin from wheat showed that it formed cation-selective ion channels in artificial lipid bilayer membranes and in the plasmalemma of rat hippocampal neurons (Hughes *et al.*, 2000). The authors suggest that the toxicity of thionins may be attributed to their ability to form such ion channels, which would result in the dissipation of ion concentration gradients essential for the maintenance of cellular homeostasis. Iwai and co-workers (2002) obtained direct evidence of the effectiveness of a thionin from oats. When rice was transformed with a gene encoding the peptide, the transformants were resistant to infection by the bacterium *Burkholderia plantarii* whereas the untransformed plants were killed (Figure 9.15).

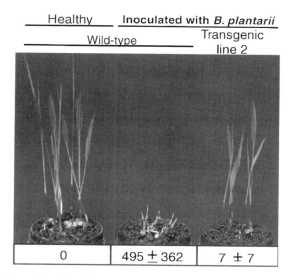

Healthy	Inoculated with *B. plantarii*	
Wild-type		Transgenic line 2
Recovered bacteria (cfu/mgFW)		
0	495 ± 362	7 ± 7

Figure 9.15 The effectiveness of transforming rice with a thionin gene from oats in protecting the plant from infection by *Burkholderia plantarii*: uninoculated controls on the left, inoculated wild-type plants in the centre, and inoculated transformants on the right; numbers below the pots are colony-forming units of the bacterium/mg fresh weight of plant material recovered from the plants (courtesy of Yuko Ohashi and the American Society of Phytopathology). A colour reproduction of this figure can be seen in the colour section

Thaumatin   A protein of about 22 100 kDa, which inhibited trypsin, was isolated from maize seed by Richardson, Valdesrodriguez and Blancolabra (1987). The protein was 50 per cent homologous with both the sweet protein thaumatin II isolated from the West African shrub *Thaumatococcus danielii* and a pathogenesis-related protein from tobacco (see Chapter 11, Section 11.3.6). More recent work has shown that such proteins are present in fruits such as apple and kiwifruit. Thaumatin-like proteins are also induced by osmotic stress and pathogen attack and are pathogenesis-related proteins (see Chapter 11, Section 11.3.6). Their anti-fungal activity is thought to be attributable to the increased permeability of hyphal membranes that they cause but more recently Grenier and co-workers (1999) have shown that several thaumatin-like proteins can hydrolyse polymeric β-1,3-glucans.

Defensins

Broekaert and co-workers (1995, 1997) have reviewed plant defensins and have shown their similarity to defensins from insects and mammals, including a defensin from human skin (Harder *et al.*, 1997). Plant defensins are 45 to 54 amino acids in length and have a net positive charge. The relative positions of eight cysteine residues, an aromatic residue and a glutamate are conserved. NMR spectroscopy showed that plant defensins, in common with those from mammals, exist as triple-stranded anti-parallel β-sheets and the primary structure of a peptide from the fruit fly *Drosophila melanogaster* also indicates that it will share this structure as well. Broekaert and co-workers (1995) suggest that these common properties provide evidence that the defensins belong to a super-family of anti-microbial peptides which predate the evolutionary divergence of plants and animals.

Two groups of plant defensins, the morphogenic and non-morphogenic have been distinguished on the basis of whether they increase hyphal branching of fungi or simply slow hyphal growth, respectively. A third group found in members of the Poaceae are referred to as the α-amylase inhibitor type as they inhibit α-amylase of insects and humans.

Defensins cause rapid membrane responses in fungi such as K^+ efflux, Ca^{++} uptake, alkalization of the medium and changes in membrane potential. Work with a defensin from *Dahlia merckii* showed that binding of the antimicrobial protein was required for activity and that binding was specific, defensins highly homologous to the dahlia defensin competing for binding sites and those more distantly related or structurally unrelated being less effective competitors (Thevissen *et al.*, 2000a,b).

There seems to be little doubt about the role of defensins as resistance factors in plants, particularly in germinating seedlings. For example, Terrass and co-workers (1995) found that sufficient defensin was produced by radish seed to protect the plant during the early stages of emergence. Moreover, as will be discussed in Chapter 11, defensins accumulate in response to pathogen attack and transgenic tobacco plants expressing a radish defensin were more resistant to infection by *Alternaria longipes* (Chapter 12).

Non-specific phospholipid transfer proteins   The discovery of these proteins as members of the anti-microbial complement of plant proteins came about through work on an α-amylase inhibitor from seed of Indian finger millet (*Eleusine coracana*) which was shown to have homology with a protein from the aleurone layer of barley. Further studies of homologies with proteins from spinach, maize and castor bean showed that they were non-specific phospholipid transfer proteins (LTPs). LTPs from different plants have a range of specificities in their activity against plant pathogenic fungi and some show antibacterial activity as well (Shewry and Lucas, 1997). Intercellular washing fluids from sugarbeet contained two proteins which, on the basis of amino acid sequence homology, including the presence of eight cysteines at conserved positions, can be classified as non-specific LTPs. The proteins showed strong anti-fungal activity *in vitro* against *Cercospora beticola*, the causal agent of leaf-spot disease in sugarbeet, inhibiting fungal growth at concentrations below 10 μg/ml (Kristensen *et al.*, 2000).

Albumins   The 2S albumin storage proteins are widely distributed in plants and consist of large and small subunits of 9 kDa and 4 kDa, respectively. However, their antifungal activity is not great but they appear to act synergistically with thionins (see above), non-inhibitory concentrations reducing the amount of thionin required to inhibit fungal growth from two- to 73-fold (Shewry and Lucas, 1997).

Ribosome-inactivating proteins   Ribosome-inactivating proteins (RIPs) were originally recognized as virus-inhibiting proteins in members of the Caryophyllales. Subsequent work showed that they are N-glucosidases which cleave the N-glycosidic bond of adenine in a specific sequence of eukaryotic ribosomes, preventing peptide elongation. Ribosomes of host plants are insensitive. There are two types: type 1 are single polypeptide chains of about 26–32 kDa, and type 2 are synthesized with a galactose-binding lectin domain which is cleaved from the rest of the molecule but remains attached by a single disulphide bond. The mechanism by which RIPs inhibit virus transmission does not seem to be known but Lodge, Kaniewski and Tumer (1993) have suggested that entry into the host cell with the virus could occur and prevent the translation of viral RNA or that entry of the virus is inhibited by binding of the molecule to the virus or the plant cell wall. A type 1 RIP from barley grains inhibits the growth of several fungi *in vitro* and there is synergy of inhibition with two other barley-seed proteins, a β-1,3 glucanase and a chitinase (Leah *et al.*, 1991).

Lysozyme

Lysozymes inhibit bacterial growth by hydrolysis of cell-wall peptidoglycans. Those from egg white and bacteriophage have been studied and there are now reports of proteins with similar activity from plants but, in addition, these enzymes have chitinase activity. Their role in resistance to disease has not been proved (Shewry and Lucas, 1997).

Proteinase inhibitors

Shewry and Lucas (1997) list 10 families of enzyme inhibitors many of which are active against trypsin and chymotrypsin, while some such as the cystatins are active against papain and other cysteine endoproteinases. These may confer resistance to nematodes or invertebrate vectors of viruses and have proved popular targets for transgenic experiments (Chapter 12).

Polygalacturonase-inhibiting proteins (PGIPs)

Although the importance of polygalacturonases in pathogenesis has been diffi-cult to prove, there is little doubt of their essential role in many diseases (Section 6.9.2). Therefore inhibitors of these enzymes would be expected to make a contribution to resistance. A number of polygalacturonase-inhibiting proteins (PGIPs) have been isolated from dicotyledonous plants and they, in common with resistance genes, possess leucine rich repeat domains. They are thought to contribute to resistance to fungal pathogens not only by inhibiting the pathogen's enzymes but also by prolonging the life of oligomers consisting of 10–13 galac-turonate residues. These are highly active elicitors of defence responses such as the hypersensitive response, lignification and the synthesis of phytoalexins (Chapter 11), whereas oligomers of higher or lower degree of polymerization are less active or inactive (De Lorenzo et al., 1990, 2001, 2002). However, Sharrock and Labavitch (1994) tested the effect of the PGIP from Bartlett pear on the action of enzymes in culture filtrates obtained from Botrytis cinerea on cell walls of pear. They found that polygalacturonide oligomers with a degree of polymerization greater than two did not accumulate but attributed this to mul-tiple polygalacturonases produced by the test pathogen.

   Not surprisingly, increased plant resistance is being sought in transgenic plants over-expressing PGIPs (Chapter 12).

10 The Genetics of Compatibility and Incompatibility

Summary

The discovery in the early years of the last century that Mendel's laws of inherit-ance were also applicable to the trait of resistance of plants to their pathogens led to an enormous effort in breeding for resistance. Unfortunately, it was soon found that the resistance conferred was, in many cases, far from permanent and, misleadingly, it was said to have 'broken down'. Owing to the classical genetic experiments of Flor and Oort, the gene-for-gene nature of such interactions was recognized. The simplest statement of this fundamental concept is that for every allele specifying resistance in the host there is a corresponding allele specifying avirulence in the pathogen. The so-called 'break down' in resistance is then readily understood as a change in the avirulence alleles of the pathogen rather than any change in the host.

The existence of alleles that specifically deny a pathogen the opportunity of infecting a host seems perverse but recent data support the view that they code for factors that enhance virulence in compatible hosts. Recognition of such viru-lence factors by the host and the subsequent mounting of an effective defence response automatically convert the factor from having a virulence function to an avirulence function and the allele which encodes it from virulence to avirulence. They may therefore be referred to as *avr/pth* alleles. The advent of molecular biology has brilliantly confirmed the validity of the gene-for-gene concept, the cloning of the first avirulence gene being reported in 1984 and the first resistance gene in 1993.

Hrp genes are another player in the field. These are genes of bacterial patho-gens that are required for the expression of the hypersensitive response in non-host plants and for pathogenicity in host plants. Understanding of this dual function came when it was realized that *hrp* genes code for elements of the type III secretory system found not only in plant pathogens but also animal patho-gens. Thus mutations in genes coding for this system would prevent the secretion

Introduction to Plant Pathology by Richard Strange
© 2003 John Wiley & Sons, Ltd   ISBN 0 470 84972 X (cased)   ISBN 0 470 84973 8 (pbk)

of elicitors of the hypersensitive response into non-host plants and also pathogenicity factors into host plants.

The rapid evolution of resistance genes of plants has allowed them to survive
despite changes in the avirulence genes of their pathogens. The means by which
plants achieve these rapid changes appears to be analogous to the hypervaria-
bility found in the major histocompatibility genes of mammals and arises from
intragenic and intergenic recombination, gene conversion and diversifying selection. However, the new concept of the guard hypothesis has explained why some
resistance genes in natural populations are ancient and has given a new impetus
to research – the identification of virulence targets in the host sought by the
products of a*vr/pth* genes.

10.1   Introduction

Compatible and incompatible are two terms which are considerable aids to the
description and understanding of plant–pathogen relations. A compatible interaction is one in which the pathogen establishes an advantageous relationship with
its host whereas an incompatible interaction describes the failure to establish such
a relationship. Compatibility and incompatibility are under the genetic control of
both host and pathogen. In biotrophic and some necrotrophic interactions the
products of host resistance genes (*R* genes) perform a surveillance function,
recognizing the advent of potential pathogens. The characteristic of the pathogen
which is recognized is also under genetic control, the genes concerned being
termed *a*virulence genes (*avr* genes) since they confer avirulence on the pathogen.
Therefore, for each resistance gene in the host there is a corresponding avirulence
gene in the pathogen. This apparently simple statement is known as the gene-for-
gene concept and has been declared a triumph of classical genetics (Fraser, 1990).

An important corollary of the gene-for-gene concept is that alteration or
deletion of an avirulence gene obviates recognition by the corresponding resistance gene. If resistance were entirely contingent on that resistance gene then the
plant would be susceptible to the pathogen with the altered or deleted avirulence
gene. Indeed, this is precisely the situation that has arisen frequently in many
agriculture crops: a cultivar which is resistant to an important pathogen is
introduced; it becomes popular and is consequently grown widely. Meantime a
variant of the pathogen with an altered or deleted avirulence gene, which may
have been present at a low incidence or may have evolved, multiplies on the
previously resistant host and comes to dominate the pathogen population, often
causing devastation. For example, ten Houten (1974) reported the demise of a
promising wheat cultivar, Heine VII, in Holland in these terms: 'When inoculated
with the known strains of stripe rust it remained resistant in our field experiments
both in 1950 and 1951; when introduced commercially in 1952, 14 per cent of the
total wheat area in the Netherlands was sown with it, and no stripe rust occurred.
In 1953 the area sown with Heine VII increased to 43 per cent of the total and

only one small locus of infection was found in a breeder's farm. In 1955 when 81 per cent of the total wheat area had been sown with Heine VII it was everywhere heavily infected with stripe rust, and in 1956 70 per cent of the winter wheat area was destroyed.' This unhappy state of affairs of many an agronomically desirable cultivar is often referred to as the 'boom-and-bust cycle'.

The hypersensitive response (HR) is fundamental to resistance in the majority of situations in which a plant is challenged with an avirulent organism (see Chapter 11, Section 11.3.1). It describes the death of a limited number of cells in the immediate vicinity of the challenge but the plant as a whole remains healthy. While it is easy to envisage that the HR is sufficient in itself to prevent colonization of a plant by obligate pathogens since, by definition, they require living host material to grow and complete their life cycles, it is not so easy to explain the association of the HR with resistance to facultative pathogens that kill the host and live off the dead remains. Here a number of active resistance responses which are set in train, either as a result of the HR or parallel with it, are responsible for resistance (see Chapter 11). One conundrum was the finding that certain mutants, which had lost their ability to elicit the HR in non-host species, were also deficient in pathogenicity for their normal hosts. The genes controlling these dual functions were given the name *hrp* genes for *h*ypersensitive *r*esponse and *p*athogenicity.

In this chapter, the pioneering experiments which led to the formalization of the gene-for-gene concept will be reviewed followed by a discussion of the cloning and nature of resistance and avirulence genes, their evolution and the natures of their products. An explanation of *hrp* genes and their function is also given. Genes controlling other pathogenicity or virulence factors, such as enzymes or toxins have already been discussed in the chapters dealing with these topics (Chapters 6 and 8).

10.2 Pioneering experiments

Biffen (1905, 1912) was the first to make a formal study of the inheritance of resistance of a plant to one of its pathogens. He found that the resistance of the wheat cultivar, Rivet, to stripe rust caused by *Puccinia striiformis* was inherited as a single recessive gene. More recent work with this pathogen and other genotypes of the host has confirmed that several genes which confer resistance are recessive (Chen and Line, 1999). In particular, genes at the *Mlo* locus of barley which confer resistance to mildew (*Blumeria graminis* = *Erysiphe graminis*) are recessive; the fact that the resistance conferred has proved, with certain exceptions, to be durable is of particular note (see Chapter 12 for a discussion of durable resistance). More often, however, resistance is inherited as a dominant rather than as a recessive character.

Plant breeders soon started to exploit resistance that was determined by one or a few genes. Although they were often extremely successful, sometimes the resistance of new cultivars was short-lived and resistance was said to have 'broken down' (see Section 10.1). This implies that a change has taken place in the plant. In fact, the change is on the part of the pathogen which has adapted to change in the host, i.e. the introduction of resistance genes by the plant breeder. Thus, host–pathogen relations can be viewed as the interaction of two genetic systems and a change on the part of the plant is likely to select for a change on the part of the pathogen. The Dutch experience described in Section 10.1 is explained by the 'resistant' Heine VII selecting for a variant of the fungus which is able to parasitize it. Such variants, which are often called physiologic races or simply races (Section 2.2), may have already existed at a low level in the population or may have arisen by recombination or mutation. The race of stripe rust which was virulent for Heine VII, on being presented with a large proportion of the Dutch wheat crop planted to this cultivar and being polycyclic (Section 3.2.1) was able to increase rapidly and cause the devastation reported by ten Houten (1974).

The complementary nature of the genetics of host and pathogen in plant diseases was originally and clearly established by Flor working with flax and its rust fungus, *Melampsora lini*. He formulated the gene-for-gene concept which has since been corroborated by studies of many host–pathogen interactions, including those of the weed, groundsel, and its mildew and the model plant *Arabidopsis thaliana* and its pathogens (Figure 10.1; Flor, 1971; Harry and Clarke, 1986). These demonstrate that gene-for-gene relationships are not simply artifacts of plant breeding but also occur in plants which have co-evolved naturally with their pathogens. Moreover, with the completion of the *Arabidopsis* genome project, it

Figure 10.1 Rust of flax caused by *Melampsora lini* in a segregating F$_2$ family of *Linum marginale*: plant on right, a fully susceptible reaction; plants at left and in centre are resistant (courtesy of Greg Lawrence). A colour reproduction of this figure can be seen in the colour section

is clear that about 1 per cent of the genome of this weed is devoted to resistance genes and their homologues (Meyers *et al.*, 1999).

10.3   Some experiments which led to the formulation of the gene-for-gene concept

The gene-for-gene concept is so fundamental to understanding the relations of plants and their pathogens that it is worth recounting some of the experiments which led to its proposal. Flor crossed 17 cultivars of the plant flax with the cultivar Bison, which was susceptible to all known races of the rust fungus, *Melampsora lini*. The progeny were selfed and the F_2 generation was tested for reaction to the pathogen. Flor obtained segregation ratios for resistance to susceptibility of 3:1, 15:1 and 63:1 showing that resistance was dominant and was governed by 1, 2 and 3 genes, respectively (Table 10.1).

Summing the resistant interactions in the data of Table 10.1, there are nine which are governed by one gene, six which are governed by two genes and two which are governed by three genes, giving a total of 27 resistance genes among the 17 resistant cultivars used as parents in the experiment. However, the experiment does not indicate if some of the genes are identical. In order to resolve this point two or more cultivars could be inoculated with a range of races of the fungus. If they gave identical reactions this would be strong evidence for identity of the resistance genes in the cultivars. Alternatively, the cultivars could be crossed, the F_1 generation selfed, and the resulting F_2 generation inoculated with a race of the fungus to which the parental types were resistant. Lack of segregation of susceptible phenotypes in the F_2 generation would again suggest that the gene conferring resistance in the two cultivars was identical.

Data in Tables 6.2 to 6.4 give the results of some of Flor's experiments. An important point about them was that Flor was able to use different physiologic races to inoculate different leaves of the same plant. In Table 10.2 the cultivar,

Table 10.1   Results of crossing 17 cultivars of flax with the susceptible cultivar Bison and screening for segregation of resistance and susceptibility in the F_2 generation to a range of races of the flax rust fungus, *Melampsora lini*

Inoculum	Race 1	Race 7	Race 24	Mixture of four races
	*(1) 3R : 1S	(5) 15R : 1S	(7) 3R : 1S	(13) 3R : 1S
	(2) 3R : 1S	(6) 15R : 1S	(8) 3R : 1S	(14) 3R : 1S
	(3) 3R : 1S		(9) 15R : 1S	(15) 3R : 1S
	(4) 3R : 1S		(10) 15R : 1S	(16) 15R : 1S
			(11) 15R : 1S	(17) 63R : 1S
			(12) 63R : 1S	

*Numbers 1–17 refer to the 17 cultivars used in the crosses.

Table 10.2   Results of crossing two cultivars of flax and screening for segregation of resistance and susceptibility in the F2 generation to three races of the flax rust fungus, *Melampsora lini* (Ellingboe, 1976)

Race	Cultivars		Types of F$_2$ Plants				R : S ratio in F$_2$
	Bison X Williston Golden						
7	S	R	R	R	R	S	15R : 1S
52	S	R	R	R	S	S	3R : 1S
16	S	R	R	S	R	S	3R : 1S
Total F$_2$ plants observed			92	26	28	10	
Approximate ratio			9	3	3	1	

Table 10.3   Results of crossing two cultivars of flax and screening for segregation of resistance and susceptibility in the F$_2$ generation to three races of the flax rust fungus, *Melampsora lini* (Ellingboe, 1976)

Race	Cultivars		Types of F$_2$ Plants				R : S ratio in F$_2$
	Ottowa X Bombay						
3	R	R	R	R	R	S	15R : 1S
22	S	R	R	S	R	S	3R : 1S
52	S	R	R	R	S	S	3R : 1S
Total F$_2$ plants observed			110	43	32	9	
Approximate ratio			9	3	3	1	

Bison, which at the time was thought to be susceptible to all known races of the fungus, was crossed with the cultivar Williston Golden. The progeny were selfed and different leaves of each plant in the F$_2$ generation were inoculated with one of three races of *M. lini* to which Williston Golden was resistant. The results show a 9:3:3:1 segregation ratio for resistance, respectively, to all three races, race 52 and not race 16, race 16 and not race 52 and to neither race. This is indicative of two independently segregating genes in Williston Golden, both of which confer resistance to race 7, one of which confers resistance to race 52 and the other to race 16.

In a similar experiment, Flor crossed the cultivars Ottawa and Bombay, which were both resistant to race 3. However, a 15:1 ratio of resistant to susceptible plants in the F$_2$ generation to this race demonstrated that resistance must have been conferred by two independently segregating genes (Table 10.3). As in the previous example, when other leaves of the F$_2$ plants were inoculated with two different races (race 52 and race 22) segregation ratios of 3:1 were obtained. Taking the data as a whole they suggest that only two genes were involved; one of these conferred resistance to race 22, the other to race 52 and both to race 3 (Table 10.3). When the cultivars Ottawa 770B and J.W.S. were crossed no plants

Table 10.4 Results of crossing two cultivars of flax and screening for segregation of resistance and susceptibility in the F_2 generation to three races of the flax rust fungus, *Melampsora lini* (from Ellingboe, 1976)

Race	Cultivars		Types of F_2 Plants				R : S ratio in F_2
	Ottowa 770B X J.W.S						
19	R	R	R	R	R	S	1R : 0S
7	S	R	R	R	S	S	3R : 1S
22	S	R	R	S	R	S	3R : 1S
Total F_2 plants observed			95	45	48	0	
Approximate ratio			2	1	1	0	

Table 10.5 Reactions of the parental, F_1 and F_2 generations of flax in a cross in which one parent is homozygous for resistance and the other for susceptibility with the corresponding cross of the fungus in which one parent is homozygous for avirulence and the other for virulence

Pathogen Genotypes	Host Genotypes					
	Parent 1 R_1R_1	Parent 2 r_1r_1	F_1 R_1r_1	F_2 R_1R_1	R_1r_1	r_1r_1
Parent 1 *Avr1Avr1*	−	+	−	−	−	+
Parent 2 *avr1avr1*	+	+	+	+	+	+
F_1 *Avr1avr1*	−	+	−	−	−	+
Avr1Avr1	−	+	−	−	−	+
F_2 *Avr1avr1*	−	+	−	−	−	+
avr1avr1	+	+	+	+	+	+

− = incompatible, i.e. the host is resistant and the pathogen avirulent; + = compatible, i.e. the host is susceptible and the pathogen virulent. Note that both resistance and avirulence are dominant traits.

were susceptible to all three races (Table 10.4). This result suggested that there were two different genes which conferred resistance to races 7 and 22 but that they were allelic. Thus, these genes did not segregate independently in the F_2 generation and therefore there was no class of plant that was susceptible to race 19.

Flor also investigated the inheritance of virulence in *M. lini*. The fungus is dikaryotic for much of its life cycle and may therefore be regarded as being somewhat similar to a conventional diploid organism. Thus an allele in one nucleus may give a phenotype which is dominant to that in the other. In the present context, Flor found that avirulence was dominant. Putting the parental, F_1, and F_2 generations of the diploid host and the dikaryotic pathogen together we therefore have the reactions shown in Table 10.5 in which, for convenience, the dominant avirulence allele is designated *Avr* in order to distinguish it from the alternative allele.

Since resistance and avirulence are dominant traits, the diploidy of the plant and the dikaryotic state of the fungus may be ignored, allowing the data of Table

Table 10.6   The quadratic check

	Host	
Pathogen	R_1	r_1
Avr1	−	+
avr1	+	+

− = incompatible, i.e. the host is resistant and the pathogen avirulent;
+ = compatible, i.e. the host is susceptible and the pathogen virulent.
Note that both resistance and avirulence are dominant traits.

10.5 to be summarized in a diagram that has been given the curious name quadratic check (Table 10.6).

We can therefore summarize the gene-for-gene concept in one simple statement: *for every allele conferring resistance in the host there is a corresponding allele conferring avirulence in the pathogen.*

A possible source of confusion in the literature is that avirulence genes of bacteria are usually prefixed by the lower case *avr* as they, in common with many plant pathogens, are not diploid. However, Oomycetes are diploid throughout almost their entire life cycle and rust fungi are usually dikaryotic. In these instances and those of fungi, which may, nevertheless, be haploid, the upper case *Avr* will be used to signify an effective avirulence gene.

10.4   Variations on the gene-for-gene concept

As initially comprehended, the gene-for-gene concept provided an excellent model for understanding race-specific resistance. In its simplest form, as exemplified by the quadratic check (Table 10.6) dominant avirulence alleles are recognized by cognate dominant resistance alleles in the plant but by no means do all plant–pathogen systems fit this simple model. In some instances there is a clear effect of gene dosage. For example, Crute and Norwood (1986), working with lettuce and the downy mildew fungus, *Bremia lactucae*, a diploid Oomycete, showed that the effectiveness of the resistance gene *R6* was dosage-dependent with a trend of decreasing colonization of the plant in the interactions *Avravr/Rr, AvrAvr/Rr, AvrAvr/RR*. Similarly, Hammond-Kosack and Jones (1994) found that tomato plants homozygous for the *Cf* genes, which confer race specific resistance to *Cladosporium fulvum*, were more effective in containing infections than plants that were heterozygous for the genes. Also in nematode interactions, homozygosity rather than heterozygosity for *R* genes appears to be more effective. For example, Tzortzakakis, Trudgill and Phillips (1998) found that tomatoes homozygous for the *Mi* gene supported less reproduction of three isolates of the nematode, *Meloidogyne javanica*, which were partially virulent, than genotypes heterozygous for the gene.

Temperature may have significant effects on the expression of resistance genes. The gene *Sr6*, which conditions resistance to races of stem rust of wheat caused by *Puccinia graminis* f.sp. *tritici* that have the corresponding avirulence gene, *P6*, is effective at 15°C, giving only a fleck, but is ineffective at 24°C, giving a 3+ reaction, i.e. medium-sized uredial pustules (Chapter 2, Table 2.1; Browder, 1985). High temperatures also negate the resistance of peppers to tomato spotted wilt virus conferred by the *Tsw* gene (Moury *et al.*, 1998) and potato (cv. Pito) to the PVYO group of potato virus Y (Valkonen, 1997). In contrast, Islam, Shepherd and Mayo (1989) showed in the flax – flax rust system that plants containing the *L2* resistance gene gave a resistant (type 1) reaction with rust strain J at 25°C but a highly susceptible reaction (type 4) at 15°C.

Genetic background may also affect the outcome of the interaction of a pathogen avirulence gene with its cognate resistance gene. For example, in stem rust of wheat, Luig and Rajaram (1972) found that the resistance conferred by the *Sr5* gene was dominant in the background of the cultivars, Kanred and Reliance, incompletely dominant in line W2928 and recessive in W2691 and W3498.

The presence of inhibitor genes is a further complication in some interactions. These are thought to interfere with the interaction of the products of resistance and avirulence genes and thus block the recognition phenomenon which leads to the resistant reaction. Furthermore, inhibitor genes may also be temperature sensitive. Islam, Shepherd and Mayo (1989) found that at higher temperatures, three strains of flax rust which possessed inhibitor genes were incompatible, suggesting interference with the expression of the inhibitor genes and the consequent unmasking of the expression of the resistance genes. In more recent experiments with oats and the crown rust fungus, *Puccinia coronata* f. sp. *avenae*, crossing a line containing the resistance gene *Pc-38* with one containing another resistance gene, *Pc-62*, gave an excess of susceptible seedlings in the segregating population. A suppressor of resistance was found on the chromosome which contained *Pc-38*. The suppressor was dosage dependent since trisomic combinations containing two doses of the *Pc-38* were found to be more effective than one in suppressing the resistance conferred by *Pc-62* (Wilson and McMullen, 1997).

In the interaction of *Arabidopsis thaliana* and *Pseudomonas syringae*, Bisgrove and co-workers (1994) found the same resistance gene conferred incompatibility to two isogenic strains of the pathogen carrying different avirulence genes. They mutagenized *A. thaliana* carrying the resistance genes *RPS3* and *RPM1* and recovered 12 mutants that were susceptible to *P. syringae* carrying *avrB* signifying that they had lost the effective resistance gene *RPS3*. All 12 mutants were also susceptible to an isogenic strain of the pathogen carrying the avirulence gene *avrRpm1* which is recognized by the resistance gene *RPM1*. Confirmation that *RPS3* and *RPM1* were indeed the same gene was obtained by genetic analysis. Four of the mutants were backcrossed to the resistant parent and the F1 generation scored for reaction to both pathogens. All plants were resistant demonstrating that the mutations were recessive. Since no complementation occurred when the four mutants were crossed either with each other or with a naturally occurring susceptible ecotype the mutations were both recessive and in the

RPS3/RPM1 resistance gene. Segregation ratios of 3 resistant to 1 susceptible in the F_2 generation of crosses with a resistant parent demonstrated that the susceptible phenotype was caused by a single mutation and this mapped to the same region of the same chromosome (chromosome 3) as *RPS3/RPM1*. Since the proteins encoded by the two avirulence genes of the pathogen contain no significant sequence similarity their relation with *RPS3/RPM1* is more accurately described as a genes-for-gene relationship.

Populations of a plant species may vary quantitatively in their resistance to a given pathogen. Such variation is often thought to be multigenic, each locus making a limited contribution to resistance. This raises the possibility that the appropriate combination of these genes would give a high level of resistance, which, owing to its multigenic nature, would be durable. One way of studying such resistance is by the use of quantitative trait loci (QTL) mapping (Young, 1996). Genetic markers are sought that are linked to resistance and, in some systems, such studies have shown that the number of genes involved may be greater than 10 but more generally it is only three to five. The use of genetic markers as an aid to breeding for resistance is discussed in Chapter 12.

In a number of other host–pathogen interactions involving fungi, Oomycetes, bacteria and viruses, resistance is inherited as a recessive character. For example, the *mlo* gene of barley confers broad spectrum resistance to several isolates of the mildew fungus *Blumeria* (= *Erysiphe*) *graminis* f. sp. *hordei*. The alternative allele, *Mlo*, is thought to encode a negative regulator of defence (see Stein and Somerville, 2002). In the absence of such regulation abnormal defence responses occur both spontaneously and during infection. Similarly, *A. thaliana* with the recessive mutation, *edr1*, exhibit a higher level of resistance to some fungi and bacteria. Again the alternative allele is thought to encode a negative regulator of defence. Deslandes and co-workers (2002) have described another recessive gene from *A. thaliana*, *RRS1-R*, which confers resistance to *Ralstonia solanacearum*. Both recessive (*RRS1-R*) and dominant (*RRS1-S*) alleles encode predicted proteins that are very similar but differ in length. The authors suggest that both are functional and may compete for a DNA target. Alternatively, they may compete for a limiting factor which is either necessary for the growth or propagation of the pathogen or for the establishment of resistance or susceptibility in the infected plant.

10.5   Molecular corroboration of the gene-for-gene concept

At the time of writing, more than 30 resistance genes have been cloned from a variety of plant species (Hulbert *et al.*, 2001) and more than 40 avirulence genes (Gabriel, 1999). Conversion of compatibility to incompatibility by transformation of susceptible plants with resistance genes and virulent pathogens with avirulence genes have amply confirmed the validity of the gene-for-gene concept. The cloning, structure, expression, function and organization of these two classes of genes are, perhaps, the most significant advances that have been made in our understanding of host–pathogen interactions during the last 20 years and have shed light on the their co-evolution. Discussion of these topics will form the bulk of the remainder of this chapter. A further class of genes which play a fundamental role in the compatibility or incompatibility of bacterial pathogens will also be discussed. These are the *hrp* genes, so named because they are required for both the *h*ypersensitive *r*esponse in resistant plants (see Chapter 11, Section 11.3.1) and *p*athogenicity in susceptible plants.

10.5.1 Cloning, structure and expression of avirulence genes

Cloning of the first avirulence gene

Among the various classes of plant pathogens bacteria, which can be cultured *in vitro* and which have comparatively small genomes, provide the most tractable material (Chapter 1). It was therefore from the bacterium *Pseudomonas syringae* pv. *glycinea*, a pathogen of soybean, that Staskawicz, Dahlbeck and Keen (1984) cloned the first avirulence gene. They mobilized a single clone from race 6 via the cosmid vector pLAFR1 into other races with the results shown in Table 10.7. This table shows that the avirulence properties of the race 6 wild type were superimposed on those of race 5, race 4 and race 1. The cosmid was mapped by restriction enzymes and two adjacent *Eco*R1 fragments unique to race 6 were shown by transposon mutagenesis to be important in determining race specificity (Staskawicz, Dahlbeck and Keen, 1984). The avirulence gene, termed *avrA*, was sequenced and found to encode a single protein product of 100 kDa (Napoli and Staskawicz, 1987).

Table 10.7 Reactions of seven soybean cultivars to three races of *Pseudomonas syringae* pv. *glycinea* and the alteration of these reactions by transformation of the pathogen with a cosmid containing an *avr* gene from race 6 (adapted from Staskawicz, Dahlbeck and Keen, 1984)

Genotypes of *P. syringae* pv. *glycinea*	Soybean cultivars						
	1	2	3	4	5	6	7
Race 5 wild type	−	−	−	+	+	−	+
Race 5 wild type + cosmid	−	−	−	+	−*	−	−*
Race 6 wild type	−	−	+	+	−	+	−
Race 4 wild type	+	+	+	+	+	+	+
Race 4 wild type + cosmid	−*	−*	+	+	−*	+	−*
Race 6 wild type	−	−	+	+	−	+	−
Race 1 wild type	+	−	+	−	−	−	+
Race 1 wild type + cosmid	−*	−	+	−	−	−	−*
Race 6 wild type	−	−	+	+	−	+	−

− Denotes an incompatible reaction; + denotes a compatible reaction; * denotes change to avirulence characteristic of race 6.

Since then, more than 40 avirulence genes have been cloned, many of which are from bacterial sources, but some are also from fungi, nematodes and viruses. Few share any sequence homology (Gabriel, 1999).

Cloning, structure and expression of other bacterial avirulence genes

Species and pathovars of *Xanthomonas* have been a popular choice of experimental material. Early work with the genes designated *avrBs1* and *avrBs2* will be discussed later when considering the evolution of avirulence genes (Section 10.6.1). In the meantime we shall be concerned with the genes *avrBs3* and *Vrxv4*.

Pioneering work with *avrBs3* from *X. campestris* pv. *vesicatoria*, a pathogen of tomatoes and bell pepper showed that it coded for a nearly identical 34-amino-acid repeat unit which was present in 17.5 copies (Herbers, Conradsstrauch and Bonas, 1992). When some of these repeats were deleted four classes of organisms were recovered:

(1) those that retained their ability to induce the hypersensitive resistance reaction (HR, see Chapter 11, Section 11.3.1) on the *Bs3* plant genotype and were therefore indistinguishable from the wild type in this respect;

(2) those that had lost their ability to induce the HR on the *Bs3* plant genotype but were avirulent (i.e. induced the HR) on an isogenic line which lacked *Bs3* (i.e. showed the converse of the specificity towards plants possessing *Bs3* and *bs3* to that of the wild type);

(3) those that induced an intermediate response on both isogenic lines (having or lacking the *Bs3* gene) in addition, these mutants induced the HR in tomato whereas the wild type was virulent and caused water-soaking;

(4) those that did not induce the HR on either isogenic line of pepper (i.e. both lines were susceptible).

Further experiments showed that these altered specificities resided in the position of the deletions rather than their length.

More recent data have shown that *avrBs3* belongs to a gene family which has members in other species and pathovars of *Xanthomonas*. The protein products of these genes share 90–97 per cent sequence identity, the differences occurring mainly within the 34-amino-acid repeats and these, as foreshadowed by the deletion experiments described above, determine specificity (Bonas and Lahaye, 2002). In addition to the repeat domain, all members of the *avrBs3* family contain nuclear localization signals and an acidic transcriptional activation domain, which are usually found only in eukaryotes. Mutations in either of these domains normally abolish recognition by plants carrying the cognate resistance gene.

Astua-Monge and co-workers (2000) have described the avirulence gene *Vrxvr*. They found that strains of tomato race 3 of *X. campestris* pv. *vesicatoria* elicited a hypersensitive response in leaves of *Lycopersicon pennellii* LA716. The avirulence gene was identified by mobilizing a library of 600 clones into a strain of the bacterium that was virulent for the tomato genotype. One cosmid clone conferred the ability to induce the hypersensitive response rendering the transconjugant avirulent. A data-base search showed that the predicted protein had homologies with two other avr proteins, avrRxv and avrBst as well as proteins from bacterial pathogens of mammals, YopJ and YopP from *Yersinia* spp. and avrA from *Salmonella typhimurium*. The authors suggest that these form part of an emerging family of bacterial proteins found in pathogens of plants and animals which have been referred to as the 'avrRxv-host interaction factor family' (Ciesiolka *et al.*, 1999).

Tsiamis and co-workers (2000) cloned the avirulence gene *avrPphF* from *Pseudomonas syringae* pv. *phaseolicola* (Pph) races 5 and 7 which confers avirulence towards bean cultivars carrying the R1 gene for halo-blight resistance, such as Red Mexican. The gene was present on a 154 kb plasmid (pAV511) and comprised two open reading frames, both of which were required for function. However, *avrPphF* in the context of other cultivars of bean displayed different properties. For example, strain RW60 of Pph, which lacks the plasmid pAV511, demonstrated a loss in virulence to a range of cultivars such as Tendergreen and Canadian Wonder. Virulence to Tendergreen was restored in RW60 by *avrPphF* alone but, in contrast, subcloned *avrPphF* in Canadian Wonder, far from restoring virulence, greatly accelerated the hypersensitive resistance reaction. *avrPphF* also conferred virulence in soybean. Thus the same gene may function as a virulence or an avirulence factor according to the genotype of both the pathogen in which it resides and the genotype of the host in which invasion is attempted (Figure 10.2 and see Section 10.6.2).

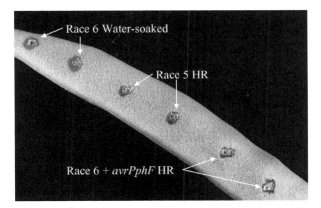

Figure 10.2 Cloning of the avirulence gene *avrPphF* from *Pseudomonas syringae* pv. *phaseolicola* race 5. Race 6 gives a susceptible water-soaked reaction whereas race 5 gives a hypersensitive response (HR). The HR was conferred on race 6 when transformed with a clone containing *avrPphF* (courtesy of John Mansfield). A colour reproduction of this figure can be seen in the colour section

Cloning, structure and expression of fungal avirulence genes

Joosten and de Wit (1999) have summarized data obtained in their laboratory over a number of years with *Cladosporium fulvum*, a pathogen of tomato. Again, this is a gene-for-gene interaction in which matching genes for avirulence and resistance give rise to resistance, expressed as a hypersensitive response (HR). This system presented an immediate technical problem: in culture the fungus produced non-specific elicitors of the HR. Perhaps the fungus behaves differently in the plant, producing specific elicitors which would be the products of avirulence genes? Clearly, such specific elicitors would be difficult to find in a plant which was reacting hypersensitively to an incompatible race of the fungus as there would be little of the pathogen present to produce the compound(s). Tomato plants were therefore infected with a race of the fungus which was compatible and the infected leaves centrifuged in order to obtain the apoplastic fluids (i.e. fluids outside the protoplasts). These were tested on a range of genotypes of tomato plants and were found to elicit the HR in a gene-for-gene specific manner. Apoplastic fluids from leaves infected with a fungus containing the avirulence gene *Avr9* were fractionated and a 28-amino-acid peptide was isolated which was specific in that it caused the HR in tomato genotypes which had the cognate resistance gene *Cf9*. The peptide was sequenced and the corresponding structural gene cloned (Van den Ackerveken, vanKan and Dewit, 1992). The gene encodes a 63-amino-acid preprotein with a signal peptide for extracellular targeting. The 40-amino-acid peptide, which is secreted, is further processed by fungal and plant proteases to the mature 28-amino-acid peptide (Van den Ackerveken, Vossen and de Wit, 1993).

Transformation of an $Avr9^-$ wild-type strain of the fungus with the *Avr9* gene rendered the transformant avirulent for plants with the *Cf9* resistance gene. Conversely, disruption of the $Avr9^+$ gene in two wild-type strains rendered them virulent to plants with the *Cf9* resistance gene. Southern blot analysis showed that wild-type strains that are virulent for *Cf9* tomatoes lacked the entire *Avr9* open reading frame.

A similar approach led to the isolation of an HR inducing protein of 12 kDa from a race with the avirulence gene *Avr4*. In contrast to the *Avr9-Cf9* interaction, genotypes of the pathogen that were virulent to *Cf4* genotypes were found to have a gene that was identical apart from one base pair. In three virulent races that were analysed the base substitution was at a different position but in each case the change was found to substitute a tyrosine residue (codon TAT) for a cysteine residue (codon TGT; Joosten, Cozijnsen and de Wit, 1994; Figure 10.3). Thus host resistance is lost owing to a single base change in the avirulence gene of the pathogen. Subsequently, de Wit's group showed that the avirulence proteins with the tyrosine substitutions were unstable in the plant, providing an explanation for their failure to trigger a resistance response (Joosten *et al.*, 1997).

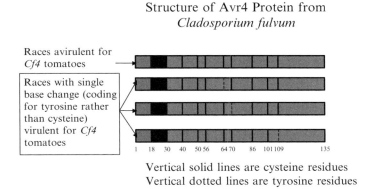

Vertical solid lines are cysteine residues
Vertical dotted lines are tyrosine residues

Figure 10.3 Structures of the Avr4 protein and variants; pathogenic races lack a cysteine residue resulting in the instability of the protein in the plant and their failure to trigger defence responses

Orbach and co-workers (2000) cloned the avirulence gene *AVR-Pita* from the rice blast fungus *Magnaporthe grisea* which confers avirulence for rice cultivars possessing the resistance gene *Pi-ta*. When introduced into virulent strains of the pathogen, the cloned gene specifically conferred avirulence toward rice cultivars that contain *Pi-ta*. The avirulence gene was a metalloprotease and was located entirely within the most distal 1.5 kb of chromosome 3, a position that provided an explanation for its frequent spontaneous loss. Moreover, the authors reported several types of mutation in the gene including point mutations, insertions, and deletions which negated its avirulence function.

Cloning, structure and expression of a nematode avirulence gene

Semblat and co-workers (2001) have cloned a putative avirulence gene from the nematode, *Meloidogyne incognita*. They used amplified fragment length polymorphisms (AFLP; see Section 2.5.5) to identify a sequence, *map-1*, present in avirulent but not virulent near-isogenic lines of the nematode for tomatoes containing the resistance gene *Mi*. The deduced protein, MAP-1, did not show any significant homology with proteins in databases but was characterized by highly conserved repetitive motifs of 58 and 13 amino acids. PCR experiments showed that *map-1*-related sequences were amplified only in nematode populations belonging to the three species against which the *Mi* gene confers resistance: *M. arenaria, M. incognita* and *M. javanica*. Evidence for the involvement of MAP-1 in the recognition of avirulent nematodes by resistant plants was obtained from polyclonal antibodies raised against a synthetic peptide deduced from the MAP-1 sequence. These strongly labelled J2 amphidial secretions in immunofluorescence microscopy assays.

Virus avirulence genes

Harrison (2002) in a review of virus variation in relation to resistance-breaking in plants has stated that several – even perhaps all – viral genes may encode avirulence factors which are recognized by cognate dominant resistance genes in the host. These include viral coat protein, RNA polymerase, movement protein (see Section 6.12) and proteinase. Some resistance-breaking virus variants may have quite subtle variations. For example, Palanichelvam and Schoelz (2002) showed that small deletions from either the 5′ or 3′ end of gene VI of cauliflower mosaic virus abolished its ability to elicit HR. More specifically, Karasawa and co-workers (1999) demonstrated that a single nucleotide change in cucumber mosaic virus strain Y was all that was required to alter the HR to a susceptible response in cowpea (*Vigna unguiculata*), recalling the same situation in the *Avr4* gene of *Cladosporium fulvum* (see above).

The role of avirulence genes in non-host resistance

The cloning of avirulence genes and their expression in heterologous pathogens has also shed light on the control of non-host resistance. Traditionally, this has been viewed as being caused by the presence of many genes that contribute to a complex of general resistance mechanisms (Heath, 1981). However, more recent evidence suggests that avirulence genes that confer race specificity towards host species may also confer avirulence to non-host species. For example, Fillingham and co-workers (1992) found that the avirulence genes of pathovars *pisi* and *phaseolicola* of *Pseudomonas syringae* which determine specificity towards cultivars of their host species, peas and beans respectively, also determined specificity towards the non-host species. This non-host specificity was characterized further. The avirulence gene, *avrPpi2* from the pea pathogen when cloned in the bean pathogen caused a rapid HR in most cultivars of bean (e.g. Canadian Wonder), but in the cv. Seafarer the reaction was slower. When Canadian Wonder and Seafarer were crossed, a ratio of 12:3:1 in the F_2 generation for plants acting with a rapid HR, the slower reaction and susceptibly was obtained. This is consistent with the two types of resistance being governed by dominant alleles at two independent loci with the faster HR response being epistatic to the slower response.

10.5.2 Function and organization of avirulence genes

The concept of a pathogen carrying avirulence genes appears counter-intuitive at first. Why should a pathogen possess genes which specifically deny it the capacity to form a parasitic relationship with its host? One answer to this conundrum is that the factors specified were, at one time, virulence factors and may still function as such in some plants (Gabriel, 1999 and see Section 10.6.1). For example, Jackson and co-workers (1999) identified a gene designated *virPphA* from *Pseudomonas syringae* pv. *phaseolicola* which acts as a virulence determinant in bean (*Phaseolus lunatus*) but as an avirulence determinant in soybean (*Glycine max*) where it caused a rapid hypersensitive response. More fundamentally, the authors of this paper present evidence to show that the gene is present on a large fragment of plasmid DNA which bears the hallmarks of a pathogenicity island. Pathogenicity islands have been recognized in bacterial pathogens of animals as DNA elements which differ in G + C content from the rest of the genome and contain domains that specify virulence and the type III secretory system (see Section 10.7) as well as mobile elements. In *P. syringae* pv. *phaseolicola*, strains cured of their plasmid were no longer virulent to bean but rather caused the hypersensitive response. Virulence was restored by complementation with cosmid clones spanning a 30 kb region of the plasmid which also contained three previously identified avirulence genes. Sequencing of 11 kb of the plasmid suggested the presence of three potential virulence genes which shared a *hrp*-box promoter (see Section 10.7). One of these partially restored virulence and was designated *virPphA*. Sequence data also showed that the G + C content was lower than expected for *P. syringae* and that homologues of two mobile elements, the insertion sequence IS*100* from *Yersinia* and transposase *Tn*501 from *P. aeruginosa* were present.

Gabriel (1999) has suggested that most microbial genes which play roles in pathogenicity are present on pathogenicity islands and that these are prone to horizontal transfer. Certainly in bacterial pathogens of animals pathogens are thought to have evolved from non-pathogenic ancestors by the acquisition of pathogenicity islands.

10.5.3 Cloning, structure and expression of resistance genes

The cloning of the first resistance gene, *Pto*, which confers resistance of tomato to *Ps. syringae* pv. *tomato* was reported in November 1993 (Martin *et al.*, 1993). Since then, more than 30 resistance genes, which are effective against fungi, bacteria, viruses and nematodes, have been cloned and sequenced. One of the nematode resistance genes is also effective against aphids.

Cloning

Singling out a single resistance gene in a plant is a formidable task since plants generally have a considerable amount of DNA. Three strategies are used: map based cloning, transposon tagging and identification by sequence homology. The last of these identifies many homologies, or resistance gene analogues as they are called but not, of course, the pathogen with the corresponding avirulence gene!

Map based cloning   The production of near isogenic lines is the starting point of map based cloning. These are plants in which the targeted resistance gene has been introgressed by back-crossing successive generations to the susceptible parent but always selecting for resistance. The result is a plant which is predominantly of the susceptible genotype (near isogenic) but contains the target resistance gene. The next step is to cross the isogenic lines and determine which markers co-segregate with resistance, the aim being to discover a marker that is as tightly linked and therefore as close to the resistance gene as possible. Clearly, selection of a resistant plant which is saturated with genetic markers is a prerequisite. A DNA library of the resistant plant is then made and usually cloned in yeast artificial chromosomes (YACs) or cosmids. The library is screened using the marker most tightly linked to the resistance gene. DNA from clones which are positive with regard to the marker are cloned in a vector which is suitable for transfer of the cloned DNA to *Agrobacterium tumefaciens*. Details of the use of *A. tumefaciens* and other techniques for transforming plants are given in Chapter 12 (Section 12.6). Suffice it to say here that sequences of interest are substituted for the tumorigenic genes in the transfer DNA (T DNA) of the Ti plasmid of *A. tumefaciens* (see Section 7.4). The transformed bacterium is then used to infect the susceptible isogenic line. Conversion of susceptibility to resistance demonstrates that the resistance gene has been transferred.

The *Pto* gene mapped to chromosome 5 of tomato and co-segregated with an RFLP marker (TG538). This was used to screen a tomato YAC library. A clone that hybridized to the marker was shown to span the *Pto* region. Preparation of a cDNA library led to the identification of another clone (CD127) which co-segregated with *Pto*. This consisted of two size classes of transcripts, one of which, when used to transform susceptible plants (via *Agrobacterium tumefaciens*-see Chapter 12, Section 12.6), conferred resistance. The 2.4 kb sequence contained an open reading frame that coded for a 321-amino-acid hydrophilic protein with similarities to the catalytic domain of protein serine-threonine kinases. This suggests that the protein may be involved in protein phosphorylation and may represent the first step in a signal transduction pathway leading to resistance responses (Martin *et al.*, 1993).

RPS2 is a gene from *Arabidopsis thaliana* which confers resistance to strains of *Ps. syringae* that express the *avrRpt2* gene. *RPS2* has been cloned independently by Staskawicz's group (Bent *et al.*, 1994) and by Ausubel's group (Mindrinos *et al.*, 1994). Both groups used map based cloning strategies and that of Staskawicz's group is summarized here. The gene was found to reside on chromosome IV and was flanked by two RFLP markers which were about 0.5 centimorgan (cM) apart. Using one of the RFLP markers as a probe, YACs and cosmid clones carrying overlapping inserts of the genome of an ecotype of *A. thaliana* containing *RPS2* were screened. RFLP markers from the ends of the YAC and cosmid inserts were generated which localized *RPS2* to a 200 kb region. A set of overlapping clones from the *RPS2* region was constructed in a vector suitable for *Agrobacterium tumefaciens*-mediated transformation of the plant. One 18 kb cosmid conferred the HR response on plants lacking *RPS2*. Five expressed sequences were found in the 18 kb sequence. When a set of overlapping cosmid clones spanning this 18 kb region was constructed and used to transform the susceptible plant, one containing a single expressed sequence conferred the resistant phenotype.

More recently a similar approach has been adopted in cloning the *Pib* gene for resistance to rice blast disease caused by *Magnaporthe grisea* (Wang *et al.*, 1999). In this case, knowledge of the nucleotide sequence of a conserved region of other resistance genes, the nucleotide binding site (NBS – see page 282) proved invaluable in pinpointing the gene. First a high-resolution linkage map was constructed. This revealed that the gene mapped to an interval of 0.5 centimorgans (cM) between two RFLP markers. The two markers were used to screen a back-crossed population for recombination between the markers and the *Pib* gene. A total of 32 recombinants were obtained out of a population of 3305 susceptible plants. These were used in a further RFLP analysis with five additional markers which allowed the gene to be mapped between two markers at a distance of 0.015 cM from one (S1916) and 0.045 cM from the other (G7030). The three other markers co-segregated with *Pib*.

A cosmid library of a resistant line of rice was constructed and screened by colony hybridizations using the five markers as probes. Six positive clones were obtained and these could be arranged in a single contig of 80 kb spanning the region between S1916 and G7030. Restriction mapping allowed the number of cosmids spanning the region to be reduced to three. The cosmids were digested with restriction enzymes, subcloned and the inserts partially sequenced from both ends. Two *Eco*R1 subclones derived from one of the cosmids had high homology with the nucleotide binding site (NBS) found in several resistance genes, suggesting that this region might code for the *Pib* gene. Sequencing of fragments from two of the cosmids revealed that each had an open reading frame (ORF) and each was more than 95 per cent similar to the other. In order to determine if either of these candidate genes were transcribed, resistant plants were inoculated and mRNA corresponding to the genes detected by primers specific to each ORF. Only one yielded a PCR product, suggesting that it was the *Pib* gene. The other did not, indicating that it might be a pseudogene. A susceptible rice cultivar was transformed with a fragment of the cosmid containing the candidate gene and from a total of 496 transformants 112 resistant plants were obtained, all of which contained the transgene.

Wang and co-workers (1999) also investigated the expression of *Pib* by Northern blot analysis. Inoculation with the pathogen induced expression although other treatments such as inoculation with a compatible race of the pathogen, mock inoculation, maintenance in the dark and temperature change also caused expression of the gene.

Transposon tagging   In this cloning strategy, plants homozygous for known resistance genes are transformed with a transposon. They are then crossed with a plant with no known resistance genes to produce F_1 progeny which are heterozygous for the resistance allele. In some instances the transposon will have inactivated the resistance allele by insertion giving rise to a susceptible phenotype. Sequences flanking the transposon may then be determined.

Jones and co-workers (1994a) isolated the *Cf-9* gene which confers resistance of tomato to races of the fungal pathogen, *Cladosporium fulvum*, with the corresponding avirulence gene *Avr9* by transposon tagging with the maize transposable element *Dissociation* (*Ds*). This element mapped to the short arm of chromosome 1 and was 3 centimorgans from the *Cf-9* locus. Plants heterozygous for both the *Ds* element and *Cf-9* were crossed with plants heterozygous also for *Cf-9* and for the stable but unlinked activator *sAc*, which is not itself capable of transposition but which activates the *Ds* element (Figure 10.4). Thus, the progeny were heterozygous for both *Ds* and *sAc* but were homozygous for *Cf-9*. These plants were crossed with transgenic tomatoes homozygous for the *Avr9* avirulence gene from *Cladosporium fulvum* but lacking the *Cf-9* gene. Plants expressing both *Cf-9* and *Avr9* became necrotic soon after germination and died. Thus, the problems associated with inoculation experiments, crucially that of disease escapes in this context, were neatly side-stepped. Out of approximately 160 00 progeny 118 survivors were recovered. Some were variegated but 33 were stable and carried *Ds*. All stable mutants were susceptible to race 5 of the fungus, which carries the *Avr9* gene, confirming the loss of *Cf-9* function. The *Cf-9* gene was isolated from one of these containing a single *Ds* element by plasmid rescue. In brief, genomic DNA was digested with a restriction enzyme, which does not cut within the *Ds* element and circularized with T4 DNA ligase. The product was used to transform *E. coli* and transformants were recovered on media containing chloramphenicol since *Ds* contains both a gene for resistance to

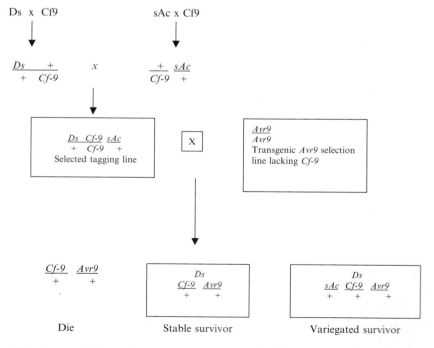

Figure 10.4 Protocol for cloning the resistance gene *Cf9* from tomato. For details see text. Reproduced with permission from *Nature*

this antibiotic and an origin of replication. Unfortunately, the open reading frame (ORF) of the gene was truncated by the restriction enzyme used for plasmid rescue as well as interrupted 93 bp 3′ to the start codon by *Ds*. In order to recover the remainder of the gene, an alternative restriction enzyme was used in plasmid rescue which cut within the *Ds* element but allowed rescue of a further 522 bp of the ORF.

Lawrence and co-workers (1995) were also able to use transposon tagging to clone resistance genes from the flax rust fungus (*Melampsora lini*), the host–pathogen system used by Flor to establish the gene-for-gene concept (Figure 10.1). They used the maize transposon *Ac* to transform the flax cultivar Forge, which is homozygous for four genes that confer resistance to flax rust. The transformed plants were crossed to the rust susceptible cultivar Hoshangabad and the progeny were screened for susceptibility. Out of 29 susceptible mutants all except one had deletions in the *L6* region. The mutant, designated X75, was determined to have the transposon at the *L6* locus by crossing to the cultivar Birio which carries the *L6* gene. Progeny that inherited *Ac* were crossed to the susceptible Hoshangabad and the progeny scored for disease reaction. Half of the progeny were susceptible demonstrating that the *Ac* insertion maps to the *L6* locus or very close to it. A DNA restriction fragment containing the transposon and flanking DNA was cloned and a fragment 459 bp from the 3′ end of *Ac* was used as a probe. This allowed the isolation of the wild-type *L6* gene from a genomic library of the cultivar Forge.

Identification of resistance genes by the polymerase chain reaction   Resistance genes share a number of structural features (Figure 10.5). In particular, a nucleotide binding site (NBS) is a characteristic domain of two of the five classes. Therefore candidate resistance genes may be amplified using degenerate oligonucleotide primers designed from conserved amino acids in the NBS motifs (Meyers *et al.*, 1999). Comparison of the sequences obtained by this procedure with those of known resistance genes and sequences from data bases allowed the NBS of *R* genes to be classified into two distinct types, one with Toll/Interleukin-1 receptor homology (TIR) and one without (see page 282 for details). However, the identification of a gene that confers resistance to a specific pathogen (i.e. recognizes an individual with a cognate *avr* gene) is difficult as genes that contain an NBS sequence are common in plants. For example, Meyers and co-workers (1999) analysed NBS domains of 14 known plant resistance genes and more than 400 homologues of the NBS domain from 26 genera of monocotyledonous, dicotyledonous and one coniferous species. Moreover, some of these may be pseudogenes (see above).

Structure of resistance genes

At the time of writing, five classes of resistance genes, based on their structure, are known (Figure 10.5). The first of these, with a single example, is the *Pto* gene which

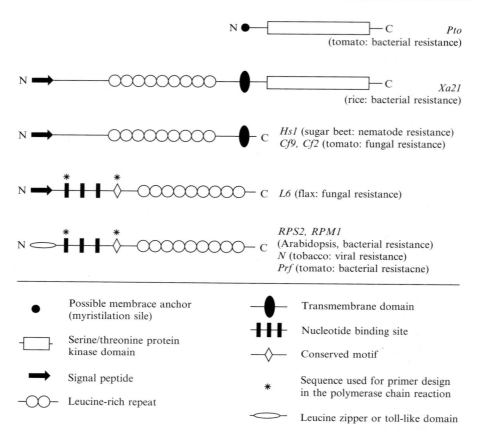

Figure 10.5 Structures of resistance genes (reproduced from Gebhardt, 1997, with permission of the author and the publisher of Trends in Plant Science)

was the first to be cloned. It has a possible membrane anchor site and a serine–threonine kinase domain but differs from the other classes since it lacks leucine-rich repeats (LRR). The second class, exemplified by the rice *Xa21* gene, possesses the serine–threonine kinase domain and, in addition, has transmembrane and LRR domains as well as a signal peptide. Transmembrane, LRR and signal peptide domains are possessed by the third class of resistance genes such as the tomato *Cf* genes but lack the serine–threonine kinase domain. The final two classes lack both the serine–threonine kinase and the transmembrane domains but have nucleotide binding sites. They differ in that one has a signal peptide, e.g. the flax-rust resistance gene *L6*, while the other has a TIR domain, so named because it has significant homology to the cytoplasmic domains of the Drosophila Toll protein and the mammalian interleukin-1 receptor (IL-R) protein, e.g. *RPS2* from *Arabidopsis*.

The significance of serine–threonine kinases is that they may function in a cascade. Thus, perception of the pathogen may activate the kinase cascade leading to the expression of genes required for resistance to the pathogen. Transmembrane domains serve to anchor the expressed protein in the membrane as, possibly, does the myristylation site of the *Pto* product. The LRR domain, by analogy with other proteins, is likely to form a coiled-coil structure which facilitates the formation of dimers or the interaction with other proteins. It is therefore a strong candidate for interaction with the products of avirulence genes, although few such interactions have been found (Section 10.5.4). The NBS is present in many ATP- and GTP-binding proteins and consists of three kinase domains. However, those of resistance genes are distinct from protein kinases. Since the TIR region has homology with signalling domains of the Toll/IL-1R superfamily it was thought that this would be its role in plant disease-resistance genes. However, work with the TIR region of alleles of the flax-rust resistance gene, *L*, have shown that this as well as the LRR region is involved in recognition (Ellis *et al.*, 1999).

Expression of resistance genes

Since plants do not have a circulating system such as that possessed by mammals, it would seem probable that for resistance genes to be effective they need to be switched on in every cell of the plant. However, there is evidence that some resistance genes are upregulated in response to pathogen attack and other stresses. As mentioned previously, the *Pib* gene for resistance to rice blast was upregulated by pathogen attack and also by alterations to the environment such as changes in temperature and darkness (Wang *et al.*, 1999). Similarly, Dinesh-Kumar and Baker (2000) showed that differential splicing of transcripts of the *N* gene occurred during infection of tobacco. The N_S transcript, which is predicted to encode the full length N protein containing the Toll-Il-1 homology region, nucleotide-binding site and LRR, is more prevalent before and for 3 h after tobacco mosaic virus infection, whereas a truncated version, the N_L transcript, predicted to encode a truncated protein lacking 13 of the 14LRR repeats is more prevalent 4–8 h after infection. Both transcripts were necessary for the expression of complete resistance to TMV.

It will be interesting to know if upregulation of resistance genes from a basal level is a constituent of systemic acquired resistance. Moreover, should this be so, a further method for cloning candidate resistance genes would be to make cDNA to upregulated genes and check them for homology with established resistance genes, a procedure which has been adopted with *NRSA-1*, a resistance gene homologue found in marigold (*Tagetes erecta*) challenged with the angiosperm pathogen *Striga asiatica* (Gowda, Riopel and Timko, 1999).

10.5.4 Function and organization of resistance genes

Resistance genes have three functions: to detect directly or indirectly *avr*-gene-specified molecules, initiate signal transduction for the activation of defence mechanisms, and to evolve rapidly in order to recognize new pathogens or variants of old ones that have lost the originally-recognized *avr* gene products (Hammond-Kosack and Jones, 1997). Recognition and the early stages of signal transduction are treated in this section while evolution of resistance genes is treated in Section 10.6.

Recognition of avr-gene-specified molecules

The simplest concept of recognition of an organism with an avirulence gene by a plant with the cognate resistance gene is that the protein products of each interact and, by doing so, set off a cascade of resistance responses which limit infection. Although about 40 avirulence genes have been cloned and perhaps rather fewer resistance genes only two avr and R protein pairs have been demonstrated to interact. These are the avrPto-Pto pair of *Pseudomonas syringae* pv. *tomato* and tomato and the avrPita-Pi-ta pair of *Magnaporthe grisea* and rice (Scofield *et al.*, 1996; Tang *et al.*, 1996; Jia *et al.*, 2000). This comparative lack of success has led to the formulation of a new model which may guide future research (Van der Hoorn, de Wit and Joosten, 2002). The hypothesis is that the R proteins only confer recognition when avr factors are complexed with host virulence targets and is discussed in Section 10.6.2.

In the meantime, considerable work has been done in order to determine those domains in resistance genes that are responsible for specificity. LRR domains are prime candidates as they play important roles in protein–protein interaction and ligand binding in a number of biological systems and in *R* genes they exhibit considerable polymorphism. For example, sequence variability in this domain of the *Cf* genes for resistance of tomato to *C. fulvum* was predicted to affect the specificity of ligand binding (Thomas *et al.*, 1998). More directly, Ellis, J.G. and co-workers (1999) found that the predicted L6 and L11 proteins from the flax *L* locus differed only in the LRR region. Further evidence was adduced from chimeric alleles constructed *in vitro* and used to transform plants. When the *L2* LRR region was combined with the *L6* or *L10* N terminal region which included the TIR and most of the NBS regions both chimeric alleles expressed the *L2* specificity showing that the LRR region was responsible. However, the N-terminal region also plays a role in specificity in other *L* alleles. First, sequence data showed that this was the only region of variation between the products of *L6* and *L7* alleles. Second, when the 5' end of *L10* was replaced by the equivalent region of *L2* the resulting gene expressed a new specificity (Ellis *et al.*, 1999). Moreover, it is clear that specificity cannot reside in an LRR region in the case of the *Pto* gene since it does not possess one but instead there is direct interaction of the product of the gene with that of the *avrPto* gene of *Pseudomonas syringae* pv. *tomato* (Scofield, 1996; Tang, 1996).

Initiation of signal transduction for the activation of defence mechanisms

At the time of writing, although the domains of some *R* genes involved in recognition are known (see previous paragraph), the mechanisms by which the recognition event generates signals for active resistance are largely unknown. However, the homologies of resistance genes give some tantalizing clues. The serine/threonine kinase homology of *Pto* and *Xa21* suggest that these products of resistance genes may initiate a phosphorylation cascade. Moreover, when Pto was used as 'bait' in the yeast two hybrid system several interacting gene products were identified which included another kinase (Pti1, *Pto*-interacting gene *1*; Sessa and Martin, 2000).

For the NBS/LRR class of resistance genes the nucleotide binding site and either the leucine zipper or the TIR homologous domains are the most likely regions to be involved in signal transduction. Through the capacity of the NBS region to bind ATP or GTP, it is thought to activate kinases or G proteins. The leucine zipper regions found in several resistance genes could allow homo-dimerization or heterodimerization with other proteins. It is speculated that this could occur on challenge by the pathogen or, alternatively, the resistance proteins may exist as dimers or oligomers which are dissociated on challenge (Hammond-Kosack and Jones, 1997). TIR regions found in some resistance genes are thought to be involved in signal transduction because of their homologies to genes involved in transduction signals in *Drosophila* and the mammalian immune response.

The tomato Cf proteins do not contain any obvious signalling domains but some work has implicated protein phosphorylation and potassium channels in signal transduction. Blatt and co-workers (1999) transformed *Nicotiana tabacum* with the *Cf-9* gene of tomato and challenged eipdermal peels with the Avr9 elicitor (see Section 10.5.1). Examination of potassium channels of intact guard cells showed almost complete suppression of the inward channel and a 2.5- to 3.0-fold stimulation of outward channel within 3–5 min of treatment. Moreover, both changes were blocked by kinase inhibitors.

Organization of resistance genes

Resistance genes are found in four arrangements in plants:

(1)   single genes with an array of distinct alleles each providing a different recognition specificity, e.g. the flax *L* locus. Thirteen specificities are known at this locus but only one is expressed in a true breeding line;

(2)   single copy genes that are present in resistant lines but absent from suscep-tible lines such as the *RPM1* gene from *Arabidopsis*;

(3)   tandem arrays of closely linked resistance gene homologues with differing
      specificities – the *M* locus of flax with seven specificities to *Melampsora lini*
      is an example;

(4)   loose clusters of genes in an *R* gene rich area of the chromosome.

10.6   The co-evolution of avirulence and resistance genes

Although the possession of avirulence genes by pathogens appears at first to be
counter-intuitive, a rational explanation becomes apparent, although remaining
speculative, when the phenomenon is viewed from an evolutionary perspective.
As stated previously, it is now clear that a number of avirulence genes may have
functioned as virulence genes or may continue to do so in appropriate hosts
(Section 10.5.2). Recognition of a virulence gene by the plant and the consequent
inception of active defence mechanisms that are effective (see Chapter 11) automatically switches the phenotype of the pathogen from virulent to avirulent. It
follows that mutation or masking of the *avr* gene will lead to the restoration
of virulence, providing such changes do not confer a significant fitness penalty.
Such changes may be slight and, in the case of the *Avr4* gene of *C. fulvum*
and cucumber mosaic virus strain Y may be confined to a single base change
(Section 10.5.1). Consequently, the survival of the host relies upon its ability to
recognize such changes. There is now evidence that several mechanisms are
available to the host for generating resistance genes with novel specificities.
Both the evolution of avirulence genes and resistance genes are discussed in the
following sections.

10.6.1   The evolution of avirulence genes

Since most plant pathogens are haploid it follows that in gene-for-gene interactions mutation of the avirulence gene will lead to virulence unless, as just stated,
such a mutation incurs a serious fitness penalty. For example, Staskawicz and co-
workers, studying race specific resistance in bacterial spot of pepper caused by
Xanthomonas campestris pv. *vesicatoria* found 13 spontaneous mutants which had
overcome the HR resistance of the host conferred by the resistance gene *Bs1*,
suggesting that the corresponding avirulence gene *avrBs1* was no longer effective
(Kearney *et al.*, 1988). When a plasmid bearing the wild-type *avrBs1* gene was
conjugated into each of the mutants and the transconjugants inoculated into
plants containing the *Bs1* gene, the plants reacted with a typical HR showing
that the avirulence gene had been transferred and was active in the transconjugants. There was therefore no block to the expression of the avirulence gene in the
mutants but rather it appeared that the mutations affected the avirulence
gene itself.

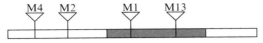

•M1 and M13 Mutants were virulent

•M2 and M4 Mutants gave intermediate reactions

Figure 10.6 Diagram showing the positions of the insertion sequence IS476 upstream of the avirulence gene, *avrBs1*, of *Xanthomonas campestris* pv. *vesicatoria* and in the open reading frame of the gene

Analysis of four of the mutants showed that the *avrBs1* gene contained an insertion sequence (IS476) of 1225 bp (Figure 10.6). In two of the mutants, M4 and M2, which gave a delayed HR and a reaction intermediate between resistance and susceptibility, respectively, the insertion sequence was found upstream to the *avrBs1* gene. The authors suggested that insertion in this region affected the regulation of the gene and may have resulted in lower protein levels. In the other two mutants, M1 and M13, which were virulent, the insertion sequence was found in the open reading frame of the gene (Figure 10.6).

The occurrence of IS476 was linked with copper resistance. Copper sprays were the normal method used to control bacterial spot of pepper in Florida and there, almost all strains of the bacterium were copper-resistant and carried IS476. Both the gene for copper resistance and *avrBs1* were found on a 200 kb plasmid (pXvCu1) which also carried three copies of IS476, at least one of which was an actively transposable element. Thus, the copper sprays used to control the disease had, ironically, resulted in the propagation of a highly mutagenic element which enabled the pathogen to overcome the genetic resistance of its host conferred by *Bs1* by disrupting the corresponding *avr* gene (Kearney *et al.*, 1988).

The possibility that loss of function of an avirulence gene incurs a fitness penalty has important implications for the durability of resistance genes and therefore disease control (see review by Leach *et al.*, 2001 and Chapter 12, Section 12.5). An early example was provided by Staskawicz's group in their work with another avirulence gene of *X. campestris* pv. *vesicatoria, avrBs2* (Kearney and Staskawicz, 1990). The gene was found in all strains of the tomato and pepper races commonly used by the authors as well as 500 other strains. When *avrBs2* was mutated by the insertion of a transposon, the mutants only grew to low titres in a susceptible host, suggesting that the gene was necessary for the fitness of the bacterium. Significantly, the corresponding resistance gene in the plant, *Bs2*, has been effective in conferring resistance for many years. Thus a resistance gene may prove durable if it recognizes an avirulence gene which has the pleiotropic effect of contributing significantly to the fitness of the pathogen. However, while this generalization remains true, unfortunately races of *X. campestris* pv. *vesicatoria* which can overcome the resistance conferred by *Bs2* are now present and are able to do significant damage to plants containing the gene (Kousik and Ritchie, 1996).

A number of other avirulence genes with fitness functions have been reviewed by Leach *et al.* (2001). For example, one member of the *avrBs3* family, the *X. citri pthA* gene, was identified initially as a pathogenicity factor in *X. citri* where it controls the ability of the pathogen to multiply intercellularly and induce cankers. However, in *X. campestris* pvs. *phaseoli* and *malvacearum* the gene confers avirulence to bean and cotton, respectively. The dual function of avirulence and pathogenicity of such genes has led to their being designated *avr/pth* genes and raises the important question as to whether such genes can evolve to lose their avirulence function while maintaining their pathogenicity function.

10.6.2   The evolution of resistance genes

While virulence may arise from loss of function of an avirulence gene in gene-for-gene pathogens (see Section 10.6.1), acquisition of resistance, being a positive phenomenon, would require a gain of function. Plants must therefore have the ability to respond to variation in the repertoire of the pathogen's avirulence genes by the acquisition of resistance genes with new specificities. A key to the mechanism by which they do this is the fact that many resistance genes are clustered or occur in tandem repeats. This suggests that resistance genes with different specificities arise by gene duplication followed by intragenic and intergenic recombination, gene conversion and diversifying selection, a situation resembling that found in the major histocompatibility genes of complex mammals.

For example, intragenic crossover events in the *L* gene of flax have given rise to recombinants with different specificities from the parental lines. Recombinants from *L2/L6* heterozygotes were found to have *L7* race specificity. Similarly, recombinants from *L9* and *suL10* gave rise to a novel specificity termed *RL10* (Luck *et al.*, 2000). In nature such events are likely to create resistance gene analogues (RGAs) randomly (Michelmore and Meyers, 1998). Most will have no function and may exist in the population as rare genes for a short time (Van der Hoorn, deWit and Joosten, 2002). In contrast, if the RGA recognizes a pathogen it becomes a resistance gene and since it confers a selective advantage to the plant it is likely to become widespread in the plant population.

Genes controlling the resistance of tomato to *Cladosporium fulvum* have provided a particularly good model system for understanding how resistant genes might have evolved. The *Cf-2, Cf-4, Cf-5* and *Cf-9* genes have been introgressed from wild tomatoes into the tomato cultivar Money Maker to give the near isogenic lines (NILs) MM-Cf2, MM-Cf4, MM-Cf5 and MM-Cf9, respectively. Mapping studies showed that the genes are present at two different loci in the genome: *Cf-4* and *Cf-9* are closely linked on the short arm of chromosome 1, while *Cf-2* and *Cf-5* map to identical positions on the short arm of chromosome 6. In addition to the functional genes originally recognized at both loci, there are a number of homologues which have been termed *Hcr* for *h*omologues of *Cladosporium fulvum* *r*esistance genes. For example, MM-Cf2 plants contain two functional *Cf-2* genes, *Cf-2.1* and *Cf-2.2* as well as a third homologue which does not appear to confer resistance to *C. fulvum*. Similarly, in MM-Cf-5 plants, beside the functional *Cf-5* gene, there are three homologues, *Hcr2-5A*, *Hcr2-5B* and *Hcr2-5D*, the functional gene being designated *Hcr2-5C*. At the *Cf-4/9* locus there are four additional homologues of the *Cf-4* gene and also four additional homologues of the *Cf-9* gene. The nomenclature of these homologues is *Hcr9-4A – E* and *Hcr9-9A – E*, the genes interacting with the Avr4 and Avr9 elicitor being *Hcr9-4D* and *Hcr9-9C*, respectively (Joosten and De Wit, 1999). Although novel resistance specificities may result from sequence exchange between tandemly repeated genes, rapid sequence exchange would lead to homogeneity rather than diversity. Sequence data offer a solution to this problem since they showed that there was a high degree of sequence rearrangements in the intergenic regions but a patchwork of sequence similarities in coding regions. Parniske and co-workers (1997) therefore proposed that the polymorphism of the intergenic regions suppresses unequal recombination in homozygotes and sister chromatids preventing sequence homogenization of the family.

Regarding the gene pool of tomato as a whole, it seems plausible that the new specificities, which are constantly being generated, enable the plant to survive changes in the factors of its pathogens that are recognized, i.e. their elicitors. Indeed, Joosten and de Wit (1999) suggest that potentially all proteins secreted by *C. fulvum* are recognized by one or more individuals in a population of tomato genotypes. In support of this they and their co-workers present data which show that extracellular proteins secreted by *C. fulvum in planta* are recognized by a collection of tomato genotypes. Two of these proteins, ECP1 and ECP2, were important virulence determinants since deletion of their corresponding genes gave rise to phenotypes with reduced pathogenic fitness. When constructs of these genes were made with potato virus X (PVX::*ECP1*, PVX::*ECP2*) and used to infect the tomato genotypes, one and four plants, respectively, were found which responded with a systemic hypersensitive response. Plants among a collection of breeding lines were also found to react to three further extracellular proteins (ECP3-5) when injected with the purified proteins or infected with PVX constructs of their genes. The possibility of using this strategy to select material with the potential for durable resistance is discussed in Chapter 12.

Returning to the more general problem as to why a direct interaction of so few products of *avr* and *R* genes has been demonstrated, three other models have been mooted (Van der Hoorn, de Wit and Joosten, 2002; Bonas and Lahayé, 2002). One of these proposes that the interaction of avr and R proteins occurs through a third protein. The second draws on the finding that several avr proteins have protease motifs and that this activity is essential for the inception of resistance responses. Here the protease is envisaged as either cleaving a protein that is required to suppress the induction of resistance responses or alternatively, and less likely, to cleave a protein to give products that interact with another protein causing the derepression of defence responses. However, a third model known as the guard model is gaining popularity. Here it is envisaged that an additional host protein, the guard protein, which is specific for each avr-R pair, is required for recognition of the avr factor. Van der Hoorn and co-workers (2002) list nine examples of putative guard proteins and show how the guard model fits the known facts concerning the origin and evolution of resistance genes. As described above, resistance genes are thought to arise by gene duplication followed by intragenic and intergenic recombination, gene conversion and diversifying selection. Correspondingly, pathogens would be selected that circumvented the newly evolved *R* genes. This situation has been described as an arms race and implies that effective *R* genes are likely to be relatively young. However, in natural populations some *R* genes are ancient and may predate speciation. The guard theory provides an explanation for this phenomenon by proposing that *R* genes are maintained by balancing selection which occurs as a consequence of the reduced fitness of pathogens which have lost the matching *avr* gene. Van der Hoorn and co-workers (2002) suggest that the function of the avr factor is to bind to a virulence target (VT) in which it induces a conformational change leading to susceptibility. In contrast, in a resistant plant the conformational change in VT is recognized by the product of the resistance gene, resulting in the triggering of defence responses. As the authors point out, because of its virulence function in some host genotypes (cf. *avrPphF*, Section 10.5.1), such a gene is likely to be maintained in the population. Besides offering a satisfying concept of our current understanding of gene-for-gene interactions the theory also points the way to future experiments which should be directed at identifying the virulence targets of avr factors.

10.7   *Hrp* genes

Studies of non-host resistance involving mutant bacteria have shown that a number of genes required to induce the HR in these heterologous plants are also required for pathogenicity in the homologous hosts. These genes are therefore denoted as *hrp* genes (*h*ypersensitive *r*esponse and *p*athogenicity; Lindgren *et al.*, 1986, 1988, 1997).

The nucleotide sequences of four *hrp* gene clusters in *Ralstonia solanacearum* (formerly *Pseudomonas solanacearum*), *Erwinia amylovora*, *Pseudomonas syringae* pv. *syringae* and *Xanthomonas campestris* pv. *vesicatoria* have been largely determined. Each contains more than 20 genes and many of them encode components of the 'type-III' protein-secretion pathway (Bogdanove *et al.*, 1996). Proteins responsible for both elicitation of defence and pathogenesis of phytopathogenic bacteria are exported via this pathway thus explaining why mutants of the genes encoding members of the secretion pathway are deficient in both attributes. Nine type III secretion genes are conserved among all four plant pathogens named at the beginning of this paragraph and also among the animal pathogens, *Salmonella typhimurium*, *Shigella flexneri*, *Yersinia enterolitica*, *Y. pestis* and *Y. pseudotuberculosis* (Salmond, 1994; Bogdanove *et al.*, 1996). Moreover, all of the predicted gene products, which comprise one outer-membrane protein, one outer-membrane-associated lipoprotein, five inner membrane proteins and two

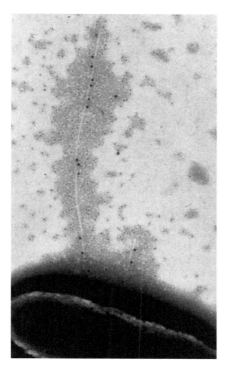

Figure 10.7 The Hrp pilus of *Pseudomonas syringae*; the pilus is a delivery system for *h*ypersensitive *r*esponse or *p*athogenicity effector molecules and is composed of HrpA subunits (large gold label) and coated with HrpZ (small gold label; courtesy of John Mansfield)

cytoplasmic proteins, one of which is possibly an ATPase, show significant similarity to components of the flagellar biogenesis complex (Blair, 1995; Bogdanove *et al.*, 1996). More recent work by Mansfield's group has shown that some of these genes are involved in the elaboration of pili (Li *et al.*, 2002). These are tubular structures that constitute the delivery system by which effector molecules are secreted into the plant cell (Figure 10.7).

There are many excellent examples demonstrating the necessity of a fully functional type III secretory system encoded by *hrp* genes for the hypersensitive response, whether this is elicited non-specifically or in gene-for-gene interactions as well as its requirement for pathogenicity. For example, Bestwick, Bennett and Mansfield (1995) showed that a *hrpD⁻* mutant of *Pseudomonas syringae* pv. *phaseolicola* was unable to elicit the hypersensitive response in the non-host, lettuce, although the mutant did induce a number of cell-wall alterations normally associated with defence. These included the formation of large paramural papillae and deposition of hydroxyproline rich glycoproteins, callose and phenolics (see Chapters 11, Sections 11.2.3 and 11.3.5). With regard to gene-for-gene interactions, it is now becoming clear that the type III secretory system is also required for the secretion of avr proteins (Leach and White, 1996). For example,

Pirhonen and co-workers (1996) have shown that phenotypic expression of *avr* genes in *P. syringae* is dependent on a functional *hrp* cluster. Their experimental system involved the transformation of *E. coli* with the plasmid-borne *hrp* cluster of *P. syringae* pv. *syringae*, Pss61 and *P. syringae* pv. *glycinea avrB*. Out of 17 cultivars of soybean tested, the transformant elicited the HR only in plants carrying the cognate resistance allele to *avrB*, *Rpg1*. Similarly, in *Arabidopsis thaliana*, only plants with the cognate *RPM1* gene responded hypersensitively. Phenotypic expression of five other avirulence genes, *avrA, avrPto, avrRpm1, avrRpt2* and *avrpPh3* in *E. coli* was also found to be dependent on the *hrp* cluster. As the authors point out the development of this system provides a powerful new tool for studying gene-for-gene recognition in plants. Similar experiments have been done by Gopalan *et al.* (1996) using the non-pathogenic bacteria *Pseudomonas fluorescens* and *E. coli*, but they were also able to demonstrate that the only site of expression of the avrB protein was within the plant cell.

The requirement of *hrp* genes for pathogenicity has been demonstrated many times. For example, Mansfield's group has cloned *hrp* genes from *Erwinia amylovora*, the causal agent of fire blight of pear (Walters *et al.*, 1990). A genomic library of *E. amylovora* isolate T was constructed in the cosmid vector pLAFR3 and maintained in *E. coli*. Clones were transferred by conjugation into the non-pathogenic isolate P66 and screened for restoration of pathogenicity by inoculation into pears. A 2.1 kb fragment was identified that restored virulence to pears and also the ability to cause the HR in *Phaseolus vulgaris*.

Beer and co-workers (1991) have cloned the *hrp* gene cluster of *E. amylovora* in several cosmids. They showed that the cluster was localized within a 25-kb region of DNA which consisted of at least eight transcriptional units. Synthesis of an intact Hrp pathway was required for the secretion of the pathogenicity factor DspE (Bogdanove, Bauer and Beer, 1998).

11 The Plant Fights Back – 2. Active Defence Mechanisms

Summary

Active resistance consists of many defence responses that interact over time and space of which the oxidative burst and the generation of nitric oxide (NO) are among the earliest. During the oxidative burst, reactive oxygen intermediates (ROI) are generated which include the superoxide anion (O_2^-), hydrogen peroxide (H_2O_2) and the hydroxyl radical ($\cdot OH$). They protect the plant in several ways: they are toxic to microorganisms and, in a remarkable analogy with the immune system of vertebrates, this toxicity is thought to be potentiated by NO; in conjunction with NO they cause the hypersensitive death of host cells, an effective resistance mechanism against obligate pathogens; hydrogen peroxide acts as a substrate for the rapid oxidative cross-linking of cell-wall proteins and lignification, making cell walls more difficult to penetrate, and it is also involved in the induction of protective genes in cells surrounding those that are directly challenged. Such genes include those coding for pathogenesis-related proteins, some of which are glucanases and chitinases. These, in combination, are potent inhibitors of fungi, presumably owing to their ability to degrade fungal cell walls. Other proteins that are expressed in response to microbial challenge are enzymes responsible for the synthesis of salicylic acid and low molecular weight antimicrobial compounds termed phytoalexins. Plants, which have been challenged with a pathogen, often develop resistance in other parts – a phenomenon referred to as systemic acquired resistance (SAR). A comparable phenomenon occurs in plants which support a population of growth promoting bacteria in the rhizosphere when it is referred to as induced systemic resistance (ISR). Tissues demonstrating induced resistance of either type are said to be protected. Characteristically, they contain elevated amounts of pathogenesis-related proteins and also respond more quickly to microbial challenge with induced defence responses such as lignification.

Induced resistance in plants occurs in response to stimuli termed elicitors which may be physical or chemical. The latter may be abiotic, such as the salts of heavy metals, or biotic if they are derived from a living source. Some biotic elicitors are specific since they are the products of avirulence genes (cf. Chapter 10) while others are general elicitors, characteristic of whole classes of

Introduction to Plant Pathology by Richard Strange
© 2003 John Wiley & Sons, Ltd   ISBN 0 470 84972 X (cased)   ISBN 0 470 84973 8 (pbk)

microorganisms, irrespective of whether they are plant pathogens or not. Examples of these are flagellins, found in bacteria, and chitin and ergosterol, constituents of fungi. In addition, a miscellaneous group of compounds has been recorded as elicitors.

Transduction of the perception of an elicitor to a defence response is mediated through calcium fluxes, mitogen-activated protein kinases, salicylic acid, jasmonates and ethylene. The central role of salicylic acid has been demonstrated by plants transformed with the *NahG* gene, which converts salicylic acid to catechol. These are susceptible to pathogens despite their possession of resistance genes which match the pathogens' avirulence genes and are incapable of developing SAR. In contrast, the signalling of ISR is mediated through the jasmonates and ethylene.

11.1 Introduction

Active defence in plants is summarized diagrammatically in Figure 11.1 but for a more complete diagram the reader is referred to the web site http://www.drastic. org.uk where the latest views of the complex interactions of plants and pathogens are displayed.

Early work centred on the elicitation of the hypersensitive response (the rapid death of challenged cells), the *de novo* production of phytoalexins (low molecular weight antimicrobial compounds) and lignification of cell walls. More recently, the generation of both reactive oxygen intermediates (ROIs), often referred to as the oxidative burst, and nitric oxide (NO) have been studied intensively. These are two of the earliest responses of the plant to challenges by potential pathogens and evidence is accumulating that they act in concert to cause the hypersensitive death of plant cells. Many other studies have established that the synthesis of pathogenesis-related proteins and the development of systemic acquired resistance (SAR) or induced systemic resistance (ISR) are important weapons of defence of plants against pathogenic organisms.

The triggers for these active defence mechanisms are termed elicitors. These are either abiotic or biotic. Abiotic elicitors include physical insults such as ultraviolet irradiation and partial freezing as well as a multitude of chemicals such as the salts of heavy metals and DNA intercalating compounds. Biotic elicitors may be specific and are, in some cases, the products of avirulence genes of pathogens (see Chapter 10). Other elicitors are characteristic of classes of microorganisms such as flagellin from bacteria and chitin and ergosterol from fungi. Still others are a miscellaneous collection of diverse compounds such as polysaccharides, enzymes, fatty acids, glycoproteins and proteins.

Whether the plant's armoury of active defence mechanisms is put into action to ward off a potential invader or not depends on detection of the pathogen through perception of its elicitors, signalling of the perceived elicitors and translation of the signal(s) into the various defence responses.

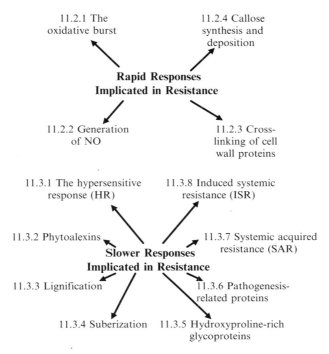

Figure 11.1   Active defence mechanisms of plants; numbers denote sections of the chapter in which the individual mechanisms are discussed

In this chapter active resistance responses and the means by which they defend the plant will first be described, followed by a consideration of elicitors and finally signal transduction.

11.2   Rapid repsonses implicated in resistance

The generation of active oxygen species and nitric oxide, the cross-linking of cell-wall proteins and the deposition of callose are resistance responses which occur rapidly – within minutes to a few hours – in plants challenged by microorganisms or treated with certain elicitors. Unlike the slower defence responses to be described later (Section 11.3) gene transcription and protein synthesis are not required for their occurrence.

11.2.1   The oxidative burst

This subject has been reviewed frequently (e.g. Lamb and Dixon, 1997; Bolwell and Wojtaszek, 1997) and much of the following is drawn from their papers.

Doke (1983) was the first to report the involvement of active oxygen species in an incompatible plant–pathogen interaction. He recorded the presence of the superoxide anion (O_2^-) and hydrogen peroxide (H_2O_2) in the hypersensitive response of potato tuber tissue to infection by an incompatible race of *Phytophthora infestans*. The superoxide anion is formed by the addition of a single electron to molecular oxygen to form O_2^- and, by analogy with the oxidative burst of neutrophils in mammalian systems, is thought to be NADPH-dependent:

$$O_2 + NADPH \longrightarrow O_2^- + NADP + H^+ \tag{11.1}$$

Certainly an NADPH-dependent O_2^--generating system is present in the microsomes of potato tubers and this was stimulated by inoculation of tuber slices with an incompatible race of *P. infestans*. Mithöfer and co-workers (1997) showed that a 1,3–1,6-β-glucan elicitor derived from the cell walls of *Phytophthora sojae* which was specifically recognized by binding proteins caused the generation of H_2O_2 and the synthesis of phytoalexins (see Section 11.3.2) in suspension cultures of soybean. Generation of H_2O_2 and O_2^- was inhibited by diphenyleneiodonium, a known inhibitor of mammalian NADPH-oxidase involved in the oxidative burst of phagocytes but elicited by the glucan suggesting the involvement of an NADPH oxidase. However, phytoalexin accumulation was not impaired by diphenyleneiodonium or ascorbate indicating that this response was independent of these active oxygen species.

Further support for the role of NADPH oxidase in the generation of O_2^- was obtained from insertion mutagenesis studies by Torres, Dangl and Jones (2002). They isolated mutants in eight genes of *Arabidopsis thaliana* that are homologues of the gene encoding the mammalian NADPH oxidase. In two of these, the normal accumulation of ROIs in response to infection with avirulent strains of *Pseudomonas syringae* and *Peronospora* was largely eliminated.

Not all plant species generate ROIs by means of NADPH oxidase. For example in French bean (*Phaseolus vulgaris*), pea (*Pisum sativum*) and cowpea (*Vigna sinensis*) cell-wall-bound peroxidases appear to be the main source (Bolwell *et al.*, 1998, 2002; Kiba *et al.*, 1997).

At low pH the superoxide anion is protonated to form the perhydroxyl radical $^{\cdot}HO_2$ and the superoxide anion and the perhydroxyl radical undergo spontaneous dismutation to produce hydrogen peroxide (Equations 11.2 and 11.3).

$$^{\cdot}HO_2 + {}^{\cdot}HO_2 \longrightarrow H_2O_2 + O_2 \tag{11.2}$$

$$^{\cdot}HO_2 + O_2^- + H_2O \longrightarrow H_2O_2 + O_2 + OH^- \tag{11.3}$$

The rapid generation of $^{\cdot}HO_2$ and H_2O_2 has now been demonstrated in many interactions of plants with fungal, bacterial or viral pathogens and the general picture has emerged that, while a weak response initially occurs in both compatible and incompatible interactions (phase 1), a second much greater and longer lasting oxidative burst occurs only in incompatible interactions 3–6 h after inoculation (phase 2).

Both O_2^- and H_2O_2 are antimicrobial and this activity, particularly in conjunction with NO (see Section 11.2.2.) may serve to check the invading organism initially. However, H_2O_2 is also involved in several other defensive measures which are discussed below.

11.2.2   The generation and role of nitric oxide (NO) in resistance

Although ROIs are capable of inducing death of plant cells and were therefore thought to be responsible for the hypersensitive response (Section 11.3.1) their concentration was found to be insufficient for this purpose. A second factor was therefore sought. NO was a candidate since, in the immune system of vertebrates, it acts in concert with ROIs in the killing of bacteria and tumour cells by macrophages (Schmidt and Walter, 1994; Nathan, 1995). Although the experimental evidence for a direct role of NO in killing plant pathogens has not been presented at the time of writing, there are data that support a role for NO in combination with ROI in the hypersensitive response (Dangl, 1998; Delledonne *et al.*, 1998; Durner, Wendehenne and Klessig, 1998). Delledonne and co-workers (2001), for example, showed in soybean suspension cultures that the efficient induction of the HR requires a balance between ROIs and NO production. Moreover, NO appeared to act in this way in the intact plant since treatment of a gene-for-gene incompatible interaction (see Chapter 10) of *Arabidopsis* and *Pseudomonas syringae* pv. *maculicola* with inhibitors of nitric oxide synthase blocked HR and caused the development of spreading lesions characteristic of the compatible interaction.

Klessig and co-workers (2000) have summarized evidence for the role of NO in defence including work from their own laboratories. They demonstrated that an increase in nitric oxide synthase-like activity occurred in resistant but not susceptible tobacco after infection with tobacco mosaic virus and that this increase was involved in the induction of the gene encoding pathogenesis-related protein 1 (*PR1*; see Section 11.3.6).

11.2.3   Cross-linking of cell-wall proteins

Two proteins, p33 and p100, from cell walls of soybean and bean suspension cultures are readily extracted by sodium dodecyl sulphate solution but rapidly become insoluble on exposure of the cultures to elicitor preparations or H_2O_2 (Bradley, Kjellbom and Lamb, 1992; Brisson, Tenhaken and Lamb, 1994). Concomitantly, the cell walls become refractory to digestion by enzymes. Moreover, protein insolubilization was detected in leaves of intact soybean plants inoculated with an incompatible but not a compatible race of *Pseudomonas syringae* pv. *glycinea* as well as tobacco expressing non-host hypersensitive resistance to *P. syringae* pv. *tabaci* (Brisson, Tenhaken and Lamb, 1994).

Studies from the laboratory of Dey have identified extensin as a protein that is subject to cross-linking. An elicitor preparation from yeast caused cross-linking of extensin from the cell walls of suspension-cultured tomato (Brownleader *et al.*, 1997). Also two forms of soluble HRGP, monomeric and cross-linked, were found in suspension-cultured potato cells and the monomeric form could be converted to the cross-linked form by incubation with tomato extensin peroxidase and H_2O_2. Such cross-linking has been likened to a self-sealing car tyre and is thought to make the cell wall more resistant to penetration by pathogens.

11.2.4 Callose synthesis and deposition

Callose is a β-1,3-linked glucan which is often synthesized and deposited as papillae – localized wall appositions – as an early response to wounding or pathogen attack. Synthesis does not require gene transcription or protein synthesis as the enzyme is held in an inactive form and only requires activation of the β 1–3 glucan synthase responsible. Some experiments have been performed with inhibitors while others have shown correlations between resistance and the responses. For example, Bayles, C.J. Ghemawat and Aist (1990) combined both approaches in a study of the resistance of barley to *Erysiphe* (=*Blumeria*) *graminis* f. sp. *hordei*. They used isogenic lines of barley, one of which contained the *mlo* mutation for resistance to powdery mildew and an inhibitor of callose formation, 2-deoxy-D-glucose. Treatment of *mlo* resistant barley coleoptiles with the inhibitor at a concentration of 10^{-5} M decreased the formation of callose-containing papillae and increased the efficiency of penetration by the fungus. Papilla formation was also delayed and those that were formed late were penetrated whereas those that formed early were not.

Perumalla and Heath (1989), investigating non-host resistance of bean, found that they could not increase haustorium production in *Phaseolus vulgaris* by the cowpea rust fungus, *Uromyces vignae* using either of two inhibitors of callose, 2-deoxy-D-glucose or the calcium chelator EGTA (ethylene glycol-bis-(aminoethyl ether)N,N,N',N'-tetraacetic acid. Perhaps, in this instance, other factors were responsible for limiting the formation of haustoria.

In other studies of barley Xi and co-workers (2000) compared penetration and cell-wall alterations of resistant and susceptible cultivars at the seedling stage challenged with *Rhynchosporium secalis*. They found that cultivars could be separated statistically into two groups that corresponded to their disease reactions. The resistant cultivars, Johnston and CDC Guardian, showed 81.2 to 99.3 per cent host cell-wall alterations (HCWA) and 0.1 to 20.1 per cent penetration at encounter sites, whereas the susceptible cultivars, Harrington, Argyle and Manley, had 30.1 to 78.3 per cent (HCWA) and 31.8 to 81.8 per cent penetration. A layer of compact osmiophilic material was deposited on the inner side of the cell of the resistant cv. Johnston indicating the formation of appositions. Cuticle at these points was not penetrated by infection pegs of the fungus. Also working with barley but with its mildew, *Blumeria graminis* f. sp. *hordei*, Bohlmann and co-workers (1988) found that thionins (Section 11.3.6) accumulated particularly in papillae and surrounding walls of resistant leaves but not in papillae of susceptible leaves.

The speed of callose deposition was emphasized in work with French bean (*Phaseolus vulgaris*) and *Xanthomonas campestris*. Here the synthesis of polymer was so rapid that the enzyme became embedded in the polysaccharide (Brown *et al.*, 1998).

11.3   Slower responses implicated in resistance

The hypersensitive response, phytoalexin synthesis, lignification, the synthesis of pathogenesis-related proteins and hydroxyproline-rich glycoproteins and systemic acquired resistance all require gene transcription and protein synthesis. As a result these responses are slower than those described in Section 11.2.

11.3.1   The hypersensitive response

The hypersensitive response (HR) is an essentially universal reaction of plants challenged with avirulent pathogens. Challenged cells and sometimes those in their immediate vicinity die rapidly (i.e. react hypersensitively) and this limited necrosis is associated with resistance of the plant as a whole (Figure 11.2). Although cell collapse does not usually occur until several hours after challenge and requires gene transcription and protein synthesis, the reaction is triggered within minutes as was originally and elegantly shown by Klement and co-workers (see Klement and Goodman, 1967, for a review). They infiltrated tobacco leaves with *Pseudomonas syringae* and followed this up with infiltration of streptomycin to which the bacteria were sensitive. If delay between the bacterial challenge and treatment with the antibiotic exceeded 20 min the HR was irreversible.

While it is easy to understand that limited necrosis might explain resistance to obligate pathogens since, by definition, they require living host cells in order to grow it is not so easy to explain why the HR is also associated with resistance to facultative pathogens since these can grow well on dead tissue. In these instances other defence phenomena, to be discussed later, must play important roles.

There has been considerable interest in determining whether the HR is a manifestation of a type of programmed cell death (pcd) which is common in animals where it is referred to as apoptosis. Apoptosis is characterized by the fragmentation of DNA into ~50 kDa pieces with 3' OH ends to produce 'ladders', membrane blebbing and nuclear and cytoplasmic condensation. In some instances, the fragmented nucleic acids may be found in membrane-bound vesicles termed apoptic bodies. Some of these symptoms have been found in plants reacting

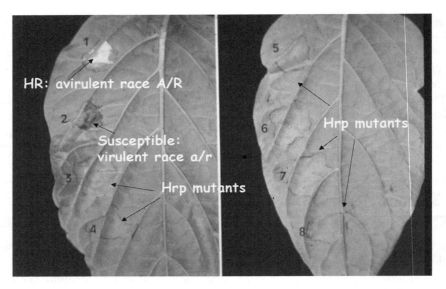

Figure 11.2 Reactions of pepper to *Xanthomonas campestris* pv *vesicatoria*: virulence, avirulence and the null reaction to *Hrp* mutants; note the water-soaked lesions of the areas of the
leaf inoculated with the virulent isolate and compare with the hypersensitive cell death caused
by the avirulent race – the *Hrp* mutants are deficient in the type III secretory system and
therefore cannot export effector molecules that cause either the susceptible reaction or the
hypersensitive response (courtesy of John Mansfield). A colour reproduction of this figure can
be seen in the colour section

$$CH_3(CH_2)_3(CH_2CH=CH)_4(CH_2)_3COOH$$

Figure 11.3 Arachidonic acid, a biotic elicitor from *Phytophthora infestans* that causes
apoptosis symptoms in tomato protoplasts and elicits terpenoid phytoalexins in potatoes

hypersensitively. For example, tomato protoplasts treated with the fatty acid
elicitor, arachidonic acid (Figure 11.3), had DNA ladders and 3' OH DNA ends.
Fragmentation of DNA has also been noted in soybean cultured cells. However,
against these observations is the finding that in tobacco the hypersensitive
response is inhibited by low oxygen pressure whereas pcd in animals is not, and
that while pcd is inhibited in animal systems by the mammalian anti-pcd protein
Bcl-X_L, in transgenic tobacco expressing this protein the HR is not inhibited
(Mittler *et al.*, 1996).

11.3.2 Phytoalexins

Introduction

Phytoalexins have been defined as low molecular weight antimicrobial compounds that are both synthesized by and accumulate in plants after exposure to

microorganisms (Paxton *et al.*, 1971). Perhaps because a role for them in plant defence was hotly disputed in some circles at that time any reference to this contentious issue was omitted! However, this is the real reason for studying them but, as we shall see, obtaining proof of such a role is not easy.

The classic experiments of Müller and Börger in 1940 were the first indication that compounds answering to the definition given above existed. Müller and Börger worked with potato and *Phytophthora infestans* and found that prior inoculation of potato tubers with an avirulent race of the pathogen protected them from disease when they were subsequently inoculated with a virulent race. They eliminated antagonism between the two races as a cause of protection when they demonstrated that the two races grew together on a potato cultivar that was susceptible to both. Moreover, when tuber cells reacting to the avirulent race were cut away, the underlying tissue was still resistant to the virulent race and other pathogens as well, such as *Fusarium caeruleum*. Müller and Börger there-fore proposed that the plants accumulated a defence substance and named it phytoalexin from the Greek phyton = plant and alexin = protecting substance.

Unfortunately, the Second World War delayed progress on this new develop-ment in plant pathology and it was not until the early 1960s that any phytoalexin was isolated and chemically characterized. The first compounds were phaseollin from beans, identified by Müller's group, and pisatin from peas, identified by Cruickshank and co-workers. Since then more than 350 phytoalexins have been characterized (Kuć, 1995).

Chemistry and distribution

Phytoalexins have been characterized from 31 plant families, mostly dicotyledon-ous, but they have also been isolated from such monocotyledonous plants as rice, sorghum, maize, wheat, barley, onions and lilies (Grayer and Harborne, 1994). To some extent the chemical class of compound is related to the plant family. For example, isoflavonoids are found in the Leguminosae but not in the Solanaceae and the converse is true of sesquiterpenoid phytoalexins. Although a single compound may be dominant, many plants produce a family of chemically related phytoalexins and some compounds that are not chemically related. Cocoa (*Theobroma cacao*) is a particularly striking instance of the latter, producing arjunolic acid, 3,4-dihydroacetophenone, 4-hydroxyacetophenone and, most surprisingly, elemental sulphur as cyclooctasulphur (Cooper *et al.*, 1996). These and a selec-tion of the many other phytoalexins from diverse chemical classes, together with the plants and families in which they are found, are presented in Figure 11.4.

Biosynthesis

Most phytoalexins are derived from one or other or a combination of three major biosynthetic pathways: shikimate, acetate-malonate and acetate-mevalonate.

(a)

H3C
CH3
H3C
H3C
COOH
HO
HO
CH3
H3C
CH2OH

(b)

S
N
N
H

(c)

OH
HO

(d)

(e)

S S
S S
S S
S S

(f)

R1
O
O
R2

(g)

O
R
CH3
HO

(h)

HO
OR
1
2
3
4
5
6
7
8
9
10

(i)

O
O
OH
O
OH

(j)

OH
O
OR

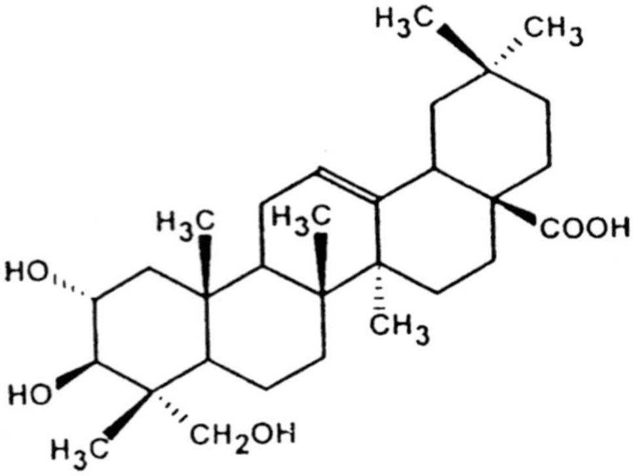

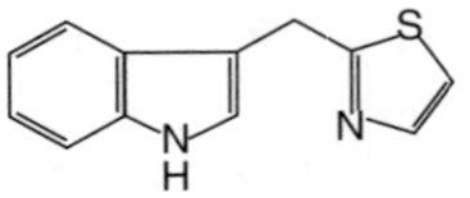

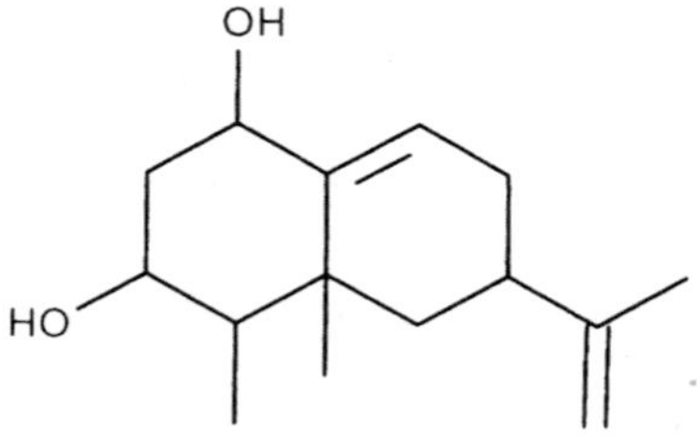

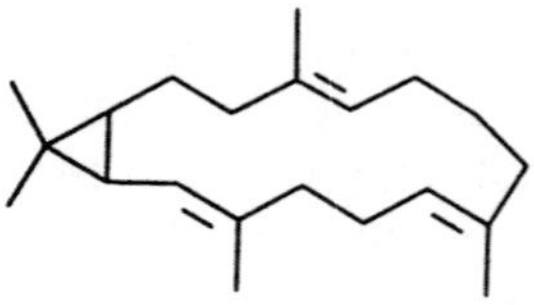

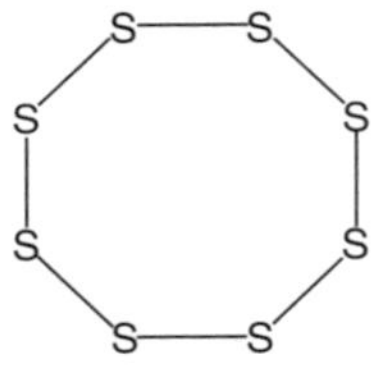

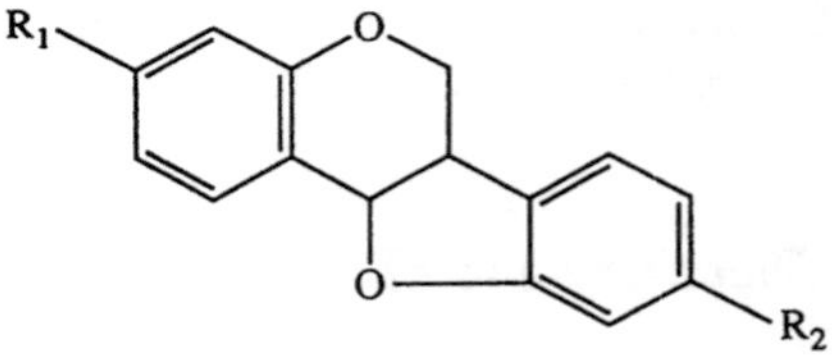

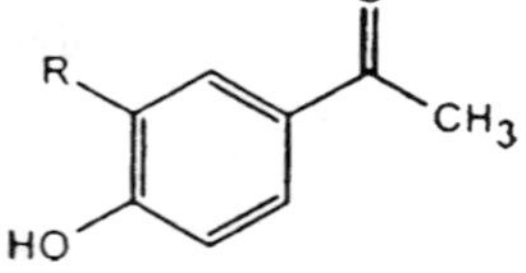

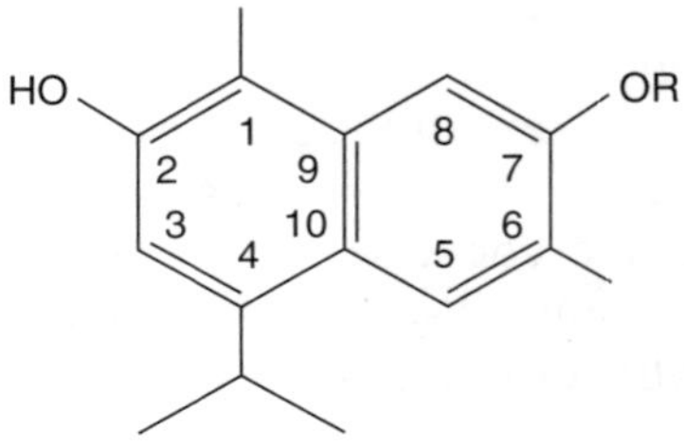

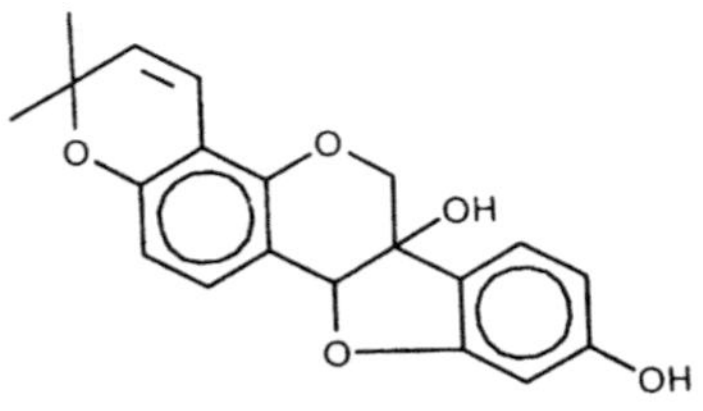

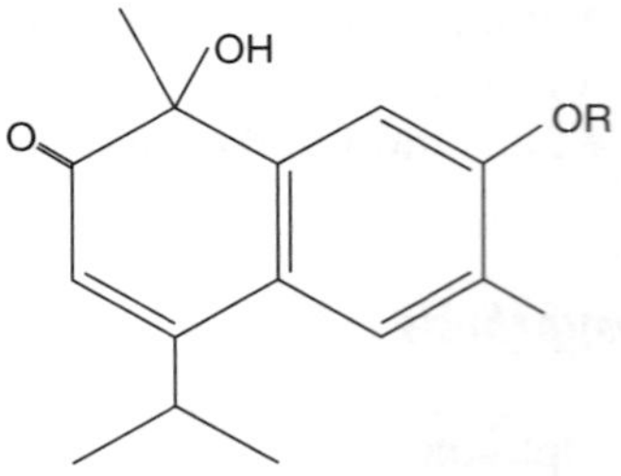

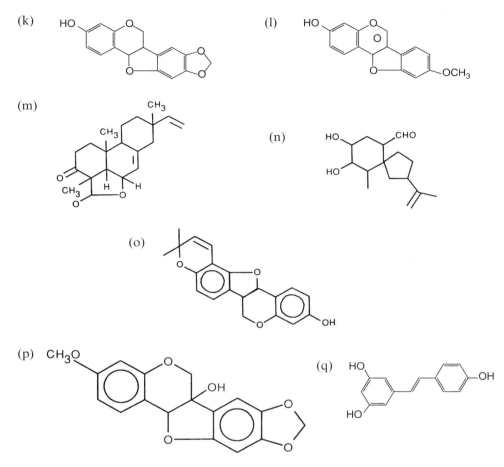

Figure 11.4 A selection of phytoalexins and the plants in which they are synthesized mentioned in the text: (a) arjunolic acid, *Theobroma cacao*, (b) camalexin, *Arabidopsis thaliana*, (c) capsidiol, *Capsicum*, (d) casbene, *Ricinus communis*, (e) cyclooctosulphur, *Theobroma cacao*, (f) demethylmedicarpin ($R_1 = R_2 = OH$), *Arachis hypogaea*, (g) 3,4-dihydroxacetophenone ($R = OH$) and 4-hydroxyacetophenone ($R = H$), *Theobroma cacao*, (h) 2,7-dihydroxycadalene ($R = H$), cotton, (i) glyceollin 1, soybean, (j) lancilene C ($R = H$), lanciline C-7-methyl ether ($R = CH_3$), cotton, (k) maackiain, *Cicer arietinum*, (l) medicarpin, *Medicago sativa*, (m) momilactone A, *Oryza sativa*, (n) phytuberin, *Solanum tuberosum*, (o) phaseollin, *Phaseolus vulgaris*, (p) pisatin, *Pisum sativum*, (q) resveratrol, *Vitis vinifera* and *Arachis hypogaea*

Inhibitors of RNA and protein synthesis usually inhibit phytoalexin production implying a requirement for both. More direct evidence has shown that there are increases in translation and transcription of genes encoding enzymes responsible for the synthesis of intermediates close to the phytoalexin end product as well as the phytoalexin itself. Moreover, such increases are also found in enzymes of the major biosynthetic pathways. For example, new transcripts were detected in elicitor-treated plant cells fed tritiated uridine in pulse-labelling experiments and when these were translated *in vitro* they gave rise to enzymes in the metabolic chains leading to phytoalexin synthesis (Chappell and Hahlbrock, 1984; Wingate, Lawton and Lamb, 1988).

Several enzymes have been studied in detail in a number of systems and these have been reviewed by Hahlbrock and Scheel (1989) and Kuć (1995). Phenylalanine ammonia lyase (PAL) is a key enzyme as it catalyses the first step of the phenylpropanoid pathway, converting phenylalanine to cinnamic acid (Figure 11.5). *De novo* synthesis of the enzyme has been confirmed by several techniques. These include the precipitation of the enzyme with specific anti-PAL sera, the *in vivo* pulse labelling of PAL after treatment with an elicitor and an increase in PAL transcripts as determined by translation *in vitro* and probing with a cDNA complementary to PAL (Edwards *et al.*, 1985; Weiergang *et al.*, 1996). PAL is tetrameric enzyme and in some systems has been found to exist as a family of genes (Hahlbrock and Scheel, 1989).

Cinnamic acid 4-hydroxylase is a mixed-function oxygenase that is dependent on cytochrome P 450 (Figure 11.5). Its rapid induction in elicitor-treated bean

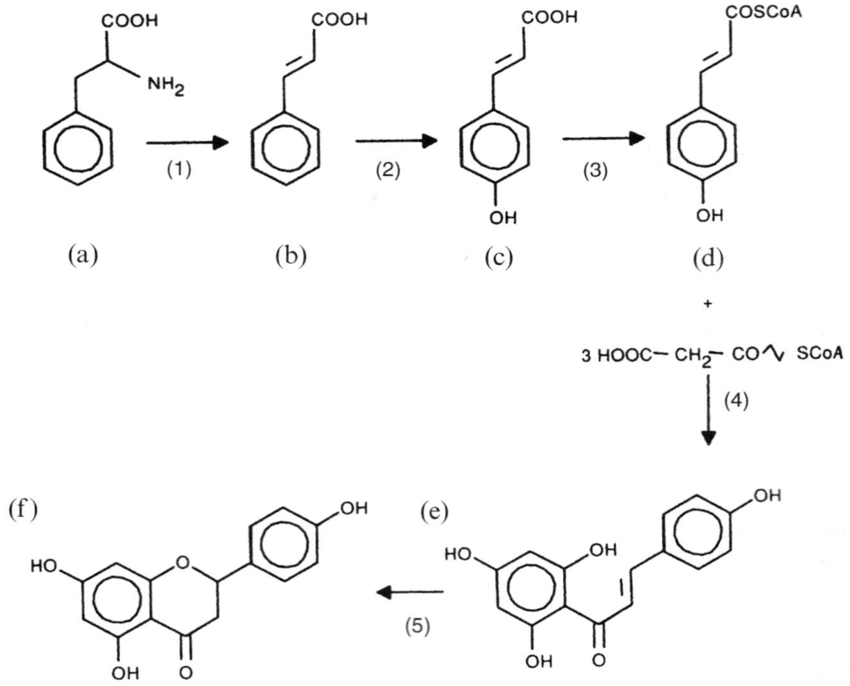

Figure 11.5 The early steps in the phenylpropanoid pathway leading to the biosynthesis of phenylpropanoid phytoalexins and lignin compounds (a) phenylalanine, (b) cinnamic acid, (c) 4-coumaric acid, (d) 4-coumaroyl CoA, (e) 4,2′,4′,6′-trihydroxychalcone, (f) naringenin; enzymes: (1) phenylalanine ammonia lyase, (2) cinnamic acid-4-hydroxylase, (3) 4-coumarate coenzyme A ligase, (4) chalcone synthase, (5) chalcone isomerase

cells was demonstrated by a monoclonal antibody raised to a highly conserved epitope of the protein. Other enzymes in the biosynthetic sequence are also dependent on cytochrome P 450 and demonstration of their regulation requires specific molecular probes.

The enzyme 4-coumarate coenzyme A ligase is monomeric and, in parsley and potato, occurs in two isoforms which are encoded by single-copy genes (Figure 11.5). Transcription of the genes encoding this enzyme and those encoding PAL is coordinated in some plants.

A number or experiments in several plants have demonstrated that the genes encoding chalcone synthase, the first committed step in the biosynthesis of flavonoids, are upregulated following treatments that induce phytoalexin synthesis (Figure 11.5). For example, messenger RNA corresponding to the enzyme rapidly accumulated in soybean cells following infection with *Phytophthora megasperma* f. sp. *sojae* or treatment with an elicitor from the fungus, as shown by a cDNA probe. Transient increases in activity of the enzyme were demonstrated in bean following elicitor treatment of cultured cells or wounded hypocotyls, as well as hypocotyls inoculated with spores of virulent or avirulent races of *Colletotrichum lindemuthianum* (Ryder *et al.*, 1987).

The second committed step in flavonoid biosynthesis is catalysed by chalcone isomerase (Figure 11.5). In bean it is a single protein and encoded by a single gene. Further modifications such as aryl migration, methylation and prenylation give rise to the many structures found among flavonoid phytoalexins. Some of these reactions have now begun to be studied. For example, Edwards, Daniell and Gregory (1997) showed that the methyl group of medicarpin was derived from methionine in suspension cultures of alfalfa treated with an elicitor from the cell wall of yeast. In a different experimental approach, Wu, Q.D. and co-workers (1997) isolated two cDNAs which encoded the terminal enzyme, (+)6a-hydroxymaackiain 3-O-methyltransferase (HMM), in the synthesis of pisatin, the major phytoalexin of pea and confirmed their HMM activity by heterologous expression in *Escherichia coli*.

Fewer studies of the biosynthesis of phytoalexins belonging to classes other than the phenylpropanoid have been made. Terpenoid phytoalexins, such as the sesquiterpene capsidiol, and the diterpene casbene are products of the acetate-mevalonate pathway. Hydroxymethylglutaryl coenzyme A reductase, the enzyme that converts hydroxy-methylglutaryl-coenzyme A to mevalonic acid, is an early regulatory enzyme in this pathway and its activity increases after wounding potato tubers or infection. However, like PAL, there is a family of genes coding for the enzyme and different members are expressed according to whether wounding or infection is the eliciting stimulus (Kuć, 1995). Mevalonate is converted to isopentenyl diphosphate which is the basic five carbon building block from which the terpenoids are derived. These may be monterpenes (C_{10}), sesquiterpenes (C_{15}) or diterpenes (C_{20}; Figure 11.6). There is considerable scope for variation among the enzymes required for the synthesis of these compounds. Not only are many enzymes involved but as in the case of hydroxymethylglutaryl coenzyme A reductase they may also exist as isozymes coded by genes which may

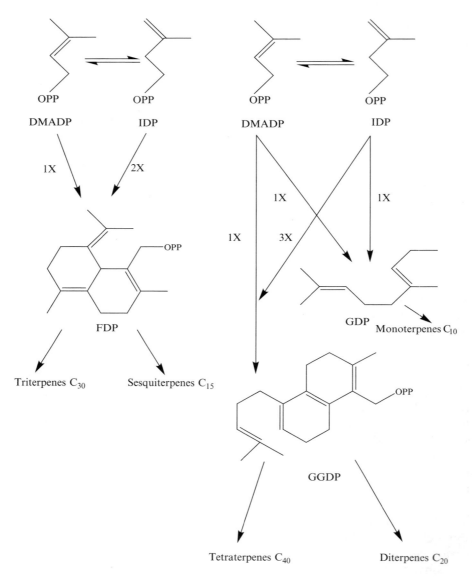

Figure 11.6 The early steps in the biosynthesis of terpenoid phytoalexins: FDP = farnesyl diphosphate, IDP = isopentenyl diphosphate, DMADP = dimethylallyl diphosphate, GDP = geranyl diphosphate and GGDP = geranyl geranyl diphosphate; sesquiterpene phytoalexins such as capsidiol (Figure 11.4) are synthesized via farnesyl diphosphate whereas diterpenoid phytoalexins such as casbene (Figure 11.4) are synthesized via geranyl geranyl diphosphate

be differentially expressed according to stimulus. For example, sesquiterpene cyclase is a key enzyme in the biosynthesis of sesquiterpene phytoalexins and, in tobacco, it is encoded by 12–15 genes (Facchini and Chappell, 1992). In some instances the terpenoid structures are modified by the addition of further components. For example, in brassicas the phytoalexins such as camalexin, found in *Arabidopsis*, are characterized by having an indole ring and at least one sulphur atom (Figure 11.4). Here, the indole ring is thought to be derived from the tryptophan pathway (Zhao and Last, 1996) and the thiazole ring from cysteine (Zook and Hammerschmidt, 1997).

Toxicity and detoxification

Phytoalexins are non-specific biocides, affecting a wide range of organisms including bacteria, fungi, nematodes, higher animals and plants themselves and some of those from legumes have the added property of being oestrogenic (Smith and Banks, 1986). Apart from Gram-positive bacteria being more sensitive than Gram-negative bacteria few generalizations can be made about the relative sensitivity of groups of organisms. Large differences in sensitivity, however, may be found among pathogens which attack plants that accumulate phytoalexins, virulent strains usually tolerating higher concentrations than avirulent ones. Such differences are usually attributable to the superior ability of virulent strains to degrade the phytoalexin.

The importance of phytoalexin detoxification in pathogenicity has been reviewed by VanEtten, Matthews and Matthews (1989). As phytoalexins belong to various classes of chemicals, plant pathogens employ various means of rendering them less toxic. Nevertheless, one common underlying trend is to render the compounds more hydrophilic. This may be achieved by oxidation, hydration and demethylation. Examples of these reactions are given in Figure 11.7.

Medicarpin is an example of a phytoalexin which is very commonly found in legumes and is subject to demethylation. Possibly significantly, demethylmedicarpin was found in leaves of groundnut susceptible to leaf rust caused by *Puccinia arachidis* but not in cultivars which were resistant (Edwards and Strange, 1991; Subba Rao, Wadia and Strange, 1996).

Although detoxification is clearly an important means by which pathogens evade phytoalexins, other mechanisms may be involved. For example, Urban, Bhargava and Hamer (1999) have reported an ATP-driven efflux pump as a novel pathogenicity factor in rice blast caused by *Magnaporthe grisea*. The pump belongs to the ATP-binding cassette (ABC) superfamily of membrane transporters. Mutants in which the ABC1 gene was disrupted by insertion in the promoter region or gene-replacement ceased to grow soon after penetration of either rice or barley epidermal cells. Moreover, ABC1 gene transcript induction by metabolic poisons was substantially reduced in the insertion mutants. The authors interpret these data as strongly suggesting that *M. grisea* requires the up-regulation of specific ABC transporters for pathogenesis in order to protect itself against plant defence mechanisms. Obvious candidates for such defence mechanisms are the rice phytoalexins, the momilactones (Figure 11.4) and the phytoanticipin, oryzalide-A (Section 9.3.1 and Figure 9.10).

(a)

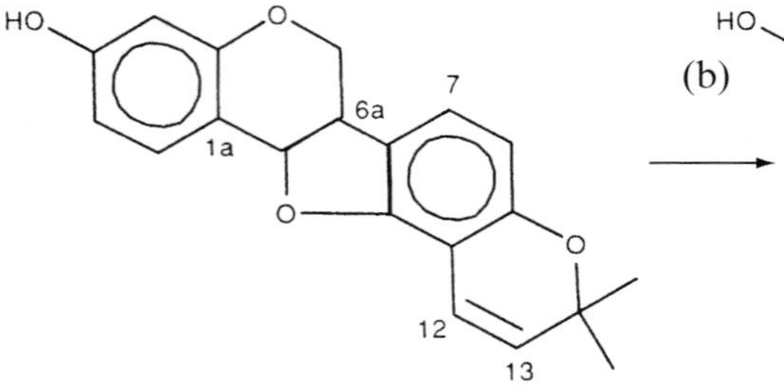
HO
O
7
6a
1a
O
O
12
13

(b)

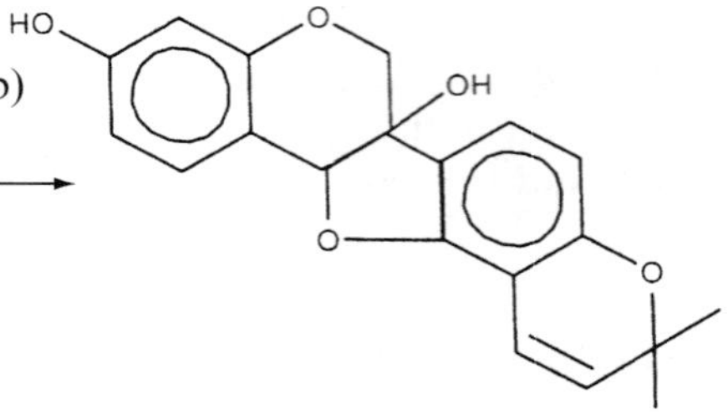
HO
O
OH
O
O

(c)

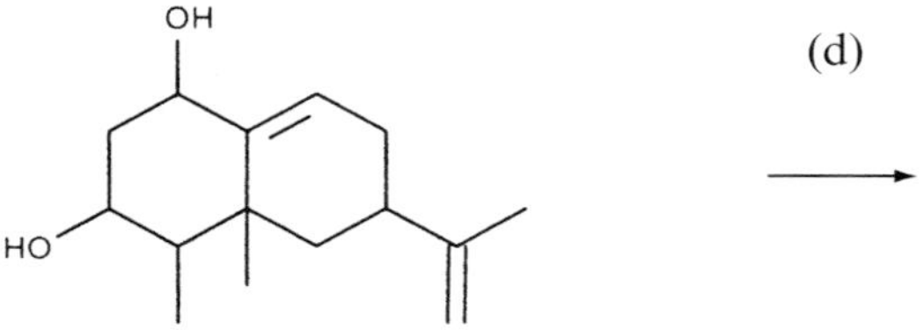
OH
HO

(d)

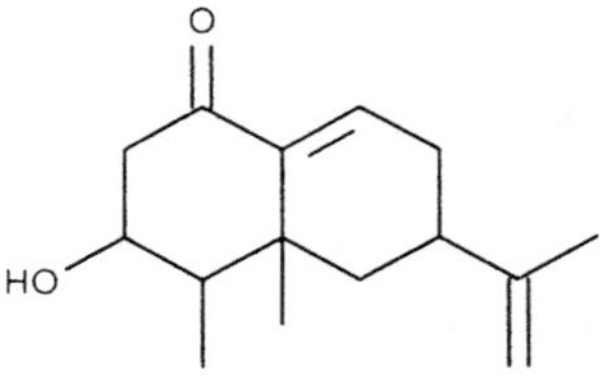
O
HO

(e)

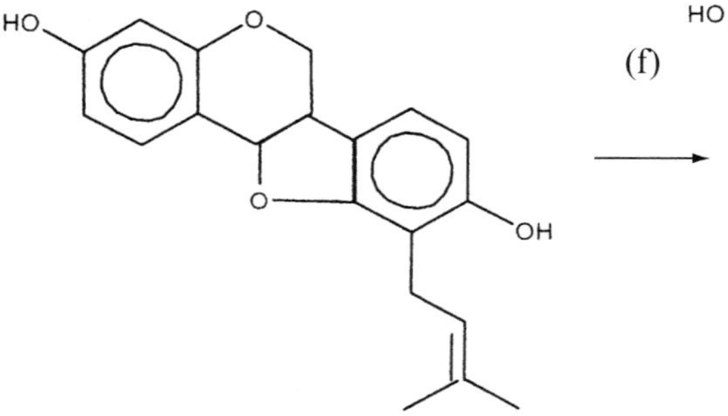
HO
O
O
OH

(f)

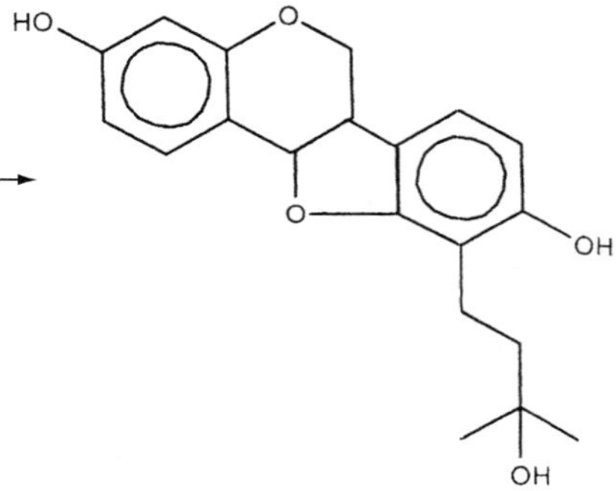
HO
O
O
OH
OH

(g)

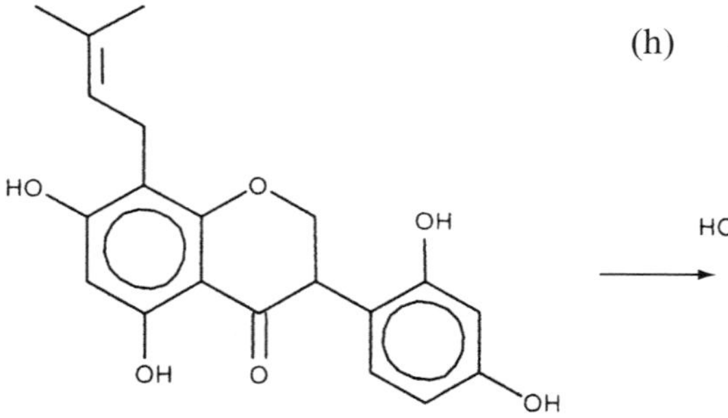
HO
O
OH
OH
O
OH

(h)

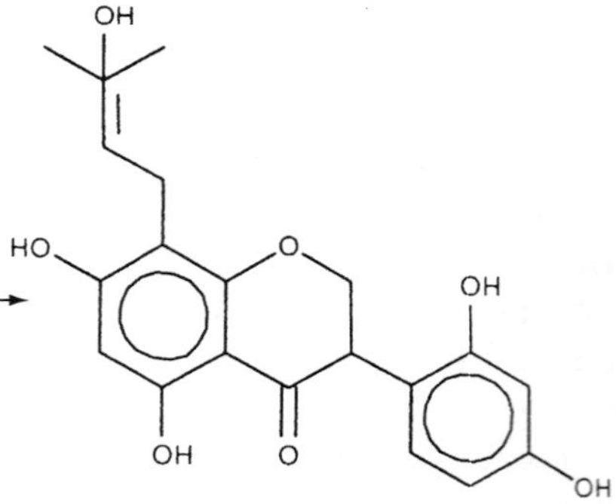
OH
HO
O
OH
OH
O
OH

(i)

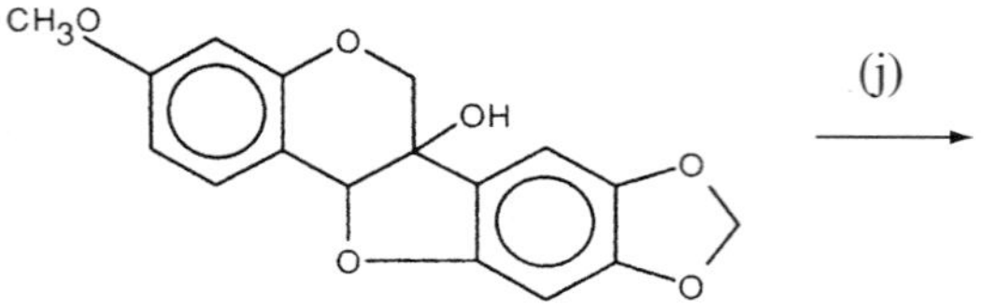
CH3O
O
OH
O
O
O

(j)

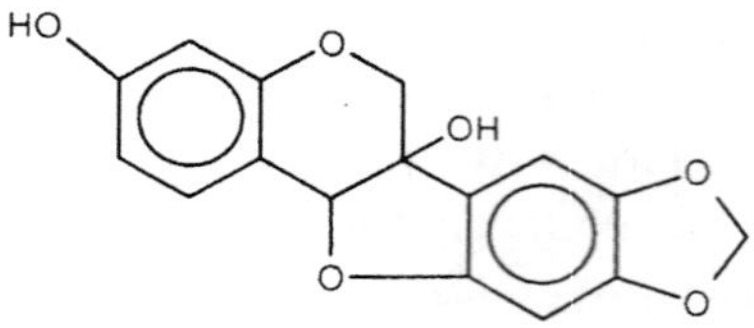
HO
O
OH
O
O
O

Determining the role of phytoalexins in resistance

Although the isolation and identification of many phytoalexins have been achieved, establishing their role in resistance has proved problematic. Various criteria have to be met of which the following are the most important.

(1) The compound must accumulate in response to infection.

(2) The compound must be inhibitory to the invading pathogen.

(3) The compound must accumulate to inhibitory concentrations in the vicinity of the pathogen at the time it ceases growing in the plant.

(4) Variation in the rate of accumulation of the phytoalexin should cause a corresponding variation in the resistance of the plant.

(5) Variation in the sensitivity of the invading organism should cause a corresponding variation in its virulence.

The first two criteria are not difficult to fulfil but demonstrating that the phytoalexin accumulates to inhibitory concentrations in the vicinity of the pathogen at the time it ceases growing is experimentally challenging. Nevertheless, all of the first three criteria are mandatory, the remaining two providing corroborative evidence.

The reason for the third criterion being so difficult to fulfil is that, if the phytoalexin is effective, it will be produced at high concentrations only locally in the vicinity of the invading pathogen. Extraction of whole tissues would therefore mean that a much larger volume of non-reacting tissue would dilute the small volume of tissue accumulating the phytoalexins. In fact, much of the early controversy about the role of phytoalexins in defence centred around this point since, in a susceptible plant, the pathogen would invade far more tissue much of which would respond by accumulating phytoalexins but to insufficient concentrations to arrest the invader.

Several careful studies have resolved this point. For example, using a radioimmune assay Hahn, Bonhoff and Grisebach (1985) were able to show that concentrations of the phytoalexin glyceollin I (Figure 11.4) exceeded the EC_{90} value (i.e. reduced radial growth of the fungus *in vitro* by 90 per cent) in resistant but not susceptible soybean roots by 8 h after inoculation. Huang and Barker (1991) also used a radioimmune assay to examine the production of the soybean phytoalexin glyceollin I. When a resistant cultivar, Centennial, and a susceptible

Figure 11.7 Detoxification of phytoalexins of various plants: monoxidation of (a) phaseollin from beans at the 6a position to give (b) 6a hydroxyphaseollin; oxidation of (c) capsidiol from peppers to give (d) capsenone; hydration of the isopentenyl groups of (e) phaseollidin and (g) kievitone, both from beans to their corresponding hydrates (f) and (h), respectively; demethylation of (i) pisatin from peas to give (j) the corresponding hydroxylated compound

Table 11.1 Comparison of concentrations of phytoalexins from *Theobroma cacao* required to inhibit the germination of conidia of *Verticillium dahliae* with those found in stems of resistant plants inoculated with the fungus (after Cooper *et al.*, 1996)

Phytoalexin	[a]ED50 (μg/ml) germination	Concentration in inoculated stems
Arjunolic acid	12.8	168
3,4-Dihydroxyacetophenone	92.5	4.9
4-Hydroxyacetophenone	7.2	2
Sulphur	3.6	51

cultivar, Ransom, were inoculated with the soybean cyst nematode, *Heterodera glycines*, they found glyceollin I in the head region of the nematode 8 h after inoculation in the resistant cultivar and the concentration increased steadily to 0.3 μmol/ml by 24 h. None was found in the susceptible cultivar.

Cooper and co-workers (1996) determined the concentrations of the cocoa phytoalexins (Figure 11.4) required to inhibit the germination of the vascular wilt fungus, *Verticillium dahliae*, and their concentrations in inoculated stems (Table 11.1). Furthermore, they were able to use coupled scanning electron microscopy and energy dispersive X-ray analysis to show that high concentrations of sulphur were present in scattered parenchyma cells which were in direct contact with xylem vessels, within vessel walls and in gels which occluded the vessels. No elemental sulphur was present in intact or wounded plants or in inoculated plants of a susceptible variety.

Other studies with cotton and the pathogen, *Xanthomonas campestris* pv. *malvacearum*, have shown that the phytoalexins 2,7-dihydroxycadalene, lancilene C and lancilene C-7-methyl ether accumulated to concentrations significantly higher than those required to inhibit the bacterium *in vitro* (Figure 11.4; Pierce *et al.*, 1996).

Indirect evidence for the importance of phytoalexins in defence comes from work on the relative sensitivity of plant pathogens to the phytoalexins of plants they either can or cannot parasitize effectively. In early work, Cruickshank (1962) found that of 50 fungal strains representing 45 species only five were tolerant of the pea phytoalexin, pisatin, and all were pathogens of pea. Only one of the 45 remaining strains was a pea pathogen. Subsequently a number of studies have supported the view that pathogens are generally more tolerant of the phytoalexins of the plants they parasitize than non-pathogens or avirulent strains of pathogens.

VanEtten, Matthews and Matthews (1989) found that only strains of *Nectria haematococca* which were virulent for pea were tolerant of its phytoalexin, pisatin, and could degrade it by demethylation (Figure 11.7). Crosses of such strains with those that were avirulent and sensitive to pisatin gave progeny that generally fell into the two parental types, i.e. those that were tolerant of pisatin, degraded it and were virulent to pea and those that were sensitive to pisatin, could not degrade it and were avirulent. Further work showed that populations of *N. haematococca* possess six genes that confer characteristic rates of pisatin demethylation (*Pda* genes) but no field isolate has been found with more than two. The phenotypes conferred by these genes fall into three classes according to their ability to demethylate pisatin; those with a short lag and high activity (Pda^{SH}), those with a short lag and moderate activity (Pda^{SM}) and those with a long lag and low activity (Pda^{LL}). Only the Pda^{SH} and Pda^{SM} phenotypes were virulent while Pda^{LL} phenotypes were no more virulent than Pda$^-$ phenotypes. Moreover, the virulence of the progeny of a cross between Pda^{SH} and Pda^{SM} was strongly correlated with their level of inducible pisatin demethylase activity. When a Pda$^-$ isolate was transformed with a fragment of DNA containing a gene which conferred the ability to demethylate pisatin, three of the transformants were more tolerant of pisatin and two were significantly more pathogenic to pea (Ciuffetti *et al.*, 1988).

Further studies determined that four genes which contributed to pathogenicity, clustered on a conditionally dispensable chromosome (Han *et al.*, 2001). The chromosome resembled bacterial pathogenicity islands in that it contained transposable elements and had differences in codon usage and GC content from other regions of the genome (cf. Section 10.5.2).

VanEtten's group has also investigated the detoxification of maackiain, a phytoalexin of chickpea by *Nectria haematococca* (Covert *et al.*, 1996; Figure 11.4). They found that detoxification was encoded by the gene, *MAK*, which, like the *pda* genes, was found to reside on a meiotically unstable chromosome that was dispensable. The gene was cloned and a comparison made of the genomic and cDNA sequences. This revealed the presence of three introns and an open reading frame encoding a protein 460 amino acids in length. The deduced amino acid sequence suggested that the gene encodes a flavin-containing mono-oxygenase.

11.3.3   Lignification

The synthesis of lignin is similar to that of many phytoalexins since it involves the derepression of the phenylpropanoid pathway, giving rise to coumaryl, coniferyl and sinapyl alcohols. Polymerization of these by a free-radical process involving H_2O_2 and peroxidase forms a very complex and resistant structure (Figure 11.8).

Early work on the induction of lignification in wheat by Ride and co-workers showed that it was specific to filamentous fungi and was caused by chitin, a polymer constituent of fungal cell walls consisting of 1-4 linked N-acetylglucosamine units (Pearce and Ride, 1980; Ride and Barber 1990; Figure 4.4). The mystery as to how such a large polymer as chitin could be perceived by the plant was solved when multiple forms of endochitinase were found in wheat leaves. It is suggested that these release soluble oligomers, those with a degree of polymerization greater than three being the most effective (Section 11.4.2). More recently, Vander and co-workers (1998) have tested a range of oligomers of chitin, partially deacetylated chitin, known as chitosan with deacetylation ranging from 1 to 60 per cent, as well as a completely deacetylated pentamer and heptamer for their ability to elicit defence responses in wheat leaves. They found that the deacetylated

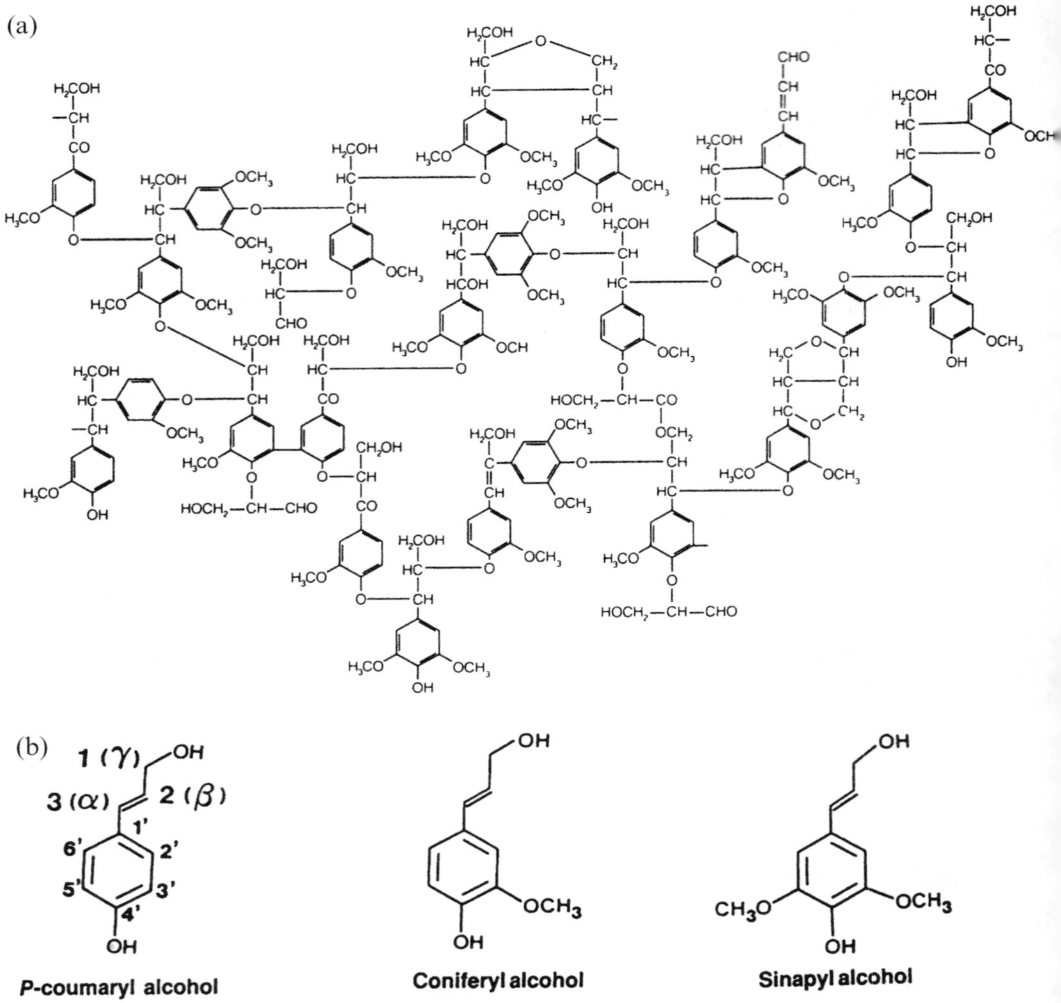

(a)

(b)

P-coumaryl alcohol   Coniferyl alcohol   Sinapyl alcohol

Figure 11.8 (a) A partial structure of a hypothetical lignin molecule from *Fagus sylvatica*; (b) the structures of the lignin monomers *p*-coumaryl, coniferyl and sinapyl alcohols ((a) reproduced with permission from Taiz and Zeiger, 1991, with permission of the authors and the Benjamin/Cumming Publishing Company Inc.)

polymers were inactive but that the acetylated polymers with a degree of polymerization of seven or greater were potent elicitors of peroxidase (POD) but not phenylalanine ammonia lyase (PAL). Partially deacetylated polymers, particularly those intermediate in the range tested, but not chitin oligomers, elicited both POD and PAL. Similarly, all chitosans, but particularly those with an intermediate degree of deacetylation but not the chitin oligomers induced the deposition of lignin, the appearance of necrotic cells, which fluoresced yellow under ultraviolet light, and necroses which were visible macroscopically.

Lignification is thought to contribute to resistance by increasing the mechanical force required for penetration, increasing the resistance of cell walls to degradation by enzymes of the pathogen and setting up impermeability barriers to the flow of nutrients and toxins. In addition, lignin precursors such as coniferyl alcohol and free radicals may be toxic to the pathogen *per se* and in some instances the hyphae of invading fungi may be lignified (Ride, 1978).

Some workers have used inhibitors to test if lignification contributes to resistance but the difficulty here is in finding an inhibitor that is sufficiently specific and does not affect the pathogen. Such experiments have involved cycloheximide (Bird and Ride, 1981) and the more specific α-amino-oxyacetic acid, which inhibits phenylalanine ammonia lyase (Tiburzy and Reisener, 1990). Usually these experiments showed that lignification was inhibited and the pathogen was more successful in colonizing the host than control plants

Antisense technology is an attractive alternative to the use of inhibitors but in practice it is fraught with difficulty owing to the production of isozymes and the multiple roles played by such enzymes. For example, Lagrimini and co-workers (1997) found that uninfected antisense tobacco mutants that were deficient in anionic peroxidase did not have reduced levels of lignin in their tissues but there was some reduction in the deposition of lignin-like polymers in response to wounding.

Another approach is to determine the correlation between lignification and resistance. In the pioneering work of Ride and his associates (Ride, 1975; Ride and Pearce, 1979; Pearce and Ride, 1980, 1982; Beardmore, Ride and Granger, 1983; Ride and Barber, 1987; Barber and Ride, 1988) wounded wheat leaves, when challenged by non-pathogens of the plant, were found to induce a ring of lignification around the inoculation site. The ring, which could be stained and quantified by densitometry, was very resistant to degradation and confined the fungus to the area of tissue within it. The lignification response was slower when pathogens such as *Septoria* species were used as inoculum and these spread from the inoculated area.

11.3.4 Suberization

Like lignin, suberin is a constituent of healthy plant tissue (Section 6.2) and also like lignin, its synthesis may be enhanced by challenge with microorganisms. As with lignin, it is difficult to prove that suberization enhances resistance although some circumstantial evidence has been obtained. Biggs (1989), for example, found that there was a positive correlation between the rate of suberin accumulation and resistance to the canker-inducing fungus, *Leucostoma persoonii*. Woodward and Pearce (1988) made a detailed study of defence in Sitka spruce and, as a result, were able to propose an integrated model, the main points of which are as follows: when a potentially pathogenic organism causes a break in the rhytidome barrier, the stilbene glycosides astringin and rhaponticin are converted to their more antifungal aglycones (Section 9.3.1; Figure 9.12). Several

days later their concentrations start to decline owing to oxidation or polymerization. At about the same time cell-wall alterations are initiated ahead of the
infection front consisting of suberization, wall thickening and the deposition of
phenolics. This results in a structural barrier which inhibits further penetration
and ultimately a periderm, which is continuous with the normal periderm barrier,
is restored. The infected area may then be sloughed off as a bark scale.

Suberization during wound-healing has long been associated with the resistance of potato tubers to infectious agents. Lulai and Corsini (1998) reported a
study of the roles of the phenolic and aliphatic domains of suberin in resistance to
a bacterial and a fungal pathogen. They found that total resistance to infection by
Erwinia carotovora subsp. *carotovora* but not to *Fusarium sambucinum* occurred
2–3 days after wounding when deposition of phenolics on the outer tangential
wall of the first layer of cells was complete. Resistance to the fungus began to
develop after deposition of the aliphatic domain of suberin was initiated and only
became total after 5–7 days when deposition within the first layer of suberizing
cells was complete. Thus it appears that the phenolic domain provides resistance
to the bacterium and the aliphatic domain to the fungus.

11.3.5 Synthesis of hydroxyproline-rich glycoproteins (HRGPs)

As stated previously (Section 6.3), HRGPs make up a significant proportion of
the protein found in plant cell walls. They consist of extensins, arabinoglactan-
proteins (AGPs), proline/hydroxyproline-rich glycoproteins (P/HPRGPs) and
solanaceous lectins (Sauer *et al.*, 1990; Showalter, 1993; Sommer-Knudsen,
Bacic and Clarke, 1998). As well as being constitutive, they are induced by
wounding and infection.

HRGPs accumulated in root cells of tomato infected by *Fusarium oxysporum*
f.sp. *radicis-lycopersici* (Benhamou *et al.*, 1990a). By use of an antiserum which
specifically recognized tomato HRGPs, they showed that they were present in
low amounts in healthy plant cell walls but that they increased in the walls of
infected tissue, especially at 96 and 120 h after inoculation. They therefore appear
to be a late response to infection in this instance. Benhamou and co-workers
(1990b) also found HRGPs associated with necrotic tissue in tobacco formed as a
result of the hypersensitive response to tobacco mosaic virus. Similarly, O'Connell and co-workers (1990) using bean leaves inoculated with *Pseudomonas
syringae* (a saprophyte) and *Pseudomonas syringae* pv. *phaseolicola* (a pathogen
of bean) and hypocotyls inoculated with *Colletotrichum lindemuthianum* showed
that HRGPs accumulated in walls of living plant cells adjoining dead hypersensitive cells during resistant reactions. HRGPs were also found in intercellular
material that encapsulated cells of *P. fluorescens* as well as in small papillae
(Section 11.2.4) adjacent to bacterial colonies. Templeton and co-workers
(1990) reported significantly greater amounts of extensin accumulation in race
specific incompatible infections of bean hypocotyls than in compatible interactions. A further mechanism by which HRGPs may confer resistance is by

providing a template for lignin deposition in papillae (Section 11.3.3; O'Connell *et al.*, 1990). Taken together, these data support the role of HRGPs in defence although much remains to be discovered about the ways in which they do so.

11.3.6 Pathogenesis-related proteins (PRPs)

Many proteins are synthesized by plants in response to wounding and microbial challenge (Bowles, 1990). Some of these are concerned with the elaboration of the defence responses already discussed but, initially, the role of a number of prominent proteins, the pathogenesis-related proteins (PRPs), was unknown.

PRPs were originally defined as acid-soluble, protease resistant, acidic proteins localized in the extracellular space (Sticher, Mauchmani and Metraux, 1997). Later, basic homologues were discovered and, in tobacco, these were found in the vacuole. However, the location of acidic PRPs in the extracellular space and basic PRPs in the vacuole does not hold true for all plant species.

PRPs may now be defined as 'proteins encoded by plants which are induced in tissue infected by pathogens as well as systemically and are associated with the development of systemic acquired resistance (SAR)' (after van Loon and van Strien, 1999). PRPs have been classified into 14 families (van Loon and van Strien, 1999; Table 11.2). PR-1 proteins in tobacco and tomato exist as multigene families. Although good evidence for their role in resistance to two Oomycetes, *Peronospora tabacina* and *Phytophthora parasitica* var. *nicotianae*,

Table 11.2 The families of pathogenesis-related proteins (after van Loon and van Strien, 1999)

Family	Type member	Properties	Gene symbols
PR-1	Tobacco PR-1a	Unknown	*Ypr1*
PR-2	Tobacco PR-2	β-1,3-glucanase	*Ypr2*, [*Gns2* ('*Gilb*')]
PR-3	Tobacco P,Q	Chitinase types I, II, III, IV, V, VI, VII	*Ypr3*, *Chia*
PR-4	Tobacco 'R'	Chitinase types I, II	*Ypr4*, *Chid*
PR-5	Tobacco S	Thaumatin-like	*Ypr5*
PR-6	Tomato Inhibitor I	Proteinase inhibitor	*Ypr6*, *Pis* ('*Pin*')
PR-7	Tomato P$_{69}$	Endoproteinase	*Ypr7*
PR-8	Cucumber chitinase	Chitinase type III	*Ypr8*, *Chib*
PR-9	Tobacco 'lignin-forming peroxidase'	Peroxidase	*Ypr9*, *prx*
PR-10	Parsley 'PR1'	'Ribonuclease-like'	*Ypr10*
PR-11	Tobacco class V chitinase	Chitinase, type I	*Ypr 11*, *Chic*
PR-12	Radish Rs-AFP3	Defensin	*Ypr12*
PR-13	*Arabidopsis* THI2.1	Thionin	*Ypr13*, *Thi*
PR-14	Barley LTP4	Lipid-transfer protein	*Ypr14*, *Ltp*

was adduced by the production of transgenic tobacco expressing PR-1a, the means by which this is achieved is not clear.

Some of the properties of the remaining PRPs are well established and in several instances clearly point to the mechanisms by which they confer resistance. For example, PRPs with β-1,3-glucanase or chitinase activity are able to inhibit fungi by attacking the glucans and chitin that make up fungal cell walls. Moreover, their antifungal activity is synergistic, being far greater than that of either enzyme acting on its own (Mauch, Mauchmani and Boller, 1988). Additionally, the action of chitinases on fungi in infected plants would be expected to release chitin oligomers that are active elicitors of lignification (Ride and Barber, 1990; Section 11.3.3). Brunner and co-workers (1998) purified 10 chitinases from tobacco leaves reacting hypersensitively to tobacco mosaic virus and tested their activity on chitin, chitosan (partially deacetylated chitin), chitin oligomers of variable length and bacterial cell walls. The enzymes were all of the endo type but differed in their action on the different substrates. For example, class I isoforms were the most active on chitin and oligomers of N-acetylglucosamine with a degree of polymerization of 4–6, whereas class III basic isoforms were the most efficient in inducing bacterial lysis. Class V and class VI chitinases hydrolysed chitin oligomers more readily than chitin. The authors suggest that the 10 tobacco chitinases represent complementary enzymes which may have synergistic effects on their substrates.

Thaumatin, defensin, thionin and lipid-transfer proteins and the probable reasons for their antimicrobial activity have already been discussed as constitutive components of plants (Section 9.3.2).

As mentioned for PR-1a, the best evidence for the role of PRPs in resistance has been obtained from their over-expression in transgenic plants and the results of several such experiments have been reviewed by Sticher, Mauchmani and Métraux (1997). For example, over-expression of tobacco PR-5 in potato delayed the onset of symptoms after inoculation with *Phytophthora infestans*. Other experiments included over-expression of a soybean β-1,3 glucanase in tobacco, a bean chitinase in brassicas and tobacco and a barley β-1,3 glucanase, chitinase and ribosome inactivating protein in tobacco. Although all these transgenic plants were more resistant to one or more pathogens a number of instances were reported in which over-expression of PRPs gave no enhanced resistance. Sticher, Mauchmani and Métraux (1997) consider that concomitant expression of several PRPs may be necessary to obtain broad resistance.

Sticher, Mauchmani and Métraux (1997) have also reviewed work with several mutants of *Arabidopsis* which provide indirect evidence for the role of PRPs in resistance. The lesion simulating disease (*lsd*) and the accelerated cell death (*acd2*) mutants, which form lesions in the absence of pathogens, have increased expression of PRPs and are resistant to pathogens that are virulent for the wild type. Similarly, a mutant (*cpr 1*) which constitutively expresses PRPs, was more resistant to *Peronospora parasitica* and *Pseudomonas syringae*. Conversely, a mutant that was incapable of expressing PRPs (*npr 1*, non expresser of PR genes) was not protected against *P. syringae* after chemical or biological elicitor treatment.

11.3.7 Systemic acquired resistance (SAR)

The phrase systemic acquired resistance (SAR) describes the resistance which develops in plants at a distance from an initial local lesion. The phenomenon was first described by Chesters in the 1930s and by Ross (1966) in tobacco inoculated with tobacco mosaic virus. However, the scientist who has championed its study over a long period is Kuć. In a conference completely devoted to the phenomenon in the year 2000 he speculated as to why it took so long to be accepted (Kuć, 2001). One reason was that up to the late 1980s 'it was greeted with curiosity and was often thought "somehow mistaken" '

It is now incontrovertible that plants given an initial inoculum which results in local necrotic lesions become resistant to a secondary inoculum given at a later date, the resistance being manifest as delayed and/or reduced symptom expression. The pathogen used as the secondary inoculum may be the same as that for the primary inoculum or an entirely different one since SAR is often effective against a wide range of plant pathogens. For example, a primary inoculum of cucumber with the fungus *Colletotrichum lagenarium* induced resistance against fungal, bacterial and viral pathogens. The period over which SAR develops varies according to the particular plant and primary inoculum and may be as short as a few hours to 2–3 weeks.

Accumulation of PR proteins in parts of plants at a distance from primary lesions is the most distinctive feature of SAR and is considered to be a marker for the phenomenon. However, another property of such plants is that they are conditioned or sensitized and are said to be protected. Such plants mount some resistance responses to a secondary inoculum more quickly and efficiently than unprotected ones.

Lignification is one of the resistance responses which appears to be primed in protected plants. For example, Dean and Kuć (1987) showed that leaf discs from protected cucumber plants incorporated [^{14}C]-labelled phenylpropanoid precursors into lignin more quickly than discs from unprotected plants. The production of callose-containing papillae is another resistance response primed in protected plants (Stumm and Gessler, 1986). Schmele and Kauss (1990) demonstrated that disruption of Ca^{2+} permeability by attempted fungal penetration activated a membrane bound β-1,3-glucan synthase which was Ca^{2+} regulated allowing rapid production of callose.

Seven criteria have been suggested for distinguishing SAR from other mechanisms that reduce disease severity or incidence, and they are:

(1) absence of toxic effects of the inducing agent on the challenging pathogen,

(2) suppression of the induced resistance by a previous application of specific inhibitors, such as actinomycin D which affect gene expression of the plant,

(3) necessity of a time interval between application of the inducer and the onset of protection,

(4)   absence of a typical dose-response correlation known for toxic compounds,

(5)   non-specificity of protection,

(6)   local as well as systemic protection,

(7)   dependence on the plant genotype.

The critical role of salicylic acid in the induction of SAR is discussed in Section 11.6.3 and the exploitation of the phenomenon to increase plant resistance in Chapter 12, Section 12.4.3.

11.3.8   Induced systemic resistance (ISR)

Plants support a considerable population of bacteria on their root surfaces and their presence often increases both growth and plant stand owing to their suppression of pathogens. Although the classical mechanisms of biocontrol, antibiosis, competition and parasitism, may be responsible (Section 5.4.2), it is now becoming clear that suppression is also mediated through the activation of host defence mechanisms. Initial experiments showed that when strain WCS417 of *Pseudomonas fluorescens* was applied to roots of carnation and the stems were inoculated 1 week later with *Fusarium oxysporum* f. sp. *dianthi* disease was reduced. Since the *P. fluorescens* was spatially separated from the pathogen the resistance induced must have been mediated through the plant (van Peer, Niemann and Schippers, 1991).

Van Loon, Bakker and Pieterse (1998) have examined ISR in the light of the criteria formulated for SAR (see Section 11.3.7). Although less work has been done with ISR all the criteria appear to be relevant with the exception of criterion 2. Here they point out that suppression of induced resistance by inhibitors of DNA-dependent RNA- or protein synthesis is difficult to apply as such inhibitors may affect many processes besides activation of defence mechanisms as well as affecting the challenge inoculum.

ISR is claimed to differ from SAR in that it is not accompanied by the accumulation of PRPs and is independent of the accumulation of salicylic acid (Knoester et al., 1999). However, DeMeyer and Hofte (1997) showed that salicylic acid production was essential for induction of resistance in bean to *Botrytis cinerea* by a strain of *Pseudomonas aeruginosa*. Transcripts of genes required for salicylic acid as well as those for the siderophore, pyochelin, were detected during the colonization of bean by the bacterium and transcription was regulated by iron supply. In contrast to SAR, the induction of ISR in some systems at least requires ethylene and jasmonic acid (see Section 11.6).

11.4   Elicitors of defence responses

The major components of active defence of plants against pathogens have been described in the previous sections of this chapter. In this section we shall be concerned with the factors that bring about these defence responses. They are known as elicitors rather than inducers since induction is a term that implies gene expression which is not part of the most rapid responses such as the oxidative burst, nitric oxide production and callose synthesis. Physical and chemical factors are usually termed abiotic elicitors while elicitors from biological sources are known as biotic elicitors.

11.4.1   Abiotic elicitors

Many physical and chemical insults are effective elicitors. Examples include ultraviolet irradiation, partial freezing, ozone, salts of heavy metals, free radicals and DNA-intercalating compounds.

Mercier, Arul and Julien (1993) tested ultraviolet-C radiation (10–280 nm), gamma radiation and heat for their ability to elicit the phytoalexin, 6-methoxymellein in carrot slices. Neither gamma irradiation nor heat was effective but ultraviolet radiation caused a maximum response with a dose of 2.20×10^5 erg cm^{-2}. Similarly, Subba Rao, Wadia and Strange (1996) found that phytoalexins were elicited in groundnut leaves by ultraviolet light with maximum emission at 254 nm and 180 μW cm^{-2} and that the treatment was more effective if the abaxial rather than the adaxial surface were exposed.

Hargreaves and Bailey (1978), in a neat experiment, bisected bean hypocotyls longitudinally, froze one half and, after thawing, bound it to the unfrozen half. Phytoalexins accumulated in the frozen half but not in the unfrozen half nor in controls in which both halves or neither were frozen. The interpretation of the experiment was that freezing allowed a signal to migrate to the live half where it elicited the phytoalexin response and the phytoalexins were exported to the frozen half.

Epperlein, Noronha-Dutra and Strange (1986) studied the elicitation of phytoalexins in cotyledons of soybean, groundnut and pea by silver nitrate which causes the production of free radicals. Since the response was suppressed by scavengers of the hydroxyl radical they suggested that the hydroxyl radical caused a chain reaction of lipid peroxidation according to the following scheme:-

$$OH^{\cdot} + RH \longrightarrow R^{\cdot} + H_2O$$

$$R^{\cdot} + O_2 \longrightarrow ROO^{\cdot}$$

$$ROO^{\cdot} + R_1H \longrightarrow ROOH + R_1^{\cdot} \tag{11.4}$$

where R and R_1 are fatty acids in membranes. Such a chain reaction is likely to result in considerable membrane damage as occurs in the hypersensitive response (Section 11.3.1) and might allow the diffusion of a signal for phytoalexin synthesis as in the experiment of Hargreaves and Bailey (1978).

The examples of the three types of abiotic elicitor given above, ultraviolet irradiation, partial freezing and the production of free radicals by silver nitrate, all involve the killing of a proportion of the cells of a plant. Thus it seems that it is the juxtaposition of live and dead cells that is crucial to effective elicitation. In contrast, it is probable that the multiplicity of DNA intercalating compounds that are effective elicitors of the phytoalexin response act directly in promoting the transcription of genes necessary for their synthesis (see Section 11.3.2).

11.4.2   Biotic elicitors

Some elicitors are said to be specific since they are the products of avirulence genes in gene-for-gene interactions (see Chapter 10). They are only effective in plants with the corresponding resistance gene. Others are general elicitors and are found as normal components of a class of microorganisms. In addition there is a range of miscellaneous compounds which are generally non-specific.

Specific elicitors

Although many avirulence genes from bacterial pathogens of plants have been cloned their products are usually cytosolic proteins which, with one exception, do not elicit the HR when introduced to plants containing the corresponding resistance gene. Recently this enigma has been solved since it is now clear that these cytosolic proteins are delivered into the plant cell by the bacterial secretory system which is encoded by *hrp* genes (see Section 10.7). The exception is the product of *avrD*. When *avrD* was cloned in *E. coli* the transformants were able to elaborate low molecular weight compounds named the syringolides which elicited the HR in soybean cultivars in a race-specific manner (Smith *et al.*, 1993; Figure 11.9).

Work from the laboratory of de Wit on the isolation and structures of the race specific elicitors of *Cladosporium fulvum* has already been described (Section 10.5.1). These are a 28-amino acid peptide, specifically produced by isolates with the *Avr9* gene, and an 86–88-amino acid peptide, the length depending on the degree of processing at the N or C terminus, specifically produced by isolates with the *Avr4* gene. These elicitors specifically induce the HR in plants containing the *Cf9* or the *Cf4* gene, respectively.

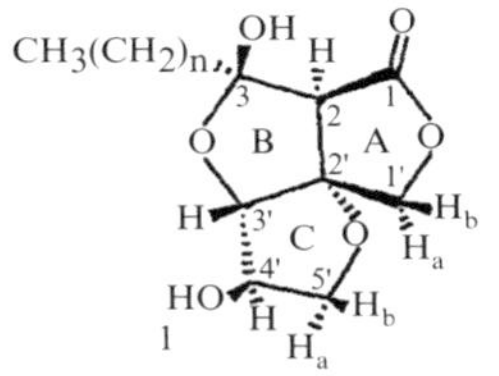

$$CH_3(CH_2)_n$$

1a, n=4, syringolide 1
1b, n=6, syringolide 2

Figure 11.9 Structures of the syringolides, race specific elicitors synthesized by strains of *Pseudomonas syringae* pv. *tomato* bearing the avirulence gene, *avrD*

General elicitors

Some elicitors are features of a class of organisms that contain plant pathogens. These include chitin and ergosterol, both constituents of fungi (Figure 4.4). Chitin is a prominent component of the cell walls of most fungi and also the exoskeleton of arthropods. It elicits lignification when injected into wheat leaves and phytoalexins when introduced to rice cell cultures. However, to be active it has to be partially depolymerized and this is achieved through the plant's chitinases. As mentioned previously (Section 11.3.3) Barber, Bertram and Ride (1989) found that only chitin oligomers with a degree of polymerization greater than 3 had activity. It is pertinent that some PR proteins are chitinases (see Section 11.3.6.).

Ergosterol is the main sterol of most higher fungi (Figure 4.4). When introduced to cell cultures of tomato it induced alkalinization of the medium with an ED_{50} concentration of 10^{-11} M. Whether other phenomena which are more clearly recognized as defence mechanisms are also elicited does not seem to have been reported.

The flagella of bacteria consist of 11 protofilaments each consisting of several thousand flagellin units. Recently, flagellin has been shown to be a potent elicitor (Felix *et al.*, 1999; Gomez-Gomez and Boller, 2002). In *Arabidopsis* it induces the oxidative burst, callose deposition and ethylene production, leading to the production of pathogenesis related proteins and the induction of other defence-related genes. However, the HR was not induced.

Harpins are proteinaceous elicitors secreted by species of *Erwinia* and *Pseudomonas*. For example, a 44-kDa protein, which is the product of the *hrpN* gene of *E. amylovora*, elicits active oxygen species in tobacco cells within minutes. Similarly, the *hrpZ* gene of *Pseudomonas syringae* pv. *syringae* encodes a 35-kDa protein which induces the hypersensitive response in tobacco and potato. The elicitors of these two bacterial species are both heat stable and sensitive to proteases but differ completely in sequence apart from a 22-amino acid domain that has 45 per cent similarity.

Elicitins are small proteins of about 10 kDa that are secreted by species of *Phytophthora* and *Pythium*. Besides inducing a necrotic response in solanaceous species, they may induce several defence responses including systemic acquired resistance (Section 11.3.7). For example, cryptogein, an elicitin from *Phytophthora cryptogaea* induces changes in membrane permeability, the production of the phytoalexin capsidiol and protein phosphorylation in cultured tobacco cells. However, as pointed out previously some elicitins are toxic (Section 8.5.2).

Various saccharide preparations of the cell walls from fungi such as the soybean pathogen, *Phytophthora megasperma* f. sp. *sojae*, are elicitors. The most effective structure for the elicitation of phytoalexins in soybean was shown by Albersheim's group to be a hepta-β-glucoside.

Other biotic elicitors

A multiplicity of other biotic elicitors have been described from a number of sources suggesting that plants have sophisticated means of detecting attributes of biological systems that are alien to itself. These may be pathogenicity or virulence factors such as enzymes or toxins or other components for which functions are not so clearly defined.

It may seem counter-intuitive to regard virulence factors as elicitors of defence reactions but this concept is explicable in terms of host–pathogen evolution, the plant evolving to recognize virulence factors of the pathogen (Sections 10.6 and 11.5). A proteinaceous elicitor from *Rhizopus stolonifer* which elicited casbene, the phytoalexin of castor bean, *Ricinus communis*, is a good example. Elicitor activity and polygalacturonase activity co-purified and both activities were destroyed by heat treatment (Lee and West, 1981a,b). However, if the elicitor/polygalacturonase were allowed to interact with plant cell walls the product was heat stable and had elicitor properties. The active component was identified as a mixture of oligogalacturonides and further work showed that a chain length nine galacturonide residues was the threshold for activity and 13 was optimal (Amin, Kurosaki and Nishi 1988; Davis *et al.*, 1986). As already mentioned, the activity of polygalacturonase-inhibiting proteins, which have structural features in common with resistance genes (Chapter 10) prolong the life of galacturonate oligomers with degrees of polymerization effective for elicitor activity (Section 9.3.2; De Lorenzo, D'Ovidio and Cervone, 2001, 2002).

Toxins are also virulence or pathogenicity factors of some plant pathogens. It would therefore be advantageous to the plant if it were able to recognize such toxins and respond with active defence mechanisms. The NIP1 peptide of the barley pathogen, *Rhynchosporium secalis*, may be an example of such a toxin since in barley plants carrying the *Rsr1* gene, defence reactions are elicited at toxin concentrations far below those required to cause necrosis (Hahn, Jungling and Knogge, 1993).

Monilicolin A, a polypeptide from the fungus *Monilinia fructicola* which causes a soft rot of stone fruit was the first elicitor to be characterized. Its action was remarkably specific, only inducing phaseollin in the non-host bean at concentrations as low as 10^{-9} M (Cruickshank and Perrin, 1968).

Kogel and co-workers (1988) have described a 67-kDa glycoprotein from *Puccinia graminis* f. sp. *tritici* that elicits hypersensitive lignification in wheat. The carbohydrate portion consisted predominantly of mannose and galactose and appeared to be the active moiety since digestion of the protein part of the molecule with trypsin or pronase had no effect.

A lipid fraction from the cystospores of *Phytophthora infestans* was an effective elicitor of the terpenoid phytoalexins of potato (Bostock, Kuć and Laine, 1981). The active components were identified as arachidonic and eicosapentaenoic acids and their activity was enhanced by fungal glucans (Figure 11.3; Maniara, Laine and Kuć, 1984). Both acids would be liable to peroxidation and perhaps this chain reaction is transferred to lipids making up the plasma membrane of the host. If so, sufficient damage to the membrane might be done to cause the hypersensitive response.

11.5 Elicitor perception

Despite the cloning of many avirulence genes and at least 30 resistance genes at the time of writing, the means by which the avirulence gene product is perceived by the resistant plant has only been established in two instances (Section 10.5.4). *A priori* one would expect the race specific product to be recognized by the product of its cognate resistance gene. One might also postulate that the resistance gene product would be located in the plasmamembrane of the plant and, on binding the avirulence gene product, initiate the defence mechanisms described in the earlier parts of this chapter. However, the finding that the avirulence gene products of bacterial pathogens are cytosolic proteins is not reconcilable with this model. On the contrary it seems that their avirulence gene products are delivered directly to the cytosol of the plant through the *hrp* secretory system.

In the case of the Avr9 protein, the product of the *Avr9* avirulence gene of the fungal pathogen of tomato, *Cladosporium fulvum*, a high affinity-binding site has been found in plants both with and without the corresponding *Cf9* gene (KoomanGersmann *et al.*, 1996). Boller (1995) has suggested that *Cf9* acts downstream from chemoperception of the Avr9 elicitor.

Greater success has attended the search for receptors of non-specific elicitors. For example, a high-affinity binding site for the heptaglucoside isolated from *Phytophthora megasperma* has been found in the membrane fraction of soybean (Schmidt and Ebel, 1987). Also a protein which binds a β-glucan elicitor fraction with an average molecular mass of 10 000 kDa was purified from a membrane fraction of soybean root cells. When a cDNA encoding the protein was expressed in tobacco cell suspension cultures or *Escherichia coli* both bound the elicitor. Furthermore, an antibody against the recombinant protein inhibited both the binding of the elicitor to a membrane fraction of soybean cotyledons and also phytoalexin accumulation. Immunogold labelling studies showed that the binding protein was located in the plasma membrane of root cells (Umemoto *et al.*,

1997). Lee, Klessig and Nurnberger (2001) have described a binding site in tobacco plasma membranes for the harpin from *Pseudomonas syringae* pv. *phaseolicola* which elicits the HR and the accumulation of pathogenesis-related gene transcripts. They found that binding to microsomal membranes and protoplasts was specific, reversible and saturable.

The 67-kDa protein from *Puccinia graminis* f. sp. *tritici* that elicits the hyper-sensitive response and lignification in wheat found by Kogel and co-workers (1988; Section 11.4.2)) bound to a highly purified plasma membrane vesicles of wheat. Binding was not specific and occurred whether or not the plant possessed the *Sr5* gene for resistance and also to plasmamembranes of barley. Plasmamembrane proteins from wheat were separated by SDS polyacrylamide gel electrophoresis and blotted onto nitrocellulose membranes. When these were overlaid with radio-labelled elicitor, binding occurred at sites corresponding to proteins of 30 and 34 kDa.

Ito, Kaku and Shibuya (1997) have described a 75-kDa protein in the plasma membrane of cultured rice cells which has high affinity for N-acetylglucosamine oligomers which elicit several cellular responses including phytoalexin accumulation.

As already mentioned, plants possess proteins which inhibit the action of polygalacturonase (PGIP proteins) with the effect that galacturonic acid oligomers of the optimal chain length for elicitor activity persist rather than being degraded to oligomers of lower activity or the inactive monomers. PGIPs contain leucine rich repeats (LRRs) and these are specific to the polygalacturonase enzymes with which they interact. For example, PGIP1 from *Phaseolus vulgaris* interacts with a polygalacturonase from *Aspergillus niger* but not with a polygalacturonase from *Fusarium moniliforme*. In contrast, PGIP2, although differing from PGIP1 by only 8-amino acid residues interacts with both enzymes (Leckie *et al.*, 1999). It is intriguing that the genes encoding these PGIPs share homology with resistance genes (i.e. LRRs). Could some resistance genes have evolved from PGIP genes? Certainly this hypothesis would be consonant with the general principle that virulence/pathogenicity genes in the context of the susceptible plant may be avirulence genes in the context of a resistant one.

11.6 Signalling

Once an elicitor has been perceived by the plant a sequence of events is initiated which culminates in the expression of one or usually more than one of the active defence responses described in the earlier part of this chapter. However, the elicitors themselves are diverse. Therefore the results of their interaction with the plant must be integrated (see Figure 11.10). At present little is known about this step but it is thought to involve fluctuations in the concentration of cytosolic calcium, alterations to the ratio of proteins with bound guanosine triphosphate (GTP) or guanosine diphosphate (GDP) – the G proteins – and kinases, in

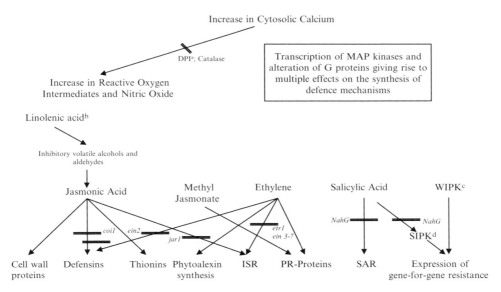

Figure 11.10 Integration of signalling pathways in the development of active resistance (letters in italics refer to mutants blocked in the pathways arrowed) An increase in cytosolic calcium is the first step in active defence, followed by an increase in reactive oxygen intermediates which may be preceded by, or occur in paralle with, the transcription of MAP kinases and alteration of G proteins (boxed). The increase in reactive oxygen intermediates is inhibited by diphenyleneiodonium (DPI), a known inhibitor of mammalian NADPH-oxidase (Figure 11.10(a)). The MAP kinases and alterations in G proteins regulate the activity of many enzymes required for the synthesis of defence mechanisms by phosphorylating or dephosphorylating them. Linolenic acid is a precursor of jasmonic acid and methyl jasmonate and is converted to these via volatile antimicrobial compounds that may be antimicrobial (Figure 11.10(b)). WIPK is a protein kinase induced by wounding (Figure 11.10(c)), and SIPK is a salicylic acid induced protein kinase. Figure 11.10(d)

particular mitogen-activated protein kinases and phosphatases, which may regulate the activity of enzymes important in the elaboration of defence mechanisms by phosphorylating or dephosphorylating them. As discussed in the sections dealing with SAR and ISR (Sections 11.3.7 and 11.3.8) three types of compound appear to be central to induced resistance: salicylic acid, jasmonates and ethylene. They and events that precede and follow their altered concentrations occurring in response to the perception by the plant of elicitors are the subjects of this section.

11.6.1 Increases in cytosolic calcium

Grant and co-workers (2000) have presented evidence for a sustained increase in cytosolic calcium which was specific in gene-for-gene interactions (see Chapter 10). In order to measure calcium fluxes they used *Arabidopsis thaliana* transformed with the Ca^{++} sensitive luminescent protein aequorin and bearing the

RPM1 gene. The product of this gene recognizes the products of two sequence unrelated avirulence genes *avrRpm1* and *avrB* which were borne by *Pseudomonas syringae* pv. *tomato* (Section 10.4). On challenge, an initial transient accumulation of Ca^{++} occurred which was non-specific and could even be triggered by *Escherichia coli*. This was followed by a second increase which was sustained and specific, depending on a functional *RPM1* gene product and delivery of the cognate avirulence gene product. Moreover, the timing of the increase differed according to the avirulence gene contained in the challenge inoculum, being earlier with *avrRpm1* (peaked at 105 ± 10 min) than *avrB* (peaked at about 137 ± 7 min).

Grant and co-workers (2000), using electron microscopy and a pharmacological approach, also addressed the question of the timing of the increase in cytosolic Ca^{++} in relation to the generation of ROIs. Plant tissues were stained with cerium chloride in order to show the presence of H_2O_2 generation, recognized in electron microscope sections as electron dense cerium perhydroxide deposits, or treated with catalase to degrade H_2O_2, diphenyleneiodonium chloride (DPI) to inhibit NADPH oxidase or lanthanum chloride to block Ca^{++} channels. Although the electron microscopy data suggested that ROI generation followed the specific Ca^{++} increase, the data were limited by the tissue-sampling techniques employed. However, when Ca^{++} channels were blocked with lanthanum chloride the increase in Ca^{++} did not occur and neither did H_2O_2 accumulation nor the hypersensitive response (Section 11.3.1). In contrast, catalase or DPI did not affect the specific increase in cytosolic Ca^{++} but greatly reduced H_2O_2 accumulation. Moreover, they only partially inhibited the hypersensitive response. The authors therefore concluded that the specific increase in cytosolic Ca^{++} preceded the generation of ROIs.

11.6.2   Mitogen-activated protein kinases (MAP kinases)

MAP kinases are important components of signalling in all eukaryotes and play a crucial role in transducing external stimuli to the machinery of the cell, ultimately bringing about a response. In animals and yeast they form cascades consisting of three members termed MAPKKK, MAPKK and MAPK in which MAPKKK phosphorylates MAPKK, which in turn phosphorylates MAPK. However, up to the end of 2001 no MAPKKK–MAPKK–MAPK module had been assembled from a plant based on *in vivo* evidence (Zhang and Klessig, 2001).

Some limited evidence showed that infection and elicitors induced the transcription of MAPKs and that this preceded or paralleled the production of H_2O_2 (Lamb and Dixon, 1997). Also a salicylic acid induced protein kinase (SIPK) and a wound induced protein kinase (WIPK) are activated in a gene-for-gene specific manner in the gene pair *Avr9/Cf9* (see Sections 10.5.1–10.5.4 for details of this gene pair) and Tobacco Mosaic Virus and the N gene (Zhang and Klessig, 2001). In suspension cultures of tobacco cells the HR could be induced by a 11.6-kDa elicitin from *Phytophthora palmivora* (Zhang, Liu and Klessig, 2000). However, experiments with inhibitors of WIPK and SIPK showed that prolonged activation of SIPK and/or delayed activation of WIPK were required for cell death to occur. The situation at the time of writing therefore seems to be that MAP kinase cascades are likely to be important in signal transduction between elicitor and effectors of plant defence but their position in the chain of events has yet to be established.

11.6.3   Salicylic acid (SA)

In 1979 White (1979) reported that treatment of the tobacco cultivar Xanthi-nc with aspirin (acetyl salicylic acid) decreased the symptoms of Tobacco Mosaic Virus. Since then it has been established that SA plays a critical role not only in systemic acquired resistance (SAR) but also genetic resistance.

The early evidence for SA involvement in SAR was that the compound elicited pathogenesis-related proteins, which are accepted markers of SAR. However, it was not until 1990 when SA concentrations were found to increase both locally and systemically in tobacco inoculated with TMV and also in the phloem of infected cucumber before the expression of SAR, that SA was postulated as the endogenous signal of SAR (Malamy *et al.*, 1990; Métraux *et al.*, 1990). More recently, experiments with tobacco and *Arabidopsis* plants transformed with the *NahG* gene (naphthalene hydroxylase G gene), which converts SA to catechol, have shown that such plants are not only unable to generate SAR but are also susceptible to pathogens for which they have R genes that are normally effective. Therefore it appears that SA is essential not only for SAR but for the expression of gene-for-gene resistance as well (Gaffney *et al.*, 1993; Delaney *et al.*, 1994). Conversely, plants expressing high constitutive concentrations of SA have a high level of constitutive resistance. For example, cultivars of rice and potatoes with the highest resistance to *Magnaporthe oryzae* and *Phytophthora infestans*, respectively, have the highest concentrations of SA (Coquoz *et al.*, 1995; Silverman *et al.*, 1995). Also, in potato, a gradient of SA from bottom to top is correlated with resistance (Silverman *et al.*, 1995).

Further evidence for the crucial role played by salicylic acid in SAR has been reported by Verberne and co-workers (2000). They transformed tobacco with two bacterial genes, which were targeted to the chloroplasts and which coded for enzymes that convert chorismate into salicylic acid by a two-step process. The transgenic plants constitutively expressed salicylic acid and salicylic acid glucoside at concentrations that were 500- to 1000-fold greater than control plants and also constitutively expressed defence genes, particularly those encoding acidic pathogenesis-related (PR) proteins but plant phenotype was not affected. Furthermore, their constitutive level of defence to viral and fungal infections were similar to those found in nontransgenic plants expressing SAR.

11.6.4 Jasmonates

Jasmonates have been implicated in resistance of plants to attack by both insects and pathogens. The principal evidence for this has been summarized in a review by Creelman and Mullet (1997) and is as follows:

(1)   Jasmonic acid (JA) accumulates in wounded plants and in plants or cell cultures treated with elicitors and may be induced by the oxidative burst.

(2)   JA activates genes encoding protease inhibitors which protect plants against insect attack.

(3)   JA modulates the expression of cell-wall proteins (cf. Yen *et al.*, 2001).

(4)   JA induces genes involved in phytoalexin synthesis such as those encoding phenylalanine ammonia lyase, chalcone synthase and hydroxymethylglu- taryl reductase.

(5)   Biosynthetic precursors of JA provide a source of volatile aldehydes and alcohols that are inhibitory to pathogens. For example, 2-hexenal is a potent inhibitor of *Pseudomonas syringae*.

(6)   Treatment of potato with jasmonate increases resistance to *Phytophthora infestans*.

Vijayan and co-workers (1998) provided strong evidence for the role of jasmo- nate in the resistance of *Arabidopsis* to *Pythium*, a ubiquitous genus of soil-borne pathogens which are often found at high inoculum levels. They studied plants with the *fad3-2 fad7-2 fad8* mutation, which are unable to accumulate jasmonate. They were highly susceptible to *Pythium mastophorum*, but could be protected by the exogenous application of methyl jasmonate. In contrast, the jasmonate insensitive mutant, *coil*, was not protected by the addition of methyl jasmonate, showing that methyl jasmonate itself was not anti-fungal but that resistance was mediated by a plant response to the compound. Wild-type plants challenged with *P. mastophorum* accumulated transcripts corresponding to the genes *PDF1.2*, *CHS*, and *LOX2* which code for a plant defensin (see Section 9.3.2), chalcone synthase (an enzyme involved in the phenylpropanoid pathway; see Section 11.3.2) and lipoxygenase 2, which is involved in the HR (see Section 11.3.1), respectively.

Methyl jasmonate and ethylene act synergistically in the induction of groups 1 and 5 of the pathogenesis proteins (Section 11.3.6). However, inhibitor studies showed that the signal transduction pathways for these two groups of PR proteins appeared to be at least partially separated (Xu *et al.*, 1994).

11.6.5 Ethylene

The role of ethylene in plant disease may seem puzzling since it has been associated with both resistance and susceptibility depending on the particular host–pathogen interaction. Induction of senescence is one reason for its ability to increase susceptibility, and this is discussed further in Chapter 7, but ethylene may increase resistance by induction of PR-proteins and the production or stimulation of aromatic biosynthesis necessary for the synthesis of phytoalexins and lignin (Boller, 1991). In order to disentangle these conflicting roles, Knoester and co-workers (1998) transformed tobacco with the gene *etr1-1*, which confers dominant insensitivity to ethylene. Expression of the basic PR-1g and PR-5c genes was markedly reduced and such plants, in contrast to non-transformed controls, were susceptible to the soil-borne Oomycete, *Pythium sylvaticum*. Significantly, these PR proteins are strongly inhibitory to Oomycetes (Woloshuk *et al.*, 1991; Niderman *et al.*, 1995).

11.6.6 Integration of signalling pathways

The evidence summarized above implicates calcium fluxes, MAP kinases and three types of signal molecule in the induction of active defence in plants, salicylic acid, jasmonates and ethylene. To what extent the effects of these compounds are additive, synergistic or inhibitory to the expression of defence has been the subject of recent research.

Although salicylic acid plays a critical role in both genetic and systemic acquired resistance (Section 11.6.3), evidence is accumulating for systemic resistance that is salicylic acid-independent. For example, Vallélian-Bindschedler, Metraux and Schweizer (1998) found that the levels of free SA and SA conjugates remained low in barley after inoculation with the biotrophs, *Blumeria* (=*Erysiphe*) *graminis* f. sp. *hordei* or *B. graminis* f. sp. *tritici* but the pathogenesis proteins PR1, PR3 (chitinase), PR5 (thaumatin-like) and PR9 (peroxidase) accumulated. Also the set of *PR* genes induced by the exogenous application of SA may differ from that induced by inoculation with pathogens. For example, SA induces the *PR-1* gene in tobacco, whereas inoculation with the soft-rot pathogen *Erwinia carotovora* induced a basic β-1,3 glucanase (*PR-2*) and a basic chitinase (*PR-3*). In this system SA antagonized the induction of *PR* genes by *Erwinia* and elicitors derived from *Erwinia* antagonized induction of *PR* gene expression by SA. Further evidence that the induction of defence genes by *Erwinia* is salicylic acid independent comes from experiments with *NahG* plants in which *PR* gene expression was not affected (Vidal *et al.*, 1998).

As discussed in Section 11.6.4 and 11.6.5, evidence is accumulating for the involvement of jasmonates and ethylene as signal molecules in induced resistance that is salicylic acid independent. Both are rapidly produced in response to pathogen attack, particularly when this results in necrosis and both induce defence genes when applied exogenously. For example, in *Arabidopsis*, the thionin gene *Thi2.1* and the plant defensin gene *PDF1.2* are activated locally and systemically after infection with a necrotizing pathogen and after treatment with methyl jasmonate but not after treatment with salicylic acid. Moreover, a similar response was evoked in *NahG* plants, which are incapable of accumulating salicylic acid. On the other hand, in some instances jasmonic acid and ethylene enhance the action of SA (Schweizer *et al.*, 1997; Lawton *et al.*, 1994). In contrast, Niki and co-workers (1998) found that JA induced basic PR genes but inhibited the induction of acidic PR genes; conversely SA induced acidic PR genes but inhibited the induction of basic PR genes. Taken together, these results imply cross-talk among the SA, ethylene and JA signalling pathways.

In order to disentangle these defence signalling pathways several workers have taken a genetic approach. For example, Penninckx and co-workers (1998) found that components of both the ethylene and jasmonate pathways were required for *PDF1.2* expression since this was blocked in the ethylene-insensitive mutant *ein2* and the jasmonic acid-insensitive mutant *coi1*. Similarly, the development of ISR was blocked by the ethylene response mutant *etr1* and the jasmonate response mutant *jar1*. Other work by Knoester and co-workers (1999) showed that five further ethylene insensitive mutants, *ein3–ein7* were all impaired with respect to the induction of ISR by a strain of *Pseudomonas fluorescens* which colonizes the rhizosphere and is non-pathogenic. In contrast, SAR was not affected.

Ton and co-workers (2002) have characterized 11 mutants of *Arabidopsis* with enhanced disease susceptibility owing to impaired SAR or ISR. Two mutants, which failed to develop SAR, were either blocked in salicylic acid synthesis or reduced in sensitivity to this molecule. With regard to ISR, three mutants were non-responsive to induction by methyl jasmonate and two to the ethylene precursor 1-aminocyclopropane-1-carboxylic acid.

In another genetic study, Rairdan and Delaney (2002) found that in *Arabidopsis*, salicylic acid accumulation was necessary for the majority of R-gene-triggered resistance. In contrast, the role of the defence regulatory protein NIM1/NPR (mutants of the gene encoding this protein are unable to accumulate pathogenesis related-proteins) in race-specific resistance was limited to resistance to *Perenospora parasitica* mediated by TIR-class R genes

Although SAR and ISR have been considered in this text as separate phenomena, SAR being dependent on SA as a signal molecule and ISR being dependent on ethylene and JA, both are regulated by the gene *npr1*. However, in *Arabidopsis* SAR and ISR are additive and since plants in which both pathways were induced did not show elevated levels of *npr1* transcripts, the constitutive level of NPR1 was deemed to be sufficient for both (van Wees *et al.*, 2000).

11.7   Overcoming induced resistance

Since induced resistance is a multifaceted phenomenon, there are many ways in which it may be overcome by successful pathogens. In the first instance they may avoid producing effective elicitors and this seems to be the mechanism in gene-for-gene interactions (see Section 10.3). Secondly, successful pathogens may interfere with the signal transduction necessary for the induction of active resistance and such interference might explain why some pathogens have wide host ranges. Thirdly, as discussed for phytoalexins (see Section 11.3.2), the pathogen might destroy the defence component itself. *Claviceps purpurea*, a fungus which is compatible with hundreds of plant species, is an example where the last two options might apply. Garre, Tenberge and Eising (1998) found that *C. purpurea* secretes a catalase during infection which migrates into the host cell wall. They suggest that not only would this negate the direct cytotoxic effect of hydrogen peroxide on the pathogen but also interfere with the cross linking of wall polymers (see Sections 11.2.3.) and the signalling mediated by hydrogen peroxide.

It seems probable that, as the signalling pathways which operate in plant defence are unravelled, other means by which pathogens interfere with them will be discovered. Already certain toxins, such as coronatine are known to interfere with defence responses (Section 8.5.1; Mittal and Davis, 1995). Perhaps one way to elucidate these pathways would be to search for components of pathogens which suppress active defence mechanisms.

12 Control of the Disease Process

Summary

The pre-penetration stage of infection is a critical one for the pathogen involving chemo-attractants, repellents and immobilizers as well as physical and chemical signals that influence germination and growth of pathogen propagules. Some of these phenomena may be susceptible to exploitation either as decoys or as a means of preventing infection. Plants also contain many antimicrobial compounds that are either constitutive (phytoanticipins) or induced (phytoalexins) and it may prove possible to vary the type and increase the concentrations of these. Systemic acquired resistance may be exploited by the use of appropriate inocula or by the use of chemicals such as benzothiadiazole. Plants may also be selected for their tolerance to enzymes and toxins secreted by the pathogen. Conventional breeding for resistance, although still the mainstay of combating plant disease, has been assisted by the availability of markers in a number of important crop plants. This has not only speeded breeding programmes but has enabled the selection of plants with more than one resistance gene

The ability to transform plants using *Agrobacterium*, electroporation or biolistics has opened the door to the rapid incorporation of defence components into plants across species barriers. Candidate genes for such transformations are those which encode proteins that inhibit pathogen enzymes or degrade their toxins, those that enhance the concentrations of saponins, antimicrobial peptides, reactive oxygen species or modify the phytoalexin response and those that switch on systemic acquired resistance. Genes derived from pathogens are also candidates since they, paradoxically, confer resistance and, less surprisingly, resistance genes may also be used to broaden the resistance of plants to a greater spectrum of pathogens.

12.1 Introduction

The aim of this chapter is to review the material of the preceding six chapters, essentially in sequence, from the standpoint of exploiting the phenomena discussed to improve the health of our crops without resorting to broad-spectrum biocides. Thus possibilities of exploiting pre-penetration and penetration events are first considered, followed by enhancement of the tolerance of plants to virulence

Introduction to Plant Pathology by Richard Strange
© 2003 John Wiley & Sons, Ltd   ISBN 0 470 84972 X (cased)   ISBN 0 470 84973 8 (pbk)

attributes of the pathogen, represented by enzymes and toxins. Promotion of resistance mechanisms of the plant – phytoanticipins, phytoalexins and acquired resistance – are then treated before discussing genetic approaches to control.

With the advent of the ability to transform plants routinely with genes of choice (Section 12.6) a new world of possibilities for the control of plant disease has become available and these are considered in Section 12.7.

12.2   Control of pre-penetration and penetration events

12.2.1   Exploiting chemo-attractants, repellents and immobilizers

Infection of plants by soil-borne pathogens is one of the more difficult areas of plant pathology, not least because soil is complex and opaque. Hawes and co-workers (1998) have drawn attention to another complication which has received surprisingly scant attention, the phenomenon of root cap cells or, as they call them, border cells (see Section 6.4). These, together with the more studied 'root exudates', represent a considerable investment by the plant in terms of energy and metabolites but their functions remain speculative. Hawes and co-workers (1998) point out that, contrary to much assumption, border cells are viable and they suggest that one of their functions might be to act as decoys and so protect the root from infection. When pea roots were inoculated with spores of a virulent strain of *Nectria haematococca*, infection was only apparent to the naked eye behind the root tip, although microscopic examination showed that the apex was covered with fungal hyphae. If the roots were placed in water, the border cells, together with the fungal hyphae fell off as a unit leaving the apex of the root free of the fungus (Section 6.4). More generally, many root pathogens initiate infection just behind the root cap in the region of elongation, possibly because this is the first place where cells remain attached to the plant.

Zhao, Schmitt and Hawes (2000) have investigated the effects of root exudates and border cells on the nematode, *Meloidogyne incognita*. When these were monitored with a quantitative assay, attraction occurred with pea and alfalfa cv. Thor, repulsion with alfalfa cv. Moapa 69, and no response with alfalfa cv. Lahonton and snap bean. With peas, juveniles (J2) accumulated rapidly in the apical region ensheathed by border cells, but not in the region of elongation. Those that accumulated on detached border cells became quiescent within 15–30 min. Neither accumulation nor quiescence occurred when border cells and exudates were removed by washing. However, when nematode responses were measured in moist sand, total root tip exudates from all three plants were repellent. Moreover, total root tip exudates from all the tested legumes as well as maize caused loss of mobility in more than 80 per cent of the J2 populations.

Border cells export a number of products, including signals for the control of mitosis, enzymes, mucilage, antibiotics, phytoalexins and proteins of unknown function. Clearly, the border cells themselves or their products as well as root exudates may have profound effects on potential pathogens. For example, some plant parasitic bacteria are motile and a number of fungal pathogens produce motile zoospores. These often respond chemotactically to root exudates (Wynn, 1981; Section 6.2). Unfortunately, there is usually limited knowledge of the chemistry of the compounds involved although there are exceptions (see Section 6.4).

Clearly we need to know more about the influence of plant roots on pathogens. If, for example, chemotaxis is a significant phenomenon in epidemiology, we need to understand the perception of the chemo-attractant and the nature of the compounds responsible. In some instances they may be constitutive but in others, as we have seen with acetosyringone, they may be induced in response to wounds (Section 6.4) or possibly challenge by other members of the soil microflora. Reciprocally, plants may possess compounds in their roots that actually repel soil-borne pathogens or, in the case of nematodes, immobilize them (Zhao, Schmitt and Hawes, 2000; Islam and Tahara, 2001). Once this information has been obtained it may prove feasible to select and breed plants with reduced concentrations of compounds that attract pathogens or enhanced concentrations of those that repel or immobilize them.

Chemo-attractants, repellents and immobilizers are also found in the aerial parts of plants. Mention has already been made of pheromones in connection with the repulsion of *Myzus persicae* by *Solanum berthaultii* (Sections 3.3.2 and 5.6.5). Trichomes may also be influential on insects which might be virus vectors (Steffens and Walters, 1991). In solanaceous plants one type of trichome (type A in *Solanum berthaultii* or type VI in *Lycopersicon*) confers resistance to small-bodied insects such as aphids and leaf-hoppers by an adhesion mechanism. On contact with the trichome, the membrane surrounding the head of the trichome ruptures, coating the insect's legs and mouthparts with its constituents. As one of these is a high concentration of polyphenol oxidase, the coating on the insect immediately begins to brown and harden. Any struggle by the insect worsens the situation and may lead to occlusion of mouthparts and adherence of the insect to the leaf. A second type of trichome (type B or type IV in *Solanum* and *Lycopersicon*, respectively) is tall and secretes continuously sucrose esters of fatty acids consisting of 4–12 carbon atoms as well as glucose. These act as feeding deterrents and entrap small pests such as spider mites and aphids. As a result, virus transmission by aphids may be reduced and this type of trichome has also been reported to inhibit fungal and bacterial growth.

Further study of other crop plants and their wild relatives for repellents and trapping agents would seem worthwhile and could perhaps be exploited through conventional breeding as well as genetic modification. For example, Coombs and co-workers (2002) investigated the effect of combining the natural resistance to the Colorado potato beetle (*Leptinotarsa decemlineata*) conferred by leptine glycoalkaloids and glandular trichomes with that conferred in plants transformed with a gene encoding the synthetic *Bacillus thuringiensis* cry3A toxin (Section 12.7.4). Although defoliation in a susceptible control was similar to that of a line with glandular trichomes, 32.3 and 32.9 per cent, respectively, defoliation in the high glycoalkaloid line was only 3.0 per cent and this was further reduced in the transgenic plants. Reciprocally, pheromones that attract insects may prove useful as decoys attracting, for example, insect vectors of pathogens away from crop plants (Bakke and Lie, 1991).

12.2.2   Exploiting chemical signals that influence germination and growth of pathogen propagules

As discussed in Section 6.6, pathogens are responsive to chemical compounds present on the surface layers of plants or secreted by them. For example, the surface waxes of plants are complex and may contain a variety of chemical cues that influence the germination and penetration of plants (Section 6.6.3). At present there is very little evidence of the specificity of these compounds although Podila, Rogers and Kolattukudy (1993) showed that the surface wax of avocado specifically triggered the conidial germination and development of infection structures of *Colletotrichum gloeosporioides* (Section 6.6.3). This would seem to be an area worth re-visiting in order to determine the scope of the specificity of such signals. Should the phenomenon prove widespread it may be possible to alter the chemistry or the quantity of such compounds so that they no longer provide the stimulus required by the pathogen.

Anthers and pollen are rich sources of nutrient which may non-specifically afford a pathogen sufficient nutrients to allow a successful attack on other tissues of the plant. In other instances, specific compounds may be involved. For example, choline and betaine were identified as the compounds in wheat anthers that promoted hyphal extension of *Fusarium graminearum* raising the question as to whether selection of plants with reduced concentrations of these compounds would be more resistant to the fungus (Section 6.6.3). Two other points are, perhaps, worth raising in connection with this host – pathogen interaction. Since wheat is self-pollinated, it may be possible to breed plants which are cleistogamous, i.e. plants in which the anthers are never exposed on the surface. These would deny the pathogen a route of entry. *F. graminearum* also causes foot-rot and this can be acute where the plant has suffered drought stress. It may be significant that accumulation of betaine is one response of wheat to drought stress. Drought tolerance in which enhanced concentrations of betaine was not a concomitant might prove to be an advantage here.

In some instances it is clear that the signals are not specific. For example, cotton, a non-host of *Striga* produces strigol, which stimulates seed germination of the pathogen. This finding led to the use of cotton as a trap crop, the principle being that the germinated seed of the pathogen, finding no host would die. However, practice does not always follow theory. Ramaiah (1987) for example, reported disappointing results when the trap crops cotton and ground-nut were grown for 2 years before a test crop of the host, sorghum. Parker (1991) suggested that the complex requirements for germination of the pathogen's seed and its distribution in the soil might explain such failures.

Where the compound that stimulates gratuitous germination of a pathogen propagule has been identified, the compound itself may be used as a soil treatment. For example, Coley-Smith (1990) reported experiments in which application of alk(en)yl-1-cysteine sulphoxides to the soil caused the germination of sclerotia of *Sclerotium cepivorum* and resulted in reduction of white rot of onions caused by the pathogen. Unfortunately, such experiments with strigol and *Striga* have not been so successful (Parker, 1991). Better success in this case has attended the use of ethylene and this has formed part of the strategy to eradicate *Striga* in the USA (Parker, 1991). However, the injection of ethylene into soils is expensive and potentially hazardous and therefore not suitable for many countries where *Striga* species are prevalent. To overcome these problems, Berner, Schaad and Volksch (1999) have investigated microorganisms as a source of ethylene. They found that ethylene producing strains of *Pseudomonas syringae* pv. *glycinea* were consistently better stimulants of seed germination of species of *Striga* than ethylene gas. Although, as the authors point out, there is a need for additional studies of the ecology and pathogenicity of the effective strains before they can be routinely deployed, this novel approach is likely to have a promising future (Figure 12.1). In the meantime, the more conventional procedure of rotation is proving effective. Carsky and co-workers (2000) found that parasitism of maize by *S. hermonthica* was significantly lower after soybean than after the sorghum control at two of three trial sites and that yield of maize at the three sites combined was increased by approximately 90 per cent.

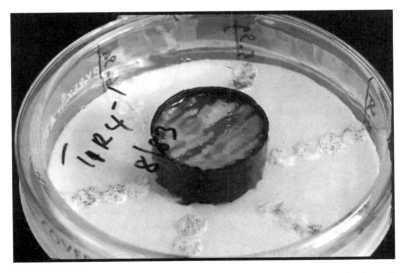

Figure 12.1   Test for stimulation of germination of *Striga* by ethylene-producing microorganisms: cultures of organisms were placed in the centre of the Petri dish and the seed of *Striga* were distributed on glass fibre filter discs radiating from the centre; germination was scored after 72 h (courtesy of Dana Berner). A colour reproduction of this figure can be seen in the colour section

12.2.3 Exploiting adhesion

Adhesion of the pathogen to the plant is essential for many pathogens if the host is to be successfully penetrated and this is particularly true for fungal pathogens that gain entry by mechanical force since otherwise their infection structures such as appressoria would be simply lifted from the plant surface (Section 6.7). However, the composition of the adhesive is variable. Compounds so far implicated are water-insoluble glycoproteins, lipids and polysaccharides. Moreover, there is variation in the environmental cues that induce the development of spore adhesiveness (Tucker and Talbot, 2001). At present there is insufficient evidence to predict if it would be feasible to exploit variation in adhesion, whether controlled by host, pathogen or environment, as a means of enhancing resistance but it is perhaps worth considering.

12.2.4 Exploiting physical cues and barriers to penetration

As discussed in Section 6.6.4, rust fungi respond thigmotropically to the surfaces of their hosts. For example *Puccinia graminis* f. sp. *tritici* orients its germ tubes at right angles to the veins of wheat leaves, increasing the chances of the growing tip encountering a stoma. The fungus also responds to multiple ridges by forming infection structures whereas *Uromyces appendiculatus* requires a single ridge 0.5 μm high to trigger the formation of its infection structures. Altering the surface topography of these hosts may therefore prevent infection and this has been demonstrated by Zekaria-Oren, Eyal and Ziv (1991). They used film forming compounds such as Bio-film and demonstrated more than 80 per cent suppression of leaf rust (*Puccinia recondita* f.sp. *tritici*) on wheat. No systematic search appears to have been reported for plants in which surface topography plays a role in preventing infection although selection of plants in which its dimensions lie outside the range of tolerance of the pathogen would seem to be a possible means of control.

It is difficult to prove that resistance of plants to pathogens is attributable to physical barriers. One clear example, however, is that of resistance of pearl millet to ergot (Willingale, Mantle and Thakur, 1986). Here constriction of the stigma caused by pollination prevented infection by the ergot fungus, *Claviceps fusiformis* (Section 9.2). Thus, plants in which protogyny is of limited duration are resistant owing to self pollination. As Willingale, Mantle and Thakur (1986) point out, it is important that this natural resistance feature is exploited in breeding ergot resistant cultivars for commercial use.

The outer layers of plants and cell walls are obstacles for vectors of plant pathogens and pathogens that penetrate directly. Differences in the ease of penetration by vectors among cultivars have been reported. For example, Montllor and Tjallingii (1989) found some lettuce cultivars were less prone to penetration by aphids than others but knowledge of the factors responsible is required so that they can be exploited to give more resistant plants.

Fungal pathogens that penetrate plants directly employ mechanical force or enzymatic action or a combination of both. Those that rely on mechanical force generally produce melanized appressoria in which sufficient turgor pressure is generated to effect penetration (Section 6.8). When melanization is inhibited by tricyclazole or other compounds penetration is prevented (Bell and Wheeler, 1986; Kim *et al.*, 1998).

Dickman and Patil (1986) demonstrated the role of cutin as a barrier to infection by showing that mutants deficient in cutinase were unable to penetrate plants. Reciprocally, transformation of *Mycosphaerella* spp., which can normally attack papaya fruit only if they are wounded, allowed the pathogen to invade intact fruit (Dickman, Podila and Kolattukudy, 1989; Section 6.9.1). Indirect evidence for the role of cutinases in the infection of host plants by fungi such as *Blumeria graminis* and *Pyrenopeziza brassicae* has been obtained by use of the ebelactones. These compounds are inhibitors of cutinase and they also inhibited infection by the fungi (Francis, Dewey and Gurr, 1996; Davies *et al.*, 2000).

Pathogens would be expected to be limited in their ability to breach the cuticle by their finite capacity to secrete the appropriate enzymes and the activity of the enzymes themselves. Bostock and co-workers (1999), in a study of genotypic variation of the resistance of peach (*Prunus persea*) to the brown rot fungus (*Monilinia fructicola*), have shown that the phenolic substances, chlorogenic and caffeic acids, prevent the secretion of cutinase rather than inhibit its action. They found that the concentrations of the phenolic acids in the epidermis and subtending cell layers of the fruit were high and especially so in peach genotypes with high levels of resistance. As the fruit ripened their concentrations declined and there was a corresponding decrease in resistance.

Thick cuticles would be expected to take longer to breach than thin ones and, on occasions, this might be sufficient for environmental conditions to change to ones that are inimical to the pathogen (e.g. warm and dry weather). Nevertheless, there seem to be few reports of conscious selection by plant breeders for plants with thick cuticles.

Similarly thick cell walls intuitively would seem to provide a greater barrier than thin ones. However, it is probable here that the success or failure of penetration resides principally in the biochemistry of the cell wall and the enzymes required for its degradation.

The cell walls of plants are complex and their degradation is correspondingly a complex process. Not only is a battery of enzymes required but also in many pathogens these seem to be present or inducible in the pathogen as multiple isozymes. Much work has been achieved in this sphere but so far, probably owing to the complexity of the phenomenon, it has not led to control. One promising aspect, however, is that the presence of inhibitors of polygalacturo-nases (PGIPs) in plants is widespread (De Lorenzo *et al.*, 1990, 2001, 2002; Sections 9.3.2 and 12.3.1).

12.3   Controlling disease by enhancing the tolerance of plants to the virulence attributes of the pathogen

Two important virulence attributes of bacterial and fungal pathogens are the production of enzymes that degrade plant cell walls (Chapter 6) and toxins (Chapter 8). Tolerance of toxins in some plants may be attributed to their capacity to metabolize or sequester them whereas tolerance to degradative enzymes may reside in inhibitors. These may have a 'knock-on' effect in that the inhibited rate of degradation may lead to the occurrence of sufficient concentrations of oligomers of cell-wall polymers to initiate defence responses (see Section 11.4.2). Undoubtedly there are other factors with unrecognized functions that are necessary for pathogenicity or virulence and an example of one of these is given at the end of the section.

12.3.1   Production of inhibitors of degradative enzymes

Evidence for cell-wall degrading enzymes as a necessity for the pathogenicity or virulence of several bacterial and fungal phytopathogens was presented in Chapter 6. It follows that inhibition of these could contribute to resistance as shown by Annis and Goodwin (1997) for a heat labile low molecular weight compound from canola acting on a polygalacturonase from *Leptosphaeria maculans*. Other polygalacturonase inhibitors are proteins with leucine-rich repeats which are structurally related to several resistance genes that have been cloned and may form part of the plant surveillance system for potential pathogens. These polygalacturonase inhibitor proteins (PGIPs) acting on polygalacturonase *in vitro* cause the accumulation of oligogalacturonides with a degree of polymerization which confers elicitor activity. Some evidence suggests that PGIPs may be induced specifically in some interactions of plants and pathogens (Nuss *et al.*, 1996) but plants with high constitutive or inducible levels of PGIPs do not appear to have been consciously selected in breeding programmes.

It would be interesting to know if inhibitors of other cell-wall degrading enzymes exist in plants and whether they can be exploited to enhance general resistance.

12.3.2   Tolerance of toxins

The recognition that the symptoms of Victoria blight of oats were caused by a toxin led to the early use of cultural filtrates of the fungus in the selection of toxin-insensitive genotypes. Such plants were also resistant to the fungus (Wheeler and Luke, 1955). More recently, the tolerance of some genotypes of grape-vine to *Eutypa lata* were shown to be related to their ability to detoxify

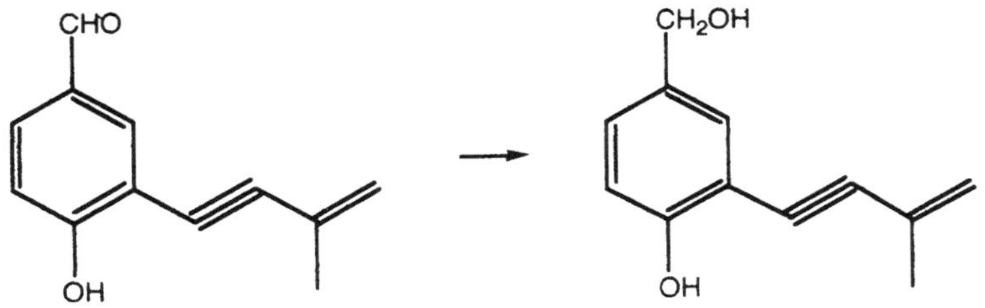

Figure 12.2   Reduction of the toxin of the grape-vine pathogen *Eutypa lata*, to the non-toxic alcohol, eutypinol, by a plant reductase

eutypine, the toxin produced by the pathogen, by reducing it to the corresponding alcohol, eutypinol (Péros and Berger, 1994; Figure 12.2). The toxin was used to screen plantlets of somaclones of *Vitis vinifera* cv. Ugni Blanc (Soulie, Roustan and Fallot, 1993). Further work showed that the reductase was not confined to species of *Vitis* and led to cloning of a gene encoding a protein with this activity. The application of this result to the production of toxin insensitive transgenes is discussed in Section 12.7.2.

Selection may also be made at the tissue-culture level (Daub, 1986). In theory, cell suspensions would be ideal for exposure to toxins since the toxin can be added in controlled amounts and survivors rescued by allowing them to grow on solid media. However, suspension cultures often contain clumps, raising the problem of diffusion and thus the possibility that some cells in the centre may escape exposure to the full toxin concentration, particularly if the compound is unstable.

Despite these difficulties there have been some successes. For example, Hartman, McCoy and Knous (1984a,b) used culture filtrates of *Fusarium oxysporum* f.sp. *medicaginis* to select alfalfa cells that were insensitive to its toxic components. Regenerants from the insensitive cells were highly resistant in the field and the resistance was both stable and heritable. Similarly, Nadel and Spiegel-Roy (1988) and Deng and co-workers (1995) were able to select lemon cells and protoplasts, respectively, that were insensitive to mal secco toxin obtained from tracheomycotic fungus *Phoma tracheiphila*, the cause of a serious disease in citrus. Regenerants were substantially more resistant than the mother cultivars from which they were derived. Vidhyasekaran and co-workers (1990) used a partially purified toxin preparation of *Cochliobolus miyabeanus* (= *Helminthosporium oryzae*) to select rice calli for resistance to brown spot of rice (cf. Section 1.3.2). The calli were shaken with toxin preparation for 48 h and then plated on medium without toxin. Of the four regenerants obtained from 360, two were highly susceptible but the remaining two showed considerably increased resistance which was stable through three generations.

Selection of calli has also been used by Venkatachalam and co-workers (1998) to obtain groundnut lines resistant to late leaf-spot disease caused by *Cercosporidium personatum.* They exposed calli to culture filtrates of the pathogen and selected those that remained viable and increased in fresh weight. Plants of the R2 generation were more resistant to the pathogen.

In some instances, however, the results have been disappointing. For example, Jin and co-workers (1996) used culture filtrates of *Fusarium solani* to select insensitive cell lines from embryonic suspension cultures of soybean but little improvement in resistance was obtained. Perhaps the policy of using whole culture filtrates rather than its constituent toxins was partly to blame.

One of the advantages of toxins is that they are not infectious. Therefore once their role in a disease syndrome has been demonstrated to be important, they can be used to screen plants in regions that the organism has not yet reached. They can also be used as convenient surrogates for the pathogen in tests of the reaction of germplasm. For example, Vanderbiezen and co-workers (1995) found that all members of the genus *Lycopersicon* with the exception of *L. cheesmanii* were moderately or highly insensitive to AAL toxin produced by *Alternaria alternata* f. sp. *lycopersici* (Section 8.4.1). In particular, *L. pennellii* and *L. peruvianum* were toxin insensitive and resistant to the pathogen. In contrast to the incomplete dominance of resistance shown by *L. esculentum*, that of *L. pennellii* was completely dominant and was conferred by a new allele at the *Asc* locus.

This would seem to be an area which is ripe for exploitation as techniques for purifying toxins and demonstrating their importance in pathogenesis or virulence have improved considerably in the last few years (cf. Chapter 8).

12.3.3 Identification of plant genes that recognize essential virulence components of pathogens

In theory, durable resistance would be expected to be a consequence of recognition of some component of the pathogen that is essential for its virulence. Progress in the application of this principle has been achieved by de Wit's group (Laugé *et al.*, 1998). Two proteins, ECP1 and ECP2, are secreted by *Cladosporium fulvum* during infection of its host, tomato, and they were shown by targeted replacement of their encoding genes to be required for full virulence (Laugé *et al.*, 1997). ECP2 was the more important since the virulence of the ECP2-deficient strain was severely impaired, the strain poorly colonizing leaf tissue, producing little emerging mycelium and few conidia. Moreover, all strains tested from a worldwide collection of *C. fulvum* were found to produce the ECP2 protein. When the gene-encoding the ECP2 protein was ligated into the potato virus X genome and the recombinant virus, which expressed the *Ecp2* gene, used to screen tomato lines, four were found that reacted with a hypersensitive resistance response. Resistance was based on a single dominant gene, designated *Cf-ECP2*, and since the factor it recognizes, ECP2, is required for virulence, the resistance conferred by the gene is expected to be durable.

12.4 Controlling disease by enhancing resistance mechanisms of the plant

12.4.1 Exploiting phytoanticipins

There is a large number of antimicrobial compounds possessed by plants which can give considerable protection against challenge by microorganisms (Chapter 9). These may be low molecular weight compound such as saponins (Section 9.3.1), low molecular weight proteins such as thionins or higher molecular weight proteins such as the virus inhibitors of the Caryophyllaceae (Section 9.3.2). Conscious selection for these seems to have been largely neglected despite the evidence for their role in resistance. In fact, some claim that, 'Development of plants for use as food crops has gradually stripped these species of their natural resistance to insects and pathogens. Consequently, most modern cultivars rely upon inputs of pesticides to produce an acceptable yield' (Steffens and Walters, 1991). The avenacins are possibly one such group of compounds. These are only found in *Avena* spp. where they confer resistance to *Gaeumannomyce graminis* and not in other cereals (Haralampidis *et al.*, 2001). One may also add that modern cultivars also rely heavily on R genes which confer specific resistance. Our current understanding of these is that they play two roles, recognition of pathogens and the instigation of a cascade of defence responses. Loss or mutation of corresponding *avr* genes leads to loss of recognition and, as a result, a failure to launch active defence mechanisms with the consequence of susceptibility. There is therefore a case to be made for examining the role of non-specific defence mechanisms in order to exploit them as a means of resistance to pathogens, although plants in which these mechanisms are well developed should be tested for any negative impact they may have on humans or animals that consume them.

As the studies of Prusky and his co-workers have shown (Prusky, 1996), effective control of disease can be obtained by exploiting the chemical defences of plants if there is a proper understanding of the processes involved. In unripe avocado fruits an antifungal diene is reponsible for resistance to anthracnose caused by *Colletotrichum gloeosporioides* (Section 9.3.1). As the fruit ripens the inhibitor is catabolized by lipoxygenase owing to the depletion of epicatechin, a natural inhibitor of lipoxygenase activity. Prusky (1988) found that treatment of fruits with 0.1 mM epicatechin or a mixture of the antioxidant, butylated hydroxy toluene (BHT, 0.1 mM) and ascorbic acid (0.5 per cent) delayed the disappearance of endogenous epicatechin and the appearance disease symptoms. Symptoms could also be delayed by enhancing the levels of carbon dioxide (Figure 12.3.).

As discussed in Chapter 9, saponins are components of many plants and for a few of them that have been studied from the point of view of defence there is strong evidence that they are effective phytoanticipins except against pathogens that have the means of detoxifying them (Section 9.3.1). Consequently it is moot

Figure 12.3 Control of anthracnose of avocado (cv. Fuerte) caused by *Colletotrichum gloeos-porioides* by subjecting freshly harvested fruit to 30 per cent CO_2 in air for 24 h (courtesy of Dov Prusky). A colour reproduction of this figure can be seen in the colour section

whether it would be feasible or desirable to select for plants with high concentrations of saponins.

Hydroxamic acids play an important role in the resistance of cereals to pathogens and pests. Gianoli and Niemeyer (1998) have investigated the content of the hydroxamic acid aglucones DIBOA (2,4-dihydroxy-1,4-benzoxazin-3-one) and DIMBOA (2,4-dihydroxy-7-methoxy-1,4-benzoxazin-3-one) in wild Poaceae belonging to the tribes Triticeae (genera *Hordeum* and *Elymus*) and Aveneae (genera *Deschampsia* and *Phalaris*). They found that the concentration of DIBOA in seedling extracts of the wild barleys *Hordeum chilense, H. brevisubulatum* subsp. *violaceum* and *H. bulbosum* was negatively correlated with the performance of the cereal aphids *Schizaphis graminum* and *Diuraphis noxia* and discuss the use of these sources of resistance in breeding for resistance against aphids.

12.4.2 Exploiting phytoalexins

The role of phytoalexins in resistance was discussed in Chapter 11 (Section 11.3.2). In plants which are capable of a phytoalexin response and yet fail to resist a specific pathogen 'too little and too late' are often the reasons rather than an innate inability to accumulate effective concentrations of the compound(s). One potential remedy for this situation would be to elicit the phytoalexins before the microbial challenge has occurred. As discussed in Chapter 11, many physical and chemical treatments do elicit phytoalexins in the absence of microorganisms and some of these have been investigated. However, a cost in terms of yield is usually the penalty for gratuitous and generalized phytoalexin elicitation, one of their properties in nature being their strictly localized accumulation in the vicinity of the pathogen. Scoparone (6,7 dimethoxycoumarin) accumulation may be elicited in citrus by gamma irradiation and phosphonic acid (Afek and

Sztejnberg, 1993; Ali, Lepoivre and Semal, 1993). In the case of phosphonic acid, the effect appeared to be mediated through the release of elicitors from the challenging fungus, *Phytophthora citrophthora*, thus maintaining the localization of the response.

12.4.3 Exploiting acquired resistance

As discussed in Chapter 11, plants respond to infection by defending themselves in many ways. When the defence is successful, the plant is often found to be more resistant to subsequent infection by the same pathogen or others. Moreover, the resistance is frequently systemic. In fact, this state of systemic acquired resistance is probably the norm for plants grown under natural conditions since they are constantly bombarded with microorganisms seeking water and nutrients. Can this natural state of plants be exploited to give enhanced resistance to some of their worst pathogens?

   As described in Chapter 11, pioneering work in this field was done by Kuć and co-workers (Kuć, 2001). However, the phenomenon was recognized much earlier by virologists who noted that infection with a mild strain of a virus often gave protection against subsequent challenge with a virulent strain. They called it cross-protection.

Cross-protection

The use of cross-protection as a means of virus control has been reviewed by Fulton (1986). Plants are treated with a mild strain of the virus which normally gives protection against a more virulent strain that may be present in the environment. However, he cautions against the large-scale application of the technique, until the following points have been carefully considered.

(1) Protection may be incomplete.

(2) Care should be taken that any experiments designed to test the validity of the techniques reflect the level of challenge inoculum prevalent in the field. Often protected plants can only be infected with inoculum that is 100-or 1000-fold greater than that needed to infect the unprotected plant.

(3) The protecting strain may spread to other hosts where it may be virulent.

(4) Vector relations of the protecting strain may differ from those of the virulent strain.

(5) There may be synergism and even hybridization with another virus such as found in African Cassava Mosaic Virus and East African Cassava Mosaic Virus (Section 1.3.10).

(6)   The protecting strain may mutate to virulence. This is a particular danger if such mutants accumulate and become dominant in stocks of virus used as the primary inoculum.

(7)   Finally, there are practical difficulties in inoculating a whole crop.

Nevertheless, cross-protection may be considered if the disease is endemic, spreading rapidly, appears impossible to eradicate and causes losses that are appreciably greater than those of the protecting strain (Posnette and Todd, 1955). Citrus tristeza virus in Brazil and papaya ringspot virus in Taiwan met these criteria and have both been controlled by cross-protection (Costa and Muller, 1980; Yeh *et al.*, 1988).

More recently, Gonsalves (1998) has described the problem of dealing with papaya ringspot virus in Hawaii. Attempts to find a mild local strain of the virus failed but two were obtained by treating a virulent strain with nitrous acid. One of these, PRSV HA 5-1, gave good protection against local strains but also produced marked symptoms itself on the fruit and leaves of some cultivars. For this reason and the extra cultural management required for cross-protection as well as the reluctance of farmers to infect their crops with virus the technique was not widely adopted. Instead, transgenic papaya was developed expressing the coat protein of the virus to control the disease (see Section 12.7.8).

Induced resistance by rhizosphere bacteria

The recognition that plant growth-promoting rhizobacteria may protect plants from disease not only by the classical mechanisms of biocontrol, i.e. antibiosis, competition and parasitism, but also by inducing systemic resistance (ISR), has opened up a new field of opportunity for enhancing plant health. Selection of such bacteria and their formulation into seed dressings promises to be an economical and environmentally appropriate way of boosting resistance in the future. Such bacteria would have to be rhizosphere competent, viable in the particular soil where the crop is grown and give effective control under the disease pressures likely to obtain during the time the crop remains in the field. As indicated in Section 11.3.8, when strain WCS417 of *Pseudomonas fluorescens* was applied to roots of carnation and the stems were inoculated 1 week later with *Fusarium oxysporum* f. sp. *dianthi* disease was reduced. As the biocontrol agent was spatially separated from the pathogen the resistance must have been mediated through the plant (van Peer, Niemann and Schippers, 1991). Other work has shown that control may be obtained in the field. Wei, Kloepper and Tuzun (1996) tested four plant growth promoting rhizobacteria (PGPR) strains of bacteria: *Pseudomonas putida* strain 89B-61, *Serratia marcesens* strain 90-166, *Flavomonas oryzihabitans* strain INR-5 and *Bacillus pumillus* strain INR-7 for protection of cucumbers against angular leaf spot caused by *Pseudomonas syringae* pv. *lachrymans*. Inoculum of the PGPR strains was applied either as a seed treatment or as a seed treatment and a soil drench. Significant reductions of disease severity occurred with three of the four bacteria in two trials and all four in the third trial.

Chemically induced resistance

Since plants in nature are under constant attack by putative pathogens, it is likely that their natural phenotype is one in which systemically acquired resistance (SAR) is expressed. Nevertheless, as demonstrated by cross-inoculation experiments, induction of SAR may be increased in such plants. SAR may also be induced by the application of several chemical compounds which themselves do not have significant antimicrobial activity. Two of the earliest of these to be described were dichloroisonicotinic acid (INA) and benzothiadiazole (BTH). Dann and co-workers (1998) showed that three or four applications of INA reduced the severity of white mould of onions, caused by *Sclerotinia cepivorum*, 20–70 per cent in highly susceptible cultivars. Similarly, severity reduction in response to two or four applications of BTH was 20–60 per cent.

BTH has been developed as Bion and has been shown to activate disease resistance in many crops to a wide variety of pathogens with consequent benefit to farmers (Hammerschmidt, Metraux and van Loon, 2001; Oostendorp *et al.*, 2001). For example, tomatoes treated with the preparation either with or without the addition of insecticide were more resistant to *Bemisia tabaci*, the whitefly vector of tomato leaf-curl virus as well as the virus itself and produced better yields than controls.

The non-amino acid β-aminobutyric acid (BABA) was shown by Oka and Cohen (2001) to induce resistance to two species of the nematode genus *Heterodera* as measured by a reduction in cyst formation and in a cereal-specific species of *Meloidogyne* as measured by a reduction in egg mass.

Oxycom TM is a combination of an active oxygen generator and fertilizer which is mildly fungicidal but enhances resistance in bean. It appears to do so by inducing defence related genes involved in phenolic metabolism and strengthening the plant cell wall (Kim *et al.*, 2001).

12.5 Genetic approaches to the control of disease

According to Day (1984), plants have been selected for resistance to disease for at least two centuries, well before its microbiological cause was understood. However, it was not until Biffen's classical experiments that disease reaction was recognized as being a character that was inherited according to Mendelian principles (Biffen, 1905, 1912; see Section 10.2).

The gene-for-gene concept of Flor and Oort was an outstanding discovery which laid the foundation for classical and molecular studies of host–pathogen interaction as well as providing a solid basis for practical plant breeding (reviewed by Flor, 1971). According to this concept as discussed in Chapter 11, incompatibility is understood to be the result of the interaction of alleles for resistance on the part of the host with alleles for avirulence on the part of the pathogen. Resistance is therefore specific and qualitative (either it is there or it is not) and the genes concerned are often referred to as race-specific or major genes. Vertical resistance is another somewhat whimsical term used to describe the same phenomenon; this was derived from the appearance of histograms depicting the resistance of cultivars challenged with a pathogen to which they either did have an effective resistance allele (high column on the resistance scale) or did not (low column).

The validity of the gene-for-gene model has been overwhelmingly substantiated by the cloning of alleles for avirulence from viruses, bacteria and fungi and the cloning of plant resistance alleles to these three classes of pathogen as well as to nematodes and aphids. Moreover, some studies have shown virulent strains may arise by modification of avirulence alleles and that these virulent strains quickly become established in pathogen populations (Section 10.6.1). Nevertheless, some cultivars of crop plants have remained resistant to noted pathogens of the species for many years. Johnson (1984) coined the term durable resistance to describe this phenomenon and remarked that it could only be recognized retrospectively.

Resistance may also be quantitative as opposed to the qualitative (major) gene resistance described above. Other terms that have been used to describe quantitative resistance are race-non-specific resistance and horizontal resistance. The latter term is complementary to the vertical resistance just described. Here the columns of the histogram depicting the resistance of a range of cultivars to a pathogen are purported to be uniformly high. Both these terms imply that the resistance conferred is to many races of the pathogen and the genes encoding this type of resistance are said to be minor. One hope for resistance controlled by such genes is that when amalgamated in an individual cultivar they would confer a good level of resistance which, owing to its polygenic nature, might also be difficult for a pathogen to overcome, i.e. such resistance might prove durable. At present there seems to be little evidence for this. Moreover, there is controversy as to whether there is a real difference between qualitative (major) and quantitative (minor) gene resistance, some scientists suggesting that qualitative resistance is merely an extreme form of quantitative resistance.

How can this incomplete knowledge of the genes involved in conferring resistance to plant disease best be used to control diseases in our crops?

Most of the plant cultivars in current use are derived from conventional plant breeding techniques using major genes for resistance. Typically, sources of resistance are searched for in the centres of origin or diversity of the plant and the pathogen. Plants that are resistant are usually unacceptable agronomically, e.g. they may be low yielding. Therefore a long process of crossing and backcrossing to the recurrent susceptible (but otherwise agronomically acceptable) parent and selecting only resistant plants among the progeny has to be undertaken. Unfortunately, the resistance that has been hard won by this laborious process may be short-lived (Section 10.1). Furthermore, genes with undesirable agronomic traits may also be carried along with the introgressed gene for resistance.

The length of time required for the introgression of resistance genes into susceptible plants has, in some cases, been sufficient for the pathogen to evolve new variants to which the resistance gene is no longer effective, rendering the whole breeding programme obsolete before the cultivar is released to the farmer. Another difficulty is that the resistance conferred by the gene may be difficult to score (cf. Chapter 4) and this is particularly true of minor genes with small effects on symptom development but which may, in aggregate, provide a good level of resistance. Also, in programmes designed to pyramid resistance genes, it is difficult to screen for genotypes that have more than one effective resistance gene since the phenotype of such plants would be similar or indistinguishable from a plant with a single resistance gene. Other problems are connected with the logistics of screening trials: inoculum may not be available, application methods may be inefficient, leading to escapes which may then be erroneously scored as resistant, and weather conditions may not favour development of the disease. Marker assisted breeding can be very helpful in such circumstances. In the first instance the number of generations and therefore the time required to recover the recurrent genotype is reduced (Ribaut and Hoisington, 1998) and secondly, there is a reduced dependence on field trials which, as just described, may not always be reliable (although, of course, the crucial test of any new cultivar is that its resistance should stand up in the field!).

A primary requirement for a marker to be of value in breeding for resistance is that it should be tightly linked to the desired resistance gene and the determination of this normally requires considerable work. The markers themselves are of various types and include Restriction Fragment Length Polymorphisms (RFLPS), Random Amplified Polymorphic DNAs (RAPDs), Amplified Fragment Length Polymorphisms (AFLPS), Inter Simple Sequence Repeats (ISSR), Sequence Tagged Microsatellite Sites (STMS) and micro- and mini-satellites. For example, in wheat, Hu, Ohm and Dweikat (1997) identified two RAPDs which co-segregated with the powdery mildew gene *Pm1*. A third RAPD was found to be 5.4 ± 1.9 cM from the resistance gene.

Bulked segregant analysis (Michelmore, Paran and Kesseli, 1991) is an efficient method to screen for markers associated with resistance genes. A single cross is made between parents which are homozygous for resistance (RR) or susceptibility (rr) and the F_2 and F_3 generations obtained. The F_3 generation enables the experimenter to determine which individuals of the F_2 generation are homozygous for the gene of interest. These homozygous F_2 individuals are then screened for markers that cosegregate with resistance. Garcia and co-workers (1996) used this technique in order to identify a RAPD marker Z3/265 in groundnut (*Arachis hypogaea*) which co-segregated with two dominant genes that conferred resistance to the nematode, *Meloidogyne arenaria*. Resistance was expressed as reduced galling and reduced egg number and the genes designated *Mag* and *Mae*, respectively. However, Z3/265 mapped some distance away from the resistance genes – 10 ± 2.5 cM from *Mag* and 14 ± 2.9 cM from *Mae*.

McClendon and co-workers (2002) were more successful in their search for a marker associated with a gene conferring resistance to wilt of pea caused by *Fusarium oxysporum* f. sp. *pisi* race 1. They found one AFLP marker which was within 1.4 cM of the gene and estimated that the probability of correctly identifying resistant lines with this marker was 96 per cent.

Where markers are very closely linked to resistance genes they have facilitated the cloning of the resistance genes themselves (see Section 10.5.3)

12.6   Plant transformation

The ability to transform plants with genes of interest and for the genes to be expressed in the transformants has been fundamental to the rapid advances that have been made in the last decade. Before discussing transformation of plants with genes which may confer resistance to pathogens the techniques for transformation that are currently available will be briefly reviewed.

The realization that plant transformation was not only a possibility but had already occurred in nature was a consequence of the discovery that the symptoms of crown gall and hairy root diseases, caused by *Agrobacterium tumefaciens* and *A. rhizogenes*, respectively, are the result of 'natural genetic engineering' (see Section 7.4.). The next question was whether the oncogenic genes of the Ti plasmid could be replaced with ones specified by the experimenter. Remarkably, this did not pose a problem, the essential components of the plasmid for successful transformation being a 25 bp border sequence at the right of the T-DNA (i.e. the DNA that is transferred to the plant) and the *vir* region.

There seem to be few restrictions on the type of DNA that is transferred although clearly there must be an upper size limit. Moreover, as the *vir* region can function *in trans*, it need not be on the same plasmid as the T-DNA. This is advantageous as the experimenter can work with two small plasmids rather than a single large one. Plasmids bearing the T-DNA are known as Ti-vectors. It is convenient to manipulate these in *Escherichia coli*, so in order for the Ti vector to be able to replicate in either *E. coli* or *Agrobacterium* spp., it must contain either a broad host range replication locus or two replication loci, one for each bacterial species.

Considerable ingenuity has gone into the construction of various Ti-vectors. One example is pGreen (Hellens *et al.*, 2000). This Ti-vector offers considerable versatility and flexibility as it is of small size (3232 bp), has the pUC replication origin permitting replication to high copy number in *E. coli* and unique restriction sites in the left- and right-hand border sequences. Moreover, the choice of restriction sites for cloning is not compromised by the experimenter's choice of selectable marker and reporter genes since the vector allows any arrangement of these. It is only able to replicate in *Agrobacterium* if the helper plasmid pSoup (!) is present in the same strain. Further information on pGreen as well as order forms may be found at the Internet site (http://www.pgreen.ac.uk).

A second method of plant transformation is to use protoplasts that have been made permeable to foreign DNA by electroporation or polyethylene glycol (PEG) treatment. However, a disadvantage of using protoplasts is that it is often difficult to regenerate plants from them as the majority of plant cells are not totipotent.

A third technique is to use biolistics. Here gold particles, coated with DNA, are fired into recipient plants (Figure 12.4). Although expression of foreign DNA is often obtained by this technique, it is usually transient and is frequently quickly lost as the DNA is seldom integrated into the host genome (Potrykus, 1991).

Despite these technical difficulties, success in obtaining transgenic plants from a range of species has been successfully achieved with all three techniques. For a fuller account of the problems and strategies for practical application, the reader is referred to the review by Birch (1997).

Questions now arise as to which crops should be genetically engineered and which genes should be introduced into them. Clearly, the cereals, because of their importance as a source of food, would be an obvious choice of recipient plant but, unfortunately, they are resistant to infection and therefore transformation by *Agrobacterium*. However, Hansen, Shillito and Chilton (1997) have reported a way around this impasse. They constructed a binary plasmid system with the Ti vector containing the gene which they wished to introduce into the plant and the other plasmid containing the genes *virD1* and *virD2* from the *vir* region of a wild-

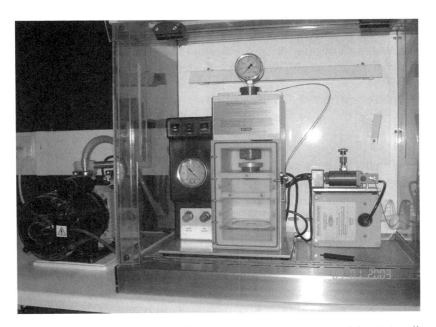

Figure 12.4 A gene gun used to transform plants by firing gold particles ($\sim$1 μ diameter) coated with DNA at plant tissue: the DNA contains the gene of interest and a selectable marker allowing only transformed cells to proliferate; these may then be regenerated into intact plants (photograph courtesy of Saul Purton)

type Ti plasmid. Successful transformation of protoplasts was achieved when they were incubated with the *Agrobacterium* containing the two types of plasmid and transformation efficiency was doubled if *virE2* were included. The legumes are also clearly candidates and here there are real possibilities for transformation by *Agrobacterium* although plants in this group too can be resistant. According to Potrykus (1991), resistance is related to the lack of an appropriate wound response, which is normally a prerequisite for successful transformation (see Section 7.4). Nevertheless, Escudero and Hohn (1997) reported transfer and integration of T-DNA into tobacco plantlets without injury.

12.7 Candidate genes for plant transformation in order to enhance resistance

With the barriers to plant transformation overcome it is now imperative to consider the ethical questions as to whether we should transform our crops and, if so, what genes should be used. This book is not the place to discuss the ethics of producing so-called genetically modified (GM) crops in detail although it would seem difficult to sustain the view that no plant should ever be transformed for two reasons: (1) every sexually reproducing plant is genetically modified in every generation, otherwise it would be a clone; (2) nature in the form of *Agrobacterium* species has been at it for thousands of years! In both cases, successful plants are selected either naturally or artificially.

Certainly there are genes which, if transferred to crop plants, hold out the promise of increasing their disease resistance and consequently reducing pollution by broad-spectrum pesticides. Moreover, in poor countries where pesticides are not available owing to cost, a genetically modified crop would seem to be a better option than starvation.

There is an increasing number of possibilities for transforming plants with genes that increase their resistance to both pests and diseases and these are discussed in the following paragraphs.

12.7.1 Transformation with genes encoding proteins that inhibit pathogen enzymes

Powell and co-workers (2000) showed that a polygalacturonase inhibiting protein (PGIP) from pear when expressed transgenically in tomato decreased the growth of *Botrytis cinerea* in ripe tomatoes and the extent of macerated tissue. Although it is not clear whether the direct inhibition of the enzyme or enhanced concentrations of breakdown products of polygalacturonic acid which are active in eliciting defence reactions were responsible, the important point is that the tomatoes were more resistant. Therefore the selection of plants with high levels of constitutive or inducible PGIPs could be a promising non-GM approach to increasing the resistance of plants to pathogens in which polygalacturonases are important virulence factors.

12.7.2   Transformation with genes encoding enzymes which degrade toxins produced by pathogens

As discussed previously (Section 8.4.2) the resistance of maize to *Helminthosporium carbonum* is conferred by a gene, *Hm1*, which reduces the side-chain of the Aeo moiety of the toxin. Similarly, the tolerance of some cultivars of grapevine to *Eutypa lata* has been attributed to conversion of its toxin, the aldehyde, eutypine to the corresponding alcohol, eutypinol (Figure 12.2). Guillén and co-workers (1998) found that mung bean was a rich source of the enzyme responsible and that there was weak homology between the gene encoding this enzyme and the *Hm1* gene. When calli of grape-vine (*Vitis vinifera*) were transformed with the gene they were far less sensitive to eutypine than the wild type.

Zhang, Xu and Birch (1999) generated transgenic sugarcane plants that expressed a gene from *Pantoea dispersa* which detoxified albicidin, a potent toxin from *Xanthomonas albilineans* (see Sections 5.4.2 and 8.5.1). Plants expressing the gene at levels of 1–10 ng of enzyme per mg of leaf protein did not develop chlorotic disease symptoms in inoculated leaves, whereas all untransformed control plants developed severe symptoms. Moreover, transgenic lines with high activity of the enzyme in young stems were also protected against systemic multiplication of the pathogen. Therefore the expression of this gene confers resistance to both disease symptoms and multiplication of the toxigenic pathogen in its host.

Toxigenic pathogens have evolved mechanisms of avoiding poisoning themselves. For example, *Pseudomonas phaseolicola* secretes phaseolotoxin, a potent inhibitor of ornithine carbamoyl transferase (OCTase) which catalyses the synthesis of citrulline. However, the reaction is not inhibited in the bacterium as one of its two OCT genes encodes an enzyme *argK* that is resistant to the toxin. Therefore transforming plants with genes that encode this toxin insensitive enzyme might render the transgenic plant insensitive. Hatziloukas and Panopoulos (1992) have adopted this approach and showed that when tobacco plants were transformed with the gene some of them were insensitive to the toxin.

12.7.3   Enhancing concentrations of saponins

A novel oxidosqualene cyclase, AsbAS1, is the first committed step in the synthesis of the triterpenoid saponins, the avenacins, which accumulate in the roots of oats (Haralampidis *et al.*, 2001). Orthologues of the gene encoding the enzyme are not present in other modern cereals and the authors suggest that it may have been lost during selection. The gene may therefore have potential for metabolic engineering for resistance in these other cereals. However, the requirement for at least six additional loci for the synthesis of avenacins suggests that cereal transformants expressing saponins may take some time to develop.

12.7.4 Transformation with genes encoding antimicrobial peptides

Early work was concerned with the transformation of plants with genes specifying factors that were injurious to plant pests rather than pathogens. For example, when tobacco was transformed with the gene specifying the peptide toxin of *Bacillus thuringiensis*, caterpillars feeding on the plants were killed. Similarly, Boulter's group at Durham University, UK have used the cowpea trypsin inhibitor to confer insect resistance (Hilder, Gatehouse and Boulter, 1989; Boulter *et al.*, 1990; Hilder and Boulter, 1999).

More recently, a number of genes encoding some of the constitutive defence peptides against pathogens, discussed in Chapter 9, have been used to transform plants and their expression has been correlated with enhanced resistance. Several other such genes but from non-plant sources such as insects and mammals have also been used. The spectrum of activity of individual peptides varies but, as a whole, those with activity against all the major classes of plant pathogens, viruses, bacteria, fungi and nematodes, can be found.

Lodge, Kanlewski and Turner (1993) transformed tobacco plants with the ribosome – inactivating protein (RIP) from pokeweed (*Phytolacca americana*) and showed that the transgenic plants were resistant to a broad spectrum of viruses. Subsequently, Smirnov, Shulaev and, Turner, (1997) showed that wild-type scions grafted onto transgenic stocks expressing the protein were also resistant to viral infection but this resistance was independent of the accumulation of salicylic acid or pathogenesis-related proteins. Oldach, Becker and Lorz (2001) expressed three cDNAs encoding the antifungal protein Ag-AFP from the fungus *Aspergillus giganteus*, a barley class II chitinase and a barley type I ribosomal inhibiting protein (RIP) in wheat under the regulation of the constitutive ubiquitin1 promoter from maize. Resistance to powdery mildew (*Erysiphe = Blumeria graminis* f. sp. *tritici*) and leaf rust (*Puccinia recondita* f. sp. *tritici*) was significantly increased in lines expressing AFP or chitinase II – but not in lines expressing RIP.

The induction of glucanases and chitinases as part of the Systemic Acquired Resistance (SAR) response has already been discussed in Chapter 11 as well as their synergistic action against fungi. Several transformation experiments involving genes encoding both types of enzyme have been performed. For example, tomato expressing glucanase and chitinase genes from tobacco were far more resistant to wilt caused by *Fusarium oxysporum* f.sp. *lycopersici* than untransformed controls (Jongedijk *et al.*, 1995). Similarly, Punja and Raharjo (1996) reported that transgenic cucumber plants expressing chitinase genes originating from petunia (acidic), tobacco (basic) or bean (basic) were more resistant to *Alternaria cucumerina*, *Botrytis cinerea*, *Colletotrichum lagenarium* and *Rhizoctonia solani*. Carrots transformed with a chitinase gene from tobacco were more resistant to *B. cinerea, R. solani* and *Sclerotium rolfsii* but no increased resistance was found in plants transformed with a chitinase gene from *Petunia*. Therefore it appears that the increase in resistance which may be obtained by transforming plants with genes encoding PR-proteins is specific to the gene concerned, the recipient plant and the challenging pathogen.

Lorito and co-workers (1998) have exploited genes encoding two endochitinases from the mycoparasite, *Trichoderma harzianum*. The genes were cloned and used in a cassette, which additionally contained a secretion sequence and the *CaMV35S* promoter, to transform tobacco and potatoes. Transformants expressed the proteins to levels of between 0.01 and 0.5 per cent total protein and 5–10 per cent of the transformants were highly resistant to three foliar pathogens, *Alternaria alternata, Alternaria solani* and *Botrytis cinerea* as well as the root pathogen *Rhizoctonia solani* (Figure 12.5).

Several possibilities for the control of nematodes by transforming crop plants with genes encoding anti-nematode proteins have been discussed in a review by Jung, Cai and Kleine (1998). For example, a gene encoding the cysteine proteinase inhibitor, oryzacystatin-I, from rice has been modified to improve its activity and the modified gene introduced into tomato and *Arabidopsis* under the control of the *CaMV35S* promoter. Female nematodes feeding on the transgenic plants were reduced in size and fecundity.

Cercropins are a family of strongly basic peptides with 35–37-amino acid residues which have a broad spectrum of activity against both gram-positive and gram-negative bacteria. They were originally found in the giant silk moth, *Hyalophora cecropia* but more recently analogues such as MB-39 have been synthesized. Huang and co-workers (1997) described the construction of a

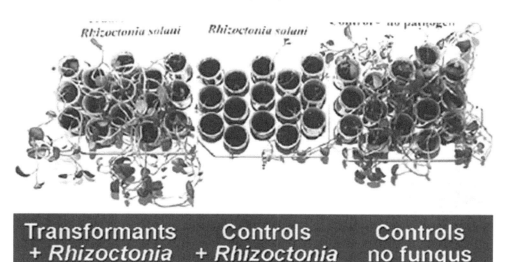

Figure 12.5 Potatoes transformed with an endochitinase gene from *Trichoderma harzianum* showing resistance to *Rhizoctonia solani* (left) compared with untransformed plants (centre) and uninoculated and untransformed plants (right) courtesy of Dr Matteo Lorito. A colour reproduction of this figure can be seen in the colour section

cassette containing MB-39. The gene for the modified cecropin product was placed under the control of the promoter and terminator of the potato proteinase inhibitor, PiII, and fused to the secretory system of barley α-amylase. When tobacco was transformed with the construct and inoculated with *Pseudomonas tabaci*, bacterial multiplication and disease symptoms were suppressed. The importance of the PiII promoter in the construct is that it is inducible by wounding and infection so that only low levels of expression of MB-39 would be expected in the intact plant with a consequent economy on the synthetic machinery of the unchallenged plant. Furthermore, fusion of the secretory system of α-amylase caused the direction of the synthesized MB-39 product to the intercellular spaces where the invading bacteria would be expected to be found.

Lactoferrin is a mammalian antibacterial protein which was recognized many years ago as a major component of infant defence systems. Tobacco has been transformed with the gene encoding the protein and its expression in the transformants monitored by an ELISA assay (Section 2.5.4). A significant positive relationship was found between expression of the protein and resistance of transformants to *Ralstonia solanacearum* (Zhang *et al.*, 1998).

Thionins were discussed in Section 9.3.2. Iwai and co-workers (2002) found that no rice thionin genes were sufficient to give resistance to *Burkholderia plantarii* and *B. glumae*. However, overexpression of a thionin gene from oats gave essentially complete protection from infection by *B. plantarii* (Figure 9.15).

12.7.5   Enhancing levels of reactive oxygen species

Wu *et al.* (1995) transformed potatoes with an H_2O_2-generating glucose oxidase gene from *Aspergillus niger*. Transgenic tubers were highly resistant to soft rot caused by *Erwinia carotovora* but the resistance was abolished by catalase confirming that H_2O_2 was responsible. Transgenic plants were also resistant to *Phytophthora infestans* and *Verticillium dahliae*

12.7.6   Modifying the phytoalexin response

The ability to detoxify a host's phytoalexins is an important virulence attribute of at least some successful pathogens (Section 11.3.2). However, phytoalexins differ widely in structure, imposing constraints on the enzymes that are able to degrade them. It follows that pathogens of plants which synthesize a particular chemical class of phytoalexin are likely to have evolved the capacity to degrade them but not others which are synthesized by other taxonomic groups of plants and differ in chemical structure. Therefore some consideration has been given to introducing genes into plants that would allow them to synthesize 'foreign' phytoalexins. This would appear to be a considerable technical challenge since the metabolic pathways leading to most phytoalexins is long. One possibility would be to introduce genes which encode enzymes that alter the last few steps in the biosynthetic sequence. However, in this approach, it is possible that the pathogen's degradative enzymes might be sufficiently non-specific to be able to degrade the novel compound. Alternatively, it might be possible to introduce the capacity to synthesize an entirely different class of compound. This approach has been adopted by Hain and co-workers (1993). They introduced a gene encoding stilbene synthase from grape-vine into tobacco, a plant that normally produces terpenoid phytoalexins. On challenge, the transgenic plant synthesized the stilbene, resveratrol (Figure 10.3), and was more resistant to attack by *Botrytis cinerea*.

12.7.7 Transformation with constructs causing the overexpression of essential regulatory genes in systemic acquired resistance

As described in Sections 11.3.6 and 11.6.6, *NPR1* (*n*on-expresser of *PR* genes) also known as *NIM1* (for *n*on-*im*munity) or *SAI1* (for *s*alicylic *a*cid-*i*nsensitive) plays a crucial role in local and systemic acquired resistance (SAR) in the model plant, *Arabidopsis thaliana*. Mutations in the gene result in loss of the ability to accumulate pathogenesis-related proteins and susceptibility to fungal and bacterial pathogens even if the plants are pre-treated with inducers of SAR. Conversely, some plants transformed with the *NPR1* gene under the control of the *CaMV35S* promoter express two- to three-fold higher levels of NPR1 protein but no increase in the expression of *PR* genes. However, on challenge with *Pseudomonas syringae* pv. *maculicola* the transgenic plants showed enhanced resistance as demonstrated by the slower rate of multiplication of the pathogen. Similarly the transgenics overexpressing *NPR1* were more resistant to the Oomycete, *Peronospora parasitica*. The enhanced resistance correlated with the greater expression of the *PR* genes *PR1, PR2* and *PR5* (Cao and Dong, 1998; Figure 12.6).

These results raise the significant possibility that broad resistance to pathogens may be conferred by alteration of signal transduction pathways leading from the eliciting agent to the final executor of the resistance response. Moreover, although over-expression of *NPR1* is required for enhanced resistance, PR proteins were not expressed until the plant was challenged by a pathogen. This represents an economy for the plant and contrasts with other transgenics where constitutive expression of genes coding for defence components such as *PR* proteins may lead to reduced plant size such as found in *cprs* mutants which are constitutive expressers of *PR* genes.

12.7.8 Pathogen-derived resistance

Pathogen-derived resistance is a term used to describe plants transformed with genes from viruses which, far from exacerbating the disease when infected with the virus, confer resistance to itself and also, in many instances, to related viruses. The

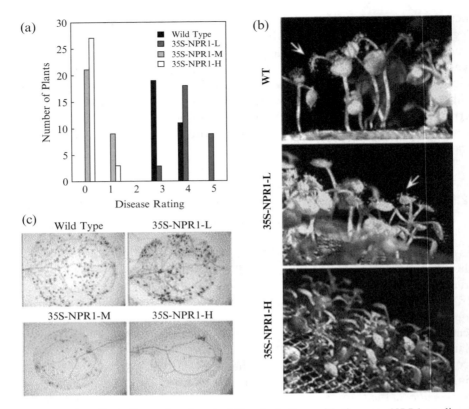

Figure 12.6 The effect of transforming *Arabidopsis thaliana* with the gene *NPR1* on disease development by *Perenospora parasitica*. (a) Disease ratings of wild-type and *NPR1* cDNA transgenic plants after infection with *P. parasitica* Noco (ratings were defined as follows: 0, no conidiophores on the plant; 1, no more than 5 conidiophores per infected plant; 2, 6–20 conidiophores on a few infected leaves; 3, 6–20 conidiophores on most of the infected leaves; 4, 5 or more conidiophores on all infected leaves; 5, 20 or more conidiophores on all infected leaves). (b) Conidiophores observed in wild-type and *NPR1* cDNA transgenic plants 7 days after inoculation with *P. parasitica* Noco. (c) Trypan blue staining of *P. parasitica*-infected leaves of wild-type and *NPR1* cDNA transgenic plants 7 days after infection; seedlings of wild-type and transgenic plants were stained with trypan blue and observed under a compound microscope (Cao and Dong, 1998 reproduced courtesy of the Proceedings of the National Academy of Sciences USA). A colour reproduction of this figure can be seen in the colour section

genes concerned are various and include those for coat proteins, replicases, movement proteins, defective interfering RNAs and DNAs and non-translated RNAs (Beachy, 1997). The mechanisms by which these transgenes confer resistance is not fully elucidated but they may be divided into two classes: those that require the production of proteins to be effective, and those that only require the accumulation of viral nucleic acids. Protein-mediated resistance can often confer resistance to a relatively broad range of plant viruses whereas expression of viral nucleic acid sequences usually provides a high level of resistance only to a specific virus strain.

Coat protein-mediated resistance

Resistance mediated by the coat protein of soybean mosaic virus (SMV), which is unable to infect tobacco, conferred resistance on tobacco to two unrelated potyviruses, potato virus Y (PVY) and tobacco etch virus (Stark and Beachy, 1989). In other instances, such as papaya ringspot virus (PRV) strain A, the resistance was only effective against the same virus strain. The coat protein of tobacco mosaic virus lies between these extremes, the degree of resistance afforded to other tobamoviruses being greater in those that have the greatest sequence similarity.

The mechanisms by which the transgenic expression of coat protein causes resistance remain unresolved in most instances although, in the case of the coat protein of TMV, interference with virus disassembly may be the means. From a practical perspective it is significant that some mutants of TMV coat protein confer much greater levels of resistance than wild-type coat protein and one of these blocked not only disassembly but also replication and local and systemic spread (Bendahmane and Beachy, 1999).

Replicase mediated resistance

Transformation of plants with genes which encode complete or partial replicase proteins confer high levels of resistance but usually only to the strain of virus from which they were derived. In one instance, however, a TMV replicase, which fortuitously contained an insertion sequence acquired from *Agrobacterium* during vector construction, conferred high levels of resistance in tobacco to a wide range of tobamoviruses but not other viruses. The mechanism by which transgenic plants expressing replicase proteins resist infection at the time of writing remains speculative.

Movement protein mediated resistance

As discussed in Section 6.12 movement proteins facilitate the spread of viruses through plasmodesmata. The exploitation of this phenomenon through trans-genes has followed a different path from those of coat protein and replicase mediated resistance since transgenic plants expressing wild-type movement protein are able to complement viruses which are defective in this protein. Therefore transgenics have been constructed which express mutant movement proteins. In a number of instances these have been shown to confer resistance. For example, tobacco plants expressing a dysfunctional movement protein of TMV were highly resistant to alfalfa mosaic virus, cauliflower mosaic virus and tobacco ringspot virus but were not resistant to cucumber mosaic virus or tobacco etch virus (Cooper *et al.*, 1995).

Nucleic acid mediated resistance

Antisense RNA technology to increase the resistance of plants to viruses was one of the first nucleic acid techniques to be tried. In some cases infection was inhibited to some extent but in others the effect was more limited or not detectable.

Other nucleic acid strategies have involved defective interfering (DI) RNAs and DNAs. These are produced during virus replication deleting some viral sequences but not those necessary for replication. As a result, the viral genome faces competition for replication by DI molecules causing depression of viral replication and symptom expression.

RNA suppression and gene silencing

Introduction of additional copies of genes in order to increase their expression often, paradoxically, results in decreased expression. One explanation of this phenomenon is that plants have a surveillance system that detects abnormally high expression of RNA sequences, possibly evolved as a defence against RNA virus infection. Thus transgenes with viral sequences may operate at this level to suppress virus replication. However, the phenomenon is not consistent and is unpredictable after the plant has undergone meiosis. Waterhouse, Graham and Wang (1998) have described experiments which not only give a satisfying explan-ation of the data already obtained but also point the way to reproducible exploitation of gene silencing as a way of controlling virus diseases. They showed that only 10 per cent of transgenic plants expressing the protease gene of Potato Virus Y in either the sense or antisense polarity gave resistance or immunity to PVY. However, if plants with the protease gene in the sense polarity were crossed with those in which the gene had been inserted in the antisense polarity, a total of 45 out of 200 plants were resistant or immune. Transcripts of both polarities were detected in the resistant plants. Moreover, when plants were transformed with tandem constructs containing the PVY protease gene in the sense and the anti-sense orientation about half were resistant or immune to the virus. Progeny of the resistant lines inherited resistance as a Mendelian character.

Waterhouse, Graham and Wang (1998) propose a model to explain their results (Figure 12.7). They suggest that sense and antisense mRNA form a duplex (step 1) which is recognized by an RNA polymerase associated with a helicase and an Rnase. The complex transcribes complementary RNA (cRNA), attaches it to the RNase and releases it from the complex (step 2). The cRNA-RNase molecules hybridize to the virus RNAs (step 3) and cleave the single-stranded regions adjacent to the hybrids (step 4) which are then degraded by other plant nucleases, giving rise to virus immunity (step 5).

One promising application of this technology is in the control of crown gall. Here it is proposed that transformation of plants with the genes that the patho-gen uses for the production of IAA and cytokinins would give durable resistance

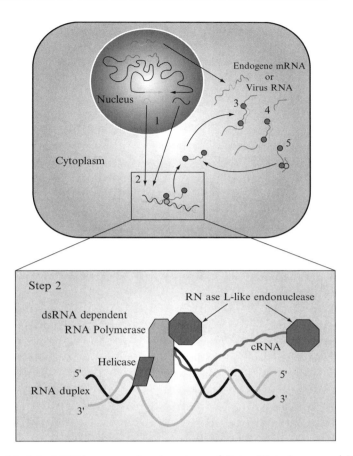

Figure 12.7 Model of RNA suppression (courtesy of Peter Waterhouse and Nature). For explanation see text

since these genes are responsible for the unorganized growth of tumours in infected plants (Figure 12.8). The metabolism of the plant's own hormones would not be affected since they are synthesized by an alternative route (see Figure 7.1; Escobar *et al.*, 2001).

12.7.9   Transformation with resistance genes

With the availability of cloned resistance genes there is now the opportunity to transfer them speedily into genotypes in which the gene is lacking. For example, there is no resistance to the nematode *Heterodera schachtii* in cultivated *Beta* species but resistance conferred by the gene *Hs1*[pro-1] is present in *B. procumbens*. The gene was introgressed into sugarbeet and cloned. Susceptible beet transformed with the gene under the control of the *CaMV35S* promoter gave an incompatible phenotype which was identical to that of resistant plants. It is likely

Figure 12.8 Control of crown gall by silencing of the genes used by the bacterium for the synthesis of IAA and cytokinins: (a) tomato transgene; (b) tomato control; (c) *Arabidopsis* transgene; (d) *Arabidopsis* control (courtesy of Escobar *et al.* and Proceedings of the National Academy of Sciences USA). A colour reproduction of this figure can be seen in the colour section

therefore that further varieties will be transformed with the gene in the future (Cai *et al.*, 1997).

The gene *Xa21* is another example of particular promise since it confers resistance of rice to 29 strains of an important pathogen, *Xanthomonas oryzae* pv. *oryzae*. The gene was cloned by Song *et al.* (1995) and susceptible plants were transformed biolistically; the gene has now been sent to many scientists throughout Europe, Africa, Asia and the USA for introduction into cultivars which are important locally (Figure 12.9). Unfortunately, however, the gene does not confer resistance to Korean strains of the pathogen (Wang *et al.*, 1996).

IRBB21   IR24   TP309   Xa21

Figure 12.9   Comparison of genotypes of rice resistant and susceptible to *Xanthomouas oryzae* pv. *oryzae* and a susceptible genotype transformed with the resistance gene *Xa21* (all plants were challenged with the pathogea); from left to right, IRBB21 (resistant), IR24 (susceptible), TP 309 (susceptible), and TP309 (transformed with *Xa21*) (courtesy of Pamela Ronald, University of California, Davis, USA). A colour reproduction of this figure can be seen in the colour section

Tang and co-workers (1999) overexpressed the *Pto* gene which confers resistance of tomato to *Pseudomonas syringae* pv. *tomato* strains with the corresponding avirulence gene *avrPto* in a gene-for-gene interaction (see Section 10.5.3). Microsocopic examination showed that the transgenes had small necrotic areas in the palisade mesophyll cells resembling the hypersensitive response induced by incompatible pathogens. Furthermore, the plants accumulated salicylic acid and increased levels of pathogenesis-related proteins and were more resistant to *Xanthomonas campestris* pv. *vesicatoria* and *Cladosporium fulvum*. It therefore seems that overexpression of a resistance gene in the absence of the corresponding avirulence gene of the pathogen can give a phenotype with broad-based resistance. How general this phenomenon is, and whether it has an acceptable cost in terms of yield of the plant, remains to be seen.

With the increasing availability of cloned resistance genes, many of the difficulties of conventional breeding may be circumvented (although, as described above, plant transformation has difficulties of its own). For instance, incompatibility of parents and allelism of R genes are no longer problems. With regard to the latter, the resistance of flax to its rust, *Melampsora lini*, has been broadened by expressing the *L2* allele in plants containing the *L6* allele (Ellis *et al.*, 1997). However, difficulties may be encountered in moving R genes into genomes that are too divergent from the donor since, although recognition of the appropriate avr product may occur, there may be problems in the transduction of the signal leading to the activation of defence responses.

Epilogue

Those new to the subject of plant pathology may have been surprised at the many different types and large numbers of pathogens that attack plants described in this book and also the devastation they cause. However, from an evolutionary perspective, this might be expected since plants are the primary producers of fixed carbon compounds and are therefore first in the food chain supplying the needs of all heterotrophs, including ourselves. An ability to parasitize the living plant puts an organism at a considerable advantage since it is in poll position in the competition to obtain nutrients from this primary food resource.

Plant pathogens are therefore our competitors and our enemies. 'Know your enemy' is a wise adage in any walk of life, particularly that of a plant pathologist! It is hoped that this book will have made a modest contribution to the greater understanding of the organisms that destroy our crops and vegetation and how we can develop the skills and knowledge to contain them – defeat is unlikely to be practicable. It is also important that this understanding is communicated to those who control budgets. Vague comments to the effect that there is a serious disease of a staple crop will not convince the increasingly ubiquitous accountants or the politician. They will want to know the extent of the damage as measured in quantity of crop destroyed or the loss in monetary terms. They may even wish to know the cause! Moreover, in a democracy, politicians might be expected to take an interest in how their constituents' lives are affected by plant diseases and what measures can be taken to combat them. Finally, those who control the purse strings must be convinced that the development of the skills and knowledge necessary to limit the ravages of plant disease comes with a price tag.

Pathogens have almost certainly exploited vegetation throughout evolutionary time. They have therefore had a long period in which to refine their mechanisms for tapping this resource. In contrast, mankind is a comparatively recent arrival. Clearly, if we are to contain these deadly enemies that threaten our food supplies, economics and environment, we have a lot of catching up to do!

Introduction to Plant Pathology by Richard Strange
© 2003 John Wiley & Sons, Ltd   ISBN 0 470 84972 X (cased)   ISBN 0 470 84973 8 (pbk)

Glossary

Abscission: the shedding of leaves, flowers or fruits.

Accession: a new member to a plant collection. In the context of plant pathology and plant breeding, accessions are often tested for disease resistance.

Actinomycetes: Gram-positive, irregularly staining filamentous bacteria with true branching that do not form spores and are non-motile.

Aecidiospores: one of the spore types of rust fungi.

Aetiology: the study of causation of disease.

Allele: (abbreviation of allelomorph) one of a series of possible alternative forms of a given gene differing in DNA sequence and affecting the function of a single product.

Amendments: the addition to soil of materials, usually with the aim of suppressing disease.

Anamorph: the imperfect (asexual) state of a fungus.

Antibody: a protein produced by lymphoid cells in response to foreign substances (antigens) and capable of coupling specifically with its homologous antigen (King and Stansfield, 1990).

monoclonal antibodies: immunoglobulins derived from a single clone of plasma cells. Since all immunoglobulins produced by a given plasma cell (or clone thereof) are chemically and structurally identical, these antibodies constitute a pure population with highly specific antigen-binding properties (King and Stansfield, 1990).

polyclonal antibodies: an antibody preparation that is the product of more than one clone of plasma cells. Such antibodies react with different components of the antigen (after King and Stansfield, 1990).

Anticodon: the triplet of nucleotides in a transfer RNA molecule which associates by complementary base pairing with a specific triplet (codon) in the messenger RNA molecule during its translation in the ribosome (King and Stansfield, 1990).

Antigen: a substance that, when injected into a vertebrate, stimulates the production of homologous antibodies.

Appressorium: a structure, formed from a fungal hypha and usually exceeding it in diameter, that serves to anchor the fungus to the host and may also assist in penetration by the exertion of mechanical force.

Basidiome: the fruiting body of a Basidiomycete.

Introduction to Plant Pathology by Richard Strange
© 2003 John Wiley & Sons, Ltd   ISBN 0 470 84972 X (cased)   ISBN 0 470 84973 8 (pbk)

Bacteriocin: proteins synthesized by bacterial species that are toxic when absorbed by bacteria belonging to sensitive strains (King and Stansfield, 1990).

Biocide: a compound that is toxic to living matter.

Biotinylated antibodies: antibodies that have been labelled with biotin. They may be recognized by avidin conjugates such as an enzyme and used in ELISA techniques.

Biotroph: an organism that in nature is entirely dependent upon another organism as a source of nutrients.

Biovar: a term used to designate subspecies of nitrogen fixing *Rhizobium* species that infect specific hosts.

bv: see biovar.

cDNA: a duplex DNA sequence complementary to an RNA molecule of interest, carried in a cloning vector (King and Stansfield, 1990).

Chlorosis: the failure to accumulate or maintain the normal complement of chlorophyll resulting in an abnormal yellow appearance.

Chemotaxis: the movement of a whole organism towards or away from a chemical stimulus.

Cognate: usually applied to resistance and avirulence genes in this book and meaning one or other of the avirulence-resistance gene pair that encode factors which allow the plant and pathogen to recognize each other. As a result of recognition a resistance reaction occurs.

Compatible: in plant pathology this term means that a pathogen is able to form a parasitic relationship with a host plant. Normally, this involves avoiding the resistance response of the host by various procedures. These may include delay or failure in triggering of the response or overcoming the response (e.g. phytoalexin degradation).

Cosmid: plasmid vectors designed for cloning large fragments of eukaryotic DNA. The vector contains *cos* site (= *co*hesive end sites) that enable it to be packaged in a phage coat *in vitro* (after King and Stansfield, 1990).

Cultivar: a variety of plant produced through selective breeding by humans and maintained by cultivation (King and Stansfield, 1990).

Differential cultivar: a cultivar that distinguishes between physiologic races of a pathogen.

Dikaryon: a fungal compartment (often called a cell but fungi are generally better regarded as being coenocytic) containing two genetically distinct haploid nuclei.

Dikaryotic: adjectival form of dikaryon.

Disease: the disturbance of the normal functions of an organism.

EC$_{50}$; EC$_{90}$: (1). the concentrations at which 50 or 90 per cent of a population reacts; (2). the concentration at which the reaction of an organism is 50 per cent or 90 per cent less than the control value.

Ecotype: a subspecific form within a true species.

Elicitor: a physical, chemical or biological stimulus that triggers defence responses in plants.

ELISA: enzyme linked immunosorbent assay.

Embolism: in plant pathology a term used to describe the disruption of water conduction, usually by an air-lock.

Epidemic: a widespread increase in disease of usually limited duration.

Epiphyte: an organism growing on a plant in a non-parasitic association.

Epinasty: the drooping of organs (usually leaves) of a plant but without loss of turgor.

Epistatic: the masking of the phenotype of a gene by that of another.

Epitope: the antigenic determinant on an antigen to which the paratope on an antibody binds.

Etiology: see aetiology.

Facultative parasite: see parasite.

Focus: a localized area of diseased plants which may serve as a source of inoculum.

Forma specialis: (usually abbreviated to f. sp.) a subspecific taxon usually denoting the host genus of which the organism is a parasite, e.g. *tritici* to denote wheat in *Puccinia graminis* f. sp. *tritici*.

Genetic pool: the total genetic information possessed by the reproductive members of a population of sexually reproducing organisms (King and Stansfield, 1990).

Geocarposphere: an area around the pod of a groundnut plant. It is usually used to describe the area in which the pod influences the microbial complement of the soil.

Genotype: the genetic constitution of an organism as distinguished from its physical appearance (its phenotype; King and Stansfield, 1990).

Gnotobiotic: adjective used to describe the condition of being free of microbial contamination, e.g. a plant that has been grown from a surface sterilized seed in a sterile environment.

Hemibiotroph: a plant pathogen that has a period of 'peaceful co-existence' with its host but later becomes necrogenic.

Haustorium: (1). a hyphal swelling that penetrates a host cell and invaginates the host plasmamembrane. In compatible associations, the plant cell is not killed; (2). a complex structure of parasitic angiosperms formed on the host that serves to anchor the parasite.

Hypha: one of the filaments of a fungal mycelium.

Horizontal gene transfer: the transfer of genes from one organism to another (usually a different organism) without sexual reproduction.

Hydathode: an epidermal structure in plants specialized for the secretion or exudation of water (Lawrence, 2000).

Hyphomycetes: fungi that bear conidia free on the mycelium.

Hypovirulence: the phenomenon by which a strain of a pathogen is less virulent than normal. In some fungi infectious dsRNA is thought to be responsible.

Incidence: disease incidence is the proportion of diseased entities within a sampling unit, e.g. the percentage of plants with at least one lesion.

Incompatible: in plant pathology this term means that a pathogen is unable to form a parasitic relationship with a host plant usually since it has triggered the resistance responses of the host.

Infection structures: a term used to describe the germ tube, appressorium, infection peg, substomatal vesicle and infection hyphae of rust fungi (Figure 7.6).

Infection type: the physical appearance of a parasitic lesion on a host plant. These are often classified as high or low and there may be a numerical system to enable further distinctions to be made (cf. Table 2.1).

Inoculum: a source of living material that may cause infection of plants or may grow on a culture medium.

Internal Transcribed Spacers: the DNA lying between ribosomal genes. In eukaryotes, such as fungi, these are ITS1, lying between the 18S and 5.8S gene, and ITS2, lying between the 5.8S gene and the 28S gene.

Isogenic: organisms of nearly identical genotype but differing in one crucial character, e.g. in the case of a pathogen, avirulence to a specific genotype of plant or, in the case of a plant, resistance to a specific genotype of pathogen.

Isolines: plants of nearly identical genotype but differing in one crucial character, e.g. resistance to a specific pathogen.

LD_{50}: the dose of a poisonous compound that causes death of 50 per cent of a population.

Lenticel: a pore in the periderm of trees and shrubs, allowing the passage of air to internal tissues (Lawrence, 2000).

Line: host plants of uniform appearance, the stability of which is maintained by selection.

Mating type: a group within a species classified on the basis of its mating behaviour. Fungi often have genes that control mating type at one, two or even three loci and there may be multiple alleles at each locus. In heterothallic species, individuals must differ in mating type alleles at each locus for mating to occur.

Meristem: the undifferentiated, mitotically active tissues of plants. The meristems at the tips of roots and shoots are referred to as apical meristems (King and Stansfield, 1990).

Microfilament: protein threads composed of globular actin subunits; a component of the cytoskeleton of eukaryotes.

Microtubule: a fine hollow protein tube composed of tubulin; a component of the cytoskeleton of eukaryotes.

Monoculture: the growing of a single-crop species continuously or in successive seasons over a wide area.

Monocyclic: see pathogen

Multiline: a cultivar made up of a mixture of isolines that differ by single major genes for resistance to a pathogen.

Mutation: (1). the process by which a gene undergoes structural change; (2). a modified gene resulting from mutation (King and Stansfield, 1990).

Mutant: an organism bearing a mutation.

Mycelium: a mass of fungal hyphae.

Mycoplasma: a bacterial genus characterized by the lack of a cell wall (King and Stansfield, 1990).

Mycotoxin: a compound formed by a fungus that is toxic to man or animals.

Necrogenic: causing necrosis.

Necrosis: the phenomenon of death usually applied to cells in a localized area.

Necrotroph: a fungus that kills host tissues in advance of growing through them.

Nectary: a gland secreting a sweet-tasting liquid usually found in flowers but also in some leaves

Nick translation: an *in vitro* procedure used to label radioactively a DNA of interest uniformly to a high specific activity. First, nicks are introduced into the unlabelled DNA by an endonuclease, generating 3' hydroxyl termini. *E. coli* DNA polymerase I is then used to add radioactive residues to the 3' hydroxy terminus of the nick, with the concomitant removal of the nucleotides from the 5' side. The resultant is an identical DNA molecule with the nick displaced further along the duplex (King and Stansfield, 1990).

Open reading frame (ORF): a DNA sequence lying between start and stop codons which is capable of transcription.

Orthologue: a gene that is homologous with one from another organism because they are derived from a common ancestral gene.

Paralogue: a gene that is homologous with one from the same organism because it has been derived from gene duplication.

Parasite: see parasitism.

    facultative: a parasite that may also exist as a saprophyte.

    obligate: a parasite that normally passes its entire life cycle in association with its host.

Parasitism: an association of two organisms in which one member, the parasite, benefits and the other (the host) is harmed (after King and Stansfield, 1990).

Paratope: the site within an immunoglobulin Fab that specifically interacts with an antigenic determinant (epitope) (King and Stansfield, 1990).

Pathogen: an organism that causes disease.

    monocyclic: one that is restricted to a single generation per cropping season.

    polycyclic: one that has more than one generation and often many generations per cropping season.

    polyetic: one in which the inoculum does not increase during a single growing season but does increase from season to season.

Pathogenicity: the ability to cause disease.

Pathovar: (usually abbreviated to pv.) a subspecific taxon of bacterial pathogens of plants usually denoting the host genus of which the organism is a parasite, e.g. *malvacearum* denotes cotton (a member of the Malvaceae) in *Xanthomonas campestris* pv. *malvacearum*.

PCR: see polymerase chain reaction.

Periderm: a three-layered tissue that replaces the epidermis in most stems and roots having secondary growth. It consists of the cork cambium which cuts off the living phelloderm on the inner surface and cork on the outer surface which is non-living at maturity.

Phenotype: the observable properties of an organism, produced by the genotype (q.v.) in conjunction with the environment.

Pheromone: a chemical perceived by an organism that affects its behaviour. Examples are alarm substances and sex attractants.

Physiologic race: a taxon of a parasite within a uniform morphologic group that differs from other members of the group in virulence towards genotypes of the host plant.

Plasmid: a small, autonomously replicating molecule of covalently closed circular DNA which is devoid of protein and which is not essential for the survival of its host.

Plasmodesmata: cytoplasmic threads running transversely through plant cell walls and connecting the cytoplasm of adjacent cells (Lawrence, 2000)

Pleiotropy: the phenomenon by which a single gene has more than one phenotypic expression.

Pleiotropic: adjectival form of pleiotropy (q.v.).

Polycyclic: see pathogen.

Polyetic: see pathogen.

Polydisperse: a colloidal sol in which the particle sizes vary within wide limits.

Polymerase chain reaction (PCR): see Section 2.5.5.

Predator: an individual of one species that attacks, kills and feeds off another.

Promoter: a region of a DNA molecule to which an RNA polymerase binds and initiates transcription.

Pv.: see pathovar.

Quarantine: regulation of the movement of living plants or their parts and plant products across political boundaries. Quarantine procedures include legislation, inspection, treatment, certification and international cooperation.

Race: see physiologic race.

Recombination: the occurrence of progeny with combinations of genes other than those that occurred in the parents, owing to independent assortment or crossing over (after King and Stansfield, 1990).

Regeneration: the process of cultivating an intact plant from tissue culture.

Resistance: the retardation of infection and growth of a parasite on or within host tissues.

    durable: a term used to describe resistance that is seldom or never circumvented by a given pathogen of the plant species under consideration.

    horizontal: resistance that operates against many races of a pathogen.

    major gene: confers a high level of resistance.

    minor gene: confers a low level of resistance but it is thought that minor genes may, when aggregated in a single genotype, give a high level of resistance that is durable (q.v.).

    race specific: see vertical resistance.

    vertical: a high level of resistance but one that may be circumvented by the pathogen and is therefore often race specific.

Restriction enzyme: an enzyme that cleaves DNA molecules at specific recognition sites (after King and Stansfield, 1990).

Rhizomorph: an aggregation of hyphae resembling a root and having a well-defined apical meristem; often differentiated into a rind of small dark cells surrounding a core of elongate colourless cells.

Rhizosphere: a region around the root of a plant, usually supporting a quantitatively different microflora from the non-rhizosphere soil.

Roguing: the removal of (usually diseased) plants from a growing crop.

Saprophyte: an organism that obtains nutrients from dead organic matter.

Sclerotium: a firm, frequently rounded mass of hyphal tissue, with or without host tissue, and bearing no spores (Holliday, 1989).

Serology: the study of the nature, production and interactions of antibodies and antigens.

Serotype: an antigenic property of a cell identified by serological methods (King and Stansfield, 1990).

Severity: disease severity is a measure of the amount of disease in a plant, e.g. the percentage leaf area occupied by lesions of a pathogen.

Siderophore: low molecular weight, virtually Fe(III)-specific ligands produced as scavenging agents in order to combat low iron stress (Neilands and Leong, 1986).

Signal peptide: a hydrophobic sequence of about 15 amino acids at the N-terminal end of secretory proteins and certain membrane proteins which is essential for their passage across or into the cell membrane (in bacteria) or the endoplasmic reticulum (in eukaryotic cells; Lawrence, 2000)

Sporodochium: a fungal structure consisting of spores and their supporting specialized hyphae.

Stele: column of primary vascular tissue consisting of xylem and phloem.

Sympatric: species inhabiting the same or overlapping geographic areas.

Teleutospore: the over-wintering spore stage produced by certain rust fungi, e.g. *Puccinia graminis* f. sp. *tritici*.

Teleomorph: the perfect (sexual) state of a fungus.

Terminator: a nucleotide sequence in DNA that causes the mRNA to detach and thus functions to stop transcription.

Totipotency: the capacity of a cell to differentiate into all of the cells of the adult organism.

Transmission: the transfer of an infectious agent (usually a virus) from one plant to another.

    horizontal: transmission between plants that are contemporaneous with each other.

    vertical: transmission to progeny.

Transposable element: a DNA sequence that can move from one chromosomal site to another (after King and Stansfield, 1990).

Transposon: a class of transposable element with characteristic flanking sequences.

Transposon mutagenesis: the production of mutants by the insertion of a transposon in a gene, thus inactivating it.

Trichome: an outgrowth of the epidermis of plants which may take any one of many forms such as branched and unbranched hairs, stinging hairs, vesicles, hooks or spines (Lawrence, 2000).

Uredinium: a lesion caused by a rust fungus bearing uredospores.

Uredium: see uredinium.

Uredospore: one of the spore types produced by rust fungi. For example, in *Puccinia graminis* f.sp. *tritici*, the cause of stem rust of wheat, it is a repeating spore stage which can quickly allow the build-up of damaging epidemics.

Vector: (1). an organism that carries and transmits a pathogen to a plant; (2). a segment of DNA that enables genes to be transferred to other organisms – the Ti plasmid of *Agrobacterium tumefaciens* may be used as a vector for transferring genes into plants (Chapter 12).

Virulence: a measure of the pathogenicity of a parasite.

Viruliferous: infected with a virus.

Volunteer: a self-sown individual of a crop plant. Volunteers when present out of the normal growing season can provide a means by which a parasite survives until the host crop is planted once more.

Water-soaked: a term used to describe a disease symptom in which an area of plant cells takes on a darker colour owing to the filling of intercellular air spaces with cell sap; sometimes indicative of the action of a toxin, particular if the symptoms are at a distance from the invading organism.

References

Abbas, H. K. *et al.*, 1999, Fumonisin B-1 from the fungus *Fusarium moniliforme* causes contact toxicity in plants: Evidence from studies with biosynthetically labeled toxin. *Journal of Natural Toxins*, **8**, 405–420.

Abdallah, R. and Black, R., 1998, Safe movement of germplasm between countries, in *Conservation and Utilization of African Plants*, R. P. Adams & J. E. Adams, eds., Missouri Botanical Garden Press, St Louis, USA, pp. 223–228.

Acuna, I. A. *et al.*, 2001, Glucosylation as a mechanism of resistance to thaxtomin A in potatoes *Plant Science*, **161**, 77–88.

Adams, P. B., 1990, The potential of mycoparasites for biological-control of plant-diseases. *Annual Review of Phytopathology*, **28**, 59–72.

Adaskaveg, J. E. and Hartin, R. J., 1997, Characterization of *Colletotrichum acutatum* isolates causing anthracnose of almond and peach in California. *Phytopathology*, **87**, 979–987.

Adhikari, T. B. *et al.*, 1999, Use of partial host resistance in the management of bacterial blight of rice. *Plant Disease*, **83**, 896–901.

Afek, U. and Sztejnberg, A., 1993, Temperature and gamma-irradiation effects on scoparone, a citrus phytoalexin conferring resistance to *Phytophthora citrophthora*. *Phytopathology*, **83**, 753–758.

Agrios, G. N., 1997, *Plant Pathology*, Fourth edition, Academic Press, San Diego, USA.

Ahn, J. H. and Walton, J. D., 1997, A fatty acid synthase gene in *Cochliobolus carbonum* required for production of HC-toxin, cyclo(D-prolyl-L-alanyl-D-alanyl-L-2-Amino-9,10-epoxi-8-oxodecanoyl). *Molecular Plant-Microbe Interactions*, **10**, 207–214.

Ahn, J. H. and Walton, J. D., 1998, Regulation of cyclic peptide biosynthesis and pathogenicity in *Cochliobolus carbonum* by TOXEp, a novel protein with a bZIP basic DNA-binding motif and four ankyrin repeats. *Molecular and General Genetics*, **260**, 462–469.

Alam, S. S. *et al.*, 1989, Chickpea blight – production of the phytotoxins solanapyrone A and C by *Ascochyta rabiei*. *Phytochemistry*, **28**, 2627–2630.

Alavi, A., Strange, R. N. and Wright, G., 1982, The relative susceptibility of some Cucurbits to an Iranian isolate of *Phytophthora drechsleri*. *Plant Pathology*, **31**, 221–227.

Alexander, N. J., McCormick, S. P. and Hohn, T. M., 1999, TRI12, a trichothecene efflux pump from *Fusarium sporotrichioides*: gene isolation and expression in yeast. *Molecular and General Genetics*, **261**, 977–984.

Alexopoulos, C. J., 1952, *Introductory Mycology*, John Wiley, New York USA.

Ali, M. K., Lepoivre, P. and Semal, J., 1993, Scoparone eliciting activity released by phosphonic acid treatment of *Phytophthora citrophthora* mycelia mimics the

Introduction to Plant Pathology by Richard Strange
© 2003 John Wiley & Sons, Ltd   ISBN 0 470 84972 X (cased)   ISBN 0 470 84973 8 (pbk)

incompatible response of phosphonic acid-treated citrus leaves inoculated with this fungus. *Plant Science*, **93**, 55–61.

Allan, R. H., Thorpe, C. J. and Deacon, J. W., 1992, Differential tropism to living and dead cereal root hairs by the biocontrol fungus *Idriella bolleyi*. *Physiological and Molecular Plant Pathology*, **41**, 217–226.

Allen, E. A. *et al.*, 1991, Appressorium formation in response to topographical signals by 27 rust species. *Phytopathology*, **81**, 323–331.

Allen, P. J., 1971, Specificity of the *cis*-isomers of inhibitors of uredospore germination in the rust fungi. *Proceedings of the National Academy of Sciences of the United States of America*, **69**, 3497–3500.

Allen, W. R. and Matteoni, J. A., 1991, Petunia as an indicator plant for use by growers to monitor for thrips carrying the tomato spotted wilt virus in greenhouses. *Plant Disease*, **75**, 78–82.

Al Rawahi, A. K. and Hancock, J. G., 1998, Parasitism and biological control of *Verticillium dahliae* by *Pythium oligandrum*. *Plant Disease*, **2**, 1100–1106.

Amin, M., Kurosaki, F. and Nishi, A., 1986, Extracellular pectinolytic enzymes of fungi elicit phytoalexin accumulation in carrot suspension-culture. *Journal of General Microbiology*, **132**, 771–777.

Amin, M., Kurosaki, F. and Nishi, A., 1988, Carrot phytoalexin alters the membrane-permeability of *Candida albicans* and multilamellar liposomes. *Journal of General Microbiology*, **134**, 241–246.

Amin, P. W., 1985, Apparent resistance of groundnut cultivar Robut 33-1 to bud necrosis disease. *Plant Disease*, **69**, 718–719.

Anagnostakis, S. L., 1987, Chestnut blight – the classical problem of an introduced pathogen. *Mycologia*, **79**, 23–37.

Andebrhan, T. *et al.*, 1995, Sensitivity of *Crinipellis perniciosa* to procyanidins from *Theobroma cacao* L.. *Physiological and Molecular Plant Pathology*, **46**, 339–348.

Anguelova, V. S., van der Westhuizen, A. J. and Pretorius, Z. A., 1999, Intercellular proteins and beta-1,3-glucanase activity associated with leaf rust resistance in wheat. *Physiologia Plantarum*, **106**, 393–401.

Annan, I. B. *et al.*, 2000, Stylet penetration activities by *Aphis craccivora* (Homoptera: Aphididae) on plants and excised plant parts of resistant and susceptible cultivars of cowpea (Leguminosae). *Annals of the Entomological Society of America*, **93**, 133–140.

Annis, S. L. and Goodwin, P. H., 1997, Production and regulation of polygalacturonase isozymes in Canadian isolates of *Leptosphaeria maculans* differing in virulence. *Canadian Journal of Plant Pathology-Revue Canadienne de Phytopathologie*, **19**, 358–365.

Antoniou, P. P., Tjamos, E. C. and Panagopoulos, C. G., 1995, Use of soil solarization for controlling bacterial canker of tomato in plastic houses in Greece. *Plant Pathology*, **44**, 438–447.

Ardi, R. *et al.*, 1998, Involvement of epicatechin biosynthesis in the activation of the mechanism of resistance of avocado fruits to *Colletotrichum gloeosporioides*. *Physiological and Molecular Plant Pathology*, **53**, 269–285.

Armentrout, V. N. and Downer, A. J., 1987, Infection cushion development by *Rhizoctonia solani* on cotton. *Phytopathology*, **77**, 619–623.

Armentrout, V. N. *et al.*, 1987, Factors affecting infection cushion development by *Rhizoctonia solani* on cotton. *Phytopathology*, **77**, 623–630.

Ashby, A. M., Watson, M. D. and Shaw, C. H., 1987, A Ti-plasmid determined function is responsible for chemotaxis of *Agrobacterium tumefaciens* towards the plant wound product acetosyringone. *FEMS Microbiology Letters*, **41**, 189–192.

Ashby, A. M. *et al.*, 1988, Ti plasmid-specified chemotaxis of *Agrobacterium tumefaciens* c58c1 toward *vir*-inducing phenolic-compounds and soluble factors from monocotyledonous and dicotyledonous plants. *Journal of Bacteriology*, **170**, 4181–4187.

Assef, G. M., Assari, K. and Vincent, E. J., 1986, Occurrence of an antifungal principle in the root extract of a bayoud – resistant date palm cultivar. *Netherlands Journal of Plant Pathology*, **92**, 43–47.

Assefa, H., van den Bosch, F. and Zadoks, J. C., 1995, Focus expansion of bean rust in cultivar mixtures. *Plant Pathology*, **44**, 503–509.

Astua-Monge, G. *et al.*, 2000 Xv4-vrxv4: A new gene-for-gene interaction identified between *Xanthomonas campestris* pv. *vesicatoria* race T3 and the wild tomato relative *Lycopersicon pennellii*. *Molecular Plant-Microbe Interactions*, **13**, 1346–1355.

Avni, A. *et al.*, 1992, Tentoxin sensitivity of chloroplasts determined by codon-83 of beta-subunit of proton-ATPase. *Science*, **257**, 1245–1247.

Aylor, D. E., 1990, The role of intermittent wind in the dispersal of fungal pathogens. *Annual Review of Phytopathology*, **28**, 73–92.

Bagga, S. and Straney, D., 2000, Modulation of cAMP and phosphodiesterase activity by flavonoids which induce spore germination of *Nectria haematococca* MP VI (*Fusarium solani*). *Physiological and Molecular Plant Pathology*, **56**, 51–61.

Baidyaroy, D. *et al.*, 2000, Transmissible mitochondrial hypovirulence in a natural population of *Cryphonectria parasitica*. *Molecular Plant-Microbe Interactions*, **13**, 88–95.

Bakke, A. and Lie, R., 1991, Mass trapping, in *Insect Pheromones in Plant Protection*, A. Bakke, ed., John Wiley and Son, Chichester, UK pp. 67–87.

Baldauf, S. L. *et al.*, 2000, A kingdom-level phylogeny of eukaryotes based on combined protein data. *Science*, **290**, 972–977.

Balesdent, M. H. *et al.*, 1998, Conidia as a substrate for internal transcribed spacer-based PCR identification of members of the *Leptosphaeria maculans* species complex. *Phytopathology*, **88**, 1210–1217.

Ball, A. M. *et al.*, 1991, Evidence for the requirement of extracellular protease in the pathogenic interaction of *Pyrenopeziza brassicae* with oilseed rape. *Physiological and Molecular Plant Pathology*, **38**, 147–161.

Ball, E. M. *et al.*, 1990, 'Polyclonal antibodies' in *Serological Methods for the Detection and Identification of Viral and Bacterial Plant Pathogens*, E. M. Ball, R. O. Hampton, & S. H. De Boer, eds., American Phytopathological Society, St Paul, Minnesota, USA, pp. 33–54.

Bandyopadhyay, R. *et al.*, 1998, Ergot: A new disease threat to sorghum in the Americas and Australia. *Plant Disease*, **82**, 356–367.

Barber, C. E. *et al.*, 1997, A novel regulatory system required for pathogenicity of *Xanthomonas campestris* is mediated by a small diffusible signal molecule. *Molecular Microbiology*, **24**, 555–566.

Barber, M. S. and Ride, J. P., 1988, A quantitative assay for induced lignification in wounded wheat leaves and its use to survey potential elicitors of the response. *Physiological and Molecular Plant Pathology*, **32**, 185–197.

Barber, M. S., Bertram, R. E. and Ride, J. P., 1989, Chitin oligosaccharides elicit lignification in wounded wheat leaves. *Physiological and Molecular Plant Pathology*, **34**, 3–12.

Barker, H., 1987, Invasion of non-phloem tissue in *Nicotiana clevelandii* by potato leafroll luteovirus is enhanced in plants also infected with Potato Y Potyvirus. *Journal of General Virology*, **68**, 1223–1227.

Barnes, I. *et al.*, 2001, Characterization of *Seiridium* spp. associated with cypress canker based on beta-tubulin and histone sequences. *Plant Disease*, **85**, 317–321.

Bartel, B., 1997, Auxin biosynthesis. *Annual Review of Plant Physiology and Plant Molecular Biology*, **48**, 49–64.

Barzic, M. R., 1999, Persicomycin production by strains of *Pseudomonas syringae* pv. *persicae*. *Physiological and Molecular Plant Pathology*, **55**, 243–250.

Bauske, E. M., Bissonnette, S. M. and Hewings, A. D., 1997, Yield loss assessment of barley yellow dwarf disease on spring oat in Illinois. *Plant Disease*, **81**, 485–488.

Bayles, C. J., Ghemawat, M. S. and Aist, J. R., 1990, Inhibition by 2-deoxy-D-glucose of callose formation, papilla deposition, and resistance to powdery mildew in an *Mlo* barley mutant. *Physiological and Molecular Plant Pathology*, **36**, 63–72.

Bayles, R. A., Clarkson, J. D. S. and Slater, S. E., 1997, 'The UK cereal pathogen virulence survey,' in *The gene for gene relationship in plant-parasite interactions*, I. R. Crute, E. B. Holub, & J. J. Burdon, eds., CAB International, Wallingford, Oxon, UK, pp. 103–118.

Beachy, R. N., 1997, Mechanisms and applications of pathogen-derived resistance in transgenic plants. *Current Opinion in Biotechnology*, **8**, 215–220.

Beardmore, J., Ride, J. P. and Granger, J. W., 1983, Cellular lignification as a factor in the hypersensitive resistance of wheat to stem rust *Physiological Plant Pathology*, **22**, 209.

Beausejour, J. *et al.*, 1999, Production of thaxtomin A by *Streptomyces scabies* strains in plant extract containing media. *Canadian Journal of Microbiology*, **45**, 764–768.

Becker, J. O. and Cook, R. J., 1988, Role of siderophores in suppression of *Pythium* species and production of increased-growth response of wheat by fluorescent Pseudomonads. *Phytopathology*, **78**, 778–782.

Beer, S. V. *et al.*, 1991, 'The *hrp* gene cluster of *Erwinia amylovora*,' in *Advances in the Genetics of Plant-microbe Interactions*, H. Hennecke & D. P. S. Verma, eds., Kluwer, Dordrecht, The Netherlands, pp. 53–60.

Beffa, R. S. *et al.*, 1996, Decreased susceptibility to viral disease of beta-1,3-glucanase-deficient plants generated by antisense transformation. *Plant Cell*, **8**, 1001–1011.

Bélair, G. and Boivin, G., 1988, Spatial pattern and sequential sampling plan for *Meloidogyne hapla* in muck-grown carrots. *Phytopathology*, **78**, 604–607.

Bell, A. A. and Wheeler, M. H., 1986, Biosynthesis and functions of fungal melanins. *Annual Review of Phytopathology*, **24**, 411–451.

Bellotti, A. C., Smith, L. and Lapointe, S. L., 1999, Recent advances in cassava pest management. *Annual Review of Entomology*, **44**, 343–370.

Bendahmane, M. and Beachy, R. N., 1999, Control of tobamovirus infections via pathogen-derived resistance. *Advances in Virus Research*, **53**, 369–386.

Bender, C. L., 1998, Bacterial phytotoxins. *Methods in Microbiology*, **27**, 169–175.

Bender, C. L., 1999, Chlorosis-inducing phytotoxins produced by *Pseudomonas syringae*. *European Journal of Plant Pathology*, **105**, 1–12.

Bender, C. L., Alarcon-Chaidez, F. and Gross, D. C., 1999, Pseudomonas syringae phytotoxins: Mode of action, regulation, and biosynthesis by peptide and polyketide synthetases. *Microbiology and Molecular Biology Reviews*, **63**, 266–292.

Bender, C. L., Malvick, D. K. and Mitchell, R. E., 1989, Plasmid-mediated production of the phytotoxin coronatine in *Pseudomonas syringae* pv. *tomato*. *Journal of Bacteriology*, **171**, 807–812.

Benhamou, N., Mazau, D. and Esquerré-Tugayé, M. T., 1990, Immunocytochemical localization of hydroxyproline-rich glycoproteins in tomato root-cells infected by *Fusarium oxysporum* f. sp. *radicis-lycopersici* – study of a compatible interaction. *Phytopathology*, **80**, 163–173.

Benhamou, N. *et al.*, 1990, Immunogold localization of hydroxyproline-rich glycoproteins in necrotic tissue of *Nicotiana tabacum* L cv. xanthi-nc infected by tobacco mosaic-virus. *Physiological and Molecular Plant Pathology*, **36**, 129–145.

Bennett, R. N. and Wallsgrove, R. M., 1994, Secondary metabolities in plant defense-mechanisms. *New Phytologist*, **127**, 617–633.

Benson, D. M., 1995, Aluminum amendment of potting mixes for control of *Phytophthora* damping-off in bedding plants. *Hortscience*, **30**, 1413–1416.

Bent, A. F. *et al.*, 1994, *Rps2* of *Arabidopsis-thaliana* – a leucine-rich repeat class of plant-disease resistance genes. *Science*, **265**, 1856–1860.

Bergamin, A. *et al.*, 1997, Angular leaf spot of *Phaseolus* beans: Relationships between disease, healthy leaf area, and yield. *Phytopathology*, **87**, 506–515.

Berger, R. D., Bergamin, A. and Amorim, L., 1997, Lesion expansion as an epidemic component. *Phytopathology*, **87**, 1005–1013.

Bermingham, S., Maltby, L. and Cooke, R. C., 1995, A critical-assessment of the validity of ergosterol as an indicator of fungal biomass. *Mycological Research*, **99**, 479–484.

Bernards, M. A. and Lewis, N. G., 1998, The macromolecular aromatic domain in suberized tissue: A changing paradigm. *Phytochemistry*, **47**, 915–933.

Berner, D. K., Schaad, N. W. and Volksch, B., 1999, Use of ethylene-producing bacteria for stimulation of *Striga* spp. seed germination. *Biological Control*, **15**, 274–282.

Bernhard, R. H., Jensen, J. E. and Andreasen, C., 1998, Prediction of yield loss caused by *Orobanche* spp. in carrot and pea crops based on the soil seedbank. *Weed Research*, **38**, 191–197.

Bertaccini, A. *et al.*, 1990, Detection of chrysanthemum yellows mycoplasmalike organism by dot hybridization and Southern blot analysis. *Plant Disease*, **74**, 40–43.

Berto, P. *et al.*, 1999, Occurrence of a lipase in spores of *Alternaria brassicicola* with a crucial role in the infection of cauliflower leaves. *FEMS Microbiology Letters*, **180**, 183–189.

Bertrand, H., 2000, Role of mitochondrial DNA in the senescence and hypovirulence of fungi and potential for plant disease control. *Annual Review of Phytopathology*, **38**, 397–422.

Bestwick, C. S., Bennett, M. H. and Mansfield, J. W., 1995, *Hrp* Mutant of *Pseudomonas syringae* pv *phaseolicola* induces cell-wall alterations but not membrane damage leading to the hypersensitive reaction in lettuce. *Plant Physiology*, **108**, 503–516.

Biffen, R. H., 1905, Mendel's laws of inheritance and wheat breeding. *Journal of Agricultural Science*, **1**, 4–48.

Biffen, R. H., 1912, Studies in inheritance in disease resistance II. *Journal of Agricultural Science*, **4**, 421–429.

Biggs, A. R., 1989, Temporal changes in the infection court after wounding of peach bark and their association with cultivar variation in infection by *Leucostoma persoonii*. *Phytopathology*, **79**, 627–630.

Birch, R. G., 1997, Plant transformation: Problems and strategies for practical application. *Annual Review of Plant Physiology and Plant Molecular Biology*, **48**, 297–326.

Bircher, U. and Hohl, H. R., 1997, Environmental signalling during induction of appressorium formation in *Phytophthora*. *Mycological Research*, **101**, 395–402.

Bird, P. M. and Ride, J. P., 1981, The resistance of wheat to *Septoria nodorum* – fungal development in relation to host lignification. *Physiological Plant Pathology*, **19**, 289.

Bisgrove, S. R. *et al.*, 1994, A disease resistance gene in *Arabidopsis* with specificity for 2 different pathogen avirulence genes. *Plant Cell*, **6**, 927–933.

Black, L. M. and Brakke, M. K., 1952, Multiplication of wound tumour virus in an insect vector. *Phytopathology*, **42**, 269–273.

Black, R., 2000 Can phytosanitary services in African countries meet the challenges of globalisation?. pp. 1167–1174. Proceedings of BCPC Pests & Diseases Conference 2000

Blair, D. F., 1995, How bacteria sense and swim. *Annual Review of Microbiology*, **49**, 489–522.

Blaker, N. S. and MacDonald, J. D., 1983, Influence of container medium pH on sporangium formation, zoospore release, and infection of rhododendron by *Phytophthora cinnamomi*. *Plant Disease*, **67**, 259–263.

Blatt, M. R. *et al.*, 1999, K+ channels of Cf-9 transgenic tobacco guard cells as targets for *Cladosporium fulvum* Avr9 elicitor-dependent signal transduction. *Plant Journal*, **19**, 453–462.

Bock, K. R., 1982, The identification and partial characterization of plant-viruses in the tropics. *Tropical Pest Management*, **28**, 399–411.

Bogdanove, A. J., Bauer, D. W., and Beer, S. V., 1998, *Erwinia amylovora* secretes DspE, a pathogenicity factor and functional *AvrE* homolog, through the *hrp* (type III secretion) pathway. *Journal of Bacteriology*, **180**, 2244–2247.

Bogdanove, A. J. *et al.*, 1996, Unified nomenclature for broadly conserved *hrp* genes of phytopathogenic bacteria. *Molecular Microbiology*, **20**, 681–683.

Bohlmann, H. and Apel, K., 1991, Thionins. *Annual Review of Plant Physiology and Plant Molecular Biology*, **42**, 227–240.

Bohlmann, H. *et al.*, 1988, Leaf-specific thionins of barley – a novel class of cell-wall proteins toxic to plant-pathogenic fungi and possibly involved in the defense-mechanism of plants. *EMBO Journal*, **7**, 1559–1565.

Boissy, G. *et al.*, 1996, Crystal structure of a fungal elicitor secreted by *Phytophthora cryptogea*, a member of a novel class of plant necrotic proteins. *Structure*, **4**, 1429–1439.

Boller, T., 1991, 'Ethylene in pathogenesis and disease resistance,' in *The Plant Hormone Ethylene*, A. K. Mattoo & J. C. Suttle, eds., CRC Press, Boca Raton, Florida, USA, pp. 293–314.

Boller, T., 1995, Chemoperception of microbial signals in plant cells. *Annual Review of Plant Physiology and Plant Molecular Biology*, **46**, 189–214.

Bolwell, G. P. and Wojtaszek, P., 1997, Mechanisms for the generation of reactive oxygen species in plant defence – a broad perspective. *Physiological and Molecular Plant Pathology*, **51**, 347–366.

Bolwell, G. P. *et al.*, 1998, Comparative biochemistry of the oxidative burst produced by rose and French bean cells reveals two distinct mechanisms. *Plant Physiology*, **116**, 1379–1385.

Bolwell, G. P. *et al.*, 2002, The apoplastic oxidative burst in response to biotic stress in plants: a three-component system. *Journal of Experimental Botany*, **53**, 1367–1376.

Bonants, P. *et al.*, 1997, Detection and identification of *Phytophthora fragariae* Hickman by the polymerase chain reaction. *European Journal of Plant Pathology*, **103**, 345–355.

Bonas, U. and Lahayé, T., 2002, Plant disease resistance triggered by pathogen-derived molecules: refined models of specific recognition. *Current Opinion in Microbiology*, **5**, 44–50.

Bonde, M. R., Micales, J. A. and Peterson, G. L., 1993, The use of isozyme analysis for identification of plant-pathogenic fungi. *Plant Disease*, **77**, 961–968.

Bordoloi, G. N. *et al.*, 2002, Potential of a novel antibiotic, 2-methylheptyl isonicotinate, as a biocontrol agent against Fusarial wilt of crucifers. *Pest Management Science*, **58**, 297–302.

Bostock, R. M., Kuć, J. A. and Laine, R. A., 1981, Eicosapentaenoic and arachidonic acids from *Phytophthora infestans* elicit fungitoxic sesquiterpenes in the potato. *Science*, **212**, 67–69.

Bostock, R. M. *et al.*, 1999, Suppression of *Monilinia fructicola* cutinase production by peach fruit surface phenolic acids. *Physiological and Molecular Plant Pathology*, **54**, 37–50.

Bouillant, M. L. *et al.*, 1994, Identification of 5-(12-heptadecenyl)-resorcinol in rice root exudates. *Phytochemistry*, **35**, 769–771.

Boulter, D. *et al.*, 1990, Genetic-engineering of plants for insect resistance. *Endeavour*, **14**, 185–190.

Bowden, J., Gregory, P. H. and Johnson, C. G., 1971, Possible wind transport of coffee leaf rust across the Atlantic Ocean. *Nature*, **229**, 500–501.

Bowers, J. H., Sonoda, R. M. and Mitchell, D. J., 1990, Path coefficient analysis of the effect of rainfall variables on the epidemiology of *Phytophthora* blight of pepper caused by *Phytophthora capsici*. *Phytopathology*, **80**, 1439–1446.

Bowers, W. S., Evans, P. H. and Katayama, M., 1987, Synthetic optimization of laetisaric acid for fungicidal activity. *Journal of Agricultural and Food Chemistry*, **35**, 1043–1046.

Bowers, W. S. *et al.*, 1986, Thallophytic allelopathy – isolation and identification of laetisaric acid. *Science*, **232**, 105–106.

Bowles, D. J., 1990, Defense-related proteins in higher-plants. *Annual Review of Biochemistry*, **59**, 873–907.

Bowyer, P. *et al.*, 1995, Host-range of a plant-pathogenic fungus determined by a saponin detoxifying enzyme. *Science*, **267**, 371–374.

Bradley, D. J., Kjellbom, P. and Lamb, C. J., 1992, Elicitor-induced and wound-induced oxidative cross-linking of a proline-rich plant-cell wall protein – a novel, rapid defense response. *Cell*, **70**, 21–30.

Bragard, C. *et al.*, 1997, *Xanthomonas translucens* from small grains: diversity and phytopathological relevance. *Phytopathology*, **87**, 1111–1117.

Brandwagt, B. F. *et al.*, 2000, A longevity assurance gene homolog of tomato mediates resistance to *Alternaria alternata* f. sp *lycopersici* toxins and fumonisin B-1. *Proceedings of the National Academy of Sciences of the United States of America*, **97**, 4961–4966.

Brandwagt, B. F. *et al.*, 2002, Overexpression of the tomato *ASC-1* gene mediates high insensitivity to AAL toxins and fumonisin B-1 in tomato hairy roots and confers resistance to *Alternaria alternata* f. sp *lycopersici* in *Nicotiana umbratica* plants *Molecular Plant-Microbe Interactions*, **15**, 35–42.

Brassett, P. R. and Gilligan, C. A., 1990, Effects of self-sown wheat on levels of the take-all disease on seedlings of winter-wheat grown in a model system. *Journal of Phytopathology-Phytopathologische Zeitschrift*, **129**, 46–57.

Braun, A. C. and Mandle, R. J., 1948, Studies on the inactivation of the tumour-inducing principle in crown gall. *Growth*, **12**, 255–269.

Braun, A. C., 1982, 'A history of the crown gall problem,' in *Molecular Biology of Plant Tumours*, J. Kahl & J. S. Schell, eds., Academic Press, New York, USA, pp. 155–210.

Brenchley, G. H., 1966, The aerial photography of potato blight epidemics. *The Royal Aeronautical Society*, **70**, 1082–1085.

Brenchley, G. H., 1968, Aerial photography for the study of plant disease. *Annual Review of Phytopathology*, **6**, 1–22.

Brisson, L. F., Tenhaken, R. and Lamb, C., 1994, Function of oxidative cross-linking of cell-wall structural proteins in plant-disease resistance. *Plant Cell*, **6**, 1703–1712.

Broek, A. V. and Vanderleyden, J., 1995, The role of bacterial motility, chemotaxis, and attachment in bacteria plant interactions. *Molecular Plant-Microbe Interactions*, **8**, 800–810.

Broekaert, W. *et al.*, 1990, Wound-induced accumulation of messenger-RNA containing a hevein sequence in laticifers of rubber tree (*Hevea brasiliensis*). *Proceedings of the National Academy of Sciences of the United States of America*, **87**, 7633–7637.

Broekaert, W. F. *et al.*, 1995, Plant defensins – novel antimicrobial peptides as components of the host-defense system. *Plant Physiology*, **108**, 1353–1358.

Broekaert, W. F. *et al.*, 1997, Antimicrobial peptides from plants. *Critical Reviews in Plant Sciences*, **16**, 297–323.

Brosch, G. *et al.*, 1995, Inhibition of maize histone deacetylases by HC toxin, the host-selective toxin of *Cochliobolus carbonum*. *Plant Cell*, **7**, 1941–1950.

Browder, L. E., 1985, Parasite – host – environment specificity in the cereal rusts. *Annual Review of Phytopathology*, **23**, 201–222.

Brown, E. B. and Sykes, G. B., 1983, Assessment of the losses caused to potatoes by the potato cyst nematodes, *Globodera rostochiensis* and *Globodera pallida*. *Annals of Applied Biology*, **103**, 271–276.

Brown, G. E. and Lee, H. S., 1993, Interactions of ethylene with citrus stem-end rot caused by *Diplodia natalensis*. *Phytopathology*, **83**, 1204–1208.

Brown, I. *et al.*, 1998, Localization of components of the oxidative cross-linking of glycoproteins and of callose synthesis in papillae formed during the interaction between non-pathogenic strains of *Xanthomonas campestris* and French bean mesophyll cells. *Plant Journal*, **15**, 333–343.

Brownleader, M. D. *et al.*, 1997, Elicitor-induced extensin insolubilization in suspension-cultured tomato cells. *Phytochemistry*, **46**, 1–9.

Brownlee, H. E. *et al.*, 1990, Antifungal effects of cocoa tannin on the witches broom pathogen *Crinipellis perniciosa*. *Physiological and Molecular Plant Pathology*, **36**, 39–48.

Brunner, F. *et al.*, 1998, Substrate specificities of tobacco chitinases. *Plant Journal*, **14**, 225–234.

Budnik, K., Laing, M. D. and daGraca, J. V., 1996, Reduction of yield losses in pepper crops caused by potato virus Y in KwaZulu-Natal, south Africa, using plastic mulch and yellow sticky traps. *Phytoparasitica*, **24**, 119–124.

Bukhalid, R. A., Chung, S. Y. and Loria, R., 1998, *nec1*, a gene conferring a necrogenic phenotype, is conserved in plant-pathogenic *Streptomyces* spp, and linked to a transposase pseudogene. *Molecular Plant-Microbe Interactions*, **11**, 960–967.

Bukhalid, R. A. *et al.*, 2002, Horizontal transfer of the plant virulence gene, *nec1*, and flanking sequences among genetically distinct *Streptomyces* strains in the diastatochromogenes cluster. *Applied and Environmental Microbiology*, **68**, 738–744.

Bulger, M. A., Stace-Smith, R. and Martin, R. R., 1990, Transmission and field spread of raspberry bushy dwarf virus. *Plant Disease*, **74**, 514–517.

Burg, S. P. and Thimann, K. V., 1959, The physiology of ethylene formation in apples. *Proceedings of the National Academy of Sciences of the United States of America*, **45**, 335–344.

Café, A. C. and Duniway, J. M., 1995, Effects of furrow irrigation schedules and host genotype on *Phytophthora* root-rot of pepper. *Plant Disease*, **79**, 39–43.

Café, A. C., Duniway, J. M. and Davis, R. M., 1995, Effects of the frequency of furrow irrigation on root and fruit rots of squash caused by *Phytophthora capsici*. *Plant Disease*, **79**, 44–48.

Cahill, D. M. and Hardham, A. R., 1994a, Exploitation of zoospore taxis in the development of a novel dipstick immunoassay for the specific detection of *Phytophthora cinnamomi*. *Phytopathology*, **84**, 193–200.

Cahill, D. M. and Hardham, A. R., 1994b, A dipstick immunoassay for the specific detection of *Phytophthora cinnamomi* in soils. *Phytopathology*, **84**, 1284–1292.

Cai, D. G. *et al.*, 1997, Positional cloning of a gene for nematode resistance in sugar beet. *Science*, **275**, 832–834.

Calonnec, A., Goyeau, H. and de Vallavielle-Pope, C., 1996, Effects of induced resistance on infection efficiency and sporulation of *Puccinia striiformis* on seedlings in varietal mixtures and on field epidemics in pure stands. *European Journal of Plant Pathology*, **102**, 733–741.

Camargo, E. P., 1999, *Phytomonas* and other trypanosomatid parasites of plants and fruit. *Advances in Parasitology*, **42**, 29–112.

Cameron, J. N. and Carlile, M. J., 1978, Fatty acids, aldehydes and alcohols as attractants for zoospores of *Phytophthora palmivora*. *Nature*, **271**, 448.

Campbell, R. N., 1996, Fungal transmission of plant viruses. *Annual Review of Phytopathology*, **34**, 87–108.

Canada, S. R. and Dunkle, L. D., 1997, Polymorphic chromosomes bearing the *Tox2* locus in *Cochliobolus carbonum* behave as homologs during meiosis. *Applied and Environmental Microbiology*, **63**, 996–1001.

Candresse, T. *et al.*, 1994, Detection of plum pox virus and analysis of its molecular variability using immuno-capture-PCR. *EPPO Bulletin*, **24**, 585–594.

Cao, H., Li, X. and Dong, X. N., 1998, Generation of broad-spectrum disease resistance by overexpression of an essential regulatory gene in systemic acquired resistance. *Proceedings of the National Academy of Sciences of the United States of America*, **95**, 6531–6536.

Capasso, R. *et al.*, 1997, Phtophorin, a phytotoxic peptide and its phytotoxic aggregates from *Phytophthora nicotianae*. *Phytopathologia Mediterranea*, **36**, 67–73.

Carlile, A. J. *et al.*, 2000, Characterization of SNP1, a cell wall-degrading trypsin, produced during infection by *Stagonospora nodorum*. *Molecular Plant-Microbe Interactions*, **13**, 538–550.

Carlton, W. M., Braun, E. J. and Gleason, M. L., 1998, Ingress of *Clavibacter michiganensis* subsp. *michiganensis* into tomato leaves through hydathodes. *Phytopathology*, **88**, 525–529.

Carrington, J. C. *et al.*, 1996, Cell-to-cell and long-distance transport of viruses in plants. *Plant Cell*, **8**, 1669–1681.

Carsky, R. J. *et al.*, 2000, Reduction of *Striga hermonthica* parasitism on maize using soybean rotation. *International Journal of Pest Management*, **46**, 115–120.

Caruso, P. *et al.*, 2002, Enrichment double-antibody sandwich indirect enzyme-linked immunosorbent assay that uses a specific monoclonal antibody for sensitive detection of *Ralstonia solanacearum* in asymptomatic potato tubers. *Applied and Environmental Microbiology*, **68**, 3634–3638.

Cavalier-Smith, T., 1998, A revised six-kingdom system of life. *Biological Reviews of the Cambridge Philosophical Society*, **73**, 203–266.

Celimene, C. C. *et al.*, 1999, Efficacy of pinosylvins against white-rot and brown-rot fungi. *Holzforschung*, **53**, 491–497.

Celimene, C. C. *et al.*, 2001, *In vitro* inhibition of *Sphaeropsis sapinea* by natural stilbenes. *Phytochemistry*, **56**, 161–165.

Chang, M. and Lynn, D. G., 1986, The haustorium and the chemistry of host recognition in parasitic Angiosperms. *Journal of Chemical Ecology*, **12**, 561–579.

Chapman, B. and Xiao, G., 2000, Inoculation of stumps with *Hypholoma fasciculare* as a possible means to control *Armillaria* root disease. *Canadian Journal of Botany-Revue Canadienne de Botanique*, **78**, 129–134.

Chappell, J. and Hahlbrock, K., 1984, Transcription of plant defense genes in response to UV-light or fungal elicitor. *Nature*, **311**, 76–78.

Chen, J. G., 2001, Dual auxin signaling pathways control cell elongation and division. *Journal of Plant Growth Regulation*, **20**, 255–264.

Chen, X. M. and Line, R. F., 1999, Recessive genes for resistance to *Puccinia striiformis* f. sp. *hordei* in barley. *Phytopathology*, **89**, 226–232.

Chen, Y. M. and Strange, R. N., 1991, Synthesis of the solanapyrone phytotoxins by *Ascochyta rabiei* in response to metal-cations and development of a defined medium for toxin production. *Plant Pathology*, **40**, 401–407.

Chen, Y. M., Tabner, B. J. and Wellburn, A. R., 1990, ACC-independent ethylene formation in brown Norway spruce needles involves organic peroxides rather than hydroperoxides as possible precursors. *Physiological and Molecular Plant Pathology*, **37**, 323–337.

Chen, Y. M., Peh, E. K. and Strange, R. N., 1991, Application of solvent optimization to the separation of the phytotoxins, solanapyrones A, B and C from culture filtrates of *Ascochyta rabiei*. *Bioseparation*, **2**, 107–113.

Cheng, Y. Q. *et al.*, 1999, C-13 labeling indicates that the epoxide-containing amino acid of HC-toxin is biosynthesized by head-to-tail condensation of acetate. *Journal of Natural Products*, **62**, 143–145.

Chérif, M. and Bélanger, R. R., 1992, Use of potassium silicate amendments in recirculating nutrient solutions to suppress *Pythium ultimum* on long English cucumber. *Plant Disease*, **76**, 1008–1011.

Childress, A. M. and Ramsdell, D. C., 1987, Bee-mediated transmission of blueberry leaf mottle virus via infected pollen in highbush blueberry. *Phytopathology*, **77**, 167–172.

Chilton, M. -D. *et al.*, 1977, Stable incorporation of plasmid DNA into higher plants: the molecular basis of crown gall tumorigenesis. *Cell*, **11**, 263–271.

Chin, A. W. *et al.*, 2001, Phenazine-1-carboxamide production in the biocontrol strain *Pseudomonas chlororaphis* PCL1391 is regulated by multiple factors secreted into the growth medium. *Molecular Plant-Microbe Interactions*, **14**, 969–979.

Chin, K. M., Wolfe, M. S. and Minchin, P. N., 1984, Host-mediated interactions between pathogen genotypes. *Plant Pathology*, **33**, 161–171.

Chittaranjan, S. and De Boer, S. H., 1997, Detection of *Xanthomonas campestris* pv. *pelargonii* in geranium and greenhouse nutrient solution by serological and PCR techniques. *European Journal of Plant Pathology*, **103**, 555–563.

Chrispeels, M. J. and Raikhel, N. V., 1991, Lectins, lectin genes, and their role in plant defense. *Plant Cell*, **3**, 1–9.

Chye, M. L. and Cheung, K. Y., 1995, Beta-1,3-glucanase is highly-expressed in laticifers of *Hevea-brasiliensis*. *Plant Molecular Biology*, **29**, 397–402.

Ciardi, J. and Klee, H., 2001, Regulation of ethylene-mediated responses at the level of the receptor. *Annals of Botany*, **88**, 813–822.

Ciesiolka, L. D. *et al.*, 1999, Regulation of expression of avirulence gene *avrRxv* and identification of a family of host interaction factors by sequence analysis of *avrBsT*. *Molecular Plant-Microbe Interactions*, **12**, 35–44.

Ciuffetti, L. M. and Tuori, R. P., 1999, Advances in the characterization of the *Pyrenophora tritici-repentis*-wheat interaction. *Phytopathology*, **89**, 444–449.

Ciuffetti, L. M., Tuori, R. P. and Gaventa, J. M., 1997, A single gene encodes a selective toxin causal to the development of tan spot of wheat. *Plant Cell*, **9**, 135–144.

Ciuffetti, L. M. *et al.*, 1988, Transformation of *Nectria haematococca* with a gene for pisatin demethylating activity, and the role of pisatin detoxification in virulence. *Journal of Cell Biochemistry*, **12C**, 278.

Clergeot, P. H. *et al.*, 2001, PLS1, a gene encoding a tetraspanin-like protein, is required for penetration of rice leaf by the fungal pathogen *Magnaporthe grisea*. *Proceedings of the National Academy of Sciences of the United States of America*, **98**, 6963–6968.

Coakley, S. M., 1988, Variation in climate and prediction of disease in plants. *Annual Review of Phytopathology*, **26**, 163–181.

Cojocaru, M. *et al.*, 1986, 5-(12-heptadecenyl)-resorcinol, the major component of the antifungal activity in the peel of mango fruit. *Phytochemistry*, **25**, 1093–1095.

Cole, R. J. *et al.*, 1985, Mean geocarposphere temperatures that induce preharvest aflatoxin contamination of peanuts under drought stress. *Mycopathologia*, **91**, 41–46.

Coley-Smith, J. R., 1990, White rot disease of *Allium* – problems of soil-borne diseases in microcosm. *Plant Pathology*, **39**, 214–222.

Collins, P. J. *et al.*, 1997, Stabilization of lignin peroxidases in white rot fungi by tryptophan. *Applied and Environmental Microbiology*, **63**, 2543–2548.

Collins, T. J., Moerschbacher, B. M. and Read, N. D., 2001, Synergistic induction of wheat stem rust appressoria by chemical and topographical signals. *Physiological and Molecular Plant Pathology*, **58**, 259–266.

Collmer, A. and Keen, N. T., 1986, The role of pectic enzymes in plant pathogenesis. *Annual Review of Phytopathology*, **24**, 383–409.

Comménil, P., Belingheri, L. and Dehorter, B., 1998, Antilipase antibodies prevent infection of tomato leaves by *Botrytis cinerea*. *Physiological and Molecular Plant Pathology*, **52**, 1–14.

Conn, K. L. *et al.*, 1998, A quantitative method for determining soil populations of *Streptomyces* and differentiating potential potato scab-inducing strains. *Plant Disease*, **82**, 631–638.

Cook, C. E. *et al.*, 1966, Germination of witchweed (*Striga lutea* Lour.): isolation and properties of a potent stimulant. *Science*, **154**, 1189–1190.

Cook, R. J. *et al.*, 1995, Molecular mechanisms of defense by rhizobacteria against root disease. *Proceedings of the National Academy of Sciences of the United States of America*, **92**, 4197–4201.

Coombs, J. J. *et al.*, 2002, Combining engineered (Bt-cry3A) and natural resistance mechanisms in potato for control of Colorado potato beetle. *Journal of the American Society for Horticultural Science*, **127**, 62–68.

Coombs, J. J. *et al.*, 2003, Field evaluation of natural, engineered, and combined resistance mechanisms in potato for control of Colorado potato beetle. *Journal of the American Society for Horticultural Science*, **128**, 219–224.

Cooper, B. *et al.*, 1995, A defective movement protein of TMV in transgenic plants confers resistance to multiple viruses whereas the functional analog increases susceptibility. *Virology*, **206**, 307–313.

Cooper, R. M., 1983, 'The mechanism and significance of enzymatic degradation of host cell walls by parasites,' in *Biochemical Plant Pathology*, J. A. Callow, ed., John Wiley and Sons Ltd., London, UK pp. 101–135.

Cooper, R. M. *et al.*, 1996, Detection and cellular localization of elemental sulphur in disease-resistant genotypes of *Theobroma cacao*. *Nature*, **379**, 159–162.

Coquoz, J. L. *et al.*, 1995, Arachidonic acid induces local but not systemic synthesis of salicylic acid and confers systemic resistance in potato plants to *Phytophthora-infestans* and *Alternaria-solani*. *Phytopathology*, **85**, 1219–1224.

Correa, A. and Hoch, H. C., 1995, Identification of thigmoresponsive loci for cell-differentiation in *Uromyces* germlings. *Protoplasma*, **186**, 34–40.

Costa, A. S. and Muller, G. W., 1980, Tristeza disease control by cross protection: a U.S. – Brazil cooperative success. *Plant Disease*, **64**, 538–541.

Cottyn, B. *et al.*, 1996, Bacterial diseases of rice. 2. Characterization of pathogenic bacteria associated with sheath rot complex and grain discoloration of rice in the Philippines. *Plant Disease*, **80**, 438–445.

Covert, S. F. *et al.*, 1996, A gene for maackiain detoxification from a dispensable chromosome of *Nectria haematococca*. *Molecular and General Genetics*, **251**, 397–406.

Creelman, R. A. and Mullet, J. E., 1997, Oligosaccharins, brassinolides, and jasmonates: nontraditional regulators of plant growth, development, and gene expression. *Plant Cell*, **9**, 1211–1223.

Crop Protection Compendium Compact Disc, 2002. CAB International, Wallingford, Oxfordshire, UK.

Cruickshank, I. A. M., 1962, Studies on phytoalexins. IV. The antimicrobial spectrum of pisatin. *Australian Journal of Biological Sciences*, **15**, 147–159.

Cruickshank, I. A. M. and Perrin, D. R., 1968, Isolation and partial characterization of monilicolin A, a polypeptide with phaseollin inducing activity from *Monilinia fructicola*. *Life Sciences*, **7**, 449–468.

Cruickshank, I. A. M., Perrin, D. R. and Mandryk, M., 1977, Fungitoxicity of duvatrienediols associated with cuticular wax of tobacco leaves. *Phytopathologische Zeitschrift-Journal of Phytopathology*, **90**, 243–249.

Crute, I. R. and Norwood, J. M., 1986, Gene-dosage effects on the relationship between *Bremia lactucae* (downy mildew) and *Lactuca sativa* (lettuce) – the relevance to a mechanistic understanding of host-parasite specificity. *Physiological and Molecular Plant Pathology*, **29**, 133–145.

Cullen, D. W. *et al.*, 2002, Detection of *Colletotrichum coccodes* from soil and potato tubers by conventional and quantitative real-time PCR. *Plant Pathology*, **51**, 281–292.

Cuppels, D. A. and Elmhirst, J., 1999, Disease development and changes in the natural *Pseudomonas syringae* pv. *tomato* populations on field tomato plants. *Plant Disease*, **83**, 759–764.

Damaan, K. E., Gardner, J. M. and Scheffer, R. P., 1974, An assay for *Helminthosporium victoriae* based on induced leakage of electrolytes from oat tissue. *Phytopathology*, **64**, 654.

Damsteegt, V. D. *et al.*, 1997, *Prunus tomentosa* as a diagnostic host for detection of plum pox virus and other *Prunus* viruses. *Plant Disease*, **81**, 329–332.

Dangl, J., 1998, Innate immunity – Plants just say NO to pathogens. *Nature*, **394**, 525.

Dann, E. *et al.*, 1998, Effect of treating soybean with 2,6-dichloroisonicotinic acid (INA) and benzothiadiazole (BTH) on seed yields and the level of disease caused by *Sclerotinia sclerotiorum* in field and greenhouse studies. *European Journal of Plant Pathology*, **104**, 271–278.

Daub, M. E., 1986, Tissue-culture and the selection of resistance to pathogens. *Annual Review of Phytopathology*, **24**, 159–186.

Davies, K. A. *et al.*, 2000, Evidence for a role of cutinase in pathogenicity of *Pyrenopeziza brassicae* on brassicas. *Physiological and Molecular Plant Pathology*, **57**, 63–75.

Davis, K. R. *et al.*, 1986, Host-pathogen interactions. 29. Oligogalacturonides released from sodium polypectate by endopolygalacturonic acid lyase are elicitors of phytoalexins in soybean. *Plant Physiology*, **80**, 568–577.

Davis, M. J., 1986, Taxonomy of plant-pathogenic Coryneform bacteria. *Annual Review of Phytopathology*, **24**, 115–140.

Davis, M. J. *et al.*, 1998, Rickettsial relative associated with papaya bunchy top disease. *Current Microbiology*, **36**, 80–84.

Day, P. R., 1984, 'Genetics of recognition systems in host-parasite interactions,' in *Cellular Interactions. Encyclopaedia of Plant Physiology New Series Vol. 17*,

H. F. Linskens & J. Heslop-Harrison, eds., Springer-Verlag, Berlin, Germany, pp. 134–147.

De Jong, J. C. *et al.*, 1997, Glycerol generates turgor in rice blast. *Nature*, **389**, 244–245.

De Laat, A. M. M. and van Loon, L. C., 1983, Regulation of ethylene biosynthesis in virus-infected tobacco-leaves. 3. The relationship between stimulated ethylene production and symptom expression in virus-infected tobacco-leaves. *Physiological Plant Pathology*, **22**, 261–273.

De Laat, P. C. A., Verhoeven, J. T. W. and Janse, J. D., 1994, Bacterial leaf rot of *Aloevera* L, caused by *Erwinia chrysanthemi* biovar-3. *European Journal of Plant Pathology*, **100**, 81–84.

De Lorenzo, G. and Ferrari, S., 2002, Polygalacturonase-inhibiting proteins in defense against phytopathogenic fungi. *Current Opinion in Plant Biology*, **5**, 295–299.

De Lorenzo, G., D'Ovidio, R. and Cervone, F., 2001, The role of polygalacturonase-inhibiting proteins (PGIPS) in defense against pathogenic fungi. *Annual Review of Phytopathology*, **39**, 313–335.

De Lorenzo, G. *et al.*, 1990, Host pathogen interactions. 37. Abilities of the polygalacturonase-inhibiting proteins from 4 cultivars of *Phaseolus vulgaris* to inhibit the endopolygalacturonases from 3 races of *Colletotrichum lindemuthianum*. *Physiological and Molecular Plant Pathology*, **36**, 421–435.

De Mackiewicz, D. *et al.*, 1998, Herbaceous weeds are not ecologically important reservoirs of *Erwinia tracheiphila*. *Plant Disease*, **82**, 521–529.

De Meyer, G. and Hofte, M., 1997, Salicylic acid produced by the rhizobacterium *Pseudomonas aeruginosa* 7NSK2 induces resistance to leaf infection by *Botrytis cinerea* on bean. *Phytopathology*, **87**, 588–593.

Dean, R. A. and Kuć, J., 1987, Rapid lignification in response to wounding and infection as a mechanism for induced systemic protection in cucumber. *Physiological and Molecular Plant Pathology*, **31**, 69–81.

Défago, G. and Kern, H., 1983, Induction of *Fusarium solani* mutants insensitive to tomatine, their pathogenicity and aggressiveness to tomato fruits and pea-plants. *Physiological Plant Pathology*, **22**, 29–37.

Défago, G., Kern, H. and Sedlar, L., 1983, Genetic-analysis of tomatine insensitivity, sterol content and pathogenicity for green tomato fruits in mutants of *Fusarium solani*. *Physiological Plant Pathology*, **22**, 39–43.

Dekhuijzen, H. M. and Overeem, J. C., 1971, Role of cytokinins in clubroot formation. *Physiological Plant Pathology*, **1**, 151–161.

Delaney, T. P. *et al.*, 1994, A central role of salicylic-acid in plant-disease resistance. *Science*, **266**, 1247–1250.

Delany, I. R. *et al.*, 2001, Enhancing the biocontrol efficacy of *Pseudomonas fluorescens* F113 by altering the regulation and production of 2,4-diacetylphloroglucinol – improved *Pseudomonas* biocontrol inoculants. *Plant and Soil*, **232**, 195–205.

Delledonne, M. *et al.*, 1998, Nitric oxide functions as a signal in plant disease resistance. *Nature*, **394**, 585–588.

Delledonne, M. *et al.*, 2001, Signal interactions between nitric oxide and reactive oxygen intermediates in the plant hypersensitive disease resistance response. *Proceedings of the National Academy of Sciences of the United States of America*, **98**, 13454–13459.

Deng, Z. N. *et al.*, 1995, Selecting lemon protoplasts for insensitivity to *Phoma trachei-phila* toxin and regenerating tolerant plants. *Journal of the American Society for Horticultural Science*, **120**, 902–905.

Dennis, J. J. C., Holderness, M. and Keane, P. J., 1992, Weather patterns associated with sporulation of *Oncobasidium theobromae* on cocoa. *Mycological Research*, **96**, 31–37.

Derie, M. L. *et al.*, 1985, *Ascochyta rabiei* on chickpeas in Idaho. *Plant Disease*, **69**, 268.

Derrick, K. S. and Timmer, L. W., 2000, Citrus blight and other diseases of recalcitrant etiology. *Annual Review of Phytopathology*, **38**, 181–205.

Desjardins, A. E., Gardner, H. W. and Weltring, K. M., 1992, Detoxification of sesquiterpene phytoalexins by *Gibberella pulicaris* (*Fusarium sambucinum*) and its importance for virulence on potato-tubers. *Journal of Industrial Microbiology*, **9**, 201–211.

Desjardins, A. E. *et al.*, 1995, Genetic-analysis of fumonisin production and virulence of *Gibberella fujikuroi* mating population A (*Fusarium moniliforme*) on maize (*Zea mays*) seedlings. *Applied and Environmental Microbiology*, **61**, 79–86.

Deslandes, L. *et al.*, 2002, Resistance to *Ralstonia solanacearum* in *Arabidopsis thaliana* is conferred by the recessive *RRS1-R* gene, a member of a novel family of resistance genes. *Proceedings of the National Academy of Sciences of the United States of America*, **99**, 2404–2409.

Dewey, F. M., Thornton, C. R. and Gilligan, C. A., 1997, Use of monoclonal antibodies to detect, quantify and visualize fungi in soils. *Advances in Botanical Research*, **24**, 275–308.

Dickie, G. A. and Bell, C. R., 1995, A full factorial analysis of 9 factors influencing invitro antagonistic screens for potential biocontrol agents. *Canadian Journal of Microbiology*, **41**, 284–293.

Dickman, M. B. and Patil, S. S., 1986, Cutinase deficient mutants of *Colletotrichum gloeosporioides* are nonpathogenic to papaya fruit. *Physiological and Molecular Plant Pathology*, **28**, 235–242.

Dickman, M. B., Podila, G. K. and Kolattukudy, P. E., 1989, Insertion of cutinase gene into a wound pathogen enables it to infect intact host. *Nature*, **342**, 446–448.

Diener, T. O., 2001, The viroid: biological oddity or evolutionary fossil? *Advances in Virus Research*, **57**, 137–184.

Dillard, H. R. and Cobb, A. C., 1993, Survival of *Colletotrichum lindemuthianum* in bean debris in New York State. *Plant Disease*, **77**, 1233–1238.

Dinesh-Kumar, S. P. and Baker, B. J., 2000, Alternatively spliced N resistance gene transcripts: their possible role in tobacco mosaic virus resistance. *Proceedings of the National Academy of Sciences of the United States of America*, **97**, 1908–1913.

Doi, Y. M. *et al.*, 1967, Mycoplasma or PLT-group-like microorganisms found in the phloem elements of plants infected with mulberry dwarf, potato witches' broom, aster yellows, or Paulownia witches' broom. *Annals of the Phytopathological Society of Japan*, **33**, 259–266.

Doke, N., 1983, Involvement of superoxide anion generation in the hypersensitive response of potato tuber tissues to infection with an incompatible race of *Phytophthora infestans* and to the hyphal wall components. *Physiological Plant Pathology*, **23**, 345–357.

Dollet, M., Sturm, N. R. and Campbell, D. A., 2001, The spliced leader RNA gene array in phloem-restricted plant trypanosomatids (*Phytomonas*) partitions into two major groupings: epidemiological implications. *Parasitology*, **122**, 289–297.

Domergue, F. *et al.*, 2000, Antifungal compounds from idioblast cells isolated from avocado fruits. *Phytochemistry*, **54**, 183–189.

Donaldson, S. P. and Deacon, J. W., 1993, Differential encystment of zoospores of *Pythium* species by saccharides in relation to establishment on roots. *Physiological and Molecular Plant Pathology*, **42**, 177–184.

Doornik, A. W., 1992, Heat-Treatment to control *Colletotrichum acutatum* on corms of *Anemone coronaria*. *Netherlands Journal of Plant Pathology*, **98**, 377–386.

Douglas, L. I. and Deacon, J. W., 1994, Strain variation in tolerance of water-stress by *Idriella* (*Microdochium*) *bolleyi*, a biocontrol agent of cereal root and stem base pathogens. *Biocontrol Science and Technology*, **4**, 239–249.

Dow, J. M. *et al.*, 1990, Extracellular proteases from *Xanthomonas campestris* pv *campestris*, the black rot pathogen. *Applied and Environmental Microbiology*, **56**, 2994–2998.

Droby, S. *et al.*, 1986, Presence of antifungal compounds in the peel of mango fruits and their relation to latent infections of *Alternaria alternata*. *Physiological and Molecular Plant Pathology*, **29**, 173–183.

Drucker, M. *et al.*, 2002, Intracellular distribution of viral gene products regulates a complex mechanism of cauliflower mosaic virus acquisition by its aphid vector. *Proceedings of the National Academy of Sciences of the United States of America*, **99**, 2422–2427.

Duffy, B. K. and Weller, D. M., 1994, A semiselective and diagnostic medium for *Gaeumannomyces graminis* var *tritici*. *Phytopathology*, **84**, 1407–1415.

Duffy, B. K., Simon, A. and Weller, D. M., 1996, Combination of *Trichoderma koningii* with fluorescent pseudomonads for control of take-all on wheat. *Phytopathology*, **86**, 188–194.

Dunkle, L. D. and Macko, V., 1995, Peritoxins and their effects on *Sorghum*. *Canadian Journal of Botany-Revue Canadienne de Botanique*, **73**, S444–S452.

Duran-Vila, N., Romero-Durban, J. and Hernandez, M., 1996, Detection and eradication of chrysanthemum stunt viroid in Spain. *EPPO Bulletin*, **26**, 399–405.

Durbin, R. A. and Uchytil, T. F., 1977, A survey of plant insensitivity to tentoxin. *Phytopathology*, **67**, 602–603.

Durner, J., Wendehenne, D. and Klessig, D. F., 1998, Defense gene induction in tobacco by nitric oxide, cyclic GMP, and cyclic ADP-ribose. *Proceedings of the National Academy of Sciences of the United States of America*, **95**, 10328–10333.

Dutton, M. V. and Evans, C. S., 1996, Oxalate production by fungi: its role in pathogenicity and ecology in the soil environment. *Canadian Journal of Microbiology*, **42**, 881–895.

Ebata, Y., Yamamoto, H. and Uchiyama, T., 1998, Chemical composition of the glue from appressoria of *Magnaporthe grisea*. *Bioscience Biotechnology and Biochemistry*, **62**, 672–674.

Edwards, C. and Strange, R. N., 1991, Separation and identification of phytoalexins from leaves of groundnut (*Arachis hypogaea*) and development of a method for their determination by reversed-phase high-performance liquid chromatography. *Journal of Chromatography*, **547**, 185–193.

Edwards, J. *et al.*, 1999, Peppermint rust in Victoria: the incidence-severity relationship and its implication for the development of an action threshold. *Australian Journal of Agricultural Research*, **51**, 91–95.

Edwards, K. *et al.*, 1985, Rapid transient induction of phenylalanine ammonia-lyase messenger RNA in elicitor-treated bean cells. *Proceedings of the National Academy of Sciences of the United States of America*, **82**, 6731–6735.

Edwards, R., Daniell, T. J. and Gregory, A. C. E., 1997, Methylation reactions and the phytoalexin response in alfalfa suspension cultures. *Planta*, **201**, 359–367.

Edwards, S. G. and Seddon, B., 2001, Mode of antagonism of *Brevibacillus brevis* against *Botrytis cinerea in vitro*. *Journal of Applied Microbiology*, **91**, 652–659.

Effertz, R. J. *et al.*, 2002, Identification of a chlorosis-inducing toxin from *Pyrenophora tritici-repentis* and the chromosomal location of an insensitivity locus in wheat. *Phytopathology*, **92**, 527–533.

Eigenbrode, S. D. *et al.*, 2002, Volatiles from potato plants infected with potato leafroll virus attract and arrest the virus vector, *Myzus persicae* (Homoptera: Aphididae). *Proceedings of the Royal Society of London Series B-Biological Sciences*, **269**, 455–460.

Ekblad, A., Wallander, H. and Nasholm, T., 1998, Chitin and ergosterol combined to measure total and living fungal biomass in ectomycorrhizas. *New Phytologist*, **138**, 143–149.

Ellingboe, A. H., 1976, Genetics of host-parasite interaction, in *Physiological Plant Pathology*, R. Heitefuss & P. H. Williams, eds., Springer-Verlag Berlin, Germany, pp. 761–778.

Ellis, J. *et al.*, 1997, Advances in the molecular genetic analysis of the flax-flax rust interaction. *Annual Review of Phytopathology*, **35**, 271–291.

Ellis, J. G. *et al.*, 1999, Identification of regions in alleles of the flax rust resistance gene L that determine differences in gene-for-gene specificity. *Plant Cell*, **11**, 495–506.

Ellis, R. J. *et al.*, 1999, Ecological basis for biocontrol of damping-off disease by *Pseudomonas fluorescens* 54/96. *Journal of Applied Microbiology*, **87**, 454–463.

Elliston, J. E., 1985, Further evidence for 2 cytoplasmic hypovirulence agents in a strain of *Endothia parasitica* from Western Michigan. *Phytopathology*, **75**, 1405–1413.

Elphinstone, J. G. *et al.*, 1996, Sensitivity of different methods for the detection of *Ralstonia solancearum* in potato tuber extracts. *EPPO Bulletin*, **26**, 663–678.

El Yamani, M. and Hill, J. H., 1990, Identification and importance of Barley Yellow Dwarf Virus in Morocco. *Plant Disease*, **74**, 291–294.

English, J. T. *et al.*, 1993, Leaf removal for control of *Botrytis* bunch rot of wine grapes in the Midwestern United States. *Plant Disease*, **77**, 1224–1227.

Epperlein, M. M., Noronha-Dutra, A. A. and Strange, R. N., 1986, Involvement of the hydroxyl radical in the abiotic elicitation of phytoalexins in legumes. *Physiological and Molecular Plant Pathology*, **28**, 67–77.

Escobar, M. A. *et al.*, 2001, RNAi-mediated oncogene silencing confers resistance to crown gall tumorigenesis. *Proceedings of the National Academy of Sciences of the United States of America*, **98**, 13437–13442.

Escudero, J. and Hohn, B., 1997, Transfer and integration of T-DNA without cell injury in the host plant. *Plant Cell*, **9**, 2135–2142.

Estabrook, E. M. & Yoder, J. I., 1998, Plant-plant communications: Rhizosphere signaling between parasitic angiosperms and their hosts. *Plant Physiology*, **116**, 1–7.

Eun, A. J. C. and Wong, S. M., 2000, Molecular beacons: A new approach to plant virus detection. *Phytopathology*, **90**, 269–275.

Eversmeyer, M. G. and Kramer, C. L., 2000, Epidemiology of wheat leaf and stem rust in the central Great Plains of the USA. *Annual Review of Phytopathology*, **38**, 491–513.

Facchini, P. J. and Chappell, J., 1992, Gene family for an elicitor-induced sesquiterpene cyclase in tobacco. *Proceedings of the National Academy of Sciences of the United States of America*, **89**, 11088–11092.

Falloon, R. E. *et al.*, 1995, Disease severity keys for powdery and downy mildews of pea, and powdery scab of potato. *New Zealand Journal of Crop and Horticultural Science*, **23**, 31–37.

Fatmi, M. and Schaad, N. W., 1988, Semiselective agar medium for isolation of *Clavibacter michiganense* subsp. *michiganense* from tomato seed. *Phytopathology*, **78**, 121–126.

Fatmi, M., Schaad, N. W. and Bolkan, H. A., 1991, Seed treatments for eradicating *Clavibacter michiganensis* subsp. *michiganensis* from naturally infected tomato seeds. *Plant Disease*, **75**, 383–385.

Fauquet, C. and Fargette, D., 1990, African Cassava Mosaic virus – etiology, epidemiology, and control. *Plant Disease*, **74**, 404–411.

Feld, S. J., Menge, J. A. and Stolzy, L. H., 1990, Influence of drip and furrow irrigation on *Phytophthora* root-rot of citrus under field and greenhouse conditions. *Plant Disease*, **74**, 21–27.

Felix, G. *et al.*, 1999, Plants have a sensitive perception system for the most conserved domain of bacterial flagellin. *Plant Journal*, **18**, 265–276.

Fernando, W. G. D. *et al.*, 1997, Head blight gradients caused by *Gibberella zeae* from area sources of inoculum in wheat field plots. *Phytopathology*, **87**, 414–421.

Feys, B. J. F. *et al.*, 1994, *Arabidopsis* mutants selected for resistance to the phytotoxin coronatine are male-sterile, insensitive to methyl jasmonate, and resistant to a bacterial pathogen. *Plant Cell*, **6**, 751–759.

Fillhart, R. C., Bachand, G. D. and Castello, J. D., 1998, Detection of infectious tobamoviruses in forest soils. *Applied and Environmental Microbiology*, **64**, 1430–1435.

Fillingham, A. J. *et al.*, 1992, Avirulence genes from *Pseudomonas syringae* pathovars *phaseolicola* and *pisi* confer specificity towards both host and nonhost species. *Physiological and Molecular Plant Pathology*, **40**, 1–15.

Finney, D. J., 1980, *Probit Analysis*, Cambridge University Press, Cambridge UK.

Flaishman, M. A. and Kolattukudy, P. E., 1994, Timing of fungal invasion using host's ripening hormone as a signal. *Proceedings of the National Academy of Sciences of the United States of America*, **91**, 6579–6583.

Flego, D. *et al.*, 1997, Control of virulence gene expression by plant calcium in the phytopathogen *Erwinia carotovora*. *Molecular Microbiology*, **25**, 831–838.

Flor, H. H., 1971, Current status of the gene-for-gene concept. *Annual Review of Phytopathology*, **9**, 275–296.

Fogliano, V. *et al.*, 1998, Characterization of a 60 kDa phytotoxic glycoprotein produced by *Phoma tracheiphila* and its relation to malseccin. *Physiological and Molecular Plant Pathology*, **53**, 149–161.

Forrer, H. R. and Zadoks, J. C., 1983, Yield reduction in wheat in relation to leaf necrosis caused by *Septoria tritici*. *Netherlands Journal of Plant Pathology*, **89**, 87–98.

Francis, S. A., Dewey, F. M. and Gurr, S. J., 1996, The role of cutinase in germling development and infection by *Erysiphe graminis* f sp *hordei*. *Physiological and Molecular Plant Pathology*, **49**, 201–211.

Franz, A., Makkouk, K. M. and Vetten, H. J., 1997, Host range of faba bean necrotic yellows virus and potential yield loss in infected faba bean. *Phytopathologia Mediterranea*, **36**, 94–103.

Fraser, R. S. S., 1990, The genetics of resistance to plant viruses. *Annual Review of Phytopathology*, **28**, 179–200.

Frederick, R. D. *et al.*, 1997, Identification of a pathogenicity locus, *rpfA*, in *Erwinia carotovora* subsp: *carotovora* that encodes a two-component sensor-regulator protein. *Molecular Plant-Microbe Interactions*, **10**, 407–415.

Friebe, A. *et al.*, 1998, Detoxification of benzoxazolinone allelochemicals from wheat by *Gaeumannomyces graminis* var. *tritici*, *G. graminis* var. *graminis*, *Graminis* var. *avenae*, and *Fusarium culmorum*. *Applied and Environmental Microbiology*, **64**, 2386–2391.

Friedrichsen, D. and Chory, J., 2001, Steroid signaling in plants: from the cell surface to the nucleus. *Bioessays*, **23**, 1028–1036.

Fry, W. E., 1982, *Principles of Plant Disease Management*, Academic Press, New York, USA.

Fulton, R. W., 1986, Practices and precautions in the use of cross protection for plant virus disease control. *Annual Review of Phytopathology*, **24**, 67–81.

Furner, I. J. *et al.*, 1986, An *Agrobacterium* transformation in the evolution of the genus *Nicotiana*. *Nature*, **319**, 422–427.

Gabriel, D. W., 1999, Why do pathogens carry avirulence genes? *Physiological and Molecular Plant Pathology*, **55**, 205–214.

Gaffney, T. *et al.*, 1993, Requirement of salicylic-acid for the induction of systemic acquired-resistance. *Science*, **261**, 754–756.

Gamba, F. M. and Lamari, L., 1998, Mendelian inheritance of resistance to tan spot (*Pyrenophora tritici-repentis*) in selected genotypes of durum wheat (*Triticum turgidum*). *Canadian Journal of Plant Pathology-Revue Canadienne de Phytopathologie*, **20**, 408–414.

Gamliel, A. and Stapleton, J. J., 1993, Characterization of antifungal volatile compounds evolved from solarized soil amended with cabbage residues. *Phytopathology*, **83**, 899–905.

Gao, S. and Shain, L., 1995, Activity of polygalacturonase produced by *Cryphonectria parasitica* in chestnut bark and its inhibition by extracts from American and Chinese chestnut. *Physiological and Molecular Plant Pathology*, **46**, 199–213.

Garcia, G. M. *et al.*, 1996, Identification of RAPD, SCAR, and RFLP markers tightly linked to nematode resistance genes introgressed from *Arachis cardenasii* into *Arachis hypogaea*. *Genome*, **39**, 836–845.

Garre, V., Tenberge, K. B. and Eising, R., 1998, Secretion of a fungal extracellular catalase by *Claviceps purpurea* during infection of rye: Putative role in pathogenicity and suppression of host defense. *Phytopathology*, **88**, 744–753.

Garrett, K. A. and Mundt, C. C., 1999, Epidemiology in mixed host populations. *Phytopathology*, **89**, 984–990.

Garrett, S. D., 1956, *Biology of Root-infecting Fungi*, Cambridge University Press, Cambridge, UK.

Gäumann, E., 1954, Toxins and plant disease. *Endeavour*, **13**, 198–204.

Gaunt, R. E., 1995, The relationship between plant-disease severity and yield. *Annual Review of Phytopathology*, **33**, 119–144.

Gebhardt, C., 1997, Plant genes for pathogen resistance – variation on a theme. *Trends in Plant Science*, **2**, 243–244.

Geering, A. D. W. and Randles, J. W., 1994, Interactions between a seed-borne strain of cucumber mosaic cucumovirus and its lupin host. *Annals of Applied Biology*, **124**, 301–314.

Gelvin, S. B., 2000, *Agrobacterium* and plant genes involved in T-DNA transfer and integration. *Annual Review of Plant Physiology and Plant Molecular Biology*, **51**, 223–256.

Ghanim, M. and Czosnek, H., 2000, Tomato yellow leaf curl geminivirus (TYLCV-Is) is transmitted among whiteflies (*Bemisia tabaci*) in a sex-related manner. *Journal of Virology*, **74**, 4738–4745.

Ghisalberti, E. L. *et al.*, 1990, Variability among strains of *Trichoderma harzianum* in their ability to reduce take-all and to produce pyrones. *Plant and Soil*, **121**, 287–291.

Giamoustaris, A. and Mithen, R., 1997, Glucosinolates and disease resistance in oilseed rape (*Brassica napus* ssp. *oleifera*). *Plant Pathology*, **46**, 271–275.

Gianoli, E. and Niemeyer, H. M., 1998, DIBOA in wild Poaceae: sources of resistance to the Russian wheat aphid (*Diuraphis noxia*) and the greenbug (*Schizaphis graminum*). *Euphytica*, **102**, 317–321.

Gibson, R. W. and Pickett, J. A., 1983, Wild potato repels aphids by release of aphid alarm pheromone. *Nature*, **302**, 608–609.

Gilbertson, R. L., Rand, R. E. and Hagedorn, D. J., 1990, Survival of *Xanthomonas campestris* pv *phaseoli* and pectolytic strains of *Xanthomonas campestris* in bean debris. *Plant Disease*, **74**, 322–327.

Gilligan, C. A., 1990, Comparison of disease progress curves. *New Phytologist*, **115**, 223–242.

Gilmer, R. N., 1965, Additional evidence of tree-to-tree transmission of sour cherry yellows virus by pollen. *Phytopathology*, **55**, 482–483.

Gindro, K. and Pezet, R., 1997, Evidence for a constitutive cytoplasmic cutinase in ungerminated conidia of *Botrytis cinerea* Pers:Fr. *FEMS Microbiology Letters*, **149**, 89–92.

Gitaitis, R. *et al.*, 1997, Bacterial streak and bulb rot of onion .1. A diagnostic medium for the semiselective isolation and enumeration of *Pseudomonas viridiflava*. *Plant Disease*, **81**, 897–900.

Glasscock, H. H., 1971, Fireblight epidemic among Kentish apple orchards. *Annals of Applied Biology*, **69**, 137–145.

Goethals, K. *et al.*, 2001, Leafy gall formation by *Rhodococcus fascians*. *Annual Review of Phytopathology*, **39**, 27–52.

Gokte, N. and Mathur, V. K., 1995, Eradication of root-knot nematodes from grapevine rootstocks by thermal therapy. *Nematologica*, **41**, 269–271.

Goldstein, P. J. and Hayes, C. E., 1977, The lectins: carbohydrate-binding proteins of plants and animals. *Advances in Carbohydrate Biochemistry and Chemistry*, **35**, 127–340.

Gomez-Gomez, L. and Boller, T., 2002, Flagellin perception: a paradigm for innate immunity. *Trends in Plant Science*, **7**, 251–256.

Gonsalves, D., 1998, Control of papaya ringspot virus in papaya: A case study. *Annual Reivew of Phytopathology*, **36**, 415–437.

Goodey, J. B., Franklin, M. T. and Hooper, D. J., 1965, *The Nematode Parasites of Plants Catalogued under their Hosts*, Third edition, CAB International, Wallingford Oxfordshire, UK.

Goor, M. C. *et al.*, 1984, The use of API systems in the identification of phytopathogenic bacteria. *Mededelingen Van De Faculteit Landbouwtenschappen Rijksuniversiteit, Gent*, **49**, 499–507.

Gopalan, S. *et al.*, 1996, Expression of the *Pseudomonas syringae* avirulence protein AvrB in plant cells alleviates its dependence on the hypersensitive response and pathogenicity (Hrp) secretion system in eliciting genotype-specific hypersensitive cell death. *Plant Cell*, **8**, 1095–1105.

Gowda, B. S., Riopel, J. L. and Timko, M. P., 1999, *NRSA-1*: a resistance gene homolog expressed in roots of non-host plants following parasitism by *Striga asiatica* (witchweed). *Plant Journal*, **20**, 217–230.

Grant, M. *et al.*, 2000, The *RPM1* plant disease resistance gene facilitates a rapid and sustained increase in cytosolic calcium that is necessary for the oxidative burst and hypersensitive cell death. *Plant Journal*, **23**, 441–450.

Grayer, R. J. and Harborne, J. B., 1994, A survey of antifungal compounds from higher-plants, 1982–1993. *Phytochemistry*, **37**, 19–42.

Greenough, D. R., Black, L. L. and Bond, W. P., 1990, Aluminum-surfaced mulch – an approach to the control of tomato spotted wilt virus in solanaceous crops. *Plant Disease*, **74**, 805–808.

Gregory, P. H., 1973, *The Microbiology of the Atmosphere*, Second edition, Leonard Hill Books, Aylesbury Buckinghamshire, UK.

Grenier, J. *et al.*, 1999, Some thaumatin-like proteins hydrolyse polymeric beta-1,3-glucans. *Plant Journal*, **19**, 473–480.

Gretenkort, M. A. and Ingram, D. S., 1993, The use of ergosterol as a quantitative measure of the resistance of cultured-tissues of *Brassica napus* ssp *oleifera* to *Leptosphaeria maculans*. *Journal of Phytopathology-Phytopathologische Zeitschrift*, **138**, 217–224.

Griep, R. A. *et al.*, 1998, Development of specific recombinant monoclonal antibodies against the lipopolysaccharide of *Ralstonia solanacearum* race 3. *Phytopathology*, **88**, 795–803.

Griffiths, H. M., Slack, S. A. and Dodds, J. H., 1990, Effect of chemical and heat therapy on virus concentrations in *in vitro* potato plantlets. *Canadian Journal of Botany-Revue Canadienne de Botanique*, **68**, 1515–1521.

Grisham, M. P., 1991, Effect of ratoon stunting disease on yield of sugarcane grown in multiple 3-year plantings. *Phytopathology*, **81**, 337–340.

Groth, J. V. and Roelfs, A. P., 1987, The concept and measurement of phenotypic diversity in *Puccinia graminis* on wheat. *Phytopathology*, **77**, 1395–1399.

Guillén, P. *et al.*, 1998, A novel NADPH-dependent aldehyde reductase gene from *Vigna radiata* confers resistance to the grapevine fungal toxin eutypine. *Plant Journal*, **16**, 335–343.

Gullino, M. L., Migheli, Q. and Mezzalama, M., 1995, Risk analysis in the release of biological-control agents – antagonistic *Fusarium oxysporum* as a case-study. *Plant Disease*, **79**, 1193–1201.

Gunnarsson, T. *et al.*, 1996, The use of ergosterol in the pathogenic fungus *Bipolaris sorokiniana* for resistance rating of barley cultivars. *European Journal of Plant Pathology*, **102**, 883–889.

Gwynne, D. C., 1984, Fire blight in perry pears and cider apples in the South West of England. *Acta Horticulturae*, **151**, 41–47.

Haber, S. *et al.*, 1995, Diagnosis of flame chlorosis by reverse transcription-polymerase chain-reaction (RT-PCR). *Plant Disease*, **79**, 626–630.

Hacker, J. and Kaper, J. B., 2000, Pathogenicity islands and the evolution of microbes. *Annual Review of Microbiology*, **54**, 641–679.

Hahlbrock, K. and Scheel, D., 1989, Physiology and molecular-biology of phenylpropanoid metabolism. *Annual Review of Plant Physiology and Plant Molecular Biology*, **40**, 347–369.

Hahn, M. and Mendgen, K., 1997, Characterization of in planta induced rust genes isolated from a haustorium-specific cDNA library. *Molecular Plant-Microbe Interactions*, **10**, 427–437.

Hahn, M., Jungling, S. and Knogge, W., 1993, Cultivar-specific elicitation of barley defense reactions by the phytotoxic peptide Nip1 from *Rhynchosporium secalis*. *Molecular Plant-Microbe Interactions*, **6**, 745–754.

Hahn, M. *et al.*, 1997, A putative amino acid transporter is specifically expressed in haustoria of the rust fungus *Uromyces fabae*. *Molecular Plant-Microbe Interactions*, **10**, 438–445.

Hahn, M. G., Bonhoff, A. and Grisebach, H., 1985, Quantitative localization of the phytoalexin glyceollin-I in relation to fungal hyphae in soybean roots infected with *Phytophthora megasperma* f sp *glycinea*. *Plant Physiology*, **77**, 591–601.

Hain, R. *et al.*, 1993, Disease resistance results from foreign phytoalexin expression in a novel plant. *Nature*, **361**, 153–156.

Hamer, J. E. *et al.*, 1989, Host species-specific conservation of a family of repeated DNA sequences in the genome of a fungal plant pathogen. *Proceedings of the National Academy of Sciences of the United States of America*, **86**, 9981–9985.

Hamid, K. and Strange, R. N., 2000, Phytotoxicity of solanapyrones A and B produced by the chickpea pathogen *Ascochyta rabiei* (Pass.) Labr. and the apparent metabolism of solanapyrone A by chickpea tissues. *Physiological and Molecular Plant Pathology*, **56**, 235–244.

Hammerschmidt, R., Métraux, J. P. and van Loon, L. C., 2001, Inducing resistance: A summary of papers presented at the First International Symposium on Induced Resistance to Plant Diseases, Corfu, May 2000. *European Journal of Plant Pathology*, **107**, 1–6.

Hammond-Kosack, K. E. and Jones, J. D. G., 1994, Incomplete dominance of tomato *Cf* genes for resistance to *Cladosporium fulvum*. *Molecular Plant-Microbe Interactions*, **7**, 58–70.

Hammond-Kosack, K. E. and Jones, J. D. G., 1997, Plant disease resistance genes. *Annual Review of Plant Physiology and Plant Molecular Biology*, **48**, 575–607.

Han, Y. N. *et al.*, 2001, Genes determining pathogenicity to pea are clustered on a supernumerary chromosome in the fungal plant pathogen *Nectria haematococca*. *Plant Journal*, **25**, 305–314.

Handelsman, J. *et al.*, 1990, Biological control of damping-off of alfalfa seedlings with *Bacillus cereus* UW85. *Applied and Environmental Microbiology*, **56**, 713–718.

Hanold, D. and Randles, J. W., 1991, Detection of coconut cadang-cadang viroid-like sequences in oil and coconut palm and other monocotyledons in the South-West Pacific. *Annals of Applied Biology*, **118**, 139–151.

Hansen, E. M. *et al.*, 2000, Incidence and impact of Swiss needle cast in forest plantations of Douglas-fir in coastal Oregon. *Plant Disease*, **84**, 773–778.

Hansen, G., Shillito, R. D. and Chilton, M. D., 1997, T-strand integration in maize protoplasts after codelivery of a T-DNA substrate and virulence genes. *Proceedings of the National Academy of Sciences of the United States of America*, **94**, 11726–11730.

Hanson, L. E., 1990, Carrot scab caused by *Streptomyces* spp in Michigan. *Plant Disease*, **74**, 1037.

Haralampidis, K. *et al.*, 2001, A new class of oxidosqualene cyclases directs synthesis of antimicrobial phytoprotectants in monocots. *Proceedings of the National Academy of Sciences of the United States of America*, **98**, 13431–13436.

Harder, J. *et al.*, 1997, A peptide antibiotic from human skin. *Nature*, **387**, 861.

Hardison, J. R., 1976, Flame and fire for plant disease control. *Annual Review of Phytopathology*, **14**, 355–379.

Hargreaves, J. A. and Bailey, J. A., 1978, Phytoalexin production by hypocotyls of *Phaseolus vulgaris* in response to constitutive metabolites released by damaged cells. *Physiological Plant Pathology*, **13**, 89–100

Harrewijn, P., Denouden, H. and Piron, P. G. M., 1991, Polymer webs to prevent virus transmission by aphids in seed potatoes. *Entomologia Experimentalis et Applicata*, **58**, 101.

Harris, L. J. *et al.*, 1999, Possible role of trichothecene mycotoxins in virulence of *Fusarium graminearum* on maize. *Plant Disease*, **83**, 954–960.

Harrison, B. D., 1977, Ecology and control of viruses with soil-inhabiting vectors. *Annual Review of Phytopathology*, **15**, 331–360.

Harrison, B. D., 2002, Virus variation in relation to resistance-breaking in plants. *Euphytica*, **124**, 181–192.

Harrison, J. G. *et al.*, 1990, Estimation of amounts of *Phytophthora infestans* mycelium in leaf tissue by Enzyme-Linked-Immunosorbent-Assay. *Plant Pathology*, **39**, 274–277.

Harry, I. B. and Clarke, D. D., 1986, Race-specific resistance in groundsel (*Senecio vulgaris*) to the powdery mildew *Erysiphe fischeri*. *New Phytologist*, **103**, 167–175.

Hart, J. H., 1981, Role of phytostilbenes in decay and disease resistance. *Annual Review of Phytopathology*, **19**, 437–458.

Hartman, C. L. and Knous, T. R., 1984, Field testing and preliminary progeny evaluation of alfalfa regenerated from cell-lines resistant to the toxins produced by *Fusarium oxysporum* f sp *medicaginis*. *Phytopathology*, **74**, 818.

Hartman, C. L., Mccoy, T. J. and Knous, T. R., 1984, Selection of alfalfa (*Medicago sativa*) cell-lines and regeneration of plants resistant to the toxin(s) produced by *Fusarium oxysporum* f sp *medicaginis*. *Plant Science Letters*, **34**, 183–194.

Haseloff, J., Mohamed, N. A. and Symons, R. H., 1982, Viroid RNAs of Cadang-Cadang disease of coconuts. *Nature*, **299**, 316–321.

Hatziloukas, E. and Panopoulos, N. J., 1992, Origin, structure, and regulation of *argk*, encoding the phaseolotoxin-resistant ornithine carbamoyltransferase in *Pseudomonas syringae* pv. *phaseolicola*, and functional expression of *argk* in transgenic tobacco. *Journal of Bacteriology*, **174**, 5895–5909.

Hauck, C., Muller, S. and Schildknecht, H., 1992, A germination stimulant for parasitic flowering plants from *Sorghum bicolor*, a genuine host plant. *Journal of Plant Physiology*, **139**, 474–478.

Hawes, M. C. *et al.*, 1998, Function of root border cells in plant health: Pioneers in the rhizosphere. *Annual Review of Phytopathology*, **36**, 311–327.

Hawn, E. J., 1963, Transmission of bacterial wilt of alfalfa by *Ditylenchus dipsaci* (Kuhn). *Nematologica*, **9**, 65–68.

Hayashi, N. *et al.*, 1990, Determination of host-selective toxin production during spore germination of *Alternaria alternata* by High Performance Liquid Chromatography. *Phytopathology*, **80**, 1088–1091.

He, H. Y. *et al.*, 1994, Zwittermicin-A, an antifungal and plant-protection agent from *Bacillus cereus*. *Tetrahedron Letters*, **35**, 2499–2502.

Healy, F. G. *et al.*, 2000, The *txtAB* genes of the plant pathogen *Streptomyces acidiscabies* encode a peptide synthetase required for phytotoxin thaxtomin A production and pathogenicity. *Molecular Microbiology*, **38**, 794–804.

Heath, M. C., 1981, Resistance of plants to rust infection. *Phytopathology*, **71**, 971–974.

Hegde, Y. and Kolattukudy, P. E., 1997, Cuticular waxes relieve self-inhibition of germination and appressorium formation by the conidia of *Magnaporthe grisea*. *Physiological and Molecular Plant Pathology*, **51**, 75–84.

Heiniger, U. and Rigling, D., 1994, Biological-control of chestnut blight in Europe. *Annual Review of Phytopathology*, **32**, 581–599.

Heinrich, M. *et al.*, 2001, Improved detection methods for fruit tree phytoplasmas. *Plant Molecular Biology Reporter*, **19**, 169–179.

Hellens, R. P. *et al.*, 2000, pGreen: a versatile and flexible binary Ti vector for *Agrobacterium*-mediated plant transformation. *Plant Molecular Biology*, **42**, 819–832.

Herbers, K., Conrads Strauch, J. and Bonas, U., 1992, Race-specificity of plant-resistance to bacterial spot disease determined by repetitive motifs in a bacterial avirulence protein. *Nature*, **356**, 172–174.

Herdina-Roget, D. K., 2000, Prediction of take-all disease risk in field soils using a rapid and quantitative DNA soil assay. *Plant and Soil*, **227**, 87–98.

Hermann, M., Zocher, R. and Haese, A., 1996, Effect of disruption of the enniatin synthetase gene on the virulence of *Fusarium avenaceum*. *Molecular Plant-Microbe Interactions*, **9**, 226–232.

Hewitt, W. B., Raski, D. J. and Goheen, A. C., 1958, Nematode vector of soil borne fan-leaf virus of grape vines. *Phytopathology*, **48**, 586–595.

Hijmans, R. J., Forbes, G. A. and Walker, T. S., 2000, Estimating the global severity of potato late blight with GIS-linked disease forecast models. *Plant Pathology*, **49**, 697–705.

Hilder, V. A. and Boulter, D., 1999, Genetic engineering of crop plants for insect resistance – a critical review. *Crop Protection*, **18**, 177–191.

Hilder, V. A., Gatehouse, A. M. R. and Boulter, D., 1989, Potential for exploiting plant genes to genetically engineer insect resistance, exemplified by the cowpea trypsin-inhibitor gene. *Pesticide Science*, **27**, 165–171.

Hiraoka, H. *et al.*, 1992, Characterization of *Bacillus subtilis RB14*, coproducer of peptide antibiotics Iturin-a and Surfactin. *Journal of General and Applied Microbiology*, **38**, 635–640.

Hoch, H. C. *et al.*, 1987, Signaling for growth orientation and cell-differentiation by surface-topography in *Uromyces*. *Science*, **235**, 1659–1662.

Hoch, H. C. and Staples, R. C., 1987, Structural and chemical-changes among the rust fungi during appressorium development. *Annual Review of Phytopathology*, **25**, 231–247.

Hodgson, R. A. J., Wall, G. C. and Randles, J. W., 1998, Specific identification of coconut tinangaja viroid for differential field diagnosis of viroids in coconut palm. *Phytopathology*, **88**, 774–781.

Hoffman, J. T. and Hobbs, E. L., 1985, Lodgepole pine dwarf mistletoe in the inter-mountain region. *Plant Disease*, **69**, 429–431.

Hoitink, H. A. J., Pelletier, R. L. and Coulson, J. G., 2003, Toxaemia of halo blight of beans. *Phytopathology*, **56**, 1062–1065.

Holcomb, G. E., 1986, Hosts of the parasitic alga *Cephaleuros virescens* in Louisiana and new host records for the continental United States. *Plant Disease*, **70**, 1080–1083.

Holcomb, G. E., Vann, S. R. and Buckley, J. B., 1998, First report of *Cephaleuros virescens* in Arkansas and its occurrence on cultivated blackberry in Arkansas and Louisiana. *Plant Disease*, **82**, 263.

Holland, N. *et al.*, 1997, Involvement of thylakoid overenergization in tentoxin-induced chlorosis in *Nicotiana* spp. *Plant Physiology*, **114**, 887–892.

Holliday, P., 1989, *A Dictionary of Plant Pathology*, Cambridge University Press, Cambridge, UK.

Holmes, G. J. and Stange, R. R., 2002, Influence of wound type and storage duration on susceptibility of sweet potatoes to *Rhizopus* soft rot. *Plant Disease*, **86**, 345–348.

Horio, T. *et al.*, 1992, A potent attractant of zoospores of *Aphanomyces cochlioides* isolated from its host, *Spinacia oleracea*. *Experientia*, **48**, 410–414.

Horsfall, J. G. and Barratt, R. W., 1945, An improved grading system for measuring plant disease. *Phytopathology*, **35**, 655.

Hoshino, T. *et al.*, 1997, Purification and characterization of a lipolytic enzyme active at low temperature from Norwegian *Typhula ishikariensis* group III strain. *European Journal of Plant Pathology*, **103**, 357–361.

Howell, C. R., 2002, Cotton seedling preemergence damping-off incited by *Rhizopus oryzae* and *Pythium* spp. and its biological control with *Trichoderma* spp. *Phytopathology*, **92**, 177–180.

Hu, G. G. and Rijkenberg, F. H. J., 1998, Ultrastructural localization of cytokinins in *Puccinia recondita* f.sp. *tritici* infected wheat leaves. *Physiological and Molecular Plant Pathology*, **52**, 79–94.

Hu, X. Y., Ohm, H. W. and Dweikat, I., 1997, Identification of RAPD markers linked to the gene *PM1* for resistance to powdery mildew in wheat. *Theoretical and Applied Genetics*, **94**, 832–840.

Huang, J. S. and Barker, K. R., 1991, Glyceollin I in soybean cyst nematode interactions – spatial and temporal distribution in roots of resistant and susceptible soybeans. *Plant Physiology*, **96**, 1302–1307.

Huang, J. W. and Kuhlman, E. G., 1991, Formulation of a soil amendment to control damping-off of slash pine-seedlings. *Phytopathology*, **81**, 163–170.

Huang, Y. *et al.*, 1997, Expression of an engineered cecropin gene cassette in transgenic tobacco plants confers disease resistance to *Pseudomonas syringae* pv *tabaci*. *Phytopathology*, **87**, 494–499.

Hughes, P. *et al.*, 2000, The cytotoxic plant protein, beta-purothionin, forms ion channels in lipid membranes. *Journal of Biological Chemistry*, **275**, 823–827.

Hugouvieux-Cotte-Pattat, N. *et al.*, 1996, Regulation of pectinolysis in *Erwinia chrysanthemi*. *Annual Review of Microbiology*, **50**, 213–257.

Hulbert, S. H. *et al.*, 2001, Resistance gene complexes: Evolution and utilization. *Annual Review of Phytopathology*, **39**, 285–312.

Humpherson-Jones, F. M., 1989, Survival of *Alternaria brassicae* and *Alternaria brassicicola* on crop debris of oilseed rape and cabbage. *Annals of Applied Biology*, **115**, 45–50.

Iacobellis, N. S. *et al.*, 1994, Pathogenicity of *Pseudomonas-syringae* subsp *savastanoi* mutants defective in phytohormone production. *Journal of Phytopathology-Phytopathologische Zeitschrift*, **140**, 238–248.

Ichihara, A., Tazaki, H. and Sakamura, S., 1983, Solanapyrones A, B and C, phytotoxic metabolites from the fungus *Alternaria solani*. *Tetrahedron Letters*, **24**, 5373–5376.

Ioannou, N., Psallidas, P. G. and Glynos, P., 2000, First record of bacterial canker (*Clavibacter michiganensis* ssp *michiganensis*) on tomato in Cyprus. *Journal of Phytopathology-Phytopathologische Zeitschrift*, **148**, 383–386.

Ishii, S., 1988, Factors influencing protoplast viability of suspension-cultured rice cells during isolation process. *Plant Physiology*, **88**, 26–29.

Islam, M. R., Shepherd, K. W. and Mayo, G. M. E., 1989, Effect of genotype and temperature on the expression of L genes in flax conferring resistance to rust. *Physiological and Molecular Plant Pathology*, **35**, 141–150.

Islam, M. T. and Tahara, S., 2001, Chemotaxis of fungal zoospores, with special reference to *Aphanomyces cochlioides*. *Bioscience Biotechnology and Biochemistry*, **65**, 1933–1948.

Isshiki, A. *et al.*, 2001, Endopolygalacturonase is essential for citrus black rot caused by *Alternaria citri* but not brown spot caused by *Alternaria alternata*. *Molecular Plant-Microbe Interactions*, **14**, 749–757.

Ito, Y., Kaku, H. and Shibuya, N., 1997, Identification of a high-affinity binding protein for N-acetylchitooligosaccharide elicitor in the plasma membrane of suspension-cultured rice cells by affinity labeling. *Plant Journal*, **12**, 347–356.

Iwai, T. *et al.*, 2002, Enhanced resistance to seed-transmitted bacterial diseases in transgenic rice plants overproducing an oat cell-wall-bound thionin. *Molecular Plant-Microbe Interactions*, **15**, 515–521.

Jackson, R. W. *et al.*, 1999, Identification of a pathogenicity island, which contains genes for virulence and avirulence, on a large native plasmid in the bean pathogen *Pseudomonas syringae* pathovar *phaseolicola*. *Proceedings of the National Academy of Sciences of the United States of America*, **96**, 10875–10880.

James, W. C., 1971, An illustrated series of assessment keys for plant diseases, their preparation and usage. *Canadian Plant Disease Survey*, **51**, 39–65.

James, W. C. and Teng, P. S., 1979, The quantification of production constraints associated with plant disease. *Applied Biology*, **4**, 201–267.

James, W. C. *et al.*, 1972, The quantitative relationship between late blight of potato and loss in tuber yield. *Phytopathology*, **62**, 92–96.

Janisiewicz, W. *et al.*, 1991, Postharvest control of blue mold and gray mold of apples and pears by dip treatment with pyrrolnitrin, a metabolite of *Pseudomonas cepacia*. *Plant Disease*, **75**, 490–494.

Janse, J. D., 1988, A *Streptomyces* species identified as the cause of carrot scab. *Netherlands Journal of Plant Pathology*, **94**, 303–306.

Jatala, P., 1986, Biological-control of plant-parasitic nematodes. *Annual Review of Phytopathology*, **24**, 453–489.

Jeffers, S. N. and Aldwinckle, H. S., 1987, Enhancing detection of *Phytophthora cactorum* in naturally infested soil. *Phytopathology*, **77**, 1475–1482.

Jetter, R. and Riederer, M., 2000, Composition of cuticular waxes on *Osmunda regalis* fronds. *Journal of Chemical Ecology*, **26**, 399–412.

Jia, Y. *et al.*, 2000, Direct interaction of resistance gene and avirulence gene products confers rice blast resistance. *EMBO Journal*, **19**, 4004–4014.

Jin, H. *et al.*, 1996, Regeneration of soybean plants from embryogenic suspension cultures treated with toxic culture filtrate of *Fusarium solani* and screening of regenerants for resistance. *Phytopathology*, **86**, 714–718.

Johansen, E., Edwards, M. C. and Hampton, R. O., 1994, Seed transmission of viruses – current perspectives. *Annual Review of Phytopathology*, **32**, 363–386.

Johnson, R., 1984, A critical analysis of durable resistance. *Annual Review of Phytopathology*, **22**, 309–330.

Johnson, R. D. *et al.*, 2000, Cloning and characterization of a cyclic peptide synthetase gene from *Alternaria alternata* apple pathotype whose product is involved in AM-toxin synthesis and pathogenicity. *Molecular Plant-Microbe Interactions*, **13**, 742–753.

Jones, A. T., 1987, Control of virus-infection in crop plants through vector resistance – a review of achievements, prospects and problems. *Annals of Applied Biology*, **111**, 745–772.

Jones, D. A. *et al.*, 1994, Isolation of the tomato *Cf-9* gene for resistance to *Cladosporiu fulvum* by transposon tagging. *Science*, **266**, 789–793.

Jones, D. A. C., *et al.*, 1994, Blackleg potential of potato seed – determination of tuber contamination by *Erwinia carotovora* subsp *atroseptica* by immunofluorescence colony staining and stock and tuber sampling. *Annals of Applied Biology*, **124**, 557–568.

Jones, J. B. *et al.*, 1995, A 3rd tomato race of *Xanthomonas campestris* pv *vesicatoria*. *Plant Disease*, **79**, 395–398.

Jones, J. D. G. *et al.*, 1986, Isolation and characterization of genes encoding 2 chitinase enzymes from *Serratia marcescens*. *EMBO Journal*, **5**, 467–473.

Jones, J. P., Engelhard, A. W. and Woltz, S. S., 1989, 'Management of Fusarium wilt of vegetables and ornamentals by macro- and microelement nutrition,' in *Soilborne Plant Pathogens: Management of Diseases with Macro- and Micro-elements*, Arthur W. Engelhard, ed., American Phytopathological Society, St. Paul, Minnesota, USA pp. 18–32.

Jones, M. J. and Epstein, L., 1989, Adhesion of *Nectria haematococca* macroconidia. *Physiological and Molecular Plant Pathology*, **35**, 453–461.

Jones, M. J. and Dunkle, L. D., 1995, Virulence gene-expression during conidial germination in *Cochliobolus carbonum*. *Molecular Plant-Microbe Interactions*, **8**, 476–479.

Jones, R. A. C. and McKirdy, S. J., 1990, Seed-borne cucumber mosaic-virus infection of subterranean clover in Western-Australia. *Annals of Applied Biology*, **116**, 73–86.

Jones, R. A. C., 2000, Determining 'threshold' levels for seed-borne virus infection in seed stocks. *Virus Research*, **71**, 171–183.

Jones, S., *et al.*, 1993, The lux autoinducer regulates the production of exoenzyme virulence determinants in *Erwinia carotovora* and *Pseudomonas aeruginosa*. *EMBO Journal*, **12**, 2477–2482.

Jones, S. W., Donaldson, S. P. and Deacon, J. W., 1991, Behavior of zoospores and zoospore cysts in relation to root infection by *Pythium aphanidermatum*. *New Phytologist*, **117**, 289–301.

Jongedijk, E. *et al.*, 1995, Synergistic activity of chitinases and beta-1, 3-glucanases enhances fungal resistance in transgenic tomato plants. *Euphytica*, **85**, 173–180.

Joosten, M. H. A. J. and de Wit, P. J. G. M., 1999, The tomato – *Cladosporium fulvum* interaction: A versatile experimental system to study plant-pathogen interactions. *Annual Review of Phytopathology*, **37**, 335–367.

Joosten, M. H. A. J., Cozijnsen, T. J. and de Wit, P. J. G. M., 1994, Host-resistance to a fungal tomato pathogen lost by a single base-pair change in an avirulence gene. *Nature*, **367**, 384–386.

Joosten, M. H. A. J., *et al.*, 1997, The biotrophic fungus *Cladosporium fulvum* circumvents *Cf-4*-mediated resistance by producing unstable AVR4 elicitors. *Plant Cell*, **9**, 367–379.

Jung, C., Cai, D. G. and Kleine, M., 1998, Engineering nematode resistance in crop species. *Trends in Plant Science*, **3**, 266–271.

Kado, C. I. and Heskett, M. G., 1970, Selective media for isolation of *Agrobacterium, Corynebacterium, Erwinia, Pseudomonas* and *Xanthomonas*. *Phytopathology*, **60**, 969–976.

Kakani, K., Sgro, J. Y. and Rochon, D., 2001, Identification of specific cucumber necrosis virus coat protein amino acids affecting fungus transmission and zoospore attachment. *Journal of Virology*, **75**, 5576–5583.

Kalinichenko, G. V. and Kalinichenko, R. I., 1983, Characteristics of fruiting of pear under bacterial infection in the Crimea. *Sel'skokhozyaistvennaya Biologiya*, **4**, 68–71.

Kaloshian, I. *et al.*, 2000, *Mi*-mediated resistance against the potato aphid *Macrosiphum euphorbiae* (Hemiptera: Aphididae) limits sieve element ingestion. *Environmental Entomology*, **29**, 690–695.

Kang, Y. W. *et al.*, 1998, Characterization of genes involved in biosynthesis of a novel antibiotic from *Burkholderia cepacia* BC11 and their role in biological control of *Rhizoctonia solani*. *Applied and Environmental Microbiology*, **64**, 3939–3947.

Karasawa, A. *et al.*, 1999. One amino acid change in cucumber mosaic virus RNA polymerase determines virulent/avirulent phenotypes on cowpea. *Phytopathology*, **89**, 1186–1192.

Katan, J., 1981, Solar heating (solarization) of soil for control of soilborne pests. *Annual Review of Phytopathology*, **19**, 211–236.

Katsuwon, J. and Anderson, A. J., 1990, Catalase and superoxide-dismutase of root-colonizing saprophytic fluorescent Pseudomonads. *Applied and Environmental Microbiology*, **56**, 3576–3582.

Kearney, B. and Staskawicz, B. J., 1990, Widespread distribution and fitness contribution of *Xanthomonas campestris* avirulence gene *avrbs2. Nature*, **346**, 385–386.

Kearney, B. *et al.*, 1988, Molecular-basis for evasion of plant host defense in bacterial spot disease of pepper. *Nature*, **332**, 541–543.

Keel, C. *et al.*, 1992, Suppression of root diseases by *Pseudomonas fluorescens* Cha0 – importance of the bacterial secondary metabolite 2,4-diacetylphloroglucinol. *Molecular Plant-Microbe Interactions*, **5**, 4–13.

Kelemu, S. and Collmer, A., 1993, Erwinia-chrysanthemi EC16 produces a 2nd set of plant-inducible pectate lyase isozymes. *Applied and Environmental Microbiology*, **59**, 1756–1761.

Kende, H. and Zeevaart, J. A. D., 1997, The five "classical" plant hormones. *Plant Cell*, **9**, 1197–1210.

Kennedy, B. S., *et al.*, 1992, Leaf surface chemicals from *Nicotiana* affecting germination of *Peronospora tabacina* (Adam) sporangia. *Journal of Chemical Ecology*, **18**, 1467–1479.

Kenyon, J. S. and Turner, J. G., 1992, The stimulation of ethylene synthesis in *Nicotiana tabacum* leaves by the phytotoxin coronatine. *Plant Physiology*, **100**, 219–224.

Kerr, A., 1969, Transfer of virulence between strains of *Agrobacterium. Nature*, **223**, 1175–1176.

Kerr, A. and Ellis, J. G., 1982, 'Conjugation and transfer of Ti plasmids in *Agrobacterium tumefaciens*', in *Molecular Biology of Plant Tumours*, J. Kahl & J. S. Schell, eds., Academic Press, New York, USA pp. 321–344.

Kerr, A. and Tate, M. E., 1984, Agrocins and the biological-control of crown gall. *Microbiological Sciences*, **1**, 1–4.

Kerry, B. R., 1990, An assessment of progress toward microbial control of plant-parasitic nematodes. *Journal of Nematology*, **22**, 621–631.

Kerry, B. R., 2000, Rhizosphere interactions and the exploitation of microbial agents for the biological control of plant-parasitic nematodes. *Annual Review of Phytopathology*, **38**, 423–441.

Kiba, A. *et al.*, 1997, Superoxide generation in extracts from isolated plant cell walls is regulated by fungal signal molecules. *Phytopathology*, **87**, 846–852.

Kim, J. C. *et al.*, 1998, Target site of a new antifungal compound KC10017 in the melanin biosynthesis of *Magnaporthe grisea. Pesticide Biochemistry and Physiology*, **62**, 102–112.

Kim, S. D., Knoche, H. W. and Dunkle, L. D., 1987, Essentiality of the ketone function for toxicity of the host-selective toxin produced by *Helminthosporium carbonum. Physiological and Molecular Plant Pathology*, **30**, 433–440.

Kim, Y. C. *et al.*, 2001, Oxycom (TM) under field and laboratory conditions increases resistance responses in plants. *European Journal of Plant Pathology*, **107**, 129–136.

Kimura, M. *et al.*, 1998, Features of *Tri101*, the trichothecene 3-O-acetyltransferase gene, related to the self-defense mechanism in *Fusarium graminearum. Bioscience Biotechnology and Biochemistry*, **62**, 1033–1036.

King, R. C. and Stansfield, W. D., 1990, *A Dictionary of Genetics*, Oxford University Press, Oxford UK.

King, R. R., Lawrence, C. H. and Calhoun, L. A., 1992, Chemistry of phytotoxins associated with *Streptomyces scabies*, the causal organism of potato common scab. *Journal of Agricultural and Food Chemistry*, **40**, 834–837.

Kinkel, L. L. *et al.*, 1998, Quantitative relationships among thaxtomin A production, potato scab severity, and fatty acid composition in *Streptomyces*. *Canadian Journal of Microbiology*, **44**, 768–776.

Kittipakorn, K. *et al.*, 1993, Strains of peanut stripe potyvirus rapidly identified by profiling peptides of the virion proteins. *Journal of Phytopathology-Phytopathologische Zeitschrift*, **137**, 257–263.

Kiyosawa, S. and Shiyomi, M., 1972, A theoretical evaluation of the effect of mixing resistant variety with susceptible variety for controlling plant diseases. *Annals of the Phytopathological Society of Japan*, **38**, 41–51.

Klement, Z. and Goodman, R. N., 1967, Hypersensitive reaction to infection by bacterial plant pathogens. *Annual Review of Phytopathology*, **5**, 17.

Klessig, D. F. *et al.*, 2000, Nitric oxide and salicylic acid signaling in plant defense. *Proceedings of the National Academy of Sciences of the United States of America*, **97**, 8849–8855.

Kloek, A. P. *et al.*, 2001, Resistance to *Pseudomonas syringae* conferred by an *Arabidopsis thaliana* coronatine-insensitive (*coi1*) mutation occurs through two distinct mechanisms. *Plant Journal*, **26**, 509–522.

Knoester, M. *et al.*, 1998, Ethylene-insensitive tobacco lacks nonhost resistance against soil-borne fungi. *Proceedings of the National Academy of Sciences of the United States of America*, **95**, 1933–1937.

Knoester, M. *et al.*, 1999, Systemic resistance in *Arabidopsis* induced by rhizobacteria requires ethylene-dependent signaling at the site of application. *Molecular Plant-Microbe Interactions*, **12**, 720–727.

Kobayashi, T. *et al.*, 2001, Detection of rice panicle blast with multispectral radiometer and the potential of using airborne multispectral scanners. *Phytopathology*, **91**, 316–323.

Kocks, C. G. *et al.*, 1998, Survival and extinction of *Xanthomonas campestris* pv. *campestris* in soil. *European Journal of Plant Pathology*, **104**, 911–923.

Kodama, R. *et al.*, 1999, The translocation-associated *Tox1* locus of *Cochliobolus heterostrophus* is two genetic elements on two different chromosomes. *Genetics*, **151**, 585–596.

Kogel, G. *et al.*, 1988, A single glycoprotein from *Puccinia graminis* f sp *tritici* cell-walls elicits the hypersensitive lignification response in wheat. *Physiological and Molecular Plant Pathology*, **33**, 173–185.

Kogel, G. *et al.*, 1991, Specific binding of a hypersensitive lignification elicitor from *Puccinia graminis* f sp *tritici* to the plasma-membrane from wheat (*Triticum aestivum L*). *Planta*, **183**, 164–169.

Kohl, J. *et al.*, 1995, Effect of *Ulocladium atrum* and other antagonists on sporulation of *Botrytis cinerea* on dead lily leaves exposed to field conditions. *Phytopathology*, **85**, 393–401.

Kohl, J. *et al.*, 1995, Suppression of sporulation of *Botrytis* spp as a valid biocontrol strategy. *European Journal of Plant Pathology*, **101**, 251–259.

Kohl, J. *et al.*, 1995, Effect of interrupted leaf wetness periods on suppression of sporulation of *Botrytis allii* and *Botrytis cinerea* by antagonists on dead onion leaves. *European Journal of Plant Pathology*, **101**, 627–637.

Kolattukudy, P. E., 1980, Biopolyester membranes of plants:cutin and suberin. *Science*, **208**, 990–1000.

Kolattukudy, P. E. *et al.*, 1995, Surface signaling in pathogenesis. *Proceedings of the National Academy of Sciences of the United States of America*, **92**, 4080–4087.

Köller, W. *et al.*, 1995, Role of cutinase in the invasion of plants. *Canadian Journal of Botany-Revue Canadienne de Botanique*, **73**, S1109–S1118.

Kono, Y. *et al.*, 1991, Structures of oryzalide-A and oryzalide-B, and oryzalic acid-A, a group of novel antimicrobial diterpenes, isolated from healthy leaves of a bacterial leaf blight-resistant cultivar of rice plant. *Agricultural and Biological Chemistry*, **55**, 803–811.

KoomanGersmann, M. *et al.*, 1996, A high-affinity binding site for the AVR9 peptide elicitor of *Cladosporium fulvum* is present on plasma membranes of tomato and other solanaceous plants. *Plant Cell*, **8**, 929–938.

Kousik, C. S. and Ritchie, D. F., 1996, Disease potential of pepper bacterial spot pathogen races that overcome the *Bs2* gene for resistance. *Phytopathology*, **86**, 1336–1343.

Kramer, R., Freytag, S. and Schmelzer, E., 1997, *In vitro* formation of infection structures of *Phytophthora infestans* is associated with synthesis of stage specific polypeptides. *European Journal of Plant Pathology*, **103**, 43–53.

Kristensen, A. K. *et al.*, 2000, Characterization of a new antifungal non-specific lipid transfer protein (nsLTP) from sugar beet leaves. *Plant Science*, **155**, 31–40.

Kuan, T. L., 1988, Inoculum thresholds of seedborne pathogens: overview. *Phytopathology*, **78**, 867–868.

Kuć, J., 1995, Phytoalexins, stress metabolism, and disease resistance in plants. *Annual Review of Phytopathology*, **33**, 275–297.

Kuć, J., 2001, Concepts and direction of induced systemic resistance in plants and its application. *European Journal of Plant Pathology*, **107**, 7–12.

Kuflu, K. M. and Cuppels, D. A. 1997, Development of a diagnostic DNA probe for xanthomonads causing bacterial spot of peppers and tomatoes. *Applied and Environmental Microbiology*, **63**, 4462–4470.

Kuijt, J., 1969, *The Biology of Parasitic Flowering Plants*, University of California, Berkeley, California USA.

Kunoh, H. *et al.*, 1988, Preparation of the infection court by *Erysiphe graminis*. 1. Contact-mediated changes in morphology of the conidium surface. *Experimental Mycology*, **12**, 325–335.

Kuo, K. C. and Hoch, H. C., 1996, Germination of *Phyllosticta ampelicida* pycnidiospores: Prerequisite of adhesion to the substratum and the relationship of substratum wettability. *Fungal Genetics and Biology*, **20**, 18–29.

Kwon, C. Y. *et al.*, 1996, A quantitative bioassay for necrosis toxin from *Pyrenophora tritici-repentis* based on electrolyte leakage. *Phytopathology*, **86**, 1360–1363.

Lacy, M. L., 1994, Influence of wetness periods on infection of celery by *Septoria-apiicola* and use in timing spray for control. *Plant Disease*, **78**, 975–979.

Lagrimini, L. M. *et al.*, 1997, Characterization of antisense transformed plants deficient in the tobacco anionic peroxidase. *Plant Physiology*, **114**, 1187–1196.

Lamari, L. *et al.*, 1995, Identification of a new race in *Pyrenophora tritici-repentis*: Implications for the current pathotype classification system. *Canadian Journal of Plant Pathology-Revue Canadienne de Phytopathologie*, **17**, 312–318.

Lamb, C. and Dixon, R. A., 1997, The oxidative burst in plant disease resistance. *Annual Review of Plant Physiology and Plant Molecular Biology*, **48**, 251–275.

Lambert, D. H., 1990, Effects of pruning method on the incidence of mummy berry and other lowbush blueberry diseases. *Plant Disease*, **74**, 199–201.

Landa, B. B. *et al.*, 2002, Differential ability of genotypes of 2,4-diacetylphloroglucinol-producing *Pseudomonas fluorescens* strains to colonize the roots of pea plants. *Applied and Environmental Microbiology*, **68**, 3226–3237.

Lane, J. A. *et al.*, 1996, Characterization of virulence and geographic distribution of *Striga gesnerioides* on cowpea in West Africa. *Plant Disease*, **80**, 299–301.

Large, E. C., 1940, *The Advance of the Fungi*, Jonathan Cape, London, UK.

Lascaris, D. and Deacon, J. W., 1994, *In vitro* growth and microcycle conidiation of *Idriella bolleyi*, a biocontrol agent of cereal pathogens. *Mycological Research*, **98**, 1200–1206.

Laugé, R. *et al.*, 1997, The *in planta*-produced extracellular proteins ECP1 and ECP2 of *Cladosporium fulvum* are virulence factors. *Molecular Plant-Microbe Interactions*, **10**, 725–734.

Laugé, R. *et al.*, 1998, Successful search for a resistance gene in tomato targeted against a virulence factor of a fungal pathogen. *Proceedings of the National Academy of Sciences of the United States of America*, **95**, 9014–9018.

Lawrence, C. H., Clark, M. C. and King, R. R., 1990, Induction of common scab symptoms in aseptically cultured potato-tubers by the vivotoxin, thaxtomin. *Phytopathology*, **80**, 606–608.

Lawrence, E. (ed.) (2000) *Henderson's Dictionary of Biological Terms*, Twelfth edition, Prentice Hall, Harlow, UK.

Lawrence, G. J. *et al.*, 1995, The L6 gene for flax rust resistance is related to the *Arabidopsis* bacterial-resistance gene *Rps2* and the tobacco viral resistance gene *N*. *Plant Cell*, **7**, 1195–1206.

Lawton, K. A. *et al.*, 1994, Acquired-resistance signal-transduction in *Arabidopsis* is ethylene independent. *Plant Cell*, **6**, 581–588.

Lazarovits, G., 2001, Management of soil-borne plant pathogens with organic soil amendments: a disease control strategy salvaged from the past. *Canadian Journal of Plant Pathology-Revue Canadienne de Phytopathologie*, **23**, 1–7.

Leach, J. E. and White, F. F., 1996, Bacterial avirulence genes. *Annual Review of Phytopathology*, **34**, 153–179.

Leach, J. E. *et al.*, 2001, Pathogen fitness penalty as a predictor of durability of disease resistance genes. *Annual Review of Phytopathology*, **39**, 187–224.

Leah, R. *et al.*, 1991, Biochemical and molecular characterization of 3 barley seed proteins with antifungal properties. *Journal of Biological Chemistry*, **266**, 1564–1573.

Leckie, F. *et al.*, 1999, The specificity of polygalacturonase-inhibiting protein (PGIP): a single amino acid substitution in the solvent-exposed beta-strand/beta-turn region of the leucine-rich repeats (LRRs) confers a new recognition capability. *EMBO Journal*, **18**, 2352–2363.

Lecomte, P. and Paulin, J. P., 1989, Disease control in apple and pear orchards. *Phytoma*, **408**, 22–26.

Lee, I. M., Davis, R. E. and Gundersen-Rindal, D. E., 2000, Phytoplasma: phytopathogenic mollicutes. *Annual Review of Microbiology*, **54**, 221–255.

Lee, J., Klessig, D. F. and Nurnberger, T., 2001, A hairpin binding site in tobacco plasma membranes mediates activation of the pathogenesis-related gene *HIN1* independent of extracellular calcium but dependent on mitogen-activated protein kinase activity. *Plant Cell*, **13**, 1079–1093.

Lee, S. C. and West, C. A., 1981, Properties of *Rhizopus stolonifer* polygalacturonase, an elicitor of casbene synthetase-activity in castor bean (*Ricinus communis* L) seedlings. *Plant Physiology*, **67**, 640–645.

Lee, Y. H. and Dean, R. A., 1994, Hydrophobicity of contact surface induces appressorium formation in *Magnaporthe grisea*. *FEMS Microbiology Letters*, **115**, 71–75.

Legg, J. P., 1999, Emergence, spread and strategies for controlling the pandemic of cassava mosaic virus disease in east and central Africa. *Crop Protection*, **18**, 627–637.

Leikin-Frenkel, A. and Prusky, D., 1998, Ethylene enhances the antifungal lipid content in idioblasts from avocado mesocarp. *Phytochemistry*, **49**, 2291–2298.

Leisola, M. S. A. *et al.*, 1984, Role of veratryl alcohol in lignin degradation by *Phanerochaete chrysosporium*. *Journal of Biotechnology*, **1**, 331–339.

Leon, R. *et al.*, 1996, Physiological and biochemical changes in shoots of coconut palms affected by lethal yellowing. *New Phytologist*, **134**, 227–234.

Leone, G., 1990, *In vivo* and *in vitro* phosphate-dependent polygalacturonase production by different isolates of *Botrytis cinerea*. *Mycological Research*, **94**, 1039–1045.

Letham, D. S., 1973, Cytokinins from *Zea mays*. *Phytochemistry*, **12**, 2445–2455.

Letham, D. S. and Palni, L. M. S., 1983, The biosynthesis and metabolism of cytokinins. *Annual Review of Plant Physiology and Plant Molecular Biology*, **34**, 163–197.

Levy, M. *et al.*, 1991, DNA fingerprinting with a dispersed repeated sequence resolves pathotype diversity in the rice blast fungus. *Plant Cell*, **3**, 95–102.

Li, C. M. *et al.*, 2002, The Hrp pilus of *Pseudomonas syringae* elongates from its tip and acts as a conduit for translocation of the effector protein HrpZ. *EMBO Journal*, **21**, 1909–1915.

Li, S. X., Spear, R. N. and Andrews, J. H., 1997, Quantitative fluorescence in situ hybridization of *Aureobasidium pullulans* on microscope slides and leaf surfaces. *Applied and Environmental Microbiology*, **63**, 3261–3267.

Li, Y. C. *et al.*, 1999, Variation in the glucosinolate content of vegetative tissues of Chinese lines of *Brassica napus* L. *Annals of Applied Biology*, **134**, 131–136.

Liang, L. Z. *et al.*, 1994, Variation in virulence, plasmid content, and genes for coronatine synthesis between *Pseudomonas syringae* pv *morsprunorum* and *P-s-syringae* from *Prunus*. *Plant Disease*, **78**, 389–392.

Lichter, A. *et al.*, 1995, The genes involved in cytokinin biosynthesis in *Erwinia herbicola* pv *gypsophilae* – characterization and role in gall formation. *Journal of Bacteriology*, **177**, 4457–4465.

Lieberei, R. *et al.*, 1989, Cyanogenesis inhibits active defense reactions in plants. *Plant Physiology*, **90**, 33–36.

Liebermann, B., Nussbaum, R. P. and Gunther, W., 2000, Bicycloalternarenes produced by the phytopathogenic fungus *Alternaria alternata*. *Phytochemistry*, **55**, 987–992.

Lim, H. S. and Kim, S. D., 1997, Role of siderophores in biocontrol of *Fusarium solani* and enhanced growth response of bean by *Pseudomonas fluorescens* GL20. *Journal of Microbiology and Biotechnology*, **7**, 13–20.

Lin, N. S., Hsu, Y. H. and Hsu, H. T., 1990, Immunological detection of plant-viruses and a mycoplasmalike organism by direct tissue blotting on nitrocellulose membranes. *Phytopathology*, **80**, 824–828.

Lin, Y. L. and Lin, C. H., 1990, Involvement of transfer-RNA bound cytokinin on the gall formation in *Zizania. Journal of Experimental Botany*, **41**, 277–281.

Lindgren, P. B., 1997, The role of *hrp* genes during plant-bacterial interactions. *Annual Review of Phytopathology*, **35**, 129–152.

Lindgren, P. B., Peet, R. C. and Panopoulos, N. J., 1986, Gene-cluster of *Pseudomonas-syringae* pv *phaseolicola* controls pathogenicity of bean-plants and hypersensitivity on nonhost plants. *Journal of Bacteriology*, **168**, 512–522.

Lindgren, P. B. *et al.*, 1988 Genes required for pathogenicity and hypersensitivity are conserved and interchangeable among pathovars of *Pseudomonas syringae. Molecular and General Genetics*, **211**, 499–506.

Link, K. P. and Walker, J. C., 1933, The isolation of catechol from pigmented onion scales and its significance in relation to disease resistance in onions. *Journal of Biological Chemistry*, **100**, 379–383.

Link, K. P., Angell, H. R. and Walker, J. C., 1929a, The isolation of protocatechuic acid from pigmented onion scales and its significance in relation to disease resistance in onions. *Journal of Biological Chemistry*, **81**, 369–375.

Link, K. P., Angell, H. R. and Walker, J. C., 1929b, Further observations on the occurrence of protocatechuic acid in pigmented onion scales and its significance in relation to disease resistance in onions. *Journal of Biological Chemistry*, **81**, 719–725.

Linke, K. H., Sauerborn, J. and Saxena, M. C., 1991, Host-parasite relationship – effect of *Orobanche crenata* seed banks on development of the parasite and yield of faba bean. *Angewandte Botanik*, **65**, 229–238.

Liu, L. X. and Shaw, P. D., 1997, Characterization of *dapB*, a gene required by *Pseudomonas syringae* pv *tabaci* BR2.024 for lysine and tabtoxinine-beta-lactam biosynthesis. *Journal of Bacteriology*, **179**, 507–513.

Liu, Y. C. and Milgroom, M. G., 1996, Correlation between hypovirus transmission and the number of vegetative incompatibility (*vic*) genes different among isolates from a natural population of *Cryphonectria parasitica. Phytopathology*, **86**, 79–86.

Löbler, M. and Klämbt, D., 1985a, Auxin-binding protein from coleoptile membranes of corn (*Zea mays* L) .1. Purification by immunological methods and characterization. *Journal of Biological Chemistry*, **260**, 9848–9853.

Löbler, M. and Klämbt, D., 1985b, Auxin-binding protein from coleoptile membranes of corn (*Zea mays* L) .2. Localization of a putative auxin receptor. *Journal of Biological Chemistry*, **260**, 9854–9859.

Locci, R., 1994, Actinomycetes as plant-pathogens. *European Journal of Plant Pathology*, **100**, 179–200.

Lodge, J. K., Kaniewski, W. K. and Tumer, N. E., 1993 Broad-spectrum virus-resistance in transgenic plants expressing pokeweed antiviral protein. *Proceedings of the National Academy of Sciences of the United States of America*, **90**, 7089–7093.

Loeffler, W. *et al.*, 1986, Antifungal effects of bacilysin and fengymycin from *Bacillus subtilis* F-29-3 a comparison with activities of other *Bacillus* antibiotics. *Journal of Phytopathology-Phytopathologische Zeitschrift*, **115**, 204–213.

Longstroth, M., 2000, The 2000 fireblight epidemic in southwest Michigan. www.msue. msu.edu/vanburen/fb2000.htm.

López-Carbonell, M., Moret, A. and Nadal, M., 1998, Changes in cell ultrastructure and zeatin riboside concentrations in *Hedera helix, Pelargonium zonale, Prunus avium*, and *Rubus ulmifolius* leaves infected by fungi. *Plant Disease*, **82**, 914–918.

López-Herrera, C. J., Verduvaliente, B. and Melerovara, J. M., 1994, Eradication of primary inoculum of *Botrytis cinerea* by soil solarization. *Plant Disease*, **78**, 594–597.

Loria, R. *et al.*, 1997 Plant pathogenicity in the genus *Streptomyces*. *Plant Disease*, **81**, 836–846.

Lorito, M. *et al.*, 1998, Genes from mycoparasitic fungi as a source for improving plant resistance to fungal pathogens. *Proceedings of the National Academy of Sciences of the United States of America*, **95**, 7860–7865.

Louws, F. J., Rademaker, J. L. W. and de Bruijn, F. J., 1999, The three Ds of PCR-based genomic analysis of phytobacteria: Diversity, detection, and disease diagnosis. *Annual Review of Phytopathology*, **37**, 81–125.

Louws, F. J. *et al.*, 1998, *rep*-PCR-mediated genomic fingerprinting: A rapid and effective method to identify *Clavibacter michiganensis*. *Phytopathology*, **88**, 862–868.

Lu, S. E., Scholz-Schroeder, B. K. and Gross, D. C., 2002, Characterization of the *salA*, *syrF*, and *syrG* regulatory genes located at the right border of the syringomycin gene cluster of *Pseudomonas syringae* pv. *syringae*. *Molecular Plant-Microbe Interactions*, **15**, 43–53.

Luck, J. E. *et al.*, 2000, Regions outside of the leucine-rich repeats of flax rust resistance proteins play a role in specificity determination. *Plant Cell*, **12**, 1367–1377.

Ludwig-Muller, J., Epstein, E. and Hilgenberg, W., 1996, Auxin-conjugate hydrolysis in Chinese cabbage: Characterization of an amidohydrolase and its role during infection with clubroot disease. *Physiologia Plantarum*, **97**, 627–634.

Luig, N. H. and Rajaram, S., 1972, The effect of temperature and genetic background on host gene expression and interaction to *Puccinia graminis tritici*. *Phytopathology*, **62**, 1171–1174.

Lukens, J. H., Mathews, D. E. and Durbin, R. D., 1987, Effect of tagetitoxin on the levels of ribulose 1,5-bisphosphate carboxylase, ribosomes, and RNA in plastids of wheat leaves. *Plant Physiology*, **84**, 808–813.

Lulai, E. C. and Corsini, D. L., 1998, Differential deposition of suberin phenolic and aliphatic domains and their roles in resistance to infection during potato tuber (*Solanum tuberosum* L.) wound-healing. *Physiological and Molecular Plant Pathology*, **53**, 209–222.

Lund, S. T., Stall, R. E. and Klee, H. J., 1998, Ethylene regulates the susceptible response to pathogen infection in tomato. *Plant Cell*, 371–382.

Lydon, J. and Patterson, C. D., 2001, Detection of tabtoxin-producing strains of *Pseudomonas syringae* by PCR. *Letters in Applied Microbiology*, **32**, 166–170.

Macko, V. *et al.*, 1992, Structure of the host-specific toxins produced by the fungal pathogen *Periconia circinata*. *Proceedings of the National Academy of Sciences of the United States of America*, **89**, 9574–9578.

MacNish, G. C. *et al.*, 1994, Anastomosis group (Ag) affinity of pectic isozyme (zymogram) groups (Zg) of *Rhizoctonia solani* from the Western-Australian cereal belt. *Mycological Research*, **98**, 1369–1375.

Madden, L. V. and Ellis, M. A., 1990, Effect of ground cover on splash dispersal of *Phytophthora cactorum* from strawberry fruits. *Journal of Phytopathology-Phytopathologische Zeitschrift*, **129**, 170–174.

Madden, L. V. and Hughes, G., 1995, Plant-disease incidence – distribution, heterogeneity, and temporal analysis. *Annual Review of Phytopathology*, **33**, 529–564.

Madrigal, C., Tadeo, J. L. and Melgarejo, P., 1991, Relationship between flavipin production by *Epicoccum nigrum* and antagonism against *Monilinia laxa*. *Mycological Research*, **95**, 1375–1381.

Mahuku, G. S., Goodwin, P. H. and Hall, R., 1995, A competitive polymerase chain-reaction to quantify DNA of *Leptosphaeria maculans* during blackleg development in oilseed rape. *Molecular Plant-Microbe Interactions*, **8**, 761–767.

Maiti, I. B. and Kolattukudy, P. E., 1979, Prevention of fungal infection of plants by specific inhibition of cutinase. *Science*, **205**, 507–508.

Malamy, J. *et al.*, 1990, Salicylic acid – a likely endogenous signal in the resistance response of tobacco to viral infection. *Science*, **250**, 1002–1004.

Manandhar, C. L. and Gill, P. S., 1984, The epidemiological role of pollen transmission of viruses. *Journal of Plant Disease Protection*, **91**, 246–249.

Manandhar, J. B., Hartman, G. L. and Wang, T. C., 1995, Semiselective medium for *Colletotrichum gloeosporioides* and occurrence of 3 *Colletotrichum* spp on pepper plants. *Plant Disease*, **79**, 376–379.

Mani, E., Hasler, T. and Charriere, J. D., 1996, How much do bees contribute to the spread of fire blight? *Schweizerische Beinen Zeitung*, **119**, 135–140.

Maniara, G., Laine, R. and Kuć, J., 1984, Oligosaccharides from *Phytophthora infestans* enhance the elicitation of sesquiterpenoid stress metabolites by arachidonic acid in potato. *Physiological Plant Pathology*, **24**, 177–186.

Mansfield, J. W., 2000, 'Antimicrobial compounds and resistance: the role of phytoalexins and phytoanticipins,' in *Mechanisms of Resistance to Plant Diseases*, A. Slusarenko, R. S. S. Fraser and L. C. van Loon, eds., Kluwer, The Netherlands, pp. 325–370.

Manulis, S., Netzer, D. and Barash, I., 1986, Structure-activity relationships as inferred from comparative phytotoxicity of stemphyloxins and betaenones. *Journal of Phytopathology-Phytopathologische Zeitschrift*, **115**, 283–287.

Manulis, S. *et al.*, 1998, Detection of *Erwinia herbicola* pv. *gypsophilae* in Gypsophila plants by PCR. *European Journal of Plant Pathology*, **104**, 85–91.

Marasas, W. F. O., Nelson, P. E. and Tousson, T. A., 1985, *Taxonomy of Toxigenic Fusaria. In Trichothecenes and other Mycotoxins*, John Wiley & Sons Ltd., Chichester, UK.

Markham, P. G. *et al.*, 1987, The acquisition of a caulimovirus by different aphid species – comparison with a potyvirus. *Annals of Applied Biology*, **111**, 571–587.

Marlatt, R. B., Pohronezny, K. and McSorley, R., 1983, Field survey of Tahiti lime, *Citrus latifolia*, for algal disease, melanose, and greasy spot in Southern Florida. *Plant Disease*, **67**, 946–949.

Martelli, G. P., 1992, Classification and nomenclature of plant-viruses – state-of-the-art. *Plant Disease*, **76**, 436–442.

Martin, G. B. A. *et al.*, 1993, Map-based cloning of a protein-kinase gene conferring disease resistance in tomato. *Science*, **262**, 1432–1436.

Martin, R. R., James, D. and Levesque, C. A., 2000, Impacts of molecular diagnostic technologies on plant disease management. *Annual Review of Phytopathology*, **38**, 207–239.

Martinez, J. P. *et al.*, 2001, Characterization of the *ToxB* gene from *Pyrenophora tritici-repentis*. *Molecular Plant-Microbe Interactions*, **14**, 675–677.

Matheussen, A. M., Morgan, P. W. and Frederiksen, R. A., 1991, Implication of gibberellins in head smut (*Sporisorium reilianum*) of *Sorghum bicolor*. *Plant Physiology*, **96**, 537–544.

Mathews, D. E. and Durbin, R. D., 1990, Tagetitoxin inhibits RNA-synthesis directed by RNA-polymerases from chloroplasts and *Escherichia coli*. *Journal of Biological Chemistry*, **265**, 493–498.

Matthews, R. E. F., 1981, *Plant Virology*, Second Edition, Academic Press, New York, USA.

Matthews, R. E. F., 1992, *Fundamentals of Plant Virology*, Academic Press, San Diego, USA.

Mauch, F., Mauch-Mani, B. and Boller, T. 1988, Antifungal hydrolases in pea tissue.2. Inhibition of fungal growth by combinations of chitinase and beta-1,3-glucanase. *Plant Physiology*, **88**, 936–942.

Maurhofer, M. *et al.*, 1994, Pyoluteorin production by *Pseudomonas fluorescens* strain Chao is involved in the suppression of *Pythium* damping-off of cress but not of cucumber. *European Journal of Plant Pathology*, **100**, 221–232.

Mayans, O. *et al.*, 1997, Two crystal structures of pectin lyase A from Aspergillus reveal a pH driven conformational change and striking divergence in the substrate-binding clefts of pectin and pectate lyases. *Structure*, **5**, 677–689.

Mazzola, M. *et al.*, 1995, Variation in sensitivity of *Gaeumannomyces graminis* to antibiotics produced by fluorescent *Pseudomonas* spp. and effect on biological-control of take-all of wheat. *Applied and Environmental Microbiology*, **61**, 2554–2559.

McClendon, M. T. *et al.*, 2002, DNA markers linked to *Fusarium* wilt race 1 resistance in pea. *Journal of the American Society for Horticultural Science*, **127**, 602–607.

Mccoy, R. E. and Martinez-Lopez, G., 1982, *Phytomonas staheli* associated with coconut and oil palm diseases in Colombia. *Plant Disease*, **66**, 675–677.

Mcgee, D. C., 1995, Epidemiologic approach to disease management through seed technology. *Annual Review of Phytopathology*, **33**, 445–466.

McGuire, R. G. *et al.*, 1991, Polygalacturonase production by *Agrobacterium tumefaciens* Biovar-3. *Applied and Environmental Microbiology*, **57**, 660–664.

McLaren, D. L. *et al.*, 1989, Ultrastructural studies on infection of sclerotia of *Sclerotinia sclerotiorum* by *Talaromyces flavus*. *Canadian Journal of Botany-Revue Canadienne de Botanique*, **67**, 2199–2205.

McLaren, N. W., 1997, Changes in pollen viability and concomitant increase in the incidence of sorghum ergot with flowering date and implications in selection for escape resistance. *Journal of Phytopathology-Phytopathologische Zeitschrift*, **145**, 261–265.

McManus, P. S. and Jones, A. L., 1994, Role of wind-driven rain, aerosols, and contaminated budwood in incidence and spatial pattern of fire blight in an apple nursery. *Plant Disease*, **78**, 1059–1066.

Meehan, F. and Murphy, H. C., 1947, Differential phytotoxicity of metabolic by-products of *Helminthosporium victoriae*. *Science*, **106**, 270–271.

Meeley, R. B. and Walton, J. D., 1991, Enzymatic detoxification of HC toxin, the host-selective cyclic peptide from *Cochliobolus carbonum*. *Plant Physiology*, **97**, 1080–1086.

Meeley, R. B. *et al.*, 1992, A biochemical phenotype for a disease resistance gene of maize. *Plant Cell*, **4**, 71–77.

Melton, R. E. *et al.*, 1998, Heterologous expression of *Septoria lycopersici* tomatinase in *Cladosporium fulvum*: Effects on compatible and incompatible interactions with tomato seedlings. *Molecular Plant-Microbe Interactions*, **11**, 228–236.

Mendgen, K., Hahn, M. and Deising, H., 1996, Morphogenesis and mechanisms of penetration by plant pathogenic fungi. *Annual Review of Phytopathology*, **34**, 367–386.

Mercier, J., Arul, J. and Julien, C., 1993, Effect of UV-C on phytoalexin accumulation and resistance to *Botrytis cinerea* in stored carrots. *Journal of Phytopathology-Phytopathologische Zeitschrift*, **139**, 17–25.

Merlot, S. and Giraudat, J., 1997, Genetic analysis of abscisic acid signal transduction. *Plant Physiology*, **114**, 751–757.

Métraux, J. P. *et al.*, 1990, Increase in salicylic-acid at the onset of systemic acquired-resistance in cucumber. *Science*, **250**, 1004–1006.

Meyer, J. R., Shew, H. D. and Harrison, U. J., 1994, Inhibition of germination and growth of *Thielaviopsis basicola* by aluminum. *Phytopathology*, **84**, 598–602.

Meyers, B. C. *et al.*, 1999, Plant disease resistance genes encode members of an ancient and diverse protein family within the nucleotide-binding superfamily. *Plant Journal*, **20**, 317–332.

Michailides, T. J. and Spotts, R. A., 1990, Postharvest diseases of pome and stone fruits caused by *Mucor piriformis* in the Pacific-Northwest and California. *Plant Disease*, **74**, 537–543.

Michelmore, R. W. and Meyers, B. C., 1998, Clusters of resistance genes in plants evolve by divergent selection and a birth-and-death process. *Genome Research*, **8**, 1113–1130.

Michelmore, R. W., Paran, I. and Kesseli, R. V., 1991, Identification of markers linked to disease-resistance genes by bulked segregant analysis – a rapid method to detect markers in specific genomic regions by using segregating populations. *Proceedings of the National Academy of Sciences of the United States of America*, **88**, 9828–9832.

Migheli, Q. *et al.*, 1998, Transformants of *Trichoderma longibrachiatum* overexpressing the beta-1,4-endoglucanase gene *egl1* show enhanced biocontrol of *Pythium ultimum* on cucumber. *Phytopathology*, **88**, 673–677.

Milholland, R. D., 1991, Muscadine grapes – some important diseases and their control. *Plant Disease*, **75**, 113–117.

Miller, C. O. *et al.*, 1955, Kinetin, a cell division factor from deoxyribonucleic acid. *Journal of the American Chemical Society*, **77**, 1392–1393.

Miller, J. D. and Arnison, P. G., 1986, Degradation of deoxynivalenol by suspension-cultures of the *Fusarium* head blight resistant wheat cultivar Frontana. *Canadian Journal of Plant Pathology-Revue Canadienne de Phytopathologie*, **8**, 147–150.

Miller, S. A. and Martin, R. R., 1988, Molecular diagnosis of plant disease. *Annual Review of Phytopathology*, **26**, 409–432.

Milne, R. G. *et al.*, 1995, Pre-embedding and post-embedding immunogold labeling and electron-microscopy in plant host tissues of 3 antigenically unrelated Mlos – primula yellows, tomato big bud and bermudagrass white leaf. *European Journal of Plant Pathology*, **101**, 57–67.

Mindrinos, M., Katagiri, F., Yu, G. L., and Ausubel, F. M. 1994, The *A.Thaliana* disease resistance gene *Rps2* encodes a protein containing a nucleotide-binding site and leucine-rich repeats. *Cell*, **78**, 1089–1099.

Mink, G. I., 1982, 'The possible role of honey bees in long-distance spread of prunus necrotic ringspot virus from California into Washington sweet cherry orchards,' in *Plant Virus Epidemiology*, R. T. Plumb & J. M. Thresh, eds., Blackwell, Oxford, UK, pp. 85–92.

Mink, G. I., 1992, Ilavirus vectors. *Advances in Disease Vector Research*, **9**, 262–281.

Mink, G. I., 1993, Pollen-transmitted and seed-transmitted viruses and viroids. *Annual Review of Phytopathology*, **31**, 375–402.

Mitchell, R. E., 1979, Bean halo blight: comparison of phaseolotoxin and N-phospho-glutamate. *Physiological Plant Pathology*, **14**, 119–128.

Mitchell, R. E., 1989, 'Biosynthesis and regulation of toxins produced by *Pseudomonas syringae* pv. *glycinea* (coronatine) and *Pseudomonas andropogonis* (rhizobitoxine),' in *Phytotoxins and Plant Pathogenesis*, A. Graniti, R. A. Durbin, & A. Ballio, eds., Springer-Verlag, Berlin, Germany.

Mitchell, R. T. and Deacon, J. W., 1986, Differential (host-specific) accumulation of zoospores of *Pythium* on roots of gramineous and non-gramineous plants. *New Phytologist*, **102**, 113–122.

Mithöfer, A. *et al.*, 1997, Involvement of an NAD(P)H oxidase in the elicitor-inducible oxidative burst of soybean. *Phytochemistry*, **45**, 1101–1107.

Mittal, S. and Davis, K. R., 1995, Role of the phytotoxin coronatine in the infection of *Arabidopsis thaliana* by *Pseudomonas-syringae* pv. *tomato*. *Molecular Plant-Microbe Interactions*, **8**, 165–171.

Mittler, R. *et al.*, 1996, Inhibition of programmed cell death in tobacco plants during a pathogen-induced hypersensitive response at low oxygen pressure. *Plant Cell*, **8**, 1991–2001.

Mizushina, Y. *et al.*, 2002, A plant phytotoxin, solanapyrone A, is an inhibitor of DNA polymerase beta and lambda. *Journal of Biological Chemistry*, **277**, 630–638.

Mo, Y. Y. *et al.*, 1995, Analysis of sweet cherry (*Prunus avium* L) leaves for plant signal molecules that activate the *syrb* gene required for synthesis of the phytotoxin, syringo-mycin, by *Pseudomonas syringae* pv *syringae*. *Plant Physiology*, **107**, 603–612.

Moire, L. *et al.*, 1999, Glycerol is a suberin monomer. New experimental evidence for an old hypothesis. *Plant Physiology*, **119**, 1137–1146.

Mok, D. W. S. and Mok, M. C., 2001, Cytokinin metabolism and action. *Annual Review of Plant Physiology and Plant Molecular Biology*, **52**, 89–118.

Möllers, C. and Sarkar, S., 1989, Regeneration of healthy plants from *Catharanthus roseus* infected with mycoplasma-like organisms through callus-culture. *Plant Science*, **60**, 83–89.

Monette, P. L., 1986, Elimination *in vitro* of 2 grapevine nepoviruses by an alternating temperature regime. *Journal of Phytopathology-Phytopathologische Zeitschrift*, **116**, 88–91.

Montllor, C. B. and Tjallingii, W. F., 1989, Stylet penetration by 2 aphid species on susceptible and resistant lettuce. *Entomologia Experimentalis et Applicata*, **52**, 103–111.

Morgan, D. P. *et al.*, 1991, Solarizing soil planted with cherry tomatoes vs solarizing fallow ground for control of *Verticillium* wilt. *Plant Disease*, **75**, 148–151.

Morjane, H. *et al.*, 1994, Oligonucleotide fingerprinting detects genetic diversity among *Ascochyta rabiei* isolates from a single chickpea field in Tunisia. *Current Genetics*, **26**, 191–197.

Morris, P. F. and Ward, E. W. B., 1992, Chemoattraction of zoospores of the soybean pathogen, *Phytophthora sojae*, by isoflavones. *Physiological and Molecular Plant Pathology*, **40**, 17–22.

Morris, P. F., Bone, E. and Tyler, B. M., 1998, Chemotropic and contact responses of *Phytophthora sojae* hyphae to soybean isoflavonoids and artificial substrates. *Plant Physiology*, **117**, 1171–1178.

Morris, R. O., 1986, Genes specifying auxin and cytokinin biosynthesis in phytopathogens. *Annual Review of Plant Physiology and Plant Molecular Biology*, **37**, 509–538.

Moses, E. *et al.*, 1996, *Colletotrichum gloeosporioides* as the cause of stem tip dieback of cassava. *Plant Pathology*, **45**, 864–871.

Moury, B. *et al.*, 1998, High temperature effects on hypersensitive resistance to Tomato Spotted wilt Tospovirus (TSWV) in pepper (*Capsicum chinense* Jacq.). *European Journal of Plant Pathology*, **104**, 489–498.

Moury, B. *et al.*, 2000, Enzyme-linked immunosorbent assay testing of shoots grown in vitro and the use of immunocapture-reverse transcription-polymerase chain reaction improve the detection of Prunus necrotic ringspot virus in rose. *Phytopathology*, **90**, 522–528.

Movahedi, S. and Heale, J. B., 1990a, Purification and characterization of an aspartic proteinase secreted by *Botrytis cinerea* Pers ex Pers in culture and in infected carrots. *Physiological and Molecular Plant Pathology*, **36**, 289–302.

Movahedi, S. and Heale, J. B., 1990b, The roles of aspartic proteinase and endo-pectin lyase enzymes in the primary stages of infection and pathogenesis of various host tissues by different isolates of *Botrytis cinerea* Pers ex Pers. *Physiological and Molecular Plant Pathology*, **36**, 303–324.

Multani, D. S. *et al.*, 1998, Plant-pathogen microevolution: Molecular basis for the origin of a fungal disease in maize. *Proceedings of the National Academy of Sciences of the United States of America*, **95**, 1686–1691.

Mumford, R. A. *et al.*, 2000, Detection of Potato mop top virus and Tobacco rattle virus using a multiplex real-time fluorescent reverse-transcription polymerase chain reaction assay. *Phytopathology*, **90**, 448–453.

MunchGarthoff, S. *et al.*, 1997, Expression of beta-1,3-glucanase and chitinase in healthy, stem-rust-affected and elicitor-treated near-isogenic wheat lines showing *Sr5*- or *Sr24*-specified race-specific rust resistance. *Planta*, **201**, 235–244.

Mundt, C. C., 1995, Models from plant pathology on the movement and fate of new genotypes of microorganisms in the environment. *Annual Review of Phytopathology*, **33**, 467–488.

Murant, A. F., Chambers, J. and Jones, A. T., 1974, Spread of raspberry bushy dwarf virus by pollination, its association with crumbly fruit, and problems of control. *Annals of Applied Biology*, **77**, 271–283.

Murphy, A. M. *et al.*, 1997, Comparison of cytokinin production in vitro by *Pyrenopeziza brassicae* with other plant pathogens. *Physiological and Molecular Plant Pathology*, **50**, 53–65.

Myers, D. F. and Fry, W. E., 1978, Enzymatic release and metabolism of the hydrogen cyanide in sorghum infected by *Gloeocercospora sorghi*. *Phytopathology*, **67**, 1717–1722.

Nadel, B. and Spiegel-Roy, P., 1987, Selection of Citrus-limon cell-culture variants resistant to the mal secco toxin. *Plant Science*, **53**, 177–182.

Nagarajan, S., 1993, Plant diseases in India and their control. *Ciba Foundation Symposia*, **177**, 208–227.

Nagarajan, S. and Singh, D. V., 1990, Long-distance dispersion of rust pathogens. *Annual Review of Phytopathology*, **28**, 139–153.

Nagtzaam, M. P. M. and Bollen, G. J., 1994, Long shelf-life of *Talaromyces flavus* in coating material of pelleted seed. *European Journal of Plant Pathology*, **100**, 279–282.

Nakajima, H. and Scheffer, R. P., 1987, Interconversions of aglycone and host-selective toxin from *Helminthosporium sacchari*. *Phytochemistry*, **26**, 1607–1611.

Napoli, C. and Staskawicz, B., 1987, Molecular characterization and nucleic-acid sequence of an avirulence gene from race 6 of *Pseudomonas syringae* pv. *glycinea*. *Journal of Bacteriology*, **169**, 572–578.

Nathan, C., 1995, Natural resistance and nitric-oxide. *Cell*, **82**, 873–876.

Nault, L. R., 1997, Arthropod transmission of plant viruses: A new synthesis. *Annals of the Entomological Society of America*, **90**, 521–541.

Nault, L. R. and Ammar, E., 1989, Leafhopper and planthopper transmission of plant-viruses. *Annual Review of Entomology*, **34**, 503–529.

Nault, L. R. and Styer, W. E., 1972, Effects of sinigrin on host selection by aphids. *Entomologia Experimentalis et Applicata*, **15**, 423–437.

Navarre, D. A. and Wolpert, T. J., 1999, Effects of light and CO_2 on victorin-induced symptom development in oats. *Physiological and Molecular Plant Pathology*, **55**, 237–242.

Navas-Cortés, J. A., Trapero-Casas, A. and Jimenez-Diaz, R. M., 1995, Survival of *Didymella rabiei* in chickpea straw debris in Spain. *Plant Pathology*, **44**, 332–339.

Neilands, J. B. and Leong, S. A., 1986, Siderophores in relation to plant growth and disease. *Annual Review of Plant Physiology and Plant Molecular Biology*, **37**, 187–208.

Nelson, E. B., 1990, Exudate molecules initiating fungal responses to seeds and roots. *Plant and Soil*, **129**, 61–73.

Nelson, E. B. and Craft, C. M., 1991, Introduction and establishment of strains of *Enterobacter cloacae* in golf course turf for the biological-control of dollar spot. *Plant Disease*, **75**, 510–514.

Nemec, S., 1995, Stress-related compounds in xylem fluid of blight-diseased citrus containing *Fusarium solani* naphthazarin toxins and their effects on the host. *Canadian Journal of Microbiology*, **41**, 515–524.

Neumann, U. *et al.*, 1999, Interface between haustoria of parasitic members of the Scrophulariaceae and their hosts: a histochemical and immunocytochemical approach. *Protoplasma*, **207**, 84–97.

Newton, A. C., 1989, Measuring the sterol content of barley leaves infected with powdery mildew as a means of assessing partial resistance to *Erysiphe graminis* f sp *hordei*. *Plant Pathology*, **38**, 534–540.

Newton, A. C., 1990, Detection of components of partial resistance to mildew (*Erysiphe graminis* f sp *hordei*) incorporated into advanced breeding lines of barley using measurement of fungal cell-wall sterol. *Plant Pathology*, **39**, 598–602.

Nicholson, P. *et al.*, 1998, Detection and quantification of *Fusarium culmorum* and *Fusarium graminearum* in cereals using PCR assays. *Physiological and Molecular Plant Pathology*, **53**, 17–37.

Nicholson, R. L. *et al.*, 1988, Preparation of the infection court by *Erysiphe graminis*. 2. Release of esterase enzyme from conidia in response to a contact stimulus. *Experimental Mycology*, **12**, 336–349.

Niderman, T. *et al.*, 1995, Pathogenesis-related Pr-1 proteins are antifungal – isolation and characterization of 3 14-kilodalton proteins of tomato and of a basic Pr-1 of tobacco with inhibitory activity against *Phytophthora infestans*. *Plant Physiology*, **108**, 17–27.

Nielsen, K. K. *et al.*, 1997, Characterization of a new antifungal chitin-binding peptide from sugar beet leaves. *Plant Physiology*, **113**, 83–91.

Niki, T. *et al.*, 1998, Antagonistic effect of salicylic acid and jasmonic acid on the expression of pathogenesis-related (PR) protein genes in wounded mature tobacco leaves. *Plant and Cell Physiology*, **39**, 500–507.

Nilsson, H. E., 1995, Remote-sensing and image-analysis in plant pathology. *Annual Review of Phytopathology*, **33**, 489–527.

Nishimura, S. and Kohmoto, K., 1983, Host-specific toxins and chemical structures from *Alternaria* species. *Annual Review of Phytopathology*, **21**, 87–116.

Norman, D. J. *et al.*, 1997, Differentiation of three species of *Xanthomonas* and *Stenotrophomonas maltophilia* using cellular fatty acid analyses. *European Journal of Plant Pathology*, **103**, 687–693.

Nuss, D. L. and Koltin, Y., 1990, Significance of DsRNA genetic elements in plant pathogenic fungi. *Annual Review of Phytopathology*, **28**, 37–58.

Nuss, L. *et al.*, 1996, Differential accumulation of PGIP (polygalacturonase-inhibiting protein) mRNA in two near-isogenic lines of *Phaseolus vulgaris* L upon infection with *Colletotrichum lindemuthianum*. *Physiological and Molecular Plant Pathology*, **48**, 83–89.

Nussbaum, R. P. *et al.*, 1999, New tricycloalternarenes produced by the phytopathogenic fungus *Alternaria alternata*. *Phytochemistry*, **52**, 593–599.

Nutter, F. W., Teng, P. S. and Royer, M. H., 1993, Terms and concepts for yield, crop loss, and disease thresholds. *Plant Disease*, **77**, 211–215.

O'Connell, R. J. *et al.*, 1990, Immunocytochemical localization of hydroxyproline-rich glycoproteins accumulating in melon and bean at sites of resistance to bacteria and fungi. *Molecular Plant-Microbe Interactions*, **3**, 33–40.

Oerke, E. C. and Schönbeck, F., 1990, Effect of nitrogen and powdery mildew on the yield formation of 2 winter barley cultivars. *Journal of Phytopathology-Phytopathologische Zeitschrift*, **130**, 89–104.

Ohkuma, K. *et al.*, 1963, Abscisin II, an abscission-accelerating substance from young cotton fruit. *Science*, **142**, 1592–1593.

Oka, Y. and Cohen, Y., 2001, Induced resistance to cyst and root-knot nematodes in cereals by DL-beta-amino-n-butyric acid. *European Journal of Plant Pathology*, **107**, 219–227.

Oldach, K. H., Becker, D. and Lorz, H., 2001, Heterologous expression of genes mediating enhanced fungal resistance in transgenic wheat. *Molecular Plant-Microbe Interactions*, **14**, 832–838.

Oliver, R. P. *et al.*, 1993, Use of fungal transformants expressing beta-glucuronidase activity to detect infection and measure hyphal biomass in infected-plant tissues. *Molecular Plant-Microbe Interactions*, **6**, 521–525.

Oostendorp, M. *et al.*, 2001, Induced disease resistance in plants by chemicals. *European Journal of Plant Pathology*, **107**, 19–28.

Opalka, N. *et al.*, 1998, Movement of rice yellow mottle virus between xylem cells through pit membranes. *Proceedings of the National Academy of Sciences of the United States of America*, **95**, 3323–3328.

Orbach, M. J. *et al.*, 2000, A telomeric avirulence gene determines efficacy for the rice blast resistance gene *Pi-ta*. *Plant Cell*, **12**, 2019–2032.

Osbourn, A., 1996, Saponins and plant defence – A soap story. *Trends in Plant Science*, **1**, 4–9.

Osbourn, A. *et al.*, 1995, Fungal pathogens of oat roots and tomato leaves employ closely related enzymes to detoxify different host plant saponins. *Molecular Plant-Microbe Interactions*, **8**, 971–978.

Osburn, R. M. *et al.*, 1995, Effect of *Bacillus cereus* UW85 on the yield of soybean at 2 field sites in Wisconsin. *Plant Disease*, **79**, 551–556.

Otten, W., Gilligan, C. A. and Thornton, C. R., 1997, Quantification of fungal antigens in soil with a monoclonal antibody-based ELISA: Analysis and reduction of soil-specific bias. *Phytopathology*, **87**, 730–736.

Oyarzun, P. J., 1993, Bioassay to assess root-rot in pea and effect of root-rot on yield. *Netherlands Journal of Plant Pathology*, **99**, 61–75.

Oyarzun, P. J., Gerlagh, M. and Zadoks, J. C., 1998, Factors associated with soil receptivity to some fungal root rot pathogens of peas. *Applied Soil Ecology*, **10**, 151–169.

Padgett, D. E. and Posey, M. H., 1993, An evaluation of the efficiencies of several ergosterol extraction techniques. *Mycological Research*, **97**, 1476–1480.

Padmanabhan, S. Y., 1973, The great Bengal famine. *Annual Review of Phytopathology*, **11**, 11–26.

Pagel, W. and Heitefuss, R., 1990, Enzyme-activities in soft rot pathogenesis of potato-tubers – effects of calcium, pH, and degree of pectin esterification on the activities of polygalacturonase and pectate lyase. *Physiological and Molecular Plant Pathology*, **37**, 9–25.

Palanichelvam, K. and Schoelz, J. E., 2002, A comparative analysis of the avirulence and translational transactivator functions of gene VI of Cauliflower mosaic virus. *Virology*, **293**, 225–233.

Panaccione, D. G. *et al.*, 1992, A cyclic peptide synthetase gene required for pathogenicity of the fungus *Cochliobolus carbonum* on Maize. *Proceedings of the National Academy of Sciences of the United States of America*, **89**, 6590–6594.

Papadopoulou, K. *et al.*, 1999, Compromised disease resistance in saponin-deficient plants. *Proceedings of the National Academy of Sciences of the United States of America*, **96**, 12923–12928.

Parke, J. L. and Gurian-Sherman, D., 2001, Diversity of the *Burkholderia cepacia* complex and implications for risk assessment of biological control strains. *Annual Review of Phytopathology*, **39**, 225–258.

Parker, C., 1991, Protection of crops against parasitic weeds. *Crop Protection*, **10**, 6–22.

Parker, C. and Riches, C. R., 1993, *Parasitic Weeds of the World: Biology and Control*, CAB International, Wallingford, Oxfordshire, UK.

Parniske, M. *et al.*, 1997, Novel disease resistance specificities result from sequence exchange between tandemly repeated genes at the *Cf*4/9 locus of tomato. *Cell*, 821–832.

Parry, D. W. and Nicholson, P., 1996, Development of a PCR assay to detect *Fusarium poae* in wheat. *Plant Pathology*, **45**, 383–391.

Patil, S. S., Hayward, S. C. and Emmons, R., 1974, An ultraviolet-induced nontoxigenic mutant of *Pseudomonas phaseolicola* of altered pathogenicity. *Phytopathology*, **64**, 590–595.

Paulitz, T. *et al.*, 2000, A novel antifungal furanone from *Pseudomonas aureofaciens*, a biocontrol agent of fungal plant pathogens. *Journal of Chemical Ecology*, **26**, 1515–1524.

Paulitz, T. C. and Bélanger, R. R., 2001, Biological control in greenhouse systems. *Annual Review of Phytopathology*, **39**, 103–133.

Paulitz, T. C., Ahmad, J. S. and Baker, R., 1990, Integration of *Pythium nunn* and *Trichoderma harzianum* isolate T-95 for the biological-control of *Pythium* damping-off of cucumber. *Plant and Soil*, **121**, 243–250.

Paxton, J., 1971, Phytoalexins – a working redefinition. *Phytopathologische Zeitschrift-Journal of Phytopathology*, **101**, 106–109.

Pearce, G. *et al.*, 2001, RALF, a 5-kDa ubiquitous polypeptide in plants, arrests root growth and development. *Proceedings of the National Academy of Sciences of the United States of America*, **98**, 12843–12847.

Pearce, M. H. and Malajczuk, N., 1990, Stump colonization by *Armillaria luteobubalina* and other wood decay fungi in an age series of cut-over stumps in Karri (*Eucalyptus diversicolor*) regrowth forests in South-Western Australia. *New Phytologist*, **115**, 129–138.

Pearce, R. B. and Ride, J. P., 1980, Specificity of induction of the lignification response in wounded wheat leaves. *Physiological Plant Pathology*, **16**, 197–204.

Pearce, R. B. and Ride, J. P., 1982, Chitin and related-compounds as elicitors of the lignification response in wounded wheat leaves. *Physiological Plant Pathology*, **20**, 119–123.

Pedras, M. S. C., 1997, Determination of the absolute configuration of the fragments composing the phytotoxin phomalide. *Canadian Journal of Chemistry-Revue Canadienne de Chimie*, **75**, 314–317.

Pedras, M. S. C., Taylor, J. L. and Morales, V. M., 1996, The blackleg fungus of rapeseed: how many species? in *Proceedings of the International Symposium on Brassicas: Ninth Crucifer Genetics Workshop*, J. S. Dias, I. Crute, & A. A. Monteiro, eds., pp. 441–446.

Penninckx, I. A. M. A. *et al.*, 1998, Concomitant activation of jasmonate and ethylene response pathways is required for induction of a plant defensin gene in Arabidopsis. *Plant Cell*, **10**, 2103–2113.

Penyalver, R., Vicedo, B. and Lopez, M. M., 2000, Use of the genetically engineered *Agrobacterium* strain K1026 for biological control of crown gall. *European Journal of Plant Pathology*, **106**, 801–810.

Perazza, D., Vachon, G. and Herzog, M., 1998, Gibberellins promote trichome formation by up-regulating *GLABROUS1* in *Arabidopsis*. *Plant Physiology*, **117**, 375–383.

Perez, V. *et al*., 1997, Mapping the elicitor and necrotic sites of *Phytophthora* elicitins with synthetic peptides and reporter genes controlled by tobacco defense gene promoters. *Molecular Plant-Microbe Interactions*, **10**, 750–760.

Pernezny, K. *et al*., 1995, An outbreak of bacterial spot of lettuce in Florida caused by *Xanthomonas campestris* pv. *vitians. Plant Disease*, **79**, 359–360.

Péros, J. P. and Berger, G., 1994, A rapid method to assess the aggressiveness of *Eutypa lata* isolates and the susceptibility of grapevine cultivars to Eutypa dieback. *Agronomie*, **14**, 515–523.

Perumalla, C. J. and Heath, M. C., 1989, Effect of callose inhibition on haustorium formation by the cowpea rust fungus in the non-host, bean plant. *Physiological and Molecular Plant Pathology*, **35**, 375–382.

Phelps, D. C. *et al*., 1990, Immunoassay for naphthazarin phytotoxins produced by *Fusarium solani. Phytopathology*, **80**, 298–302.

Picard, K., Tirilly, Y. and Benhamou, N., 2000, Cytological effects of cellulases in the parasitism of *Phytophthora parasitica* by *Pythium oligandrum. Applied and Environmental Microbiology*, **66**, 4305–4314.

Pierce, M. L. *et al*., 1996, Adequacy of cellular phytoalexin concentrations in hypersensitively responding cotton leaves. *Physiological and Molecular Plant Pathology*, **48**, 305–324.

Pinkerton, J. and Strobel, G. A., 1976, Serinol as an activator of toxin production in attenuated cultures of *Helminthosporium sacchari. Proceedings of the National Academy of Sciences of the United States of America*, **73**, 4007.

Piotte, C. *et al*., 1994, Cloning and characterization of 2 satellite DNAs in the low-C-value genome of the nematode *Meloidogyne* spp. *Gene*, **138**, 175–180.

Piotte, C. *et al*., 1995, Analysis of a satellite DNA from *Meloidogyne hapla* and its use as a diagnostic probe. *Phytopathology*, **85**, 458–462.

Pirhonen, M. U. *et al*., 1996, Phenotypic expression of *Pseudomonas syringae* avr genes in *E. coli* is linked to the activities of the hrp-encoded secretion system. *Molecular Plant-Microbe Interactions*, **9**, 252–260.

Pirone, T. P. and Blanc, S., 1996, Helper-dependent vector transmission of plant viruses. *Annual Review of Phytopathology*, **34**, 227–247.

Pitkin, J. W., Panaccione, D. G. and Walton, J. D., 1996, A putative cyclic peptide efflux pump encoded by the *TOXA* gene of the plant-pathogenic fungus *Cochliobolus carbonum. Microbiology-UK*, **142**, 1557–1565.

Ploeg, A. T., Zoon, F. C. and Maas, P. W. T., 1996, Transmission efficiency of five tobravirus strains by *Paratrichodorus teres. European Journal of Plant Pathology*, **102**, 123–126.

Podila, G. K., Rogers, L. M. and Kolattukudy, P. E., 1993, Chemical signals from avocado surface wax trigger germination and appressorium formation in *Colletotrichum gloeosporioides. Plant Physiology*, **103**, 267–272.

Pohronezny, K. *et al.*, 1990, Dispersal and management of *Xanthomonas campestris* pv *vesicatoria* during thinning of direct-seeded tomato. *Plant Disease*, **74**, 800–805.

Poplawsky, A. R. and Ellingboe, A. H., 1989, Take-all suppressive properties of bacterial mutants affected in antibiosis. *Phytopathology*, **79**, 143–146.

Posnette, A. F. and Todd, J. M., 1955, Virus diseases of cocao in West Africa. IX. Strain variation and interference in virus 1A. *Annals of Applied Biology*, **43**, 433–453.

Potrykus, I., 1991, Gene-transfer to plants – assessment of published approaches and results. *Annual Review of Plant Physiology and Plant Molecular Biology*, **42**, 205–225.

Powell, A. L. T. *et al.*, 2000, Transgenic expression of pear PGIP in tomato limits fungal colonization. *Molecular Plant-Microbe Interactions*, **13**, 942–950.

Pradhanang, P. M., Elphinstone, J. G. and Fox, R. T. V., 2000, Sensitive detection of *Ralstonia solanacearum* in soil: a comparison of different detection techniques. *Plant Pathology*, **49**, 414–422.

Pradhanang, P. M., Elphinstone, J. G. and Fox, R. T. V., 2000, Identification of crop and weed hosts of *Ralstonia solanacearum* biovar 2 in the hills of Nepal. *Plant Pathology*, **49**, 403–413.

Proctor, R. H., Hohn, T. M. and McCormick, S. P., 1995, Reduced virulence of *Gibberella zeae* caused by disruption of a trichothecene toxin biosynthetic gene. *Molecular Plant-Microbe Interactions*, **8**, 593–601.

Prusky, D., 1988, The use of antioxidants to delay the onset of anthracnose and stem end decay in avocado fruits after harvest. *Plant Disease*, **72**, 381–384.

Prusky, D., 1996, Pathogen quiescence in postharvest diseases. *Annual Review of Phytopathology*, **34**, 413–434.

Prusky, D., Wattad, C. and Kobiler, I., 1996, Effect of ethylene on activation of lesion development from quiescent infections of *Colletotrichum gloeosporioides* in avocado fruits. *Molecular Plant-Microbe Interactions*, **9**, 864–868.

Prusky, D. *et al.*, 1990, Induction of the antifungal diene in unripe avocado fruits – effect of inoculation with *Colletotrichum gloeosporioides*. *Physiological and Molecular Plant Pathology*, **37**, 425–435.

Punja, Z. K. and Raharjo, S. H. T., 1996, Response of transgenic cucumber and carrot plants expressing different chitinase enzymes to inoculation with fungal pathogens. *Plant Disease*, **80**, 999–1005.

Quigley, N. B. and Gross, D. C., 1994, Syringomycin production among strains of *Pseudomonas syringae* pv. *syringae* – conservation of the *SyrB* and *SyrD* genes and activation of phytotoxin production by plant signal molecules. *Molecular Plant-Microbe Interactions*, **7**, 78–90.

Raaijmakers, J. M. and Weller, D. M., 1998, Natural plant protection by 2,4-diacetylphloroglucinol – producing *Pseudomonas* spp. in take-all decline soils. *Molecular Plant-Microbe Interactions*, **11**, 144–152.

Rairdan, G. J. and Delaney, T. P., 2002, Role of salicylic acid and *NIM1/NPR1* in race-specific resistance in Arabidopsis. *Genetics*, **161**, 803–811.

Raju, B. C. and Wells, J. M., 1986, Diseases caused by fastidious xylem-limited bacteria and strategies for management. *Plant Disease*, **70**, 182–186.

Ramaiah, K. V., 1987, 'Control of *Orobanche* and *Striga* species – a review' in *Proceedings of the 4th International Symposium on Parasitic Flowering Plants*, Marburg, Germany, pp. 637–644.

Rangaswamy, V. *et al.*, 1997, Expression and analysis of coronafacate ligase, a thermo-regulated gene required for production of the phytotoxin coronatine in *Pseudomonas syringae*. *FEMS Microbiology Letters*, **154**, 65–72.

Rasmussen, J. B. and Scheffer, R. P., 1988, Isolation and biological activities of 4 selective toxins from *Helminthosporium carbonum*. *Plant Physiology*, **86**, 187–191.

Rasmussen, J. B. and Scheffer, R. P., 1988, Effects of selective toxin from *Helminthosporium carbonum* on chlorophyll synthesis in maize. *Physiological and Molecular Plant Pathology*, **32**, 283–291.

Read, N. D. *et al.*, 1997, Role of topography sensing for infection-structure differentiation in cereal rust fungi. *Planta*, **202**, 163–170.

Reddy, M. V., Singh, K. B. and Nene, Y. L., 1981, 'Screening techniques for *Ascochyta* blight of chickpea,' in *Proceedings of a Workshop on Ascochyta Blight and Winter Sowing of Chickpea*, M. C. Saxena & K. B. Singh, eds., Martinus Nijhoff/Dr. W. Junk, Aleppo, Syria.

Reeves, P. J. *et al.*, 1993, Molecular-cloning and characterization of 13 *out* genes from *Erwinia carotovora* subspecies *carotovora* – genes encoding members of a general secretion pathway (gsp) widespread in gram-negative bacteria. *Molecular Microbiology*, **8**, 443–456.

Reifschneider, F. J. B. and Lopes, C. A., 1982, Bacterial top and stalk rot of maize in Brazil. *Plant Disease*, **66**, 519–520.

Rekah, Y., Shtienberg, D. and Katan, J., 2001, Population dynamics of *Fusarium oxysporum* f. sp. *radicis-lycopersici* in relation to the onset of *Fusarium* crown and root rot of tomato. *European Journal of Plant Pathology*, **107**, 367–375.

Reuveni, R., Raviv, M. and Bar, R. 1989, Sporulation of *Botrytis cinerea* as affected by photoselective polyethylene sheets and filters. *Annals of Applied Biology*, **115**, 417–424.

Reuveni, M. *et al.*, 1987, Removal of duvatrienediols from the surface of tobacco-leaves increases their susceptibility to blue mold. *Physiological and Molecular Plant Pathology*, **30**, 441–451.

Reverchon, S. *et al.*, 1997, The cyclic AMP receptor protein is the main activator of pectinolysis genes in *Erwinia chrysanthemi*. *Journal of Bacteriology*, **179**, 3500–3508.

Reynolds, K. M., Madden, L. V. and Ellis, M. A., 1988, Effect of weather variables on strawberry leather rot epidemics. *Phytopathology*, **78**, 822–827.

Rezzonico, E. *et al.*, 1988, Transcriptional down-regulation by abscisic acid of pathogenesis-related beta-1,3-glucanase genes in tobacco cell cultures. *Plant Physiology*, **117**, 585–592.

Riahi, H. *et al.*, 1990, A quantitative scale for assessing chickpea reaction to *Ascochyta rabiei*. *Canadian Journal of Botany-Revue Canadienne de Botanique*, **68**, 2736–2738.

Ribaut, J. M. and Hoisington, D., 1998, Marker-assisted selection: new tools and strategies. *Trends in Plant Science*, **3**, 236–239.

Richards, D. E. *et al.*, 2001, How gibberellin regulates plant growth and development: A molecular genetic analysis of gibberellin signaling. *Annual Review of Plant Physiology and Plant Molecular Biology*, **52**, 67–88.

Richardson, M., Valdes-Rodriguez, S. and Blanco-Labra, A., 1987, A possible function for thaumatin and a TMV-induced protein suggested by homology to a maize inhibitor. *Nature*, **327**, 432–434.

Ride, J. P., 1975, Lignification in wounded wheat leaves in response to fungi and its possible role in resistance. *Physiological Plant Pathology*, **5**, 125–134.

Ride, J. P., 1978, The role of cell wall alteration in resistance to fungi. *Annals of Applied Biology*, **89**, 302–306.

Ride, J. P. and Barber, M. S., 1987, The effects of various treatments on induced lignification and the resistance of wheat to fungi. *Physiological and Molecular Plant Pathology*, **31**, 349–360.

Ride, J. P. and Barber, M. S., 1990, Purification and characterization of multiple forms of endochitinase from wheat leaves. *Plant Science*, **71**, 185–197.

Ride, J. P. and Drysdale, R. B., 1972, A rapid method for the chemical estimation of filamentous fungi in plants. *Physiological Plant Pathology*, **2**, 7–15.

Ride, J. P. and Pearce, R. B., 1979, Lignification and papilla formation at sites of attempted penetration of wheat leaves by non-pathogenic fungi. *Physiological Plant Pathology*, **15**, 79–92.

Riley, D. G. and Pappu, H. R., 2000, Evaluation of tactics for management of thrips-vectored Tomato spotted wilt virus in tomato. *Plant Disease*, **84**, 847–852.

Ristaino, J. B. and Gumpertz, M. L., 2000, New frontiers in the study of dispersal and spatial analysis of epidemics caused by species in the genus *Phytophthora*. *Annual Review of Phytopathology*, **38**, 541–576.

Roberts, A. G. *et al.*, 1997, Phloem unloading in sink leaves of *Nicotiana benthamiana*: Comparison of a fluorescent solute with a fluorescent virus. *Plant Cell*, **9**, 1381–1396.

Roberts, D. P. and Lumsden, R. D., 1990, Effect of extracellular metabolites from *Gliocladium virens* on germination of sporangia and mycelial growth of *Pythium ultimum*. *Phytopathology*, **80**, 461–465.

Roberts, S. J., *et al.*, 1995, Effect of pea bacterial-blight (*Pseudomonas syringae* pv. *pisi*) on the yield of spring sown combining peas (*Pisum sativum*). *Annals of Applied Biology*, **126**, 61–73.

Robinson, S. P., Jacobs, A. K. and Dry, I. B., 1997, A class IV chitinase is highly expressed in grape berries during ripening. *Plant Physiology*, **114**, 771–778.

Robson, G. D., Trinci, A. P. J. and Wiebe, M. G., 1991, Phosphatidylinositol 4,5 bisphosphate (Pip2) is present in *Fusarium graminearum*. *Mycological Research*, **95**, 1082–1084.

Robson, G. D., Wiebe, M. G. and Trinci, A. P. J., 1994, Betaine transport in *Fusarium graminearum*. *Mycological Research*, **98**, 176–178.

Robson, G. D. *et al.*, 1992, Choline transport in *Fusarium graminearum* A3/5. *FEMS Microbiology Letters*, **92**, 247–252.

Robson, G. D. *et al.*, 1995, Choline-induced and acetylcholine-induced changes in the morphology of *Fusarium graminearum* – evidence for the involvement of the choline transport-system and acetylcholinesterase. *Microbiology-UK*, **141**, 1309–1314.

Rodriguez-Palenzuela, P., Burr, T. J. and Collmer, A., 1991, Polygalacturonase is a virulence factor in *Agrobacterium tumefaciens* Biovar-3. *Journal of Bacteriology*, **173**, 6547–6552.

Roelfs, A. P. and Martens, J. W., 1988, An international system of nomenclature for *Puccinia graminis* f sp *tritici*. *Phytopathology*, **78**, 526–533.

Rogers, L. M., Flaishman, M. A. and Kolattukudy, P. E., 1994, Cutinase gene disruption in *Fusarium solani* f sp *pisi* decreases its virulence on pea. *Plant Cell*, **6**, 935–945.

Rogers, L. M. *et al.*, 2000, Requirement for either a host- or pectin-induced pectate lyase for infection of *Pisum sativum* by *Nectria hematococca*. *Proceedings of the National Academy of Sciences of the United States of America*, **97**, 9813–9818.

Romantschuk, M. and Bamford, D. H., 1986, The causal agent of halo blight in bean, *Pseudomonas syringae* pv. *phaseolicola*, attaches to stomata via its pili. *Microbial Pathogenesis*, **1**, 139–148.

Ross, A., 1966, 'Systemic effects of local lesion formation', In *Viruses of Plants*, A. Beemster & S. Dykstra, eds., North Holland, Amsterdam, pp. 127–150.

Ross, J. J. *et al.*, 1995, Genetic-regulation of gibberellin deactivation in *Pisum*. *Plant Journal*, **7**, 513–523.

Rossman, A. Y., Palm, M. E. and Spieman, L. J., 1987, *A Literature Guide to the Identification of Plant Pathogenic Fungi*, American Phytopathological Society, St. Paul, Minnesota, USA.

Roth, U., Friebe, A. and Schnabl, H., 2000, Resistance induction in plants by a brassi-nosteroid-containing extract of *Lychnis viscaria* L. *Zeitschrift fur Naturforschung C-A Journal of Biosciences*, **55**, 552–559.

Rouse, D. I., 1988, Use of crop growth-models to predict the effects of disease. *Annual Review of Phytopathology*, **26**, 183–201.

Roux, C. *et al.*, 1998, Identification of new early auxin markers in tobacco by mRNA differential display. *Plant Molecular Biology*, **37**, 385–389.

Rouxel, T., Kollmann, A. and Bousquet, J. F., 1990, Zinc suppresses Sirodesmin PL toxicity and protects *Brassica napus* plants against the blackleg disease caused by *Leptosphaeria maculans*. *Plant Science*, **68**, 77–86.

Rovira, A. D., 1990, The impact of soil and crop management-practices on soil-borne root diseases and wheat yields. *Soil Use and Management*, **6**, 195–200.

Roy, M. A., 1988, Use of fatty-acids for the identification of phytopathogenic bacteria. *Plant Disease*, **72**, 460.

Rush, C. M. and Heidel, G. B., 1995, Furovirus diseases of sugar-beets in the United-States. *Plant Disease*, **79**, 868–875.

Ryder, T. B. *et al.*, 1987, Organization and differential activation of a gene family encoding the plant defense enzyme chalcone synthase in *Phaseolus vulgaris*. *Molecular and General Genetics*, **210**, 219–233.

Sah, D. N., 1991, Influence of environmental factors on infection of maize (*Zea mays* L.) by *Erwinia chrysanthemi* pv. *zeae*. *Journal of the Institute of Agriculture and Animal Science*, **12**, 41–45.

Salmond, G. P. C., 1994, Secretion of extracellular virulence factors by plant-pathogenic bacteria. *Annual Review of Phytopathology*, **32**, 181–200.

Sauer, N. *et al.*, 1990, Cloning and characterization of a wound-specific hydroxyproline-rich glycoprotein in *Phaseolus vulgaris*. *Plant Cell and Environment*, **13**, 257–266.

Sawada, H., Takeuchi, T. and Matsuda, I., 1997, Comparative analysis of *Pseudomonas syringae* pv. *actinidiae* and *pv. phaseolicola* based on phaseolotoxin-resistant ornithine carbamoyltransferase gene (*argK*) and 16S–23S rRNA intergenic spacer sequences. *Applied and Environmental Microbiology*, **63**, 282–288.

Saxena, S. C. and Lal, S., 1984, Use of meteorological factors in prediction of *Erwinia* stalk rot of maize. *Tropical Pest Management*, **30**, 82–85.

Schaad, N. W., Sitterly, W. R. and Humaydan, H., 1980, Relationship of incidence of seedborne *Xanthomonas campestris* to black rot of crucifers. *Plant Disease*, **64**, 91–92.

Schaad, N. W. *et al.*, 1995, A combined biological and enzymatic amplification (Bio-Pcr) technique to detect *Pseudomonas syringae* pv. *phaseolicola* in bean seed extracts. *Phytopathology*, **85**, 243–248.

Schäfer, J. F. and Roelfs, A. P., 1985, Estimated relation between numbers of urediniospores of *Puccinia-graminis* f. sp. *tritici* and rates of occurrence of virulence. *Phytopathology*, **75**, 749–750.

Scheer, J. M. and Ryan, C. A., 2002, The systemin receptor SR 160 from *Lycopersicon peruvianum* is a member of the LRR receptor kinase family. *Proceedings of the National Academy of Sciences of the United States of America*, **99**, 9585–9590.

Schilling, A. G., Moller, E. M. and Geiger, H. H., 1996, Polymerase chain reaction-based assays for species-specific detection of *Fusarium culmorum, F graminearum*, and *F avenaceum. Phytopathology*, **86**, 515–522.

Schisler, D. A. and Slininger, P. J., 1997, Microbial selection strategies that enhance the likelihood of developing commercial biological control products. *Journal of Industrial Microbiology and Biotechnology*, **19**, 172–179.

Schmele, I. and Kauss, H., 1990, Enhanced activity of the plasma-membrane localized callose synthase in cucumber leaves with induced resistance. *Physiological and Molecular Plant Pathology*, **37**, 221–228.

Schmidt, H. H. H. W. and Walter, U., 1994, NO at Work. *Cell*, **78**, 919–925.

Schmidt, W. E. and Ebel, J., 1987, Specific binding of a fungal glucan phytoalexin elicitor to membrane-fractions from soybean, *Glycine max. Proceedings of the National Academy of Sciences of the United States of America*, **84**, 4117–4121.

Scholz-Schroeder, B. K. *et al.*, 2001, A physical map of the syringomycin and syringopeptin gene clusters localized to an approximately 145-kb DNA region of *Pseudomonas syringae* pv. *syringae* strain B301D. *Molecular Plant-Microbe Interactions*, **14**, 1426–1435.

Scholz-Schroeder, B. K. *et al.*, 2001, The contribution of syringopeptin and syringomycin to virulence of *Pseudomonas syringae* pv. *syringae* strain B301D on the basis of *sypA* and *syrB1* biosynthesis mutant analysis. *Molecular Plant-Microbe Interactions*, **14**, 336–348.

Schoulties, C. L. *et al.*, 1987, Citrus canker in Florida. *Plant Disease*, **71**, 388–395.

Schuler, R. T., Rodakowski, N. N. and Kucera, H. L., 1977, 'Small grain harvesting loss in Dakota', in *Grain and Forage Harvesting. Proceedings of the First International Grain and Forgae Conference*, St Joseph, Michigan, USA.

Schultz, T. P. and Nicholas, D. D., 2000, Lignin influence on angiosperm sapwood susceptibility to white-rot fungal colonization: A hypothesis. *Lignin: Historical, Biological, and Materials Perspectives*, **742**, 205–213.

Schultz, T. P. *et al.*, 1990, Role of stilbenes in the natural durability of wood – fungicidal structure-activity-relationships. *Phytochemistry*, **29**, 1501–1507.

Schweizer, P. *et al.*, 1997, Jasmonate-inducible genes are activated in rice by pathogen attack without a concomitant increase in endogenous jasmonic acid levels. *Plant Physiology*, **114**, 79–88.

Scofield, S. R. *et al.*, 1996, Molecular basis of gene-for-gene specificity in bacterial speck disease of tomato. *Science*, **274**, 2063–2065.

Scott-Craig, J. S. *et al.*, 1990, Endopolygalacturonase is not required for pathogenicity of *Cochliobolus carbonum* on maize. *Plant Cell*, **2**, 1191–1200.

Seem, R. C., 1984, Disease incidence and severity relationships. *Annual Review of Phytopathology*, **22**, 133–150.

Selmar, D., Irandoost, Z. and Wray, V., 1996, Dhurrin-6'-glucoside, a cyanogenic diglucoside from Sorghum bicolor. *Phytochemistry*, **43**, 569–572.

Semblat, J. P. *et al.*, 2001, Molecular cloning of a cDNA encoding an amphid-secreted putative avirulence protein from the root-knot nematode *Meloidogyne incognita*. *Molecular Plant-Microbe Interactions*, **14**, 72–79.

Serrano, M. G., Camargo, E. P. and Teixeira, M. M. G., 1999, *Phytomonas*: Analysis of polymorphism and genetic relatedness between isolates from plants and phytophagous insects from different geographic regions by RAPD fingerprints and synapomorphic markers. *Journal of Eukaryotic Microbiology*. **46**, 618–625.

Sessa, G. and Martin, G. B., 2000, Signal recognition and transduction mediated by the tomato Pto kinase: a paradigm of innate immunity in plants. *Microbes and Infection*, **2**, 1591–1597.

Shah, D., Bergstrom, G. C. and Ueng, P. P., 1995, Initiation of *Septoria nodorum* blotch epidemics in winter-wheat by seed-borne *Stagonospora nodorum*. *Phytopathology*, **85**, 452–457.

Shaner, E. and Finney, R. E., 1977, The effect of nitrogen fertilization on the expression of slow-mildewing resistance in Knox wheat. *Phytopathology*, **67**, 1051–1056.

Sharrock, K. R. and Labavitch, J. M., 1994, Polygalacturonase inhibitors of Bartlett pear fruits – differential-effects on *Botrytis cinerea* polygalacturonase isozymes, and influence on products of fungal hydrolysis of pear cell-walls and on ethylene induction in cell-culture. *Physiological and Molecular Plant Pathology*, **45**, 305–319.

Shaw, D. A., Adaskaveg, J. E. and Ogawa, J. M., 1990, Influence of wetness period and temperature on infection and development of shot-hole disease of almond caused by *Wilsonomyces carpophilus*. *Phytopathology*, **80**, 749–756.

Shaw, M. W., 1994, Modeling stochastic processing in plant pathology. *Annual Review of Phytopathology*, **32**, 523–544.

Shepherd, T. *et al.*, 1999, Epicuticular wax composition in relation to aphid infestation and resistance in red raspberry (Rubus idaeus L.). *Phytochemistry*, **52**, 1239–1254.

Sherriff, C. *et al.*, 1994, Ribosomal DNA-sequence analysis reveals new species groupings in the genus *Colletotrichum*. *Experimental Mycology*, **18**, 121–138.

Sherriff, C. *et al.*, 1995, rDNA sequence-analysis confirms the distinction between *Colletotrichum graminicola* and *C sublineolum*. *Mycological Research*, **99**, 475–478.

Shewry, P. R. and Lucas, J. A., 1997, Plant proteins that confer resistance to pests and pathogens. *Advances in Botanical Research*, **26**, 135–192.

Shieh, M. T. *et al.*, 1997, Molecular genetic evidence for the involvement of a specific polygalacturonase, P2c, in the invasion and spread of *Aspergillus flavus* in cotton bolls. *Applied and Environmental Microbiology*, **63**, 3548–3552.

Showalter, A. M., 1993, Structure and function of plant-cell wall proteins. *Plant Cell*, **5**, 9–23.

Shtienberg, D. *et al.*, 1990, Development and evaluation of a general-model for yield loss assessment in potatoes. *Phytopathology*, **80**, 466–472.

Siame, B. A. *et al.*, 1993, Isolation of strigol, a germination stimulant for *Striga asiatica*, from host plants. *Journal of Agricultural and Food Chemistry*, **41**, 1486–1491.

Siddiqi, M. R., 1986, *The Tylenchidae*, CAB International, Wallingford, Oxfordshire, UK.

Silverman, P. *et al.*, 1995, Salicylic-acid in rice – biosynthesis, conjugation, and possible role. *Plant Physiology*, **108**, 633–639.

Simpson, A. J. G. *et al.*, 2000, The genome sequence of the plant pathogen *Xylella fastidiosa*. *Nature*, **406**, 151–157.

Singh, P. *et al.*, 1999, Purification and biological characterization of host-specific SV-toxins from *Stemphylium vesicarium* causing brown spot of European pear. *Phytopathology*, **89**, 947–953.

Singh, P. *et al.*, 2000, Effects of host-selective SV-toxin from *Stemphylium vesicarium*, the cause of brown spot of European pear plants, on ultrastructure of leaf cells. *Journal of Phytopathology-Phytopathologische Zeitschrift*, **148**, 87–93.

Sivan, A. and Chet, I., 1989, Degradation of fungal cell-walls by lytic enzymes of *Trichoderma harzianum*. *Journal of General Microbiology*, **135**, 675–682.

Smirnov, S., Shulaev, V. and Tumer, N. E., 1997, Expression of pokeweed antiviral protein in transgenic plants induces virus resistance in grafted wild-type plants independently of salicylic acid accumulation and pathogenesis-related protein synthesis. *Plant Physiology*, **114**, 1113–1121.

Smith, D. A. and Banks, S. W., 1986, Biosynthesis, elicitation and biological activity of isoflavonoid phytoalexins. *Phytochemistry*, **25**, 979–995.

Smith, M. J. *et al.*, 1993, The syringolides – bacterial c-glycosyl lipids that trigger plant-disease resistance. *Tetrahedron Letters*, **34**, 223–226.

Smith, V. L., Wilcox, W. F. and Harman, G. E., 1990, Potential for biological-control of *Phytophthora* root and crown rots of apple by *Trichoderma* and *Gliocladium* spp. *Phytopathology*, **80**, 880–885.

Sommer-Knudsen, J., Bacic, A. and Clarke, A. E., 1998, Hydroxyproline-rich plant glycoproteins. *Phytochemistry*, **47**, 483–497.

Song, W. Y. *et al.*, 1995, A receptor kinase-like protein encoded by the rice disease resistance gene, *Xa21*. *Science*, **270**, 1804–1806.

Sonnenbichler, J. *et al.*, 1989, Secondary fungal metabolites and their biological-activities. 1. Isolation of antibiotic compounds from cultures of *Heterobasidion annosum* synthesized in the presence of antagonistic fungi or host plant-cells. *Biological Chemistry Hoppe-Seyler*, **370**, 1295–1303.

Sorensen, K. N., Kim, K. H. and Takemoto, J. Y. 1998, PCR detection of cyclic lipodepsinonapeptide-producing *Pseudomonas syringae* pv *syringae* and similarity of strains. *Applied and Environmental Microbiology*, **64**, 226–230.

Soulie, O., Roustan, J. P. and Fallot, J., 1993, Early *in vitro* selection of eutypine-tolerant plantlets – application to screening of *Vitis vinifera* cv. Ugni Blanc somaclones. *Vitis*, **32**, 243–244.

Spanu, P. and Boller, T., 1989, Ethylene biosynthesis in tomato plants infected by *Phytophthora infestans*. *Journal of Plant Physiology*, **134**, 533–537.

Spendley, P. J. and Ride, J. P., 1984, Fungitoxic effects of 2 alka-2, 4-dienals, alpha-triticene and beta-triticene, isolated from wheat. *Transactions of the British Mycological Society*, **82**, 283–288.

Spendley, P. J. *et al.*, 1982, 2 novel antifungal alka-2,4-dienals from *Triticum-aestivum*. *Phytochemistry*, **21**, 2403–2404.

Sperry, J. S. and Tyree, M. T., 1988, Mechanism of water stress-induced xylem embolism. *Plant Physiology*, **88**, 581–587.

Sposato, P., Ahn, J. H. and Walton, J. D., 1995, Characterization and disruption of a gene in the maize pathogen *Cochliobolus carbonum* encoding a cellulase lacking a cellulose-binding domain and hinge region. *Molecular Plant-Microbe Interactions*, **8**, 602–609.

Stace-Smith, R. and Hamilton, R. I., 1988, Inoculum thresholds of seedborne pathogens – viruses. *Phytopathology*, **78**, 875–880.

Stahl, D. J. and Schäfer, W., 1992, Cutinase is not required for fungal pathogenicity on pea. *Plant Cell*, **4**, 621–629.

Staples, R. C., 2001, Nutrients for a rust fungus: the role of haustoria. *Trends in Plant Science*, **6**, 496–498.

Staples, R. C. and Hoch, H. C., 1987, Infection structures – form and function. *Experimental Mycology*, **11**, 163–169.

Stapleton, J. J. and Grant, R. S., 1992, Leaf removal for nonchemical control of the summer bunch rot complex of wine grapes in the San-Joaquin valley. *Plant Disease*, **76**, 205–208.

Stark, D. M. and Beachy, R. N., 1989, Protection against potyvirus infection in transgenic plants – evidence for broad-spectrum resistance. *Bio-Technology*, **7**, 1257–1262.

Staskawicz, B. J., Dahlbeck, D. and Keen, N. T., 1984, Cloned avirulence gene of *Pseudomonas syringae* pv. *glycinea* determines race-specific incompatibility on *Glycine max* (L) Merr. *Proceedings of the National Academy of Sciences of the United States of America-Biological Sciences*, **81**, 6024–6028.

Steffens, J. C. and Walters, D. S., 1991, Biochemical aspects of glandular trichome-mediated insect resistance in the Solanaceae. *ACS Symposium Series*, **449**, 136–149.

Stein, M. and Somerville, S. C., 2002, MLO, a novel modulator of plant defenses and cell death, binds calmodulin. *Trends in Plant Science*, **7**, 379–380.

Stewart, G. R. and Press, M. C., 1990, The physiology and biochemistry of parasitic angiosperms. *Annual Review of Plant Physiology and Plant Molecular Biology*, **41**, 127–151.

Sticher, L., Mauch-Mani, B. and Métraux, J. P., 1997, Systemic acquired resistance. *Annual Review of Phytopathology*, **35**, 235–270.

Strange, R. N., 1993, *Plant Disease Control: Towards Environmentally Acceptable Methods*, Chapman & Hall, London UK.

Strange, R. N. and Smith, H., 1971, A fungal growth stimulant in anthers which predisposes wheat to attack by *Fusarium graminearum*. *Physiological Plant Pathology*, **1**, 141–150.

Strange, R. N. and Smith, H., 1978, The effects of choline, betaine and wheat-germ extract on growth of cereal pathogens. *Transactions of the British Mycological Society*, **70**, 193–199.

Strange, R. N., Deramo, A. and Smith, H., 1978, Virulence enhancement of *Fusarium graminearum* by choline and betaine and of *Botrytis cinerea* by other constituents of wheat germ. *Transactions of the British Mycological Society*, **70**, 210–217.

Strange, R. N., Majer, J. R. and Smith, H., 1974, The isolation and identification of choline and betaine as the two major components in anther and wheat germ that stimulate *Fusarium graminearum in vitro*. *Physiological Plant Pathology*, **4**, 277–290.

Strelkov, S. E., Lamari, L. and Ballance, G. M., 1998, Induced chlorophyll degradation by a chlorosis toxin from *Pyrenophora a tritici-repentis*. *Canadian Journal of Plant Pathology-Revue Canadienne de Phytopathologie*, **20**, 428–435.

Stumm, D. and Gessler, C., 1986, Role of papillae in the induced systemic resistance of cucumbers against *Colletotrichum lagenarium*. *Physiological and Molecular Plant Pathology*, **29**, 405–410.

Subba-Rao, P. V., Wadia, K. D. R. and Strange, R. N., 1996, Biotic and abiotic elicitation of phytoalexins in leaves of groundnut (*Arachis hypogaea* L). *Physiological and Molecular Plant Pathology*, **49**, 343–357.

Summerell, B. A. and Burgess, L. W., 1989, Factors influencing survival of *Pyrenophora tritici-repentis* – stubble management. *Mycological Research*, **93**, 38–40.

Sutton, B. C., 1992, 'The genus *Glomerella* and its anamorph' in *Colletotrichum – Biology, Pathology and Control*, J. A. Bailey & M. J. Jeger, eds., CAB International, Wallingford, Oxfordshire, UK, pp. 1–26.

Suzuki, Y. *et al.*, 1996, Specific accumulation of antifungal 5-alk(en)ylresorchinol homologs in etiolated rice seedlings. *Bioscience Biotechnology and Biochemistry*, **60**, 1786–1789.

Suzuki, Y. *et al.*, 1998, Identification of 5-N-(2'-oxo)alkylresorcinols from etiolated rice seedlings. *Phytochemistry*, **47**, 1247–1252.

Suzuki, Y. *et al.*, 1998, A potent antifungal benzoquinone in etiolated sorghum seedlings and its metabolites. *Phytochemistry*, **47**, 997–1001.

Swart, S. *et al.*, 1994, Rhicadhesin-mediated attachment and virulence of an *Agrobacterium tumefaciens Chvb* mutant can be restored by growth in a highly osmotic medium. *Journal of Bacteriology*, **176**, 3816–3819.

Swart, S. *et al.*, 1994, Several phenotypic changes in the cell-envelope of *Agrobacterium tumefaciens chvb* mutants are prevented by calcium limitation. *Archives of Microbiology*, **161**, 310–315.

Sykes, P. J. *et al.*, 1992, Quantitation of targets for PCR by use of limiting dilution. *Biotechniques*, **13**, 444–449.

Taiz, E. and Zeiger, E., 1991, *Plant Physiology*, The Benjamin Cummings Publishing Company, Redwood City, California USA.

Taiz, E. and Zeiger, E., 2002, *Plant Physiology*, Third edition Sinauer Associates Inc., Sunderland, Massachusetts USA.

Tanaka, A. *et al.*, 1999, Insertional mutagenesis and cloning of the genes required for biosynthesis of the host-specific AK-toxin in the Japanese pear pathotype of *Alternaria alternata*. *Molecular Plant-Microbe Interactions*, **12**, 691–702.

Tanaka, S., 1933, Studies on black spot disease of Japanese pear (*Pyrus serotina* Rehd.). *Memoirs of the College of Agriculture, Kyoto Imperial University*, **28**, 1–31.

Tang, X. Y. *et al.*, 1996, Initiation of plant disease resistance by physical interaction of AvrPto and Pto kinase. *Science*, **274**, 2060–2063.

Tang, X. Y. *et al.*, 1999, Overexpression of *Pto* activates defense responses and confers broad resistance. *Plant Cell*, **11**, 15–29.

Tariq, V. N. and Jeffries, P., 1987, Cytochemical-localization of lipolytic enzyme-activity during penetration of host tissues by *Sclerotinia sclerotiorum*. *Physiological and Molecular Plant Pathology*, **30**, 77–91.

Taylor, A. L., 1971, 'Estimating nematode densities in soil and roots' in *Crop Loss Assessment Methods*, L. Chiarappa, ed., FAO and CAB, Oxford, UK, p. 3.1.4/1–3.1.4/5.

Taylor, L. R., 1971, 'The measurement of arthropod numbers and activity by sampling with sweep nets and traps' in *Crop Loss Assessment Methods*, L. Chiarappa, ed., FAO and CAB, Oxford, UK, p. 3.1.3/1–3.1.3/9.

Templeton, M. D. *et al.*, 1990, Hydroxyproline-rich glycoprotein transcripts exhibit different spatial patterns of accumulation in compatible and incompatible interactions between *Phaseolus vulgaris* and *Colletotrichum-lindemuthianum. Plant Physiology*, **94**, 1265–1269.

Tenhaken, R. *et al.*, 1997, Characterization and cloning of cutinase from *Ascochyta rabiei. Zeitschrift fur Naturforschung C-A Journal of Biosciences*, **52**, 197–208.

Ten Have, A. *et al.*, 1998, The endopolygalacturonase gene *Bcpg1* is required for full virulence of *Botrytis cinerea. Molecular Plant-Microbe Interactions*, **11**, 1009–1016.

Ten Houten, J. G., 1974, Plant pathology: changing agricultural methods and human society. *Annual Review of Phytopathology*, **12**, 1–11.

Terhune, B. T. and Hoch, H. C., 1993, Substrate hydrophobicity and adhesion of *Uromyces* urediospores and germlings. *Experimental Mycology*, **17**, 241–252.

Terhune, B. T., Bojko, R. J. and Hoch, H. C., 1993, Deformation of stomatal guard-cell lips and microfabricated artificial topographies during appressorium formation by *Uromyces. Experimental Mycology*, **17**, 70–78.

Termorshuizen, A. J. *et al.*, 1998, Interlaboratory comparison of methods to quantify microsclerotia of *Verticillium dahliae* in soil. *Applied and Environmental Microbiology*, **64**, 3846–3853.

Terras, F. R. G. *et al.*, 1995, Small cysteine-rich antifungal proteins from radish – their role in host-defense. *Plant Cell*, **7**, 573–588.

Thakur, R. P. and Williams, R. J., 1980, Pollination effects on pearl millet ergot. *Phytopathology*, **70**, 80–84.

Thevissen, K. *et al.*, 2000, A gene encoding a sphingolipid biosynthesis enzyme determines the sensitivity of *Saccharomyces cerevisiae* to an antifungal plant defense from dahlia (*Dahlia merckii*). *Proceedings of the National Academy of Sciences of the United States of America*, **97**, 9531–9536.

Thevissen, K. *et al.*, 2000, Specific binding sites for an antifungal plant defensin from dahlia (*Dahlia merckii*) on fungal cells are required for antifungal activity. *Molecular Plant-Microbe Interactions*, **13**, 54–61.

Thomas, C. M., *et al.*, 1998, Genetic and molecular analysis of tomato *Cf* genes for resistance to *Cladosporium fulvum. Philosophical Transactions of the Royal Society of London Series B-Biological Sciences*, **353**, 1413–1424.

Thomas, J. and John, V. T., 1981, Effect of gibberellic-acid and indole-3-acetic-acid on the infection of rice plants by rice tungro virus. *Phytopathologische Zeitschrift-Journal of Phytopathology*, **101**, 168–174.

Thomashow, L. S. and Weller, D. M., 1990, Role of antibiotics and siderophores in biocontrol of take-all disease of wheat. *Plant and Soil*, **129**, 93–99.

Thornbury, D. W. and Pirone, T. P., 1983, Helper components of 2 potyviruses are serologically distinct. *Virology*, **125**, 487–490.

Thrane, C. *et al.*, 2000, Viscosinamide-producing *Pseudomonas fluorescens* DR54 exerts a biocontrol effect on *Pythium ultimum* in sugar beet rhizosphere. *FEMS Microbiology Ecology*, **33**, 139–146.

Thresh, J. M. and Owusu, G. K., 1986 The control of cocoa swollen shoot disease in Ghana – an evaluation of eradication procedures. *Crop Protection*, **5**, 41–52.

Thurston, H. D., 1990, Plant-disease management-practices of traditional farmers. *Plant Disease*, **74**, 96–102.

Tiburzy, R. and Reisener, H. J., 1990, Resistance of wheat to *Puccinia graminis* f sp *tritici* – association of the hypersensitive reaction with the cellular accumulation of lignin-like material and callose. *Physiological and Molecular Plant Pathology*, **36**, 109–120.

Tieman, D. V. *et al.*, 2000, The tomato ethylene receptors NR and LeETR4 are negative regulators of ethylene response and exhibit functional compensation within a multigene family. *Proceedings of the National Academy of Sciences of the United States of America*, **97**, 5663–5668.

Tien, M. and Kirk, T. K., 1983, Lignin-degrading enzyme from the hymenomycete *Phanerochaete chrysosporium* Burds. *Science*, **221**, 661–662.

Timms-Wilson, T. M., Bryant, K. and Bailey, M. J., 2001, Strain characterization and 16S–23S probe development for differentiating geographically dispersed isolates of the phytopathogen *Ralstonia solanacearum*. *Environmental Microbiology*, **3**, 785–797.

Timms-Wilson, T. M. *et al.*, 2000, Chromosomal insertion of phenazine-1-carboxylic acid biosynthetic pathway enhances efficacy of damping-off disease control by *Pseudomonas fluorescens*. *Molecular Plant-Microbe Interactions*, **13**, 1293–1300.

Tinland, B., 1996, The integration of T-DNA into plant genomes. *Trends in Plant Science*, **1**, 178–184.

Tjamos, E. C. and Fravel, D. R., 1995, Detrimental effects of sublethal heating and *Talaromyces flavus* on microsclerotia of *Verticillium dahliae*. *Phytopathology*, **85**, 388–392.

Tomas, A. *et al.*, 1990, Purification of a cultivar-specific toxin from *Pyrenophora tritici-repentis*, causal agent of tan spot of wheat. *Molecular Plant-Microbe Interactions*, **3**, 221–224.

Tomlinson, J. A., 1960, Crook rot of watercress. A review of research. *National Agricultural Advisory Service Quarterly Review*, **49**, 13–19.

Tomlinson, J. A. and Hunt, J., 1987, Studies on watercress chlorotic leaf-spot virus and on the control of the fungus vector (*Spongospora subterranea*) with zinc. *Annals of Applied Biology*, **110**, 75–88.

Ton, J. *et al.*, (2002), Characterization of *Arabidopsis* enhanced disease susceptibility mutants that are affected in systemically induced resistance. *Plant Journal*, 29, 11–21.

Torres, M. A., Dangl, J. L. and Jones, J. D. G., 2002, Arabidopsis *gp91* (phox) homologues *AtrbohD* and *AtrbohF* are required for accumulation of reactive oxygen intermediates in the plant defense response. *Proceedings of the National Academy of Sciences of the United States of America*, **99**, 517–522.

Trapero-Casas, A. and Kaiser, W. J., 1992, Development of *Didymella rabiei*, the teleomorph of *Ascochyta rabiei*, on chickpea straw. *Phytopathology*, **82**, 1261–1266.

Trebitsh, T., Goldschmidt, E. E. and Riov, J., 1993, Ethylene induces *de novo* synthesis of chlorophyllase, a chlorophyll degrading enzyme, in citrus-fruit peel. *Proceedings of the National Academy of Sciences of the United States of America*, **90**, 9441–9445.

Trutmann, P. and Pyndji, M. M., 1994, Partial replacement of local common bean mixtures by high-yielding angular leaf-spot resistant varieties to conserve local genetic diversity while increasing yield. *Annals of Applied Biology*, **125**, 45–52.

Trutmann, P., Paul, K. B. and Cishabayo, D., 1992, Seed treatments increase yield of farmer varietal field bean mixtures in the central African highlands through multiple disease and beanfly control. *Crop Protection*, **11**, 458–464.

Tsiamis, G. *et al.*, 2000, Cultivar-specific avirulence and virulence functions assigned to *avrPphF* in *Pseudomonas syringae* pv. *phaseolicola*, the cause of bean halo-blight disease. *EMBO Journal*, **19**, 3204–3214.

Tucker, S. L. and Talbot, N. J., 2001, Surface attachment and pre-penetration stage development by plant pathogenic fungi. *Annual Review of Phytopathology*, **39**, 385–417.

Turner, J. G., 1986, Activities of ribulose-1,5-bisphosphate carboxylase and glutamine-synthetase in isolated mesophyll-cells exposed to tabtoxin. *Physiological and Molecular Plant Pathology*, **29**, 59–68.

Turner, J. G. and Mitchell, R. E., 1985, Association between symptom development and inhibition of ornithine carbamoyltransferase in bean-leaves treated with phaseolotoxin. *Plant Physiology*, **79**, 468–473.

Turner, S. J., 1993, Soil sampling to detect potato cyst-nematodes (*Globodera* spp). *Annals of Applied Biology*, **123**, 349–357.

Tyagi, S. and Kramer, F. R., 1996, Molecular beacons: Probes that fluoresce upon hybridization. *Nature Biotechnology*, **14**, 303–308.

Tyler, B. M. *et al.*, 1996, Chemotactic preferences and strain variation in the response of *Phytophthora sojae* zoospores to host isoflavones. *Applied and Environmental Microbiology*, **62**, 2811–2817.

Tzeng, D. D. and DeVay, J. E., 1985, Physiological-responses of *Gossypium hirsutum* L to infection by defoliating and nondefoliating pathotypes of *Verticillium dahliae* Kleb. *Physiological Plant Pathology*, **26**, 57–72.

Tzfira, T. *et al.*, 2000, Nucleic acid transport in plant-microbe interactions: The molecules that walk through the walls. *Annual Review of Microbiology*, **54**, 187–219.

Tzortzakakis, E. A., Trudgill, D. L. and Phillips, M. S., 1998, Evidence for a dosage effect of the *Mi* gene on partially virulent isolates of *Meloidogyne javanica*. *Journal of Nematology*, **30**, 76–80.

Uchiyama, T. and Okuyama, K., 1990, Participation of *Oryza sativa* leaf wax in appressorium formation by *Pyricularia oryzae*. *Phytochemistry*, **29**, 91–92.

Umemoto, N. *et al.*, 1997, The structure and function of a soybean beta-glucan-elicitor-binding protein. *Proceedings of the National Academy of Sciences of the United States of America*, **94**, 1029–1034.

Urban, M., Bhargava, T. and Hamer, J. E., 1999, An ATP-driven efflux pump is a novel pathogenicity factor in rice blast disease. *EMBO Journal*, **18**, 512–521.

Valkonen, J. P. T., 1997, Novel resistances to four potyviruses in tuber-bearing potato species, and temperature-sensitive expression of hypersensitive resistance to potato virus Y. *Annals of Applied Biology*, **130**, 91–104.

Vallélian-Bindschedler, L., Métraux, J. P. and Schweizer, P., 1998, Salicylic acid accumulation in barley is pathogen specific but not required for defense-gene activation. *Molecular Plant-Microbe Interactions*, **11**, 702–705.

Van Alfen, N. K., Mcmillan, B. D. and Dryden, P., 1987, The multicomponent extracellular polysaccharide of *Clavibacter michiganense* subsp. *insidiosum*. *Phytopathology*, **77**, 496–501.

Van Alfen, N. K., Mcmillan, B. D. and Wang, Y., 1987, Properties of the extracellular polysaccharides of *Clavibacter michiganense* subsp. *insidiosum* that may affect pathogenesis. *Phytopathology*, **77**, 501–505.

Van den Ackerveken, G. F. J. M., vanKan, J. A. L. and de Wit, P. J. G. M., 1992, Molecular analysis of the avirulence gene *Avr9* of the fungal tomato pathogen *Cladosporium fulvum* fully supports the gene-for-gene hypothesis. *Plant Journal*, **2**, 359–366.

Van den Ackerveken, G. F. J. M., Vossen, P. and de Wit, P. J. G. M., 1993, The *Avr9* race-specific elicitor of *Cladosporium fulvum* is processed by endogenous and plant proteases. *Plant Physiology*, **103**, 91–96.

Van den Bosch, F., Metz, J. A. J. and Zadoks, J. C., 1999, Pandemics of focal plant disease, a model. *Phytopathology*, **89**, 495–505.

Van den Bosch, F., Zadoks, J. C. and Metz, J. A. J., 1988, Focus expansion in plant-disease.1. The constant rate of focus expansion. *Phytopathology*, **78**, 54–58.

Van den Bosch, F., Zadoks, J. C. and Metz, J. A. J., 1988, Focus expansion in plant-disease. 2. Realistic parameter-sparse models. *Phytopathology*, **78**, 59–64.

Van den Bosch, F. *et al.*, 1988, Focus expansion in plant-disease. 3. 2 Experimental examples. *Phytopathology*, **78**, 919–925.

Van den Bosch, F. *et al.*, 1990, Focus expansion in plant-disease. 4. Expansion rates in mixtures of resistant and susceptible hosts. *Phytopathology*, **80**, 598–602.

Van der Hoorn, R. A. L., de Wit, P. J. G. M. and Joosten, M. H. A. J., 2002, Balancing selection favors guarding resistance proteins. *Trends in Plant Science*, **7**, 67–71.

Van der Plank, J. E., 1963, *Plant Diseases: Epidemics and Control*, Academic Press, New York USA.

Van Dijk, K. and Nelson, E. B., 2000, Fatty acid competition as a mechanism by which *Enterobacter cloacae* suppresses *Pythium ultimum* sporangium germination and damping-off. *Applied and Environmental Microbiology*, **66**, 5340–5347.

Van Dijk, P., Vandermeer, F. A. and Piron, P. G. M., 1987, Accessions of Australian *Nicotiana* species suitable as indicator hosts in the diagnosis of plant virus diseases. *Netherlands Journal of Plant Pathology*, **93**, 73–85.

Van Loon, L. C. and Van Strien, E. A., 1999, The families of pathogenesis-related proteins, their activities, and comparative analysis of PR-1 type proteins. *Physiological and Molecular Plant Pathology*, **55**, 85–97.

Van Loon, L. C., Bakker, P. A. H. M. and Pieterse, C. M. J., 1998, Systemic resistance induced by rhizosphere bacteria. *Annual Review of Phytopathology*, **36**, 453–483.

Van Peer, R., Niemann, G. J. and Schippers, B., 1991, Induced resistance and phytoalexin accumulation in biological-control of *Fusarium*-wilt of carnation by *Pseudomonas* sp strain WCS417R. *Phytopathology*, **81**, 728–734.

Van Wees, S. C. M. *et al.*, 2000, Enhancement of induced disease resistance by simultaneous activation of salicylate- and jasmonate-dependent defense pathways in *Arabidopsis thaliana*. *Proceedings of the National Academy of Sciences of the United States of America*, **97**, 8711–8716.

Vander, P. *et al.*, 1998, Comparison of the ability of partially N-acetylated chitosans and chitooligosaccharides to elicit resistance reactions in wheat leaves. *Plant Physiology*, **118**, 1353–1359.

Vanderbiezen, E. A. *et al.*, 1995, Inheritance and genetic-mapping of resistance to *Alternaria alternata* f sp *lycopersici* in *Lycopersicon pennellii*. *Molecular and General Genetics*, **247**, 453–461.

VanEtten, H. D., Matthews, D. E. and Matthews, P. S., 1989, Phytoalexin detoxification – importance for pathogenicity and practical implications. *Annual Review of Phytopathology*, **27**, 143–164.

Van Kan, J. A. L. *et al.*, 1997, Cutinase A of *Botrytis cinerea* is expressed, but not essential, during penetration of *Gerbera* and tomato. *Molecular Plant-Microbe Interactions*, **10**, 30–38.

Van Parijs, J. *et al.*, 1992, Effect of the *Urtica dioica* agglutinin on germination and cell-wall formation of *Phycomyces blakesleeanus* Burgeff. *Archives of Microbiology*, **158**, 19–25.

Veness, R. G. and Evans, C. S., 1989, The role of hydrogen-peroxide in the degradation of crystalline cellulose by basidiomycete fungi. *Journal of General Microbiology*, **135**, 2799–2806.

Venkatachalam, P. *et al.*, 1998, Regeneration of late leaf spot-resistant groundnut plants from *Cercosporidium personatum* culture filtrate-treated callus. *Current Science*, **74**, 61–65.

Verbeek, M., Vandijk, P. and Vanwell, P. M. A., 1995, Efficiency of eradication of 4 viruses from garlic (*Allium sativum*) by meristem-tip culture. *European Journal of Plant Pathology*, **101**, 231–239.

Verberne, M. C. *et al.*, 2000, Overproduction of salicylic acid in plants by bacterial transgenes enhances pathogen resistance. *Nature Biotechnology*, **18**, 779–783.

Vereecke, D. *et al.*, 2000, The *Rhodococcus fascians*-plant interaction: morphological traits and biotechnological applications. *Planta*, **210**, 241–251.

Vidal, S. *et al.*, 1998, Cell wall-degrading enzymes from *Erwinia carotovora* cooperate in the salicylic acid-independent induction of a plant defense response. *Molecular Plant-Microbe Interactions*, **11**, 23–32.

Vidhyasekaran, P. *et al.*, 1990, Selection of brown spot-resistant rice plants from *Helminthosporium oryzae* toxin-resistant calluses. *Annals of Applied Biology*, **117**, 515–523.

Vietmeyer, N. D., 1986, Lesser-known plants of potential use in agriculture and forestry. *Science*, **232**, 1379–1384.

Vijayan, P. *et al.*, 1998, A role for jasmonate in pathogen defense of *Arabidopsis*. *Proceedings of the National Academy of Sciences of the United States of America*, **95**, 7209–7214.

Visser, P. B. and Bol, J. F., 1999, Nonstructural proteins of Tobacco rattle virus which have a role in nematode-transmission: expression pattern and interaction with viral coat protein. *Journal of General Virology*, **80**, 3273–3280.

Vloutoglou, I., Fitt, B. D. L. and Lucas, J. A., 1995, Periodicity and gradients in dispersal of *Alternaria linicola* in linseed crops. *European Journal of Plant Pathology*, **101**, 639–653.

Voegele, R. T. *et al.*, 2001, The role of haustoria in sugar supply during infection of broad bean by the rust fungus *Uromyces fabae*. *Proceedings of the National Academy of Sciences of the United States of America*, **98**, 8133–8138.

Voisard, C. *et al.*, 1989, Cyanide Production by *Pseudomonas fluorescens* helps suppress black root-rot of tobacco under gnotobiotic conditions. *EMBO Journal*, **8**, 351–358.

Vos, P. *et al.*, 1995, AFLP – a new technique for DNA fingerprinting. *Nucleic Acids Research*, **23**, 4407–4414.

Wada, H. *et al.*, 1996, Occurrence of the strawberry pathotype of *Alternaria alternata* in Italy. *Plant Disease*, **80**, 372–374.

Walker, D. S., Reeves, P. J. and Salmond, G. P. C., 1994, The major secreted cellulase, celV, of *Erwinia carotovora* subsp *carotovora* is an important soft-rot virulence factor. *Molecular Plant-Microbe Interactions*, **7**, 425–431.

Wallen, V. R. and Galway, D. A., 1977, Studies on the biology and control of *Ascochyta fabae* on faba bean. *Canadian Plant Disease Survey*, **57**, 31–35.

Walters, K. *et al.*, 1990, Gene for pathogenicity and ability to cause the hypersensitive reaction cloned from *Erwinia amylovora*. *Physiological and Molecular Plant Pathology*, **36**, 509–521.

Walton, J. D., 1996, Host-selective toxins: Agents of compatibility *Plant Cell*, **8**, 1723–1733.

Walton, J. D., Earle, E. D. and Gibson, B. W., 1982, Purification and structure of the host-specific toxin from *Helminthosporium carbonum* race 1. *Biochemical and Biophysical Research Communications*, **107**, 785–794.

Wang, E. M. *et al.*, 2001, Suppression of a P450 hydroxylase gene in plant trichome glands enhances natural-product-based aphid resistance. *Nature Biotechnology*, **19**, 371–374.

Wang, G. L. *et al.*, 1996, The cloned gene, *Xa21*, confers resistance to multiple *Xanthomonas oryzae* pv. *oryzae* isolates in transgenic plants. *Molecular Plant-Microbe Interactions*, **9**, 850–855.

Wang, H. *et al.*, 1996, Fumonisins and *Alternaria alternata lycopersici* toxins: Sphinganine analog mycotoxins induce apoptosis in monkey kidney cells. *Proceedings of the National Academy of Sciences of the United States of America*, **93**, 3461–3465.

Wang, Z. X. *et al.*, 1999, The *Pib* gene for rice blast resistance belongs to the nucleotide binding and leucine-rich repeat class of plant disease resistance genes. *Plant Journal*, **19**, 55–64.

Wangai, A. W., 1990, 'Effects of barley yellow dwarf virus on cereals in Kenya', in *Barley Yellow Dwarf*, CIMMYT, ed., Mexico, pp. 391–393.

Wardlaw, C. W., 1972, *Banana Diseases Including Plantains and Abaca*, Second edition, Longman, London UK.

Waterhouse, P. M., Gerlach, W. L. and Miller, W. A., 1986, Serotype-specific and general luteovirus probes from cloned cDNA sequences of barley yellow dwarf virus. *Journal of General Virology*, **67**, 1273–1281.

Waterhouse, P. M., Graham, H. W. and Wang, M. B., 1998, Virus resistance and gene silencing in plants can be induced by simultaneous expression of sense and antisense RNA. *Proceedings of the National Academy of Sciences of the United States of America*, **95**, 13959–13964.

Waterworth, H., 1993, Processing foreign plant germ plasm at the National-Plant-Germplasm Quarantine Center. *Plant Disease*, **77**, 854–860.

Wattad, C., Dinoor, A. and Prusky, D., 1994, Purification of pectate lyase produced by *Colletotrichum gloeosporioides* and its inhibition by epicatechin – a possible factor involved in the resistance of unripe avocado fruits to anthracnose. *Molecular Plant-Microbe Interactions*, **7**, 293–297.

Webb, S. E. and Linda, S. B., 1992, Evaluation of spunbonded polyethylene row covers as a method of excluding insects and viruses affecting fall-grown squash in Florida. *Journal of Economic Entomology*, **85**, 2344–2352.

Wei, G., Kloepper, J. W. and Tuzun, S., 1996, Induced systemic resistance to cucumber diseases and increased plant growth by plant growth-promoting rhizobacteria under field conditions. *Phytopathology*, **86**, 221–224.

Weiergang, I., Hipskind, J. D. and Nicholson, R. L., 1996, Synthesis of 3-deoxyanthocyanidin phytoalexins in sorghum occurs independent of light. *Physiological and Molecular Plant Pathology*, **49**, 377–388.

Weiergang, I. *et al.*, 1996, Morphogenic regulation of pathotoxin synthesis in *Cochliobolus carbonum*. *Fungal Genetics and Biology*, **20**, 74–78.

Weiler, E. W., 1997, Octadecanoid-mediated signal transduction in higher plants. *Naturwissenschaften*, **84**, 340–349.

Weller, S. A. *et al.*, 2000, Detection of *Ralstonia solanacearum* strains with a quantitative, multiplex, real-time, fluorogenic PCR (TaqMan) assay. *Applied and Environmental Microbiology*, **66**, 2853–2858.

Wells, J. M., Vanderzwet, T. and Butterfield, J. E., 1994, Differentiation of *Erwinia* species in the *herbicola* group by class analysis of cellular fatty-acids. *Journal of Phytopathology-Phytopathologische Zeitschrift*, **140**, 39–48.

Weste, G. and Marks, G. C., 1987, The Biology of *Phytophthora cinnamomi* in Australasian Forests. *Annual Review of Phytopathology*, **25**, 207–229.

Wheeler, H. and Luke, H. H., 1955, Mass screening for disease-resistant mutants in oats. *Science*, **122**, 1229.

Whenham, R. J. and Fraser, R. S. S., 1981, Effect of systemic and local lesion-forming strains of tobacco mosaic-virus on abscisic-acid concentration in tobacco-leaves-consequences for the control of leaf growth. *Physiological Plant Pathology*, **18**, 267–278.

Whipps, J. M., 2001, Microbial interactions and biocontrol in the rhizosphere. *Journal of Experimental Botany*, **52**, 487–511.

White, R. F., 1979, Acetysalicylic acid (aspirin) induces resistance to tobacco mosaic virus in tobacco. *Virology*, **99**, 410–412.

Wiebe, M. G., Robson, G. D. and Trinci, A. P. J., 1989, Effect of choline on the morphology, growth and phospholipid-composition of *Fusarium graminearum*. *Journal of General Microbiology*, **135**, 2155–2162.

Wilcox, J. R. and St Martin, S. K., 1998, Soybean genotypes resistant to *Phytophthora sojae* and compensation for yield losses of susceptible isolines. *Plant Disease*, **82**, 303–306.

Wilhite, S. E., Lumsden, R. D. and Straney, D. C., 2001, Peptide synthetase gene in *Trichoderma virens*. *Applied and Environmental Microbiology*, **67**, 5055–5062.

Willingale, J., Mantle, P. G. and Thakur, R. P., 1986, Postpollination stigmatic constriction, the basis of ergot resistance in selected lines of pearl-millet. *Phytopathology*, **76**, 536–539.

Wilson, W. A. and McMullen, M. S., 1997, Dosage dependent genetic suppression of oat crown rust resistance gene *Pc-62*. *Crop Science*, **37**, 1699–1705.

Wingate, V. P. M., Lawton, M. A. and Lamb, C. J., 1988, Glutathione causes a massive and selective induction of plant defense genes. *Plant Physiology*, **87**, 206–210.

Wolfe, M. S., 1985, The current status and prospects of multiline cultivars and variety mixtures for disease resistance. *Annual Review of Phytopathology*, **23**, 251–273.

Woloshuk, C. P. *et al.*, 1991, Pathogen-induced proteins with inhibitory activity toward *Phytophthora infestans. Plant Cell*, **3**, 619–628.

Wolpert, T. J. and Macko, V. 1991, Immunological comparison of the *in vitro* and *in vivo* labeled victorin binding-protein from susceptible oats. *Plant Physiology*, **95**, 917–920.

Wolpert, T. J. *et al.*, 1994, Identification of the 100-Kd victorin binding-protein from oats. *Plant Cell*, **6**, 1145–1155.

Woo, E. J. *et al.*, 2000, Crystallization and preliminary X-ray analysis of the auxin receptor ABP1. *Acta Crystallographica Section D-Biological Crystallography*, **56**, 1476–1478.

Woo, S. *et al.*, 2002, Synergism between fungal enzymes and bacterial antibiotics may enhance biocontrol. *Antonie Van Leeuwenhoek International Journal of General and Molecular Microbiology*, **81**, 353–356.

Woodward, S. and Pearce, R. B., 1988, The role of stilbenes in resistance of Sitka spruce (*Picea sitchensis* (Bong) Carr) to entry of fungal pathogens. *Physiological and Molecular Plant Pathology*, **33**, 127–149.

Woodward, S. and Pearce, R. B., 1988, Responses of Sitka spruce callus to challenge with wood decay fungi. *European Journal of Forest Pathology*, **18**, 217–229.

Wotton, H. R. and Strange, R. N., 1985 Circumstantial evidence for phytoalexin involvement in the resistance of peanuts to *Aspergillus flavus. Journal of General Microbiology*, **131**, 487–494.

Wotton, H. R. and Strange, R. N., 1987, Increased susceptibility and reduced phytoalexin accumulation in drought-stressed peanut kernels challenged with *Aspergillus flavus. Applied and Environmental Microbiology*, **53**, 270–273.

Wu, G. S. *et al.*, 1995, Disease resistance conferred by expression of a gene encoding H_2O_2-generating glucose-oxidase in transgenic potato plants. *Plant Cell*, **7**, 1357–1368.

Wu, Q. D., Preisig, C. L. and VanEtten, H. D., 1997, Isolation of the cDNAs encoding (+)6a-hydroxymaackiain 3-O-methyltransferase, the terminal step for the synthesis of the phytoalexin pisatin in *Pisum satvium. Plant Molecular Biology*, **35**, 551–560.

Wu, S. C. *et al.*, 1997, Deletion of two endo-beta-1,4-xylanase genes reveals additional isozymes secreted by the rice blast fungus. *Molecular Plant-Microbe Interactions*, **10**, 700–708.

Wynn, W. K., 1981, Tropic and taxic responses of pathogens to plants. *Annual Review of Phytopathology*, **19**, 237–255.

Wyss, U., 1982, Virus-transmitting nematodes – feeding-behavior and effect on root-cells. *Plant Disease*, **66**, 639–644.

Xi, K. *et al.*, 2000, Histopathological study of barley cultivars resistant and susceptible to *Rhynchosporium secalis. Phytopathology*, **90**, 94–102.

Xiao, J. Z. *et al.*, 1994, Studies on cellular-differentiation of *Magnaporthe grisea* – physicochemical aspects of substratum surfaces in relation to appressorium formation. *Physiological and Molecular Plant Pathology*, **44**, 227–236.

Xu, X. M., Guerin, L. and Robinson, J. D., 2001, Effects of temperature and relative humidity on conidial germination and viability, colonization and sporulation of *Monilinia fructigena. Plant Pathology*, **50**, 561–568.

Xu, Y. *et al.*, 1994, Plant defense genes are synergistically induced by ethylene and methyl jasmonate. *Plant Cell*, **6**, 1077–1085.

Yakoby, N. *et al.*, 2001, *Colletotrichum gloeosporioides pelB* is an important virulence factor in avocado fruit-fungus interaction. *Molecular Plant-Microbe Interactions*, **14**, 988–995.

Yang, X. B., Berggren, G. T. and Snow, J. P., 1990, Effects of free moisture and soybean growth stage on focus expansion of *Rhizoctonia* aerial blight. *Phytopathology*, **80**, 497–503.

Yao, C. L., Conway, W. S. and Sams, C. E., 1996, Purification and characterization of a polygalacturonase produced by *Penicillium expansum* in apple fruit. *Phytopathology*, **86**, 1160–1166.

Yeh, S. D. *et al.*, 1988, Control of Papaya ringspot virus by cross protection. *Plant Disease*, **72**, 375–380.

Yen, S. K. *et al.*, 2001, Environmental and developmental regulation of the wound-induced cell wall protein WI12 in the halophyte ice plant. *Plant Physiology*, **127**, 517–528.

Young, D. H. and Sequeira, L., 1986, Binding of *Pseudomonas solanacearum* fimbriae to tobacco leaf cell-walls and its inhibition by bacterial extracellular polysaccharides. *Physiological and Molecular Plant Pathology*, **28**, 393–402.

Young, J. M. *et al.*, 2001, A revision of Rhizobium Frank 1889, with an emended description of the genus, and the inclusion of all species of *Agrobacterium* Conn 1942 and *Allorhizobium undicola* de Lajudie *et al.* 1998 as new combinations: *Rhizobium radiobacter, R. rhizogenes, R. rubi, R. undicola* and *R. vitis. International Journal of Systematic and Evolutionary Microbiology*, **51**, 89–103.

Young, N. D., 1996, QTL mapping and quantitative disease resistance in plants. *Annual Review of Phytopathology*, **34**, 479–501.

Yun, S. H., Turgeon, B. G. and Yoder, O. C., 1998, REMI-induced mutants of *Mycosphaerella zeae-maydis* lacking the polyketide PM-toxin are deficient in pathogenesis to corn. *Physiological and Molecular Plant Pathology*, **52**, 53–66.

Zadoks, J. C., 1985, On the conceptual basis of crop loss assessment – the threshold theory. *Annual Review of Phytopathology*, **23**, 455–473.

Zadoks, J. C. and Van den Bosch, F., 1994, On the spread of plant-disease – a theory on foci. *Annual Review of Phytopathology*, **32**, 503–521.

Zawolek, M. W., 1993, Shaping a focus – wind and stochasticity. *Netherlands Journal of Plant Pathology*, **99**, 241–255.

Zekaria-Oren, J., Eyal, Z. and Ziv, O., 1991, Effect of film-forming compounds on the development of leaf rust on wheat seedlings. *Plant Disease*, **75**, 231–234.

Zeringue, H. J., Bhatnagar, D. and Cleveland, T. E., 1993, $C_{15}H_{24}$ volatile compounds unique to aflatoxigenic strains of *Aspergillus flavus. Applied and Environmental Microbiology*, **59**, 2264–2270.

Zhang, L. H. and Birch, R. G., 1997, The gene for albicidin detoxification from *Pantoea dispersa* encodes an esterase and attenuates pathogenicity of *Xanthomonas albilineans* to sugarcane. *Proceedings of the National Academy of Sciences of the United States of America*, **94**, 9984–9989.

Zhang, L. H., Xu, J. L. and Birch, R. G., 1999, Engineered detoxification confers resistance against a pathogenic bacterium. *Nature Biotechnology*, **17**, 1021–1024.

Zhang, S. Q. and Klessig, D. F., 1998, Resistance gene *N*-mediated *de novo* synthesis and activation of a tobacco mitogen-activated protein kinase by tobacco mosaic virus infection. *Proceedings of the National Academy of Sciences of the United States of America*, **95**, 7433–7438.

Zhang, S. Q. and Klessig, D. F., 2001, MAPK cascades in plant defense signaling. *Trends in Plant Science*, **6**, 520–527.

Zhang, S. Q., Liu, Y. D. and Klessig, D. F., 2000, Multiple levels of tobacco WIPK activation during the induction of cell death by fungal elicitins. *Plant Journal*, **23**, 339–347.

Zhang, Z. Y. *et al.*, 1998, Expression of human lactoferrin cDNA confers resistance to *Ralstonia solanacearum* in transgenic tobacco plants. *Phytopathology*, **88**, 730–734.

Zhao, J. M. and Last, R. L., 1996, Coordinate regulation of the tryptophan biosynthetic pathway and indolic phytoalexin accumulation in *Arabidopsis*. *Plant Cell*, **8**, 2235–2244.

Zhao, X. W., Schmitt, M. and Hawes, M. C., 2000, Species-dependent effects of border cell and root tip exudates on nematode behavior. *Phytopathology*, **90**, 1239–1245.

Zhou, X. P. *et al.*, 1997, Evidence that DNA-A of a geminivirus associated with severe cassava mosaic disease in Uganda has arisen by interspecific recombination. *Journal of General Virology*, **78**, 2101–2111.

Zhu, Y. Y. *et al.*, 2000, Genetic diversity and disease control in rice. *Nature*, **406**, 718–722.

Zimmermann, W. and Seemuller, E., 1984, Degradation of raspberry suberin by *Fusarium solani* f sp *pisi* and *Armillaria mellea*. *Phytopathologische Zeitschrift-Journal of Phyto-pathology*, **110**, 192–199.

Zook, M. and Hammerschmidt, R., 1997, Origin of the thiazole ring of camalexin, a phytoalexin from Arabidopsis thaliana. *Plant Physiology*, **113**, 463–468.

Zweimuller, M. *et al.*, 1997, Biotransformation of the fungal toxin fomannoxin by conifer cell cultures. *Biological Chemistry*, **378**, 915–921.

Index